Dynamic Stochastic Models from Empirical Data

This is Volume 122 in
MATHEMATICS IN SCIENCE AND ENGINEERING
A Series of Monographs and Textbooks
Edited by RICHARD BELLMAN, *University of Southern California*

The complete listing of books in this series is available from the Publisher upon request.

Dynamic Stochastic Models from Empirical Data

R. L. Kashyap

School of Electrical Engineering
Purdue University
West Lafayette, Indiana

A. Ramachandra Rao

School of Civil Engineering
Purdue University
West Lafayette, Indiana

ACADEMIC PRESS New York San Francisco London 1976

A Subsidiary of Harcourt Brace Jovanovich, Publishers

ACADEMIC PRESS, INC.
111 Fifth Avenue, New York, New York 10003

United Kingdom Edition published by
ACADEMIC PRESS, INC. (LONDON) LTD.
24/28 Oval Road, London NW1

Library of Congress Cataloging in Publication Data

Kashyap, Rangasami Laksminarayana, (date)
Dynamic stochastic models from empirical data.

(Mathematics in science and engineering)
Bibliography: p.
1. Time-series analysis. 2. Stochastic processes.
3. Estimation theory. 4. System analysis. I. Rao, Adiseshappa Ramachendra, (date) joint author. II. Title. III. Series.
QA280.K37 519.2 75-13093
ISBN 0–12–400550–0

PRINTED IN THE UNITED STATES OF AMERICA

To our many teachers

Scientific research is not a clamor of affirmation and denial. Theories and hypotheses are modified more often than they are discredited. A realistic methodology must be one that allows for repair as readily as for refutation.

P. B. MEDAWAR

Contents

Preface

The construction of stochastic dynamic models from empirical time series is practiced in a variety of disciplines, including engineering, ecology, and applied statistics, with specific forecasting aims. However, there have been few systematic expositions of the major problems facing model builders: determination of the plausible classes of models for the given time series by inspection of the series and examination of its characteristics, detailed comparison of the various classes of models, the role and methods of model validation, etc. Of course, there have been a number of books discussing some techniques for developing models for time series data, as mentioned in the text.

The central problems in model building are, in our view, the choice of the appropriate class of models and the validation or checking for adequacy of the best fitting models from the selected class. Even though optimal parameter estimation methods are used, the performance of the best fitting model from an inappropriate class of models may be poor in comparison with the performance of the corresponding member of the appropriate class. Moreover, detailed validation tests bring out the limitations of the selected class and may suggest a more appropriate class, if one exists. In earlier expositions of the subject, the class comparison and validation problem, if considered at all, was discussed entirely in terms of the classical theory of hypothesis testing and other similar decision theoretic methods. The comparison of many important classes of models can be demonstrated to be beyond the scope of the theory of hypothesis testing because of the difficulty in finding the probability distribution of the test statistic. Furthermore, a validation program in which only the residuals are tested by using the methods of hypothesis testing is often inconclusive. Thus the development of various approaches to comparison of different classes and subsequent validation of the final models are the major themes of this book. In addition, relatively standard topics such as parameter estimation methods and estimability are covered in some detail.

The validity of the methodology developed in the text is demonstrated by presenting detailed case studies of model development for about 15 univariate and multivariate time series. In these case studies, all the important numerical details of parameter estimation, class selection, and validation are included. The rainfall and riverflow series, the animal population series, the U.S. population series, and sales figures of a certain company are some of the data sets treated here. The potential application of the model for forecasting, generation of synthetic data, and verification of certain causal hypotheses about environmental processes is discussed at some length. In particular, stochastic models are demonstrated to be superior to deterministic models even though the latter are popular.

The plan of the book is as follows: Chapter I is an introduction. Chapter II contains a brief discussion of a number of topics, such as prediction and other prerequisite materials that will be needed in later chapters. Chapter III is a qualitative discussion of the dominant features of time series obeying various schema for models, including autoregressive moving average (ARMA) models, integrated ARMA models, covariance stationary models, etc., with suggested guidelines for choice of the possible classes of models for any given series. Chapters IV and V deal with the problem of model multiplicity and estimability, i.e., the conditions needed on a class of models to ensure that there is at most one model in the class for the given series. Chapters VI and VII deal with various parameter estimation methods and the corresponding tradeoff involved between estimation accuracy and computational complexity. Chapters VIII and IX deal with methods of comparison of the various classes of models and with validation of the chosen model. Case studies of modeling are discussed in Chapters X and XI.

This book, developed from our teaching and research at Purdue University for the past four years on the topic of system identification, has been designed to be used by practicing engineers, ecologists, and applied statisticians interested in constructing models and by graduate students as a textbook in a course on time series analysis or system identification. The readers of the book are assumed to have some knowledge, albeit elementary, of statistics and random processes; otherwise it is self-contained. Some of the more complicated derivations are postponed to appendixes so that the text reads smoothly. This book has been used as a textbook for a one-semester first-year graduate course with Chapter V and parts of Chapter VII omitted.

In writing a book on a subject that is being actively investigated, we can perforce describe only a few of the various methods proposed for estimation and model comparison. Our guideline in this selection has been the availability of empirical support of the methods. It is entirely possible that a number of methods that may have been successful in practice are not included here. We take comfort from the Bhagavad Gita, "All actions are tainted with some blemish."

Acknowledgments

We would like to thank Dr. Richard Bellman for encouraging us to undertake this monograph. We would like to thank Dr. Robert E. Nasburg who carefully read the manuscript and offered quite a few suggestions for its improvement. We are pleased to acknowledge the interest in our project shown by Dean J. C. Hancock, Prof. C. L. Coates, and Prof. K. S. Fu of the School of Electrical Engineering and Prof. J. F. McLaughlin of the School of Civil Engineering at Purdue University. We are grateful to Mr. P. S. Ramakrishna, Dr. R. G. Srinivasa Rao, Mr. M. C. Mittal, Mr. R. McBride, and Dr. L. K. Rastogi for helping us with the numerical work. We would like to thank Mrs. Barbara Pounds, Terry L. Brown, Terry L. Sanders, Mary Ann Harrison, and Wanitta Booth who have typed this book from a marginally legible manuscript.

The authors are grateful to the National Science Foundation and to the Office of Water Research and Technology of the U.S. Department of the Interior for partial support of the research work reported here.

Notation and Symbols

All bold-faced letters are either vectors or matrices. All upper case bold-faced Latin letters, the bold-faced Greek letter $\boldsymbol{\rho}$, and all bold-faced script letters with the exception of $\boldsymbol{\mathcal{Y}}$ are matrices. The remaining bold-faced letters represent column vectors. Row vectors can be obtained by transposition indicated by superscript T.

Fixed Notation

$\mathbf{A}_i$, $\mathbf{B}_j$	$m \times m$ coefficient matrices in the difference equation.
$\mathbf{A}(D)$, $\mathbf{B}(D)$	$m \times m$ polynomial matrices in D made up of $\mathbf{A}_i$, $i = 1, \ldots, m_1$ and $\mathbf{B}_j$, $j = 1, \ldots, m_2$, respectively
D	Unit delay operator.
det	Determinant.
$\mathscr{E}$	Equivalence relation defined in Chapter V.
$\mathscr{H}$	Set of values that can be assumed by $\boldsymbol{\theta}$.
$\mathscr{I}$	$n \times n$ information matrix.
l_1	Dimension of $\mathbf{u}$.
l_2	Dimension of $\boldsymbol{\psi}$.
$\mathscr{M}$	A set of multivariate equations.
n_0	The number of all unknown coefficients in the difference equation.
n	n_0, if $m = 1$.
$O(x)$	$O(x)/x \to K \neq 0$ as $x \to 0$.
$o(x)$	$o(x)/x \to 0$ as $x \to 0$.
p	Probability density function.
$\mathscr{s}$	Denotes a single equation.
t	Time variable, discrete or continuous.
T (superscript)	Transposition sign for vectors and matrices.
$\mathbf{u}$	l_1-vector of inputs.
$\mathbf{w}$	m-vector of white noise disturbances.
$\mathbf{W}(t-1)$	Matrix made up of $\mathbf{w}(t-1), \ldots, \mathbf{w}(t-m_2)$ used in Chapter VII.
$\mathbf{y}$	m-vector of outputs.
$\boldsymbol{\mathcal{Y}}$	Vector obtained by stacking the y-vectors at various instants.
$\boldsymbol{\theta}$	Vector of all the unknown coefficients in the difference equation.
$\boldsymbol{\xi}(t-1)$	History of all the observations until time t.
$\boldsymbol{\rho}$	Noise covariance.
σ	Standard deviation.

$\boldsymbol{\psi}$	l_2-vector of deterministic trend terms.
ω	Frequency variable.
Ω	Set of allowed values of $\boldsymbol{\rho}$.
$\in$	Belongs to.
$\subseteq$	A subset of or all of.

Variable Symbols

The most often used meaning of the following symbols is given here. In a few cases, these symbols are used for other purposes as indicated in the text.

$\mathbf{C}_j$	Time averaged correlation matrix.
$\mathbf{F}$	$(m \times l_2)$-dimensional coefficient matrix in the difference equation, used in F-test.
f_i	Functions of $y(\cdot)$.
$\mathbf{G}_i$	$m \times l_1$ coefficient matrices in the difference equation.
$\mathbf{G}(D)$	A matrix polynomial in D constructed from $\mathbf{G}_i$, $i = 1, \ldots, m_3$.
H_0	Null hypothesis.
H_1	Alternative hypothesis.
i, j, k	Dummy integer variables.
i	$\sqrt{-1}$
J	Criterion function.
n_α	Dimension of $\boldsymbol{\alpha}$.
n_β	Dimension of $\boldsymbol{\beta}$.
$\mathbf{P}$	Covariance matrix.
$\mathbf{R}_j$	Correlation function.
$\mathbf{S}_{yy}$	Spectral density of variable $\mathbf{y}$.
$\mathbf{S}_{yw}$	Cross spectral density of y and w.
$\mathbf{S}(t)$	$\sum_{j=1}^{t-1} \mathbf{z}(t-1)\mathbf{z}^{\mathrm{T}}(t-1)$
$\mathbf{T}(D)$	Transfer function matrix whose elements are rational functions in D.
$\mathbf{v}$	m-dimensional disturbance.
$\mathbf{x}$	State vector, noisy observation.
$\mathbf{X}(t-1)$	$[\mathbf{Z}(t-1), \mathbf{W}(t-1)]$.
$Y(\omega)$	Fourier transform of $\{y(\cdot)\}$.
$\mathbf{z}_i(t-1)$	An n_i-dimensional vector made up of $y_k(t-j)$, $\psi_i(t-j)$, and $u_i(t-j), j = 1, 2, \ldots; k = 1, 2, \ldots$; etc.
$\mathbf{Z}(t-1)$	A matrix constructed from $\mathbf{z}_1(t-1), \ldots, \mathbf{z}_m(t-1)$.
$\mathscr{Z}$	A matrix obtained by stacking the matrices $\mathbf{Z}(t-1)$ at various instants.
$\boldsymbol{\alpha}$	Part of vector $\boldsymbol{\theta}$ which does not involve moving average coefficients.
$\boldsymbol{\beta}$	Part of vector $\boldsymbol{\theta}$ which involves only moving average coefficients; $\boldsymbol{\theta} = (\boldsymbol{\alpha}^{\mathrm{T}}, \boldsymbol{\beta}^{\mathrm{T}})^{\mathrm{T}}$.
$\boldsymbol{\zeta}(\cdot)$	m-dimensional disturbance vector.
$\boldsymbol{\phi}$	$(\boldsymbol{\theta}, \boldsymbol{\rho})$.
$\boldsymbol{\nabla}$	Derivative.
∇	Differencing operator $[\nabla y(t) = y(t) - y(t-1)]$.

Chapter I | Introduction to the Construction of Models

The word "model" is used in many situations to describe the system at hand. Consequently, there are strong differences of opinion as to the appropriate use of the word "model." It may suggest a photographic replication of the system under study which reflects all its ramifications so that the model may adequately represent the original system. This type of replication is seldom achieved in practice. Every model may have a few specific purposes, such as forecasting and control, and the model need only have just enough significant detail to satisfy these purposes. Thus the basic premise in model building is that complicated systems—all real systems are usually complicated—do not always need complicated models. For instance, the complete phenomenological description of river flow processes is very complicated. Still one can get relatively simple models of river flow processes which yield adequate performance in forecasting and control. Models with a degree of complexity beyond a certain level often perform poorly in comparison with some simpler models. Often, if a model for a given process involves a large number of parameters, it is a good indication that we have to consider an entirely different family of models for the given process. Thus it is advisable to fit relatively simple models to the given data and to increase the complexity of the model only if the simpler model is not satisfactory. In this regard, the methodology used in model building is not different from the usual practice in other branches of science.

Systems are traditionally regarded as deterministic or stochastic. One of the principal disadvantages of the deterministic models of systems is the absence of an effective method of comparison of various possible models that can be constructed by using the same empirical data such that the comparison is relevant to the ultimate purpose of the models. For instance, a criterion such as least squares (between the output of the model and the observed output) can be used to compare the models. But the result of the comparison may not always be very relevant for forecasting or control, as systems that satisfy the least squares criterion may yield poor predictions. Furthermore, the utility of a model is intimately related to the so-called *generalization capability* of the model. A model can be said to have generalization capability if the model constructed by using n observations of the system does not differ very much from a model constructed by using a subsequent set of n observations. The construction of models in a probabilistic framework allows us to compare different models and to give a precise meaning to words such as "significant" and "negligible," which is lacking in models constructed in a deterministic context. Since the class of stochastic models encompasses the class of deterministic models, searching for the best stochastic model does not entail any loss of generality.

While dealing with mechanical or electromechanical systems, such as turbogenerators and aircraft, the variables whose time histories are available to us are divided into two groups, the so-called inputs and outputs. The usual assumption in such a division is that the output variables are those whose behavior is of particular interest to us. However, they cannot be manipulated directly. Instead, they can be influenced by means of the input variables or the independent variables which can be directly manipulated. If $y(\cdot)$ denotes the output and $u(\cdot)$ the input, then $y(t)$ is a function of past values of $y(\cdot)$ and $u(\cdot)$. However, this does not rule out the possibility of the dependence of $u(t)$ on the past values of $y(\cdot)$ and $u(\cdot)$ since there may be hidden feedbacks in the system. When dealing with meteorological or economic systems, the causal relationship between the variables may not be immediately apparent and in such cases the arbitrary division of variables into inputs and outputs may not be quite useful. All variables are then treated as output variables in the construction of a satisfactory model, and the causal relations among the variables are inferred from the model.

1a. Nature and Goals of Modeling

Our main purpose in constructing a model for the given input–output data is to obtain an understanding of the process. Typically, any analytical expression that explains the nature and extent of the dependency of the present observation on past history can be said to increase our understanding of the process. Another way of looking at a model of an empirical process is that the model is a convenient way of summarizing the entire available set of observations; i.e., the important characteristics of the data can be recovered from the model by analyzing the model or simulating it. An excellent historical illustration of modeling an empirical process is Kepler's model of planetary motion to explain the observational history of the position of planets. Kepler's laws can be used to predict the future state of a planet knowing its current state. They also elegantly summarize the entire history of observations of the positions of the various planets. Other classical illustrations of the modeling process are the second-order autoregressive model for the annual sunspot sequence and the first-order stationary stochastic differential equation model for turbulent flows.

But it is not enough that a model be consistent with the numerical observations. We want the model to be "simple" or "pleasing to the mind." A model that has too many parameters is usually considered unsatisfactory. A classic example of a model of this kind is the Ptolemy theory of cycles and epicycles for explaining the planetary motions. This model is quite clumsy even though it could explain all the available observations available in its time; i.e., the difference between the forecast given by the model and the actual position of a planet was of the same order as the error of the observation. Kepler's laws explain the same set of observations with relatively few parameters. Of course, the clumsy model is accepted if a better one is not available.

When we are dealing with complex processes, we cannot hope to obtain a satisfactory deterministic model. Hence we regard the process $\mathbf{y}(t)$, at time t, as a sum of two parts: (i) a part that is a function of the observation history until the time $t - 1$, and, (ii) a purely random component $\mathbf{w}(\cdot)$, i.e., a variable from a zero mean and independent, identically distributed (IID) sequence. The corresponding equation for $\mathbf{y}(t)$ can be represented as

$$\mathbf{f}_0(\mathbf{y}(t)) = \mathbf{g}(\boldsymbol{\xi}(t-1), \boldsymbol{\theta}) + \mathbf{w}(t), \tag{1a.1.1}$$

where $\boldsymbol{\xi}(t-1)$ is the observation history until $(t - 1)$, $\{\mathbf{w}(\cdot)\}$ is the zero mean IID sequence, $\mathbf{f}_0$ and $\mathbf{g}$ are functions of the indicated arguments, and $\boldsymbol{\theta}$ is a finite vector of parameters.

Regarding the functions $\mathbf{f}_0$ and $\mathbf{g}$, we need to consider two cases: (i) The functions $\mathbf{f}_0$ and $\mathbf{g}$ are known except that the parameter vector $\boldsymbol{\theta}$ is unknown. (ii) The functions $\mathbf{f}_0$ and $\mathbf{g}$ are unknown. In case (i), the modeling problem is completely solved if the parameter vector $\boldsymbol{\theta}$ is estimated from the given data. Such a situation occurs in dealing with mechanical or electromechanical systems such as aircraft and turbogenerators, whose dynamics are clearly known except for a few parameters. But case (ii) is typical of the modeling problems occurring in economics, hydrology, meteorology, population studies, etc. Here, either by physical reasoning or by conjecture, we first restrict the function pair $(\mathbf{f}_0, \mathbf{g})$ to a finite number of candidates $\{(\mathbf{f}_0{}^i, \mathbf{g}^i), i = 1, \ldots, n_1\}$, where the function $\mathbf{g}^i$ depends on $\boldsymbol{\xi}(t-1)$ and $\boldsymbol{\theta}^i$, a finite vector of parameters which is unknown. We first have to choose the appropriate function pair among the candidates and then estimate the corresponding parameter $\boldsymbol{\theta}$ in the chosen pair. We will discuss this aspect later.

Once we have a good model, it can be used for a variety of purposes, such as forecasting the variables one or several units of time ahead, and the control of the system. It can be used for generating data that have characteristics similar to those of the empirical data. For example, in the case of river flow studies, the data obtained by simulating the model for the river flow are called synthetic flow data and are used extensively in reliability studies connected with storage designs, studies of the operating policies of reservoirs, and so forth. Still another use of the model is in data transmission. When we are interested in transmitting speech data, we first digitize the speech waveform. We have two options for transmission. We can either transmit the digitized data directly, or first fit an autoregressive model to a batch of speech data involving about, say, 40 msec, and transmit only the coefficients of the autoregressive process. By using these coefficients, the speech can be synthesized at the receiving end by simulating the autoregressive model on a computer.

We now turn to the topic of forecasting. It is often said that the main aim in modeling is prediction or forecasting. This statement is not completely true. It should be noted that forecasts can often be obtained by using plain regression methods in which the past history of the process is used. An excellent historical example of obtaining good forecasts by only inspecting the data and not using any theory at all is the Babylonian method of forecasting the first visibility of the new moon and lunar eclipses, etc.

In many cases, two entirely different models may lead to more or less the same one-step-ahead prediction error. As such, the forecasting ability alone cannot be taken to be the final judge of the quality of the model, unless we want a model solely for forecasting. If forecasting is our only aim, then we may not need a sophisticated model. For example, in population studies, a deterministic model is quite satisfactory for forecasting the population one year ahead with about 5% error. But such a model cannot be validated by the methods to be discussed; i.e., such a model does not preserve all the principal statistical characteristics of the data.

1b. Description of Models

The observable variables will be of two kinds, an m-dimensional output vector $\mathbf{y}$ and an l_1-dimensional input vector $\mathbf{u}$. We will consider only stochastic difference equation models such as

$$\mathbf{f}_0(\mathbf{y}(t)) = \sum_{j=1}^{m_1} \mathbf{A}_j \mathbf{f}_j(\mathbf{y}(t-j)) + \sum_{j=1}^{m_3} \mathbf{G}_j \mathbf{g}_j(\mathbf{u}(t-j)) + \mathbf{F}\boldsymbol{\psi}(t-1) + \boldsymbol{\zeta}(t), \tag{1b.1.1}$$

where the dynamical behavior of $\mathbf{y}(t)$ is explained in terms of its own past values, the past values of $\mathbf{u}$, and a disturbance vector $\boldsymbol{\zeta}(t)$ of dimension n which is independent of the past history of both $\mathbf{y}$ and $\mathbf{u}$. We will show later that the family (1b.1.1) is versatile enough to handle a variety of stochastic sequences. The stochastic difference equations are of finite order everywhere except in some models occurring in Chapters II and X, the so-called fractional noise models, which are described by infinite-order difference equations.

By and large, the elements of coefficient matrices $\mathbf{A}_j$, $\mathbf{G}_j$ will be treated as unknown constants. However, in modeling certain natural processes, such an assumption is inappropriate. In such cases, we consider the variables $\mathbf{A}_j$, $\mathbf{G}_j$ to be *slowly varying random variables*. The details can be found in Chapter VI and the method is illustrated in Chapter X.

The l_2-dimensional vector $\boldsymbol{\psi}(t)$ in (1b.1.1) is made up of deterministic trend functions such as $\psi_1(t) = 1$, $\psi_2(t) = t$, and $\psi_3(t) = \sin \omega_1 t$. The appropriate deterministic function of time is usually chosen by inspection of the empirical data and its characteristics in conjunction with some hypothesis testing. The functions $\mathbf{f}_j$, $j = 1, \ldots, m_1$, and $\mathbf{g}_j$, $j = 1, \ldots, m_3$, are vectors of dimension m and l_1, respectively,

$$\mathbf{f}_j = (f_{j1}, \ldots, f_{jm})^{\mathrm{T}}, \qquad j = 1, \ldots, m_1$$
$$\mathbf{g}_j = (g_{j1}, \ldots, g_{jl_1})^{\mathrm{T}}, \qquad j = 1, \ldots, m_3.$$

When the functions $\mathbf{f}_j$, $\mathbf{g}_j$ are unknown, our first guess is that they are all linear. This may be too restrictive. A more relaxed assumption is

$$\mathbf{f}_j = \mathbf{f}, \quad j = 1, \ldots, m_1; \qquad \mathbf{g}_j = \mathbf{g}, \quad j = 1, \ldots, m_3. \tag{1b.1.2}$$

In such a case, Eq. (1b.1.1) can be rewritten in a compact form by utilizing the unit delay operator D defined by the relation $D(\mathbf{f}(y(t))) = \mathbf{f}(y(t-1))$:

$$\mathbf{A}(D)\mathbf{f}(\mathbf{y}(t)) = \mathbf{G}(D)\mathbf{g}(u(t-1)) + \mathbf{F}\boldsymbol{\psi}(t-1) + \boldsymbol{\zeta}(t), \tag{1b.1.3}$$

where

$$\mathbf{A}(D) = \mathbf{I} - \sum_{j=1}^{m_1} \mathbf{A}_j D^j, \qquad \mathbf{G}(D) = \sum_{j=1}^{m_3} \mathbf{G}_j D^j.$$

We next discuss the disturbance $\boldsymbol{\zeta}(\cdot)$ appearing in (1b.1.1) and (1b.1.3). The simplest assumption on the disturbance $\boldsymbol{\zeta}(t)$ would be that it is a white noise. However, such an assumption would be too restrictive. Instead, we assume that $\boldsymbol{\zeta}(t)$ has zero mean and finite memory:

$$E[\boldsymbol{\zeta}(t)] = 0, \qquad E[\boldsymbol{\zeta}(t)\boldsymbol{\zeta}^{\mathrm{T}}(t-j)] = 0, \qquad \forall j \geq (m_2 + 1).$$

In cases where $\boldsymbol{\zeta}(t)$ has a finite memory it possesses the moving average representation

$$\boldsymbol{\zeta}(t) = \mathbf{w}(t) + \sum_{j=1}^{m_2} \mathbf{B}_j \mathbf{w}(t-j) = \mathbf{B}(D)\mathbf{w}(t), \tag{1b.1.4}$$

where

$$\mathbf{B}(D) = \mathbf{I} + \sum_{j=1}^{m_2} \mathbf{B}_j D^j; \tag{1b.1.5}$$

$$E[\mathbf{w}(t)] = 0, \qquad E[\mathbf{w}(t)\mathbf{w}^{\mathrm{T}}(j)] = \boldsymbol{\rho}\delta_{tj}, \tag{1b.1.6}$$

where $\mathbf{w}(\cdot)$ is a zero mean white noise sequence, with $\mathbf{w}(t)$ independent of $\mathbf{w}(j)$ for $t \neq j$. It is also implicitly assumed that the noise $\mathbf{w}(t)$ is independent of the past values $\mathbf{y}(t-j)$, $\mathbf{u}(t-j)$, $j \geq 1$. Alternatively, the elementary disturbance $\mathbf{w}(\cdot)$ is nonanticipatory. The sequence $\boldsymbol{\zeta}(\cdot)$ can also be represented as an autoregressive process, i.e., $\boldsymbol{\zeta}(t)$ as a linear function of its past values and $\mathbf{w}(t)$. However, it is easy to show that the system (1b.1.3) with an autoregressive representation for $\boldsymbol{\zeta}(\cdot)$ does not possess any greater flexibility than the system given by (1b.1.1) or (1b.1.3) with $\boldsymbol{\zeta}$ having a moving average representation.

Thus, (1b.1.3) has the final form

$$\mathbf{A}(D)\mathbf{f}(\mathbf{y}(t)) = \mathbf{G}(D)\mathbf{g}(\mathbf{u}(t-1)) + \mathbf{F}\boldsymbol{\psi}(t-1) + \mathbf{B}(D)\mathbf{w}(t), \tag{1b.1.7}$$

where $\{\mathbf{w}(\cdot)\}$ obeys (1b.1.6) and is independent.

The family of models in (1b.1.1) and (1b.1.7) is characterized by two sets of parameters that may be called the primary and secondary parameters. The order parameters m_1, m_2, and m_3, and similar other parameters to be defined later, constitute the set of primary parameters, and these can assume only nonnegative integer values. The elements of the coefficient matrices $\mathbf{A}_j$, $\mathbf{B}_j$, $\mathbf{G}_j$, $\mathbf{F}$, $\boldsymbol{\rho}$, which assume only real values, constitute the secondary parameters.

We emphasize the fact that the model (1b.1.7) is linear only in the coefficient matrices $\mathbf{A}_j$, $\mathbf{B}_j$, ... It is not necessarily linear in the variable $\mathbf{y}$. Consequently, the probability distribution of $\mathbf{y}$ may be non-Gaussian even if $\mathbf{w}(\cdot)$ is Gaussian. This feature is important because naturally occurring stochastic processes such as the monthly temperature and rainfall observed at a station are often highly non-Gaussian.

The model in Eq. (1b.1.7) is versatile. One can handle a variety of processes by a proper choice of the functions $\mathbf{f}$ and $\mathbf{g}$. For instance, consider the following multiplicative process:

$$y(t) = \prod_{j=1}^{m_1} (y(t-j))^{A_j} \prod_{j=1}^{m_3} (u(t-j))^{G_j} w(t). \tag{1b.1.8}$$

It can be expressed in the form (1b.1.7) by choosing the functions $\mathbf{f}$ and $\mathbf{g}$ to be log functions.

Another important consideration in modeling is the role of the error in the observations. Often the observations are erroneous because of the low accuracy of the measuring instruments and schemes. This is especially true when the instruments are in open air as in rainfall measurement. Let us label the theoretical value of the output variable by the usual symbol $\mathbf{y}(t)$ and its observation by $\mathbf{y}'(\cdot)$. Note that $\mathbf{y}(\cdot)$ is not directly observable. We can represent the relation between $\mathbf{y}$ and $\mathbf{y}'$ as

$$\mathbf{y}'(t) = \mathbf{y}(t) + \boldsymbol{\eta}(t), \tag{1b.1.9}$$

where $\boldsymbol{\eta}(\cdot)$ is a zero mean IID noise, the so-called measurement noise. If necessary, we can postulate more complicated relationships between $\mathbf{y}$ and $\mathbf{y}'$. One of the purposes of modeling is to determine the need for the introduction of the measurement noise $\boldsymbol{\eta}$ into the model; i.e., we have to investigate whether the given observation sequence can be explained adequately with the aid of disturbance sequence $\mathbf{w}(\cdot)$ alone or whether the measurement noise sequence $\boldsymbol{\eta}$ has to be introduced. In aerospace applications such as satellite tracking, the dynamics of the system (namely, Kepler's laws) are well known. In such cases, the introduction of the measurement noise $\boldsymbol{\eta}$ can be justified because the state estimators and predictors obtained from the more detailed model give better results than the ones obtained with the simpler model, in which the noise $\boldsymbol{\eta}$ is omitted. On the other hand, in a typical economic system, the dynamics of the system are unknown and the coefficients $\mathbf{A}_j$, $\mathbf{B}_j$ have to be estimated from the data. It is not obvious that the state estimators and predictors obtained from the complicated models which allow for the noise $\boldsymbol{\eta}$ are necessarily superior to the estimators and predictors derived from the simpler model in which $\boldsymbol{\eta}$ is omitted.

1c. Choice of a Model for the Given Data

The choice of an appropriate model in the family (1b.1.7) for the given empirical time series $\{\mathbf{y}(\cdot), \mathbf{u}(\cdot)\}$ can be discussed by using the following three steps: (i) the choice of a structure; (ii) the choice of the primary parameters; (iii) the choice of the secondary parameters.

1c.1. Choice of the Structure and the Primary Parameters

In single output systems, by the choice of the structure we mean the choice of the functions $\mathbf{f}$ and $\mathbf{g}$ [or the functions $\mathbf{f}_j$, $\mathbf{g}_j$ as in (1b.1.1)] and the choice of the trend functions ψ_i, if they are found to be necessary. In the multiple output systems we have to consider the nature of the interconnections between the various equations for y_i, $i = 1, 2, \ldots, m$.

If we do not have any prior information about the process that is being modeled, there is little to be gained by fitting a complicated model to the given data without first exhausting the possibilities of fitting a simple model. In particular, there is no need to begin with an interconnected system of equations if each variable y_i can be adequately described in terms of its own past history and the random disturbance. The interconnected structure is expressed by the structure of the matrices $\mathbf{A}(D)$, $\mathbf{B}(D)$ and covariance matrix $\boldsymbol{\rho}$. We present the following possible structures for $\mathbf{A}$, $\mathbf{B}$, and $\boldsymbol{\rho}$ beginning with the structure having the least complexity:

Case (i). $\mathbf{A}(D)$ = diagonal, $\mathbf{B}(D) = \mathbf{I}$, $\boldsymbol{\rho}$ = diagonal.
Case (ii). $\mathbf{A}(D)$ = nondiagonal, $\mathbf{B}(D) = \mathbf{I}$, $\boldsymbol{\rho}$ = diagonal.
Case (iii). $\mathbf{A}(D)$ = nondiagonal, $\mathbf{B}(D) = \mathbf{I}$, $\boldsymbol{\rho}$ = nondiagonal.
Case (iv). $\mathbf{A}(D)$ = nondiagonal, $\mathbf{B}(D)$ = diagonal, $\boldsymbol{\rho}$ = diagonal.
Case (v). $\mathbf{A}(D)$ = nondiagonal, $\mathbf{B}(D)$ = diagonal, $\boldsymbol{\rho}$ = nondiagonal.
Case (vi). $\mathbf{A}(D)$ = nondiagonal, $\mathbf{B}(D)$ = nondiagonal, $\boldsymbol{\rho}$ = nondiagonal.

Cases (i)–(iii) are the so-called autoregressive (AR) models and Cases (iv)–(vi) are the autoregressive moving average (ARMA) models. Case (i) corresponds to a system of autoregressive equations that are not connected with one another, Case (iii) is the familiar multivariate autoregressive models, and Case (vi) is the multivariate autoregressive moving average representation in its generality. Thus, if we cannot fit an AR model to the data, there is no need to consider the ARMA family in Case (vi) without having exhausted Cases (iv) and (v). We will show in Chapter V that the family of models given in Case (iv) is as rich as the family of models in Case (vi). Consequently, we may as well approximate the given empirical process by a model in the family of Case (iv) instead of one in family (vi), since the computation complexity involved in the former case is much less than that in the latter.

To choose the appropriate structure and the primary parameters for the given data, we narrow down our choice to a finite *number* of different *classes* of models. All the models in each class have the same primary parameters, or the different models in the same class differ only in their secondary parameters which come from the set, say, $\mathscr{H}_i$, for the *i*th class. Some guidelines for this preliminary choice are discussed in Chapters III and V. If the constant coefficient models are not appropriate, models with time-varying parameters will be considered. We have to compare the different classes by using the given data. This topic is discussed in detail in Chapter VIII.

1c.2. Choice of the Secondary Parameters

As mentioned earlier, the secondary parameters are the various coefficients in Eq. (1b.1.1) and the covariance matrix $\boldsymbol{\rho}$, all of which assume real values from a suitable set. The secondary parameters are estimated only after fixing or at least tentatively considering the structure and the primary parameters in the model. The question of estimability must be considered first, and may be stated as follows: "Assuming that the given data have been generated from a model, with its secondary parameters drawn from the specified set, is it always possible to recover the values of secondary parameters even if we are given a semi-infinite sequence of observations?" The answer to this question is not always affirmative, especially if there exists another set of secondary parameters which in conjunction with the model can give rise to an observation sequence that is statistically indistinguishable from that of the earlier sequence. The problem of recovery of the true values of the parameters is referred to as *estimability* and is discussed in Chapters IV and V. Various methods of estimation of the secondary parameters and the corresponding accuracy of the estimates are discussed in Chapters IV–VI.

1d. Validation

In the preceding sections, methods were discussed for the choice of a model in the family of models (1b.1.1) along with a set of values for the primary and secondary parameters. However, the question of determining whether the given model satisfactorily represents the given data still remains. This problem is referred to as the validation of the model. If the model is adequate for the purposes for which it was originally designed (such as forecasting), then the model can be considered as validated. Of course, the validity of a model can be specified only in relative terms. We can only say that a model is better than some others according to some criterion. It may be argued that if the primary and secondary parameters are chosen in a reasonable manner, then the final model must be satisfactory in a relative sense. However, the primary parameters are often chosen in an arbitrary manner in view of the reasons briefly mentioned earlier. Consequently, the problem of validating the model still remains. We can consider at least two different approaches to the problem of validation. The first of these has some analytical basis behind it whereas the second approach is based on simulation.

In the first approach, the validity of the assumptions behind the model is checked. In many cases, the only important assumption is that the disturbance sequence has zero mean and is independent. By using the given model and the available observations, the estimates of the disturbances, the so-called residuals, are obtained and the independence assumption is checked by using the usual theory of hypothesis testing, after reformulating the problem as a choice between two hypotheses H_0 and H_1. The hypothesis H_0, usually called the null hypothesis, declares the residuals to be zero mean and independent. The hypothesis H_1, usually called the alternative, declares the successive residuals to be dependent and to obey an autoregressive process of order m_2. The autoregressive model of dependence is chosen for H_1 because of the computational ease. In addition, the residuals are assumed to be normally distributed. By using the likelihood ratio testing procedure, a statistic having an approximate F or χ^2 distribution can be obtained for discriminating between the two hypotheses. If the hypothesis H_0 is accepted at a suitable prespecified level of significance, then the corresponding model is also accepted. If H_1 is accepted, then of course the model under consideration is unsatisfactory. To obtain further guidelines for modifying the model, the residuals are analyzed further to understand the nature of the serial dependence among them. For instance, there may be a sinusoidal trend component in the residuals that would cause the acceptance of H_1 over H_0. In such a case, one has to add a sinusoidal trend component to the model. The details of this method of validation are given in Chapter VIII.

The acceptance of hypothesis H_0 in the test described above is a necessary condition for the validity of the model but is not sufficient. One can easily construct examples in which the model passes the above test (i.e., the acceptance of H_0 in the hypothesis testing problem), and yet the model may not make any sense whatsoever. Moreover, the choice of the significance level in hypothesis testing is completely subjective. Sometimes, H_0 may be preferred to H_1 at the 95% significance level and H_1 preferred to H_0 at the 98% level. This discrepancy is considered later.

To confirm the validity of the model, we have to directly compare the characteristics of the model output such as the correlograms, spectral densities, and extreme value characteristics, with the corresponding characteristics of the empirical data. In order to compare these characteristics, the various statistical characteristics of the model output can be obtained either by analysis or by simulation. We accept the model if the discrepancy between the two sets of characteristics is within one or two standard deviations of the corresponding characteristics, which is inversely proportional to $1/\sqrt{N}$, N being the length of the observation history.

It is only fair to remark that sometimes we may not be able to construct a model with the desired level of accuracy and generality. For instance, in modeling river flows, we cannot construct a single model which gives both good one-day-ahead forecasts and one-year-ahead forecasts. In such cases, we construct separate models for each frequency domain. Our inability to construct models with the required level of generality and accuracy prompts us to look for quite different representations of the process. Similarly, linear interactive models constructed for some economic time series do not appear to possess the pseudo-oscillatory behavior which is supposedly present in the empirical time series. There are some suggestions that the presence of pseudocyclical behavior in economic time series is fictitious and that given any random sequence, one can detect any frequency in it depending on one's psychological makeup. This controversy again suggests the need for developing new representations of the data and new methods of validation, which are applicable for problems with limited data. These aspects of class selection and validation are discussed at length in Chapter VIII.

Notes

We have briefly hinted at the relationship between the concepts of model building and the methodology in science. Toulmin (1957) gives a concise introduction to this topic. We have been strongly influenced by the work of Popper (1934) on the importance of validating models and hypotheses. The quotation on the first page of the book is taken from Medawar (1969).

A detailed review of the various books on time series and such related topics as system identification is beyond the scope of this book. Some of the references are mentioned at various places in the other chapters. There is an annotated bibliography of papers in time series prior to 1959 compiled by Wold (1965). The special issue on system identification and time series analysis of the *IEEE Transactions on Automatic Control* (December 1974) has a number of survey and other papers dealing with estimation methods, validation, and applications.

Chapter II | Preliminary Analysis of Stochastic Dynamical Systems

Introduction

Beginning with the concepts of weak stationarity, covariance stationarity, and invertibility, several important concepts and properties connected with the processes obeying stochastic difference equations with additive noise are discussed in this chapter. The spectral representation of such processes, including the problems of estimation of the covariances and spectral densities, are also discussed. The problem of prediction is considered in some detail. Finally, the multiplicative noise models and fractional noise models are considered with some emphasis on the methods of forecasting by using such systems.

Although many of the topics discussed below are treated in detail in various other books and monographs they are included here because they will be used in subsequent chapters. In order to avoid unnecessary duplication the discussion is brief.

2a. Assumptions and Discussion

We will list the principal assumptions made on the basic difference equation

$$\mathbf{A}(D)\mathbf{y}(t) = \mathbf{G}(D)\mathbf{u}(t-1) + \mathbf{F}\boldsymbol{\psi}(t-1) + \mathbf{B}(D)\mathbf{w}(t). \qquad (2a.1.1)$$

In each problem, we will choose a subset of these assumptions to ensure the uniqueness of the parameters characterizing the process $\mathbf{y}$ obeying the difference equation and to ensure their estimability using the observed histories of $\mathbf{y}$ and $\mathbf{u}$. Recapitulating, $\mathbf{A}(D)$, $\mathbf{G}(D)$, and $\mathbf{B}(D)$ are matrix polynomials of degree m_1, m_3, and m_2, respectively. The vectors $\mathbf{y}$, $\mathbf{w}$, $\mathbf{u}$, and $\boldsymbol{\psi}$ are of dimensions m, m, l_1, and l_2. $\mathbf{F}$ is an $m \times l_2$ matrix. The assumptions can be easily specialized for the case of $m = 1$, i.e., single equation systems. All of the assumptions stated below are *not* used simultaneously. In each section or chapter, the particular assumptions used therein are specifically mentioned.

Assumptions on the Inputs w, u, and $\boldsymbol{\psi}$

A1. The sequence $\{\mathbf{w}(\cdot)\}$ is made up of zero mean independent and identically distributed m-dimensional random vectors having a nonsingular covariance matrix $\boldsymbol{\rho}$. $\mathbf{w}(t)$ is independent of $\mathbf{u}(t-j)$ and $\mathbf{y}(t-j)$ for all $j \geq 1$, $\mathbf{u}(\cdot)$ obeying (A2).

A2. The l_1-dimensional observable vector input $\mathbf{u}(\cdot)$ is a nondeterministic, covariance stationary process with a positive-definite covariance matrix. $\mathbf{u}(t)$ is independent of $\mathbf{w}(t-j)$ for all $j \geq 1$.

A3. $\mathbf{u}$ and $\mathbf{w}$ are mutually independent.

A4. The l_2-dimensional vector of deterministic trend function $\boldsymbol{\psi}(\cdot)$ obeys the persistency condition

$$\lim_{N\to\infty} \frac{1}{N} \sum_{t=1}^{N} \boldsymbol{\psi}(t)\boldsymbol{\psi}^{\mathrm{T}}(t) \text{ exists and is positive definite.}$$

A4′. The vector of trends $\boldsymbol{\psi}$ obeys the condition:

$$\sum_{t=1}^{\infty} \sum_{i=1}^{l_2} (\alpha_i \psi_i(t))^2 = \infty,$$

for any nonzero vector $\boldsymbol{\alpha} = (\alpha_1, \ldots, \alpha_{l_2})^{\mathrm{T}}$.

Discussion. Assumptions (A1), (A2), and (A4) are used throughout. Assumption (A3) will be used almost everywhere except in parts of Chapters IV and V. Assumption (A4) and its weaker version (A4′) are discussed in Chapter IV.

Assumptions on the Matrices A, B, and G

A5. All zeros of the determinant $\mathbf{A}(D)$ lie outside the unit circle and $\mathbf{A}(D)$ is nonsingular.

A6. All zeros of the determinant $\mathbf{B}(D)$ are outside the unit circle and $\mathbf{B}(0) = \mathbf{I}$.

A7. The greatest common polynomial left factor matrix of **A**, **B**, and **G** is unimodular (i.e., has constant determinant) or, equivalently, the Smith form of $[\mathbf{A}, \mathbf{B}, \mathbf{G}]$ is $[\mathbf{I}, \mathbf{0}]$.

A7′. For every i, $i = 1, \ldots, m$, the greatest common divisor of the polynomials in the set $\{b_{ii}(D), a_{i1}(D), \ldots, a_{im}(D), g_{i1}(D), \ldots, g_{il_1}(D)\}$ has degree zero.

Only in the multivariate case do we need *one* of the following assumptions, which characterize the corresponding canonical form or pseudocanonical form:

A8. The matrix $[\mathbf{A}_{m_1} \vdots \mathbf{B}_{m_2} \vdots \mathbf{G}_{m_3}]$ has rank m.

A9. $\mathbf{A}(D)$ is lower triangular with $\mathbf{A}(0) = \mathbf{I}$. Degree of $A_{ij}(D) \leq$ degree of $A_{jj}(D)$ for all $i = j+1, \ldots, m$, i.e.,

$$\mathbf{A} = \begin{bmatrix} A_{11} & & & \\ & \ddots & & \\ & & A_{ii} & \\ & & A_{ji} & \\ & & \vdots & \ddots \\ A_{m1} & & A_{mi} & & A_{mm} \end{bmatrix}.$$

A10. $\mathbf{B}(D)$ is diagonal, and $\mathbf{A}(0) = \mathbf{I}$.

A11. $\mathbf{B}(D)$ is diagonal, $\boldsymbol{\rho}$ is diagonal, and $\mathbf{A}(0)$ is lower triangular with $(A(0))_{ii} = 1, \forall i$.

Discussion. Assumptions (A5)–(A6) are used throughout. One of the assumptions (A8)–(A11) is necessary only for multivariate systems for characterizing the canonical forms of the difference equation. Assumption (A5) guarantees the asymptotic covariance stationarity of process $\mathbf{y}$ in (2a.1.1) as discussed in this chapter. Assumption (A6) ensures the invertibility of (2a.1.1) and is discussed later in this chapter. Some of the principal results of Chapters IV and V are valid even when (A6) is replaced by (A6′).

A6′. The zeros of det $\mathbf{B}$ are either on or outside the unit circle.

But assumption (A6′) is not sufficient for the computation of maximum likelihood (ML) estimates in Chapter IV or VII. As such we have retained (A6) throughout the book. (A7′) and (A7) ensure the noncancellation of *poles* and zeros in univariate systems and in their multivariate counterpart. Assumptions (A7)–(A11) are discussed in Chapter V.

2b. Stationarity

A process $\mathbf{y}$ is said to be covariance stationary if it obeys

$$E[\{\mathbf{y}(t) - E(\mathbf{y}(t))\}\{\mathbf{y}(t-j) - E(\mathbf{y}(t-j))\}^{\mathrm{T}}] = \mathbf{R}_j, \qquad \forall j = 0, \pm 1, \pm 2, \ldots, \tag{2b.1.1}$$

$E[\mathbf{y}(\cdot)]$ being arbitrary, not necessarily a constant.

In (2b.1.1), $\mathbf{R}_j, j = 0, \pm 1, \ldots,$ are known as autocovariance matrices. Whenever the covariance matrices $\mathbf{R}_j$ are introduced, the corresponding process $\mathbf{y}(\cdot)$ is implicitly assumed to be covariance stationary. If (2b.1.1) is valid only as j tends to infinity, then the corresponding process is said to be only asymptotically covariance stationary.

If a process is covariance stationary and in addition $E[\mathbf{y}(t)]$ is a constant vector for all t, then the process is called weakly stationary.

Necessary and sufficient conditions for the *asymptotic* covariance stationarity of a process $\mathbf{y}$ obeying (2a.1.1) are (A2) and (A5). However, they will not ensure covariance stationarity for all t in view of the arbitrariness of the initial condition for the difference equation. To ensure covariance stationarity, an additional condition on the initial conditions is needed. The required additional condition is indicated in

$$E[\{\mathbf{y}(t) - E(\mathbf{y}(t))\}\{\mathbf{y}(j) - E(\mathbf{y}(j))\}^{\mathrm{T}}] = \mathbf{R}_{|t-j|},$$
$$\forall t; \quad j = 0, 1, \ldots, l_3 - 1, \qquad l_3 = \max(m_1 + 1, m_2).$$

A covariance stationary process is completely described by the mean, possibly time varying, and the set of covariance matrices $(\mathbf{R}_j, j = 0, 1, \ldots)$ or the corresponding spectral density $\mathbf{S}_{yy}(\omega)$ and mean:

$$\mathbf{S}_{yy}(e^{-i\omega}) = \sum_{k=-\infty}^{\infty} \mathbf{R}_k e^{-ik\omega}, \tag{2b.1.2}$$

or

$$\mathbf{R}_k = \frac{1}{2\pi}\int_{-\pi}^{\pi} e^{ik\omega}\, d\mathbf{\Phi}(\omega). \tag{2b.1.3}$$

Where ω is the frequency operator, $\mathbf{\Phi}(\omega)$ is the spectral distribution function, and $\mathbf{S}_{yy}(\cdot)$ is the derivative of $\mathbf{\Phi}(\omega)$.

The process $\mathbf{y}$ itself possesses a spectral representation

$$\mathbf{y}(t) = \frac{1}{2\pi}\int_{-\pi}^{\pi} e^{it\omega}\, d\boldsymbol{\phi}(\lambda), \tag{2b.1.4}$$

where $\boldsymbol{\phi}(\cdot)$ is a process whose increments are independent:

$$E[d\boldsymbol{\phi}(\lambda)\, d\boldsymbol{\phi}^{\mathrm{T}}(u)] = d\mathbf{\Phi}(\lambda)\delta_{\lambda u},$$

where $\delta_{\lambda u} = 0$ if $\lambda \neq u$; $\delta_{\lambda u} = 1$ if $\lambda = u$. More informally, let $\mathbf{Y}(\omega)$ denote the Fourier series representation of a given realization $[\mathbf{y}(k), k = -\infty, \ldots, \infty]$ of the process $\mathbf{y}$,

$$\mathbf{Y}(\omega) = \sum_{t=-\infty}^{\infty} \mathbf{y}(t)e^{-it\omega}.$$

Then the relation between the Fourier series representation $\mathbf{Y}(\omega)$ and the spectral density $\mathbf{S}_{yy}$ can be written as

$$\mathbf{S}_{yy}(e^{-i\omega}) = \int_{-\pi}^{\pi} E[\mathbf{Y}(\omega)\mathbf{Y}^*(\omega - \omega_1)]\, d\omega_1$$

where $\mathbf{Y}^*(\omega)$ is the complex conjugate of $\mathbf{Y}(\omega)$.

2c. Invertibility

The concept of invertibility forms the basis for the parameter estimation and prediction in systems with moving average terms. A process $\mathbf{y}(\cdot)$ obeying (2a.1.1) is said to be invertible if the disturbances $\mathbf{w}(t)$ can be recovered with probability 1 or in the mean square sense from the semi-infinite history of observations of $\mathbf{y}$ and $\mathbf{u}$. If we denote an estimate of $\mathbf{w}(t)$ by $\overline{\mathbf{w}}(t)$, Eq. (2a.1.1) suggests this scheme for computing subsequent estimates:

$$\mathbf{B}(D)\overline{\mathbf{w}}(t) = \mathbf{A}(D)\mathbf{y}(t) - \mathbf{G}(D)\mathbf{u}(t-1) - \mathbf{F}\boldsymbol{\psi}(t-1). \tag{2c.1.1}$$

In order to find out how $\overline{\mathbf{w}}(t)$ converges to $\mathbf{w}(t)$, subtract (2c.1.1) from (2a.1.1), to obtain

$$\mathbf{B}(D)(\mathbf{w}(t) - \overline{\mathbf{w}}(t)) = 0. \tag{2c.1.2}$$

It is easy to see that the necessary and sufficient condition for the error $(\mathbf{w}(t) - \overline{\mathbf{w}}(t))$ to tend to zero asymptotically is the asymptotic stability of Eq. (2c.1.2), which in turn is guaranteed by assumption (A6). Invertibility is automatically present in a system with no moving average terms. The invertibility property is completely characterized by the coefficient matrices in the moving average part of the system and the noise covariance matrix.

If a system (2a.1.1) is not invertible, then one can replace the moving average part in it by another moving average representation such that the new representation is invertible. The technique can be illustrated by an example. Let the moving average part be denoted by $\boldsymbol{\zeta}(\cdot)$.

Let

$$\mathbf{B}(D) = \mathbf{I} - \mathbf{B}_1 D, \qquad \boldsymbol{\zeta}(t) = \mathbf{w}(t) - \mathbf{B}_1\mathbf{w}(t-1), \qquad E[\mathbf{w}(t)\mathbf{w}^{\mathrm{T}}(t)] = \boldsymbol{\rho}. \tag{2c.1.3}$$

If the polynomial $\mathbf{B}(D)$ does not obey (A6), the moving average representation in (2c.1.4) can be considered for $\boldsymbol{\zeta}(\cdot)$:

$$\boldsymbol{\zeta}(t) = \mathbf{w}'(t) + \mathbf{B}_1'\mathbf{w}'(t-1), \qquad E[\mathbf{w}'(t)(\mathbf{w}'(t))^{\mathrm{T}}] = \boldsymbol{\rho}_1'. \tag{2c.1.4}$$

Let

$$\mathbf{R}_j' \triangleq E[\boldsymbol{\zeta}(t)\boldsymbol{\zeta}^{\mathrm{T}}(t-j)] = 0, \qquad \forall j \geq 2.$$

Hence both the pairs $(\mathbf{B}_1, \boldsymbol{\rho}_1)$ and $(\mathbf{B}_1', \boldsymbol{\rho}_1')$ must be solutions of

$$\begin{aligned} \mathbf{R}_0' &= \boldsymbol{\rho}_1 + \mathbf{B}_1\boldsymbol{\rho}_1\mathbf{B}_1^{\mathrm{T}} = \boldsymbol{\rho}_1' + \mathbf{B}_1'\boldsymbol{\rho}_1(\mathbf{B}_1')^{\mathrm{T}}, \\ \mathbf{R}_1' &= \mathbf{B}_1\boldsymbol{\rho}_1 = \mathbf{B}_1'\boldsymbol{\rho}_1' \end{aligned} \tag{2c.1.5}$$

which is obtained by equating the first two covariance matrices of $\boldsymbol{\zeta}(\cdot)$ obtained with the aid of (2c.1.3) and (2c.1.4).

Alternatively, one can show that there exists a pair $(\mathbf{B}_1', \boldsymbol{\rho}_1')$ as a solution of (2c.1.5) such that the matrix $\mathbf{B}(\cdot)$ constructed from $\mathbf{B}_1'$ obeys assumption (A6).

When $\zeta(\cdot)$ is a scalar one can explcitly display the solution for B_1' in terms of R_0 and R_1:

$$B_1, B_1' = \frac{1}{2}\left[\frac{R_0'}{R_1'} \mp \left(\left(\frac{R_0'}{R_1'}\right)^2 + 4\right)^{1/2}\right], \qquad \rho_1 = \frac{R_1'}{B_1}, \quad \rho_1' = \frac{R_1'}{B_1'}.$$

One can easily verify that $|B_1'| < 1$ since $B_1'B_1 = -1$. Further, it should be noted that $\rho_1' > \rho_1$, i.e., the noise covariance in the invertible system is *larger* than that in the noninvertible system. In other words, given a noninvertible system, we can construct an invertible system whose input noise variance is more than that of the noninvertible system.

2d. Covariance Functions and Correlograms

A preliminary analysis of the empirical data in terms of the covariances and the correlogram is very common. We will briefly consider some of the basic ideas underlying this analysis, and make the appropriate assumptions among those discussed in Section 2a. The graphical introduction to stochastic processes given by Wold (1965) is highly informative on the estimation of spectrum and correlogram.

2d.1. Relationships among the Covariances

Consider an autoregressive system obeying (2a.1.1) with $\mathbf{B}_j = 0$ and $\mathbf{G}_j = 0$ for all j. An equation for $\mathbf{R}_j$ can be obtained by multiplying the equation for $\mathbf{y}(t)$ throughout by $\mathbf{y}^{\mathrm{T}}(t-j)$ and taking expectation:

$$\mathbf{R}_i = \sum_{i=1}^{m_1} \mathbf{A}_j \mathbf{R}_{|j-i|}, \qquad \forall j \geq 1. \tag{2d.1.1}$$

For all $j \geq m_1$, (2d.1.1) constitutes a linear difference equation in $\mathbf{R}_j$ with $\mathbf{R}_0$, $\mathbf{R}_1, \ldots, \mathbf{R}_{m_1-1}$ as the initial conditions. These initial conditions can be determined as functions of the parameter matrices $\mathbf{A}_j, j = 1, 2, \ldots$, and $\boldsymbol{\rho}$ by means of the $(m_1 - 1)$ linear equations given by (2d.1.1) for $j = 1, 2, \ldots, m_1 - 1$ and the following equation:

$$\mathbf{R}_0 = \sum_{i,j=1}^{m_1} \mathbf{A}_i \mathbf{R}_{|i-j|} \mathbf{A}_j^{\mathrm{T}} + \boldsymbol{\rho}, \tag{2d.1.2}$$

which is obtained by considering the equation for $\mathbf{y}(\cdot)$, multiplying it by its transpose, and taking expectation.

Since (2d.1.2) is linear, the solution for $\mathbf{R}_j$ as a function of j can only be a geometrically decaying function of j or a linear combination of such functions. This conclusion is not altered even if there are moving average terms. Thus, the stationary stochastic processes whose correlation function $\mathbf{R}_j$ behaves like $1/(j+1)$ cannot be exactly represented by linear difference equations.

When there are moving average terms in the difference equation, the covariance matrices $\mathbf{R}_j$ still obey (2d.1.1) for all $j \geq m_1$, without involving the moving average coefficient matrices $\mathbf{B}_j$. However, the initial conditions for the difference equation, namely $\mathbf{R}_0$, $\mathbf{R}_1, \ldots, \mathbf{R}_{m_1-1}$, involve the coefficients $\mathbf{B}_j$, $\mathbf{A}_j$, and $\boldsymbol{\rho}$.

We will give two univariate examples to illustrate the preceding discussion.

Example 2d.1. Let y be a scalar process:

$$y(t) = A_1 y(t-1) + w(t) + B_1 w(t-1). \tag{2d.1.3}$$

The equation for R_j is given by

$$R_j = A_1 R_{j-1}, \qquad j \geq 2. \tag{2d.1.4}$$

To obtain expressions for the initial conditions R_0 and R_1 in terms of A_1, B_1 and ρ, rewrite (2d.1.3) as

$$\zeta(t) = w(t) + B_1 w(t-1) = y(t) - A_1 y(t-1). \tag{2d.1.5}$$

By using (2d.1.5), we can obtain

$$\begin{aligned} E[\zeta(t)\zeta(t-1)] &= B_1\rho = R_1 - A_1 R_0 - R_2 A_1 + A_1^2 R_1 \\ &= R_1 - A_1 R_0, \qquad \text{since} \quad R_2 = A_1 R_1, \end{aligned} \tag{2d.1.6}$$

$$E[\zeta^2(t)] = \rho(1 + B_1^2) = (1 + A_1^2)R_0 - 2A_1 R_1, \tag{2d.1.7}$$

by following the indicated operations. Solve these for R_0 and R_1 in terms of ρ, B_1, and A_1:

$$R_0 = \frac{\rho(1 + 2A_1B_1 + B_1^2)}{1 - A_1^2}, \qquad R_1 = A_1R_0 + B_1\rho.$$

Example 2d.2

$$y(t) = A_1y(t-1) + A_2y(t-2) + w(t). \tag{2d.1.8}$$

The equation for R_j is given by

$$R_j = A_1R_{j-1} + A_2R_{j-2}, \qquad j \geq 2. \tag{2d.1.9}$$

$$R_0 = \frac{\rho(1 - A_2)}{1 + A_2}((1 - A_2)^2 - A_1^2), \qquad R_1 = \frac{A_1R_0}{1 - A_2}.$$

We can simplify the solution by considering two separate conditions.

Case (*i*). Let $-1 < A_2 < 0$, $A_1^2 + 4A_2 < 0$. Let $\cos\omega_2 = A_1/2(-A_2)^{1/2}$. The solution for R_j can be rewritten as

$$R_j = R_0(-A_2)^{j/2}\cos(\omega_2 j + \phi)/\cos\phi, \tag{2d.1.10}$$

where

$$\tan\phi = -\frac{A_1}{2(-A_2)^{1/2}}\frac{1 + A_2}{1 - A_2}\frac{1}{(1 + A_1^2/4A_2)^{1/2}}.$$

Case (*ii*). $A_1^2 + 4A_2 > 0$, $-1 < A_2 < 1$, $-2 < A_1 < 2$. The solution is

$$R_j = \left(\frac{R_1 - R_0\lambda_2}{\lambda_1 - \lambda_2}\right)\lambda_1^j + \left(\frac{R_0\lambda_1 - R_1}{\lambda_1 - \lambda_2}\right)\lambda_2^j,$$

where $\lambda_1, \lambda_2 = A_1 \pm (A_1^2 + 4A_2)^{1/2}$.

Case (i) is referred to in subsequent chapters. The interesting feature in Case (i) is the damped sinusoidal nature of the covariances even though no explicit sinusoidal terms are present in the system equation. Case (i) is a possible model for empirical time series displaying periodicity. The comparison of the performance of the system equation corresponding to Case (i) with another model having sinusoidal terms is interesting and can be handled by the methods discussed in Chapter VIII.

We will give the general formula for evaluating the quantities $\{R_j\}$ in terms of the coefficients $A_j, B_j, \ldots,$ for the single output ARMA system (2a.1.1) with $m_2 \leq m_1 - 1$ and $G_j = 0$ for all j:

$$R_k = \sum_{j=1}^{m_1} A_jR_{k-j}, \qquad k \geq m_1.$$

The remaining quantities $R_0, \ldots, R_{m_1-1}$ are obtained by solving the following system of simultaneous equations; the proof will be left as an exercise:

$$[\mathbf{I} - \mathscr{A}_1]\begin{bmatrix} R_0 \\ \vdots \\ R_{m_1-1} \end{bmatrix} - \mathscr{A}_2\begin{bmatrix} R_1 \\ \vdots \\ R_{m_1} \end{bmatrix} = \mathscr{B}(\mathbf{I} - \mathscr{A}_1)^{-1}\begin{bmatrix} B_0 \\ B_1 \\ \vdots \\ B_{m_1-1} \end{bmatrix}\rho$$

where $R_{m_1} = \sum_{j=1}^{m_1} A_j R_{m_1-j}$, $B_0 = 1$ and the $m_1 \times m_1$ matrices $\mathscr{A}_1$, $\mathscr{A}_2$, and $\mathscr{B}$ have the form

$$\mathscr{A}_1 = \begin{bmatrix} 0 & & & & \\ A_1 & 0 & & 0 & \\ A_2 & A_1 & 0 & & \\ \vdots & \vdots & & 0 & \\ A_{m_1-1} & A_{m_1-2} & \cdots & A_1 & 0 \end{bmatrix}, \qquad \mathscr{A}_2 = \begin{bmatrix} A_1 & A_2 & A_3 & \cdots & A_{m_1} \\ A_2 & A_3 & A_4 & & \\ \vdots & \vdots & A_{m_1} & & \\ & A_{m_1} & & & 0 \\ A_{m_1} & & & & \end{bmatrix},$$

$$\mathscr{B} = \begin{bmatrix} B_0 & B_1 & \cdots & B_{m_1-1} \\ B_1 & B_2 & & \\ \vdots & \vdots & & \\ & B_{m_1-1} & 0 & \\ B_{m_1-1} & & & \end{bmatrix}.$$

2d.2. Time-Averaged Covariance Matrices

Here we introduce another set of covariance matrices obtained by time averaging. These are denoted by $\mathbf{C}_j$. We use the symbol $\hat{E}$ to denote the time averaging operator:

$$\hat{E}[\mathbf{y}(t)] \triangleq \lim_{N\to\infty} \frac{1}{N}\sum_{t=1}^{N} \mathbf{y}(t)$$

$$\mathbf{C}_j \triangleq \hat{E}[\{\mathbf{y}(t) - \hat{E}(\mathbf{y}(t))\}\{\mathbf{y}(t-j) - \hat{E}(\mathbf{y}(t-j))\}^{\mathrm{T}}]$$

$$\triangleq \lim_{N\to\infty} \frac{1}{N}\sum_{t=1}^{N} [\{\mathbf{y}(t) - \hat{E}(\mathbf{y}(t))\}\{\mathbf{y}(t-j) - \hat{E}(\mathbf{y}(t-j))^{\mathrm{T}}\}]. \qquad (2d.2.1)$$

The existence of the limits mentioned above is implicitly assumed. When the process $\mathbf{y}(\cdot)$ is weakly stationary, $\mathbf{R}_j = \mathbf{C}_j$ for all j. Otherwise, they may not be the same. To illustrate the difference between $\mathbf{R}_j$ and $\mathbf{C}_j$ consider the following example.

Example 2d.3. Consider the following single output ARMA process

$$A(D)y(t) = B(D)w(t) + C_1 \cos \omega_1 t + C_2 \sin \omega_1 t.$$

Let $y(t) = x_1(t) + x_2(t)$, where $x_1(\cdot)$ is a weakly stationary process and $x_2(\cdot)$ is a deterministic process. By substituting $x_1 + x_2$ for y in the difference equation for y and noting $\hat{E}(x_1(t) \cos \omega_1 t) = 0$ we get the following relations for x_1 and x_2:

$$A(D)x_1(t) = B(D)w(t), \tag{2d.2.2}$$

$$A(D)x_2(t) = C_1 \cos \omega_1 t + C_2 \sin \omega_1 t \tag{2d.2.3}$$

The relation between C_j and R_j is given by

$$C_j = R_j + \lim_{N\to\infty} \frac{1}{N} \sum_{t=1}^{N} x_2(t)x_2(t-j). \tag{2d.2.4}$$

Let $F_j = \hat{E}[x_2(t)x_2(t-j)]$. By using (2d.2.3) we obtain the following difference equation for F_j:

$$A(D)F_j = \frac{1}{2}\frac{C_1^2 + C_2^2}{D_1^2 + D_2^2}(D_1 \cos \omega_1 j - D_2 \sin \omega_1 j),$$

where $D_1 = \text{Re}[A(e^{-i\omega_1})]$ and $D_2 = \text{Im}[A(e^{-i\omega_1})]$. When j is large, F_j will have the approximate solution

$$F_j \approx \frac{1}{2}\frac{C_1^2 + C_2^2}{D_1^2 + D_2^2} \cos \omega_1 j.$$

Alternatively, if j is large, $C_j \approx F_j$.

2d.3. Estimation of $\mathbf{C}_j$ and $\mathbf{R}_j$

If the available history of $\mathbf{y}$ is $\{\mathbf{y}(t), t = 1, 2, \ldots, N\}$, then $\hat{\mathbf{C}}_j(N)$ is a consistent estimate of $\mathbf{C}_j$. This estimate follows naturally from the definition of $\mathbf{C}_j$:

$$\hat{\mathbf{C}}_j(N) = \frac{1}{N-j} \sum_{t=j+1}^{N} [(\mathbf{y}(t) - \hat{\mathbf{E}}(\mathbf{y}(t)))(\mathbf{y}(t-j) - \hat{\mathbf{E}}(\mathbf{y}(t-j)))^{\mathrm{T}}].$$

If the process $\mathbf{y}$ is weakly stationary, then $\hat{\mathbf{C}}_j(N)$ is also a consistent estimate of $\mathbf{R}_j$.

When $\mathbf{y}$ is a scalar, the graph of the normalized estimates $\hat{C}_j(N)$ [i.e., normalized to make $\hat{C}_0(N) = 1$] versus j is usually called the correlogram of the process.

We will obtain an expression for the mean square error of the estimate $\hat{C}_j$ given above which is an indication of the accuracy of the estimate (Hannan, 1960). Let y be a weak stationary, scalar Gaussian process with an autoregressive representation. Let $\hat{C}_j(N)$ denote an estimate of R_j based on N observations:

$$\begin{aligned} E[(\hat{C}_j(N) - R_j)^2] &= E(\hat{C}_j(N))^2 - R_j^2 \\ &= \frac{1}{N^2} E\left[\sum_{k,i=1}^{N} y(k)y(k-j)y(i)y(i-j)\right] - R_j^2 \\ &= \frac{1}{N^2} \sum_{k,i=1}^{N} [R_{|i-k|}^2 + R_j^2 + R_{k-i+j}R_{k-j+1}] - R_j^2, \end{aligned}$$

by the normality assumption. Since R_j exponentially decays with j, the dominant term in the expression above is

$$E[(\hat{C}_j(N) - R_j)^2] = \frac{\sum_{j=-\infty}^{\infty} R_j^2}{N}. \qquad (2d.3.1)$$

To obtain further understanding of the expression given above let us consider the system in Example 2d.2, Case (i). Let $A_1 = 1.41$, $A_2 = -0.77$. The summation $\sum R_j^2$ can be evaluated easily since we are dealing with an exponential series

$$\left\{\frac{E[\hat{C}_j(N) - R_j]^2}{R_0}\right\}^{1/2} \approx \frac{3.75}{\sqrt{N}}.$$

When N is small, the estimates $\hat{C}_j$ can be very inaccurate. The expression (2d.3.1) is useful for comparing the goodness of fit between the correlogram of a model proposed for the data and the correlogram of the data itself. The discrepancy between the two is not considered excessive if it is not more than two standard deviations.

2e. Spectral Analysis

2e.1. Definition and Examples

Spectral analysis is routinely used in the preliminary analysis of empirical data. However, it is also used as a standard characteristic of a process, especially in the comparison of the different models to fit the same data. In particular, a model may be judged to be superior to another if its spectral density fits that of the given data better than the spectral density of another model. However, accurate estimation of spectral density is difficult, especially when the sample sizes are small. Sufficient care must be exercised in obtaining inferences based only on the spectrum. The inaccuracies in the estimates of the spectral densities are not unexpected considering the possible error that can occur even in the estimation of covariances in a single output process.

If $\mathbf{y}(\cdot)$ and $\mathbf{u}(\cdot)$ are two processes connected by the transfer matrix $\mathbf{T}$, $\mathbf{y}(t) = \mathbf{T}(D)\mathbf{u}(t)$, then $\mathbf{S}_{yy}(\cdot)$, the spectral density of $\mathbf{y}$, and $\mathbf{S}_{yu}(\cdot)$, the cross spectral density, can be written as

$$\mathbf{S}_{yy}(e^{-i\omega}) = \mathbf{T}(e^{-i\omega})\mathbf{S}_{uu}(e^{-i\omega})\mathbf{T}^{\mathrm{T}}(e^{i\omega})$$
$$\mathbf{S}_{yu}(e^{-i\omega}) = \mathbf{T}(e^{-i\omega})\mathbf{S}_{uu}(e^{-i\omega}).$$

An alternative representation for $\mathbf{S}_{yu}(\cdot)$ is

$$\mathbf{S}_{yu}(e^{-i\omega}) = \sum_{k=-\infty}^{\infty} e^{-i\omega k} E[\mathbf{y}(j)\mathbf{u}^{\mathrm{T}}(j-k)].$$

If the input $\mathbf{u}(\cdot)$ is an m-vector white noise sequence $\mathbf{w}(\cdot)$ with covariance matrix $\boldsymbol{\rho}$, then the corresponding representation for $\mathbf{S}_{yy}(e^{-i\omega})$ is $\mathbf{S}_{yy}(e^{-i\omega}) = \mathbf{T}(e^{-i\omega})\boldsymbol{\rho}\mathbf{T}^{\mathrm{T}}(e^{i\omega})$. Some examples illustrating these expressions are given below.

Example 2e.1. Consider the process in Example 2d.2, Case (i):

$$S_{yy}(e^{-i\omega}) = \rho\left[(1 + A_2)^2\left(1 + \frac{A_1^2}{4A_2}\right) - 4A_2(\cos\omega - \cos\omega_1)^2\right]^{-1}$$

$$\cos\omega_1 = A_1(A_2 - 1)/4A_2.$$

The interesting feature of this example is the sharp peak at the frequency $\omega = \omega_1$. ω_1 is different from ω_2, the frequency of the oscillation associated with the correlations $R_j, j = 0, 1, 2, \ldots$.

It is relevant to point out that we often have problems in discriminating between models such as the one in Example 2e.1 which show sharp maxima in their spectral densities and the models that have deterministic sinusoidal components of the corresponding frequency in them.

2e.2. Estimation of Spectral Density in Weak Stationary Processes

In this section we consider the estimation of the spectral density of a scalar weakly stationary process. Let $y(1), \ldots, y(N)$ be the N available observations of the process which have been normalized so that their sample mean is zero. The estimate of the covariance $\hat{C}_j$ of the process is

$$\hat{C}_j = \frac{1}{N-j}\sum_{t=j+1}^{N} y(t)y(t-j). \tag{2e.2.1}$$

A common estimate of $S_{yy}(e^{-i\omega})$ is

$$\hat{S}_{yy}(e^{-i\omega}) = \sum_{k=-N+1}^{N-1} \hat{C}_k e^{-ik\omega}. \tag{2e.2.2}$$

Another common estimate of S_{yy} is the periodogram estimate which is obtained by direct Fourier transformation of data. Let

$$Y_N(\omega) = \sum_{k=1}^{N} y(k)e^{-ik\omega}. \tag{2e.2.3}$$

Then the periodogram estimate of S_{yy} is denoted by $\bar{S}_{yy}$:

$$\bar{S}_{yy}(e^{-i\omega}) = \frac{1}{N}[Y_N(\omega)Y_N^*(\omega)] = \frac{1}{N}\left\|\sum_{k=1}^{N} y(k)e^{-ik\omega}\right\|^2. \tag{2e.2.4}$$

To relate the estimate $\bar{S}_{yy}$ to $\hat{S}_{yy}$ we will first simplify the expression for $\bar{S}_{yy}$ in (2e.2.4):

$$\begin{aligned}
\bar{S}_{yy}(e^{-i\omega}) &= \frac{1}{N}\sum_{k=1}^{N}\sum_{j=1}^{N} y(k)y(j)e^{-ik\omega}e^{ij\omega} \\
&= \frac{1}{N}\sum_{l=-N}^{N}\left(\sum_{k=j+1}^{N} y(k)y(k-l)\right)e^{-il\omega} \\
&= \frac{1}{N}\sum_{l=-N}^{N}(N-j)\hat{C}_l e^{-il\omega}, \qquad \text{by (2e.2.1)} \\
&= \sum_{l=-N}^{N}\left(1 - \frac{j}{N}\right)\hat{C}_l e^{-il\omega}.
\end{aligned}$$

Thus the difference between the two estimates $\bar{S}_{yy}$ and $\hat{S}_{yy}$ lies in the different weightings attached to the empirical covariances $\hat{C}_j$, $j = 0, 1, 2, \ldots$.

The Fourier transform $Y_N(\omega)$ does not converge in any sense as N tends to infinity. Further, the estimate $\bar{S}_{yy}$ has the asymptotic properties

$$\lim_{N\to\infty} E[\bar{S}_{yy}(e^{-i\omega})] = S_{yy}(\omega) + O\left(\frac{\log N}{N}\right)$$

$$\lim_{N\to\infty} \operatorname{cov}[\bar{S}_{yy}(e^{-i\omega})] = S_{yy}^2(e^{-i\omega}), \qquad \omega \neq 0, \pm\pi$$

$$= 2S_{yy}^2(e^{-i\omega}), \qquad \omega = 0, \pm\pi.$$

Thus the periodogram estimate is not a good estimate of $S_{yy}(e^{-i\omega})$ since its standard deviation is of the same magnitude as the quantity to be estimated. Similarly, we can show that the estimate $\hat{S}_{yy}$ is also biased.

To obtain better estimates, we can consider two alternatives. The given data can be divided into a number of batches. The spectral density of each batch of data can be separately estimated and the various estimates averaged. This procedure greatly improves the quality of the estimate because the averaging effectively removes the bias. The second method is to smooth the estimate obtained ("raw" estimate) by means of suitable window functions. To illustrate the use of one such window function let us consider the Bartlett window. Let $\bar{S}_{yy}(e^{-i\omega})$ denote the estimated spectral density obtained above from the averaging process, with each batch of data having M observations. Then the smoothed estimate S_{yy}^* has the form

$$S_{yy}^*(e^{-i\omega}) = \int_{-\pi}^{\pi} W_B(\omega - \omega_1)\bar{S}_{yy}(e^{-i\omega_1})\, d\omega_1$$

where

$$W_B(\omega) = M\left(\frac{\sin \omega M}{\omega M}\right)^2.$$

The estimate $\bar{S}_{yy}$ is smoothed by using the Bartlett window function W_B. The function W_B is symmetric about the origin and has zeros at $\omega = 2\pi/M$, $4\pi/M, \ldots$. There are other window functions defined by Hamming, Tukey, Parzen, and others [Jenkins and Watts (1968)].

An entirely new approach has been suggested by Parzen (1969) which does away with the use of the spectral window. One of the main criticisms of the window methods is their ad hoc nature. Even though we are dealing with an estimation problem, a well-defined criterion is lacking. Parzen uses the least squares criterion, fits an autoregressive process to the data, and considers the spectral density of the fitted AR process to be an estimate of the unknown spectral density of the process. Numerical results obtained from this method appear to be good.

2e.3. Estimation of Spectral Density with Covariance Stationary Processes

Let us illustrate the problems involved by considering the following scalar process $y(\cdot)$ which is a sum of a weakly stationary part $x(\cdot)$ and the sinusoidal trend term of frequency $\omega_1 : y(t) = x(t) + \gamma e^{i\omega_1 t}$. If we do not know the structure of the process $y(\cdot)$ beforehand, then we can only estimate the time-averaged covariances $\hat{C}_j$ by using the given observations of $y(\cdot)$ and obtain an estimate of the spectral density in the usual manner: $\hat{S}_{yy}(e^{-i\omega}) = \sum_{j=-N}^{N} \hat{C}_j e^{-ij\omega}$. Even if we ignore the sampling problem (which causes the estimates $\hat{C}_j$ to be inaccurate), the estimate $\hat{S}_{yy}(\cdot)$ is not consistent since it tends to a sum of the true spectral density of y and $\gamma^2 \, \delta(\omega - \omega_1)$, where δ is the delta function. If we consider the cumulative spectral function, it should show a sharp jump at $\omega = \omega_1$.

Hence, consistent estimation is possible if we remove the trend component and repeat the operation. This procedure implies that we first have to test the given observations for the presence of trends. Although this is not an easy task some tests for determining whether a given set of observations possesses a sinusoidal trend are developed in Chapter VIII.

2f. Prediction

We have limited ourselves to the theory of prediction relevant for the stochastic difference equation models. A more general theory of prediction in linear dynamical systems can be found in the papers by Kalman (1963) and Kailath (1968). A discussion of the various methods of forecasting used in sales literature can be found in Kendall (1973).

2f.1. The General Form of the Predictor in Multivariate Systems

Using the probability distribution of $\mathbf{y}(t)$ only, we can obtain a suitable forecast of $\mathbf{y}(t)$. It is reasonable to say that a forecast of $\mathbf{y}(t)$ which utilizes both the observation history $\boldsymbol{\xi}(t-1)$ until time $(t-1)$ and the probability distribution function of $\mathbf{y}(t)$ should be better than the forecast based only on the probability distribution of $\mathbf{y}(t)$:

$$\boldsymbol{\xi}(t-1) = \{\mathbf{y}(t-1), \ldots, \mathbf{y}(-m_1), \mathbf{u}(t-2), \ldots, \mathbf{u}(-m_3)\}.$$

However, one can incorporate the information contained in $\boldsymbol{\xi}(t-1)$ into the forecast of $\mathbf{y}(\cdot)$ only if the functional relationship between $\mathbf{y}(t)$ and $\boldsymbol{\xi}(t-1)$ is known. All ad hoc schemes for forecasting imply such a relationship whether or not it is explicitly stated. The functional relationship between $\mathbf{y}(t)$ and $\boldsymbol{\xi}(t-1)$ can be represented as

$$\mathbf{y}(t) = \mathbf{h}_t[\boldsymbol{\xi}(t-1), \boldsymbol{\theta}, \mathbf{w}(t)], \qquad (2f.1.1)$$

where $\{\mathbf{w}(\cdot)\}$ is an IID sequence that is not directly observable, $\boldsymbol{\theta}$ is a vector of parameters, and $\mathbf{h}_t$ is a known function. If the function $\mathbf{h}_t$ is unknown,

(2f.1.1) does not give us any new information, and consequently knowledge of $\boldsymbol{\xi}(t-1)$ is not of any use in forecasting $\mathbf{y}(t)$. The parameter $\boldsymbol{\theta}$ is usually unknown. When $\boldsymbol{\theta}$ is not known we can regard it as a random variable whose value is constant throughout the experiment, unlike $\mathbf{w}(\cdot)$, whose value continuously changes throughout the experiment.

The second component in forecasting is the criterion function. A forecast will rarely coincide with the actual value; a loss function or criterion function reflects the degree of importance attached to the difference between the forecast and the actual value. A commonly used loss function is the quadratic function $J_1 = \|\mathbf{y}(t) - \hat{\mathbf{y}}(t|t-1)\|^2$, where $\hat{\mathbf{y}}(t|t-1)$ is a predictor of $\mathbf{y}(t)$. The quadratic loss function is analytically easy to handle. Moreover, the predictors obtained from the quadratic function are often satisfactory. The expected value of the loss function is called the risk function Q_1. $Q_1 = E[\|\mathbf{y}(t) - \hat{\mathbf{y}}(t|t-1)\|^2]$. It is customary to choose the forecast to minimize Q_1. Let $\hat{\mathbf{y}}(t|t-1) = \mathbf{g}(\boldsymbol{\xi}(t-1))$, where $\mathbf{g}$ is any function of the indicated arguments. Then we choose the function $\mathbf{g}$ to minimize Q_1. The optimal predictor that minimizes the quadratic risk function Q_1 has the form

$$\mathbf{y}^*(t|t-1) = E[\mathbf{y}(t)|\boldsymbol{\xi}(t-1)]. \qquad (2f.1.2)$$

An important property of the optimal predictor $\mathbf{y}^*$ is its orthogonality property

$$\tilde{\mathbf{y}}(t) = \mathbf{y}(t) - \mathbf{y}^*(t|t-1), \qquad E[(\tilde{\mathbf{y}}(t))^{\mathrm{T}}\mathbf{h}(\boldsymbol{\xi}(t-1))] = 0, \qquad (2f.1.3)$$

where $\mathbf{h}(\cdot)$ is any m-vector function of the arguments, and $\tilde{\mathbf{y}}(\cdot)$ denotes the error in the optimal predictor. Often (2f.1.3) is stated as follows: The optimal prediction error is orthogonal to all observations used in the predictor. The use of geometric language is prompted by the interesting geometrical interpretation of the predictor given below.

Let us regard the random variables $\mathbf{y}(t)$, $\mathbf{y}(t-1)$, and $\mathbf{y}(t-2)$ as elements of a Hilbert space. If $\mathbf{y}'$, $\mathbf{y}''$ are any two elements of the space, the distance function between them is the quadratic function $E[\|\mathbf{y}' - \mathbf{y}''\|^2]$. The vectors corresponding to $\mathbf{y}(t)$, $\mathbf{y}(t-1)$, and $\mathbf{y}(t-2)$ are represented in Fig. 2f.1.1 as OC, OB, and OA. Then $\mathbf{y}^*(t|t-1) = E[\mathbf{y}(t)|\boldsymbol{\xi}(t-1)] \triangleq OD$, the orthogonal

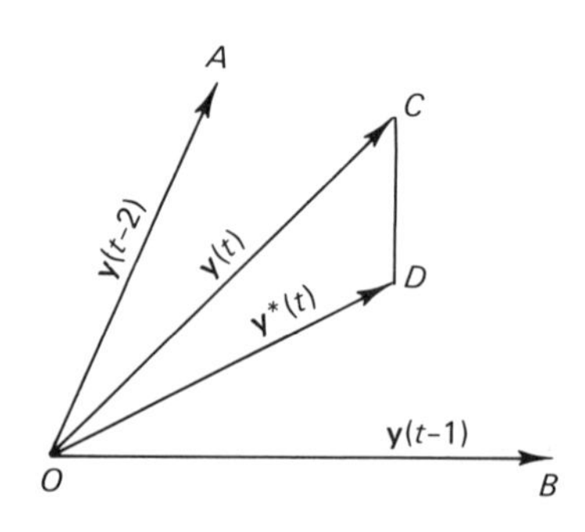

projection of the vector $\mathbf{y}(t)$ on the space spanned by $\mathbf{y}(t-1)$ and $\mathbf{y}(t-2)$, and $CD = OC - OD = \tilde{\mathbf{y}}(t)$, the prediction error. By definition, CD is perpendicular to the space spanned by OA and OB, and hence CD is perpendicular to any vector in that space. This is the orthogonality property in (2f.1.3), which can be proved as follows:

$$\begin{aligned} E[(\tilde{\mathbf{y}}(t))^{\mathrm{T}}\mathbf{h}(\boldsymbol{\xi}(t-1))] &= E[E\{(\tilde{\mathbf{y}}(t))^{\mathrm{T}}\mathbf{h}(\boldsymbol{\xi}(t-1))|\boldsymbol{\xi}(t-1)\}]\dagger \\ &= E[\{E(\mathbf{y}(t)|\boldsymbol{\xi}(t-1)) - \mathbf{y}^*(t|t-1)\}^{\mathrm{T}}\mathbf{h}(\boldsymbol{\xi}(t-1))] \\ &= 0, \end{aligned} \tag{2f.1.4}$$

by the definition of $\mathbf{y}^*$ and since $\mathbf{h}(\boldsymbol{\xi}(t-1))$ is a function of $\boldsymbol{\xi}(t-1)$ alone. To prove the optimality of the predictor (2f.1.2), let $\hat{\mathbf{y}}$ denote any *other* predictor:

$$\begin{aligned} &E[\|\mathbf{y}(t) - \hat{\mathbf{y}}(t|t-1)\|^2] \\ &\quad = E[\|\mathbf{y}(t) - \mathbf{y}^*(t|t-1) + \mathbf{y}^*(t|t-1) - \hat{\mathbf{y}}(t|t-1)\|^2], \\ &\quad\quad \text{by adding and subtracting } \mathbf{y}^*, \\ &\quad = E[\|\mathbf{y}(t) - \mathbf{y}^*(t|t-1)\|^2] + E[\|\mathbf{y}^*(t|t-1) - \hat{\mathbf{y}}(t|t-1)\|^2] \\ &\quad\quad + 2E[(\mathbf{y}(t) - \mathbf{y}^*(t|t-1))^{\mathrm{T}}(\mathbf{y}^*(t|t-1) - \hat{\mathbf{y}}(t|t-1))]. \end{aligned} \tag{2f.1.5}$$

The third term of (2f.1.5) is zero by the orthogonality property of (2f.1.3) since the term $[\mathbf{y}^*(t|t-1) - \hat{\mathbf{y}}(t|t-1)]$ is a function of $\boldsymbol{\xi}(t-1)$ alone. Since the second term in (2f.1.5) is nonnegative, the left-hand side of (2f.1.5) cannot be less than the first term of the right-hand side, which proves the optimality of the predictor in (2f.1.2). The predictor in (2f.1.2) is also optimal with respect to any loss function that is a convex function of $(\mathbf{y} - \hat{\mathbf{y}})$.

2f.2. Determination of the Predictor

To obtain explicit expressions for the predictor, specific assumptions must be made about $\mathbf{y}(t)$ and $\boldsymbol{\xi}(t-1)$. Let the relationship between $\mathbf{y}(t)$ and $\boldsymbol{\xi}(t-1)$ be the difference equation (2a.1.1), and let $\boldsymbol{\theta}$ be the vector of the coefficients in it, i.e., the coefficients of the polynomial matrices **A**, **B**, **G**, and **F**.

Case (i): $\boldsymbol{\theta}$ is known. We need to assume that the $\mathbf{w}(\cdot)$ sequence is zero mean, independent, and that the system (2a.1.1) is invertible. No other assumptions are needed about the probability distribution of $\mathbf{w}$. By definition

$$E[\mathbf{w}(t)|\boldsymbol{\xi}(t-j)] = 0, \qquad j = 1, 2, \ldots. \tag{2f.2.1}$$

By (2a.1.1), $\mathbf{w}(t)$ is a function of $\mathbf{y}(t-j)$, $j \geq 0$, and $\mathbf{u}(t-k)$, $k \geq 1$, or an implicit function of $\boldsymbol{\xi}(t)$. Since the system (2a.1.1) is invertible

$$E[\mathbf{w}(t)|\boldsymbol{\xi}(\tau)] = \mathbf{w}(t) \qquad \text{for} \quad \forall\tau \geq t. \tag{2f.2.2}$$

Taking conditional expectation of all the terms in (2a.1.1) conditioned on

† This is a standard technique in stochastic systems. The inner expectation is a conditional expectation and is a function of $\boldsymbol{\xi}(t-1)$. The outer expectation is taken over $\boldsymbol{\xi}(t-1)$.

$\boldsymbol{\xi}(t-1)$ and using (2f.2.1) and (2f.2.2), we obtain the following equation for $\mathbf{y}^*(t|t-1)$:

$$\mathbf{y}^*(t|t-1) = \sum_{j=1}^{m_1} \mathbf{A}_j\mathbf{y}(t-j) + \sum_{j=1}^{m_3} \mathbf{G}_j\mathbf{u}(t-j) + \sum_{j=1}^{m_2} \mathbf{B}_j\mathbf{w}(t-j) + \mathbf{F}\boldsymbol{\psi}(t-1). \tag{2f.2.3}$$

A method of estimating $\mathbf{w}(t-j)$, $j = 1, 2, \ldots$, from $\boldsymbol{\xi}(t-1)$ is needed in order to use the predictor in (2f.2.3). Hence, we define the following estimate $\bar{\mathbf{w}}(t)$ of $\mathbf{w}(t)$, which is recursively computed from (2f.2.3) which in turn is obtained from (2a.1.1) by replacing $\mathbf{w}$ by $\bar{\mathbf{w}}$:

$$\bar{\mathbf{w}}(t) = \mathbf{y}(t) - \sum_{j=1}^{m_1} \mathbf{A}_j\mathbf{y}(t-j) - \sum_{j=1}^{m_2} \mathbf{B}_j\bar{\mathbf{w}}(t-j) - \sum_{j=1}^{m_3} \mathbf{G}_j\mathbf{u}(t-j) - \mathbf{F}\boldsymbol{\psi}(t-1). \tag{2f.2.4}$$

The right-hand side of (2f.2.4) involves only the knowledge available at time t. It is easy to show that $\bar{\mathbf{w}}(t)$ tends to $\mathbf{w}(t)$ as t tends to infinity. Replacing $\mathbf{w}(t)$ in (2f.2.3) by $\bar{\mathbf{w}}(t)$ in (2f.2.4) yields the required one-step-ahead predictor. Such a replacement results in a slight error since $\bar{\mathbf{w}}(t)$ is equal to $\mathbf{w}(t)$ only asymptotically. But this error decreases exponentially since (2f.2.4) is an exponentially stable linear system.

We will also obtain an expression for the accuracy of the predictor $\mathbf{y}^*$. A measure of the accuracy of the predictor is the mean square value of the error. Since $\tilde{\mathbf{y}}(t) \triangleq \mathbf{y}(t) - \mathbf{y}^*(t|t-1) = \mathbf{w}(t)$ asymptotically,

$$E[(\tilde{\mathbf{y}}(t))(\tilde{\mathbf{y}}(t))^{\mathrm{T}}] = \mathbf{P} = \text{error covariance matrix} = \boldsymbol{\rho}.$$

The invertibility assumption is important. If this assumption is invalid, (2f.2.2) is not true and the predictor in (2f.2.3) will not be optimal. In such a case, we have to replace the noninvertible system by an equivalent invertible system and repeat the process of prediction.

Case (ii): $\boldsymbol{\theta}$ is unknown. This case is clearly more realistic than Case (i). It is also more complex since it involves the estimation of the variable $\boldsymbol{\theta}$. As such, it will be discussed in detail in Chapter VI. To illustrate the method, consider a first-order autoregressive system $\mathbf{y}(t) = \mathbf{A}_1\mathbf{y}(t-1) + \mathbf{w}(t)$. From the Bayesian point of view, the unknown quantity $\mathbf{A}_1$ may be regarded as a random matrix that assumes a constant value throughout this experiment. Even though such an interpretation may create philosophical problems, its practical utility is considerable. By taking conditional expectation throughout, conditioning on $\boldsymbol{\xi}(t-1)$, we obtain $\hat{\mathbf{y}}(t|t-1) = E[\mathbf{A}_1|\boldsymbol{\xi}(t-1)]\mathbf{y}(t-1)$. We need to evaluate the estimate $E[\mathbf{A}_1|\boldsymbol{\xi}(t-1)]$, the posterior estimate of $\mathbf{A}_1$ given the history until time $t-1$. Evaluation of $E[\mathbf{A}_1|\boldsymbol{\xi}(t-1)]$ requires knowledge of the probability density of the disturbances $\mathbf{w}(\cdot)$. We will treat this problem later.

2g. Prediction in Multiplicative Systems

2g.1. Assumptions and Motivations

We will consider the analysis and forecasting of multiplicative systems such as

$$y(t) = e^{A_0} \prod_{j=1}^{m_1} (y(t-j))^{A_j} \eta(t), \tag{2g.1.1}$$

where $\{\eta(\cdot)\}$ is a white noise sequence obeying assumption (A1) and $y(\cdot)$ is a univariate process.

Such models are needed for analyzing a variety of time series arising in many different disciplines such as animal population studies, and economics. Some examples are considered in Chapter XI.

By applying the log transform to Eq. (2g.1.1), we obtain the following, which is linear in $\ln y$:

$$\ln y(t) = A_0 + \sum_{j=1}^{m_1} A_j \ln y(t-j) + \zeta(t), \tag{2g.1.2}$$

where $\zeta(t) = \ln \eta(t)$. Equation (2g.1.2) can be generalized by adding moving average or deterministic trend terms to it. It is also convenient for the estimation of the coefficients A_j since the noise $\zeta(t)$ enters the system additively. However, the problem of prediction is more involved since we want to forecast the variable y and not $\ln y$.

We will introduce some assumptions on the noise process $\zeta(\cdot)$ or $\eta(\cdot)$. Let the variable $\zeta(t)$ be normally distributed with zero mean and variance ρ:

$$\zeta(t) \triangleq \ln \eta(t) \sim N(0, \rho), \qquad \eta(t) \sim \Lambda(0, \rho)$$

where Λ is the log-normal distribution whose density is defined as

$$p(\eta) = \frac{1}{\eta(2\pi\rho)^{1/2}} \exp\left[\frac{-(\ln \eta)^2}{2\rho}\right].$$

The reproducible property is valid here; i.e., if $y(0)$ is log-normal, then $y(t)$, $t = 1, 2, \ldots$, are all log-normal. The steady state (normal) distribution of $x(t) = \ln y(t)$ can be found and so can the steady state (log-normal) distribution of $y(\cdot)$. If the order of the system in (2g.1.2) is unity ($m_1 = 1$), then $x(\cdot)$ has the steady state distribution $N(\mu_1, \mu_2)$ or $y(t) \sim \Lambda(\mu_1, \mu_2)$, where $E[x(t)] = \alpha_0/(1-\alpha_1) \triangleq \mu_1$, $V[x(t)] = \rho/(1-\alpha_1^2) \triangleq \mu_2$.

The following formula will be of repeated use in the ensuing discussion:

$$\text{If} \quad x \sim N(\mu_1, \mu_2), \qquad \text{then} \quad E[e^{tx}] = \exp[t\mu_1 + \tfrac{1}{2}t^2\mu_2]. \tag{2g.1.3}$$

2g.2. Prediction

We will consider a one-step-ahead predictor of $y(t)$ based on the past history $\boldsymbol{\xi}(t-1)$ (Teekens, 1972). We will find the optimal predictor according to the quadratic criterion function in the preceding section,

$$Q_1 = E[(y(t) - \hat{y}(t|t-1))^2].$$

The optimal predictor $y^*(t|t-1)$ has the form

$$y^*(t|t-1) = E[y(t)|\xi(t-1)]$$
$$= e^{A_0} \prod_{j=1}^{m_1} (y(t-j))^{A_j} E(e^{\zeta(t)}), \qquad \text{from} \quad (2g.1.1)$$
$$= e^{A_0} \prod_{j=1}^{m_1} (y(t-j))^{A_j} \exp(\rho/2), \tag{2g.2.1}$$

using (2g.1.3) since $\zeta(t) \sim N(0, \rho)$.

Let us find the mean square value of the optimal one-step-ahead prediction error:

$$E[(y(t) - y^*(t|t-1))^2] = E\left[\left(e^{A_0} \prod_{j=1}^{m_1} y(t-j)^{A_j}\right)^2 (e^{\rho/2} - e^{\zeta(t)})^2\right]. \tag{2g.2.2}$$

For simplicity, we will compute the mean square value of the optimal one-step-ahead prediction error only when $m_1 = 1$. Notice that

$$E[(e^{\rho/2} - e^{\zeta(t)})^2] = E[e^{\rho} - 2e^{\rho/2}e^{\zeta(t)} + e^{2\zeta(t)}] = e^{2\rho} - e^{\rho}, \tag{2g.2.3}$$

$$E[(y(t-1))^{2A_1}] = E\{\exp[2A_1 x(t-1)]\}, \qquad \text{since} \quad x(t) = \ln y(t)$$
$$= \exp\left[(2A_1)\frac{A_0}{1-A_1} + \frac{2A_1^2}{1-A_1^2}\rho\right], \qquad \text{by (2g.1.3).} \tag{2g.2.4}$$

Substituting (2g.2.3) and (2g.2.4) into (2g.2.2), we obtain

$$E[(y(t) - y^*(t|t-1))^2] = E[e^{2A_0}(y(t-1))^{2A_1}]E[(e^{\rho/2} - e^{\zeta(t)})^2]$$
$$= [e^{2A_0}]\left[\exp\left(\frac{2A_1A_0}{1-A_1} + \frac{2A_1^2\rho}{1-A_1^2}\right)\right][e^{\rho}(e^{\rho} - 1)].$$

Note that the mean square value of the optimal one-step-ahead prediction error is dependent on the coefficients A_j of the dynamical system.

We will find the predictor according to the risk function Q_2 as well:

$$Q_2 = E[(\ln y(t) - \ln \hat{y}(t|t-1))^2] = E[\{\ln(y(t)/\hat{y}(t|t-1))\}^2].$$

The risk function Q_2 is appropriate when the ratio (or the percentage) of the actual value of the variable to that of its estimate is of interest. The problem may be dealt with in terms of the variable x:

$$x(t) \triangleq \ln y(t); \qquad \hat{x}(t|t-1) \triangleq \ln y(t|t-1)$$
$$Q_2 = E[\{x(t) - \hat{x}(t|t-1)\}^2].$$

As shown earlier,

$$\ln y^*(t|t-1) = x^*(t|t-1) = A_0 + \sum_{j=1}^{m_1} A_j x(t-j)$$
$$= A_0 + \sum_{j=1}^{m_1} A_j \ln y(t-j). \tag{2g.2.5}$$

The mean square error of the optimal estimate x^* is ρ. The main difference between the optimal predictor in (2g.2.5) and that considered in (2g.2.1) is the factor $\exp[\rho/2]$. The factor may be quite significant if ρ is large. Typically, if $\rho = 4$, the multiplying factor is e^2, which is 9.

2h. Prediction in Systems with Noisy Observations

2h.1. The Form of the Predictor

In the preceding section, the signal $\mathbf{y}(t)$ was predicted based on its past values, assuming that the past values could be measured exactly. In many cases, this assumption is unwarranted and may be incorrect. The observed variable, say $\mathbf{x}(\cdot)$, is a sum of the signal $\mathbf{y}(\cdot)$ and a zero mean white noise $\boldsymbol{\eta}$. We would like to predict $\mathbf{y}(t)$ and $\mathbf{x}(t)$ based on the history $\mathbf{x}(j), j \leq t - 1$.

We will consider a system with only autoregressive and moving average terms, and will assume $m_1 > m_2$:

$$\mathbf{A}(D)\mathbf{y}(t) = \mathbf{B}(D)\mathbf{w}(t), \qquad \mathbf{x}(t) = \mathbf{y}(t) + \boldsymbol{\eta}(t). \tag{2h.1.1}$$

The sequence $\{\boldsymbol{\eta}(\cdot)\}$ is zero mean and IID with a covariance matrix $\mathbf{R}_\eta$, and is independent of $\{\mathbf{w}(\cdot)\}$. We will denote the covariance matrix of $\mathbf{w}$ by $\mathbf{R}_w$ instead of the usual symbol $\boldsymbol{\rho}$.

We can express Eq. (2h.1.1) in state variable form and hence obtain the requisite predictor by means of the standard Kalman (1963) theory. This method is not given here since it is available in many books. Instead we will develop another approach which gives further insight into the structure of the process $\mathbf{x}(\cdot)$ in addition to giving us the required predictors.

The basic idea (Kashyap, 1970a) is to obtain an ARMA equation for $\mathbf{x}(t)$:

$$\mathbf{A}(D)\mathbf{x}(t) = \mathbf{B}'(D)\mathbf{v}(t) = \sum_{j=0}^{m_5} \mathbf{B}_j'\mathbf{v}(t-j), \tag{2h.1.2}$$

where $m_5 = \max[m_1, m_2]$, $\mathbf{B}_0' = \mathbf{I}$, and $\mathbf{v}(\cdot)$ is a zero mean IID noise sequence with covariance matrix $\mathbf{R}_v$. Later we will develop explicit expressions for $\mathbf{B}_j'$, $j = 1, \ldots, m_5$, and $\mathbf{R}_v$, in terms of $\mathbf{A}_j$, $\mathbf{B}_j$, $\mathbf{R}_w$, and $\mathbf{R}_\eta$. Using (2h.1.1), one can obtain an expression for the one-step-ahead predictor $\hat{\mathbf{x}}(t|t-1)$ in terms of $\mathbf{x}(t-j), j \geq 1$. By using the techniques of the earlier sections and Eq. (2h.1.1), one can obtain the following one-step-ahead predictor of $\mathbf{y}(t)$ based on $\mathbf{x}(t-j)$, $j \geq 1$: $\hat{\mathbf{y}}(t|t-j) = \hat{\mathbf{x}}(t|t-j), j \geq 1$.

2h.2. The ARMA Equation for the Noise-Corrupted Output

We will assume that $m_1 = m_2 - 1$. Substitute for $\mathbf{y}$ in (2h.1.1) in terms of $\mathbf{x}$,

$$\mathbf{A}(D)\mathbf{x}(t) = \mathbf{B}(D)\mathbf{w}(t) - \mathbf{A}(D)\boldsymbol{\eta}(t). \tag{2h.2.1}$$

Let

$$\boldsymbol{\zeta}(t) = \mathbf{B}(D)\mathbf{w}(t) - \mathbf{A}(D)\boldsymbol{\eta}(t). \tag{2h.2.2}$$

The first term in (2h.2.1), $\mathbf{B}(D)\mathbf{w}(t)$, is a moving average expression of order m_2. The second term in (2h.2.1), $\mathbf{A}(D)\boldsymbol{\eta}(t)$, is a moving average representation of order m_1. Hence the right-hand side of (2h.2.1) can be represented by a moving average representation of order $m_5 = \max[m_1, m_2] = m_1$ by the earlier assumption. Let us call this new moving average representation $\mathbf{B}'(D)\mathbf{v}(t)$, where $\{\mathbf{v}(\cdot)\}$ is a zero mean white noise with covariance matrix $\mathbf{R}_v$. Hence,

$$\boldsymbol{\zeta}(t) \triangleq \mathbf{B}'(D)\mathbf{v}(t) = \left(\sum_{j=0}^{m_1} \mathbf{B}_j' D^j\right)\mathbf{v}(t). \tag{2h.2.3}$$

To find the matrix coefficients $\mathbf{B}_j'$ in terms of $\mathbf{A}_j$, $\mathbf{B}_j$, $\mathbf{R}_w$, and $\mathbf{R}_\eta$, we need to equate the corresponding correlations of the process $\boldsymbol{\zeta}(\cdot)$ as given by (2h.2.3) and (2h.2.2). Let

$$\boldsymbol{\zeta}_1(t) = \mathbf{B}(D)\mathbf{w}(t), \qquad \boldsymbol{\zeta}_2(t) = \mathbf{A}(D)\boldsymbol{\eta}(t).$$

Recall

$$\mathbf{B}(D) = \mathbf{B}_0 + \mathbf{B}_1 D + \cdots + \mathbf{B}_{m_2} D^{m_2}$$
$$\mathbf{A}(D) = \mathbf{A}_0 - \mathbf{A}_1 D - \cdots - \mathbf{A}_{m_1} D^{m_1} = \mathbf{A}_0' + \mathbf{A}_1' D + \cdots + \mathbf{A}_{m_1}' D^{m_1},$$

i.e.,

$$\mathbf{B}_0 = \mathbf{A}_0 = \mathbf{B}_0' = \mathbf{A}_0' = \mathbf{I} \quad \text{and} \quad \mathbf{A}_j' \triangleq -\mathbf{A}_j, \quad j \geq 1$$

$$E[\boldsymbol{\zeta}_1(t)\boldsymbol{\zeta}_1^{\mathrm{T}}(t-k)] = E\left[\left\{\sum_{j=0}^{m_1} \mathbf{B}_j \mathbf{w}(t-j)\right\}\left\{\sum_{j=0}^{m_1} \mathbf{w}^{\mathrm{T}}(t-j-k)\mathbf{B}_j^{\mathrm{T}}\right\}\right]$$
$$= \sum_{j=k}^{m_1} \mathbf{B}_j \mathbf{R}_w \mathbf{B}_{j-k}^{\mathrm{T}}. \tag{2h.2.4}$$

Similarly,

$$E[\boldsymbol{\zeta}_2(t)\boldsymbol{\zeta}_2^{\mathrm{T}}(t-k)] = \sum_{j=k}^{m_1} \mathbf{A}_j' \mathbf{R}_\eta (\mathbf{A}_{j-k}')^{\mathrm{T}}. \tag{2h.2.5}$$

Since $\mathbf{w}$ and $\boldsymbol{\eta}$ are independent processes, Eq. (2h.2.4) yields

$$E[\boldsymbol{\zeta}(t)\boldsymbol{\zeta}^{\mathrm{T}}(t-k)] = E[\boldsymbol{\zeta}_1(t)\boldsymbol{\zeta}_1^{\mathrm{T}}(t-k)] + E[\boldsymbol{\zeta}_2(t)\boldsymbol{\zeta}_2^{\mathrm{T}}(t-k)]$$
$$= \sum_{j=k}^{m_1} \mathbf{B}_j \mathbf{R}_w \mathbf{B}_{j-k}^{\mathrm{T}} + \sum_{j=k}^{m_1} \mathbf{A}_j' \mathbf{R}_\eta (\mathbf{A}_{j-k}')^{\mathrm{T}}$$
$$= 0 \qquad \forall k > m_1. \tag{2h.2.6}$$

Next, we can use the expression for $\boldsymbol{\zeta}(\cdot)$ in (2h.2.3) and obtain expressions for the correlations,

$$E[\boldsymbol{\zeta}(t)\boldsymbol{\zeta}^{\mathrm{T}}(t-k)] = \sum_{j=k}^{m_1} \mathbf{B}_j' \mathbf{R}_v (\mathbf{B}_{j-k}')^{\mathrm{T}}$$
$$= 0, \qquad \forall k > m_1. \tag{2h.2.7}$$

By equating the corresponding correlations from Eqs. (2h.2.6) and (2h.2.7) for $k = m_1, m_1 - 1, \ldots, 0$, we get the following $(m_1 + 1)$ matrix equations in $(m_1 + 1)$ unknowns $\mathbf{B}_j', j = 1, \ldots, m_1$, and $\mathbf{R}_v$:

$$\begin{aligned}
\mathbf{B}'_{m_1}\mathbf{R}_v &= \mathbf{A}'_{m_1}\mathbf{R}_\eta \\
\mathbf{B}'_{m_1-1}\mathbf{R}_v + \mathbf{B}'_{m_1}\mathbf{R}_v(\mathbf{B}_1')^{\mathrm{T}} &= \mathbf{A}'_{m_1-1}\mathbf{R}_\eta + \mathbf{A}'_{m_1}\mathbf{R}_\eta(\mathbf{A}_1')^{\mathrm{T}} + \mathbf{B}_{m_1-1}\mathbf{R}_w \\
&\vdots \\
\sum_{j=0}^{m_1} \mathbf{B}_j'\mathbf{R}_v(\mathbf{B}_j')^{\mathrm{T}} &= \sum_{j=0}^{m_1} \mathbf{A}_j'\mathbf{R}_\eta(\mathbf{A}_j')^{\mathrm{T}} + \sum_{j=0}^{m_1-1} \mathbf{B}_j\mathbf{R}_w\mathbf{B}_j^{\mathrm{T}}.
\end{aligned} \tag{2h.2.8}$$

Equation (2h.2.8) may appear formidable, but it can be solved for $\mathbf{R}_v$ and $\mathbf{B}_j', j = 1, 2, \ldots, m_1$. Usually more than one solution is possible and we should pick that solution whose corresponding polynomial $\mathbf{B}'(D)$ obeys assumption (A6), i.e., its determinant has all its zeros outside the unit circle. We will illustrate the technique by an example.

Example 2h.1. Let the scalar signal y obey an AR$(\cdot)$ process

$$y(t) = A_1 y(t-1) + w(t), \qquad x(t) = y(t) + \eta(t). \tag{2h.2.9}$$

Then the ARMA equation for x will be

$$x(t) - A_1 x(t-1) = v(t) + B_1' v(t-1). \tag{2h.2.10}$$

From (2h.2.8) we obtain two nonlinear simultaneous equations for B_1' and R_v, noting that $A_1' = -A_1$:

$$B_1' R_v = -A_1 R_\eta \tag{2h.2.11}$$

$$(1 + (B_1')^2)R_v = (1 + A_1^2)R_\eta + R_w. \tag{2h.2.12}$$

Dividing (2h.1.12) by (2h.1.11) we obtain a quadratic equation for B_1':

$$(B_1')^2 + B_1' \frac{\{(1 + A_1^2)R_\eta + R_w\}}{A_1 R_\eta} + 1 = 0. \tag{2h.2.13}$$

We can solve for B_1' from (2h.2.13). Between the two solutions, we prefer the one satisfying $|B_1'| < 1$ so that the invertibility condition (A6) is satisfied. Using this solution and (2h.2.11) we can obtain R_v. Since B_1' and R_v are known, $x(t)$ can be predicted by using (2h.2.10).

2i. Rescaled Range–Lag Characteristic

In many cases the standard second-order characterization of a process given by the spectral density or correlogram is not sufficient for analyzing the process, especially when we are dealing with its extreme values. For example, in designing a dam on a river we are interested in knowing the expected value of the maximum of the cumulative flows in a certain period of time. Another example is the inventory control problem where we want to know the cumulative demand in a certain period of time. In studying such problems involving a scalar sequence of random variables, the rescaled range–lag characteristic is useful; the information given by this characteristic complements the information given by the spectral density. The importance of this characteristic was first discussed by Hurst *et al.* (1965) in analyzing a variety of "real life" time series such as those of tree ring thicknesses, sunspots, river flows, and rainfall.

Let us define the range $R_1(s)$ associated with a scalar sequence of zero mean random variables $\{y(\cdot)\}$:

$$R_1(s) = \max_{1 \le t \le s} \sum_{i=1}^{t} y(i) - \min_{1 \le t \le s} \sum_{i=1}^{t} y(i). \tag{2i.1.1}$$

Note that the range $R_1(s)$ is nonnegative for all s. The qualitative behavior of $R_1(s)$ with s can be ascertained by considering the following variable $R_2(s)$, which is easier to analyze:

$$R_2(s) = \max_{1 \le t \le s} \sum_{i=1}^{t} y(i). \tag{2i.1.2}$$

By inspection, we see that $R_2(s)$ obeys the difference equation

$$R_2(s+1) = \max[R_2(s), R_2(s) + y(s+1)], \tag{2i.1.3}$$

from which we can obtain expressions for the mean and variance of $R_2(s)$. Taking expectation throughout (2i.1.3), we obtain

$$E[R_2(s+1)] = \int_{-\infty}^{\infty} dR \left[\int_{-\infty}^{0} dy R p(y, R) + \int_{0}^{\infty} dy (R + y) p(y, R) \right], \tag{2i.1.4}$$

where $p(y, R)$ is the joint probability density of $y(k+1)$ and $R_2(s)$. Simplifying (2i.1.4), we obtain

$$\begin{aligned} E[R_2(s+1)] &= \int_{-\infty}^{\infty} dR \left[\int_{-\infty}^{0} dy R p(y, R) + \int_{0}^{\infty} y p(y, R)\, dy \right] \\ &= E[R_2(s)] + \int_{-\infty}^{\infty} dR \int_{0}^{\infty} y p(y, R)\, dy \\ &= E[R_2(s)] + \int_{0}^{\infty} y p(y)\, dy \\ &\triangleq E[R_2(s)] + c_1, \qquad c_1 > 0, \end{aligned} \tag{2i.1.5}$$

where $p(y)$ is the probability density of $y(s+1)$. Equation (2i.1.5) indicates that $E[R_2(s)]$, and hence $E[R_1(s)]$, varies *linearly* with the lag s. This result may be a little surprising considering that $(\sum_{t=1}^{N} y(t)/\ln \ln N)$ tends to ± 1 with probability 1 as N tends to infinity in view of the law of iterated logarithm.

Next let us find the mean square value of $R_2(s)$ and hence its variance. Squaring both sides of (2i.1.3) and taking expectation, we obtain

$$\begin{aligned}E[R_2{}^2(s+1)] &= \int_{-\infty}^{\infty} dk\left[\int_{-\infty}^{0} dyR^2p(y,R) + \int_0^{\infty} dy(R+y)^2p(y,R)\right]\\ &= \int_{-\infty}^{\infty} dR\int_{-\infty}^{\infty} dyR^2p(y,R) + \int_0^{\infty} y^2p(y)\,dy\\ &\quad + 2\int_{-\infty}^{\infty} dR\int_0^{\infty} dyRyp(y,R). \qquad (2i.1.6)\end{aligned}$$

Let us assume $\{y(\cdot)\}$ is IID. Then $y(s+1)$ and $R_2(s)$ are independent. In that case, the last term on the right-hand side of (2i.1.6) is equal to $2E(R_2(s))c_1$:

$$\begin{aligned}E[R_2{}^2(s+1)] &= E[R_2{}^2(s)] + \int_0^{\infty} y^2p(y)\,dy + 2E[R_2(s)]c_1\\ \mathrm{var}[R_2(s+1)] &\triangleq E[R_2{}^2(s+1)] - (E(R_2(s+1)))^2\\ &= E[R_2{}^2(s)] + \int_0^{\infty} y^2p(y)\,dy + 2E[R_2(s)]c_1 - (E(R_2(s)) + c_1)^2\\ &= \mathrm{var}[R_2(s)] + \int_0^{\infty} y^2p(y)\,dy - c_1{}^2\\ &\triangleq \mathrm{var}[R_2(s)] + c_3,\end{aligned}$$

or

$$\mathrm{var}[R_2(s)] = sc_3, \qquad (2i.1.7)$$

where

$$c_3 = \int_0^{\infty} y^2p(y)\,dy - \left(\int_0^{\infty} yp(y)dy\right)^2.$$

According to (2i.1.7) $\mathrm{var}[R_2(s)]$ grows linearly with s and so also does the variance of $R_1(s)$. In the derivation of (2i.1.7), $\{y\}$ is assumed to be IID. Equation (2i.1.7) is also asymptotically valid if $y(\cdot)$ obeys an asymptotically stable ARMA process with zero mean.

In the foregoing analysis, we have assumed that the mean of the process $y(\cdot)$ is exactly zero. In practice one does not know the mean, and the process may even have a time-varying mean. To handle such processes, we define the variable $R_3(t, s)$. Consider any sequence $\{y(\cdot)\}$ not necessarily zero mean. Define $x(t, s, v) = \sum_{j=1}^{v} y(t+j) - (v/s)\sum_{j=1}^{s} y(t+j)$. Note that $x(t, s, s) \triangleq 0$,

$$R_3(t, s) = \max_{1 \le v \le s} x(t, s, v) - \min_{1 \le v \le s} x(t, s, v).$$

We can illustrate the role of the variable x by an example. Suppose $y(t)$ represents the flow in a river at a location in the tth month and we are interested in designing a reservoir at the location. Suppose at the tth month, we know in advance the flows $y(t+1), \ldots, y(t+s)$ in the next s months. If we adopt a policy of uniform discharge of $(1/s) \sum_{j=1}^{s} y(t+j)$ during the entire s months, then actual water content in the reservoir at the $(t+v)$th month is $x(t, s, v)$. The variable $R_3(t, s)$ tells us the reservoir capacity that is necessary to provide the storage for the s months.

It is convenient to normalize R_3 by dividing it by the standard deviation σ_3:

$$\begin{aligned}\sigma_3{}^2(t, s) &= \text{sample variance of the process } y \text{ in the interval } (t, t+s) \\ &= \frac{1}{s} \sum_{u=1}^{s} y^2(t+v) - \left\{\frac{1}{s} \sum_{u=1}^{s} y(t+v)\right\}^2.\end{aligned}$$

The rescaled range is the ratio $R_3(t, s)/\sigma_3(t, s)$.

Suppose we postulate a suitable probabilistic dynamic model for the given sequence $y(\cdot)$, such as a stochastic difference equation. Then we can consider the following expectations:

$$E[R_3(t, s)|s] = R_4(s), \qquad E[\sigma_3{}^2(t, s)|s] = \sigma_4{}^2(s), \tag{2i.1.8}$$

where the right-hand sides have been implicitly assumed to be free of t. The rescaled range is estimated by computing $R_4(s)$ and $\sigma_4(s)$ values for various values of s, from the observed values $\{y(\cdot)\}$. This is accomplished by estimating $R_3(t, s)$ and $\sigma_3(t, s)$ values for various t for the given s, and averaging over t. Since computational considerations preclude the evaluation of $R_3(t, s)$ for every t and s, $R_3(t, s)$ and $\sigma_3(t, s)$ are estimated for any given s at $t = t_1, \ldots, t_n$, where t_i are prespecified. The estimated rescaled range which is an estimate of $R_4(s)/\sigma_4(s)$ is denoted by

$$R(s)/\sigma(s) = \frac{1}{n} \sum_{i=1}^{n} [R_3(t_i, s)/\sigma_3(t_i, s)]. \tag{2i.1.9}$$

It is not possible to obtain an analytical expression for a suitable measure of the accuracy of the estimate $R(s)/\sigma(s)$, such as the mean square error $E[(R(s)/\sigma(s) - R_4(s)/\sigma_4(s))^2]$, in terms of the postulated model for the process $y(\cdot)$. Hence, one has to study the accuracy of the estimate on an empirical basis. Consequently, $\mu(s)$, the estimated variance of the estimate $R(s)/\sigma(s)$, is defined:

$$\mu^2(s) = \left(\frac{1}{n}\right) \sum_{i=1}^{n} \left[\frac{R_3(t_i, s)}{\sigma_3(t_i, s)} - \frac{R(s)}{\sigma(s)}\right]^2. \tag{2i.1.10}$$

The proximity of the estimated mean square error $\mu^2(s)$ to the true value of the mean square error $E[(R(s)/\sigma(s)) - (R_4(s)/\sigma_4(s))^2]$ is unknown. Still, the standard deviation $\mu(s)$ gives a good idea of the variability of the estimate $R(s)/\sigma(s)$. Thus, if $R(s)/\sigma(s)$ estimates are computed for two independent sequences of the same stochastic process, then it is reasonable to expect the two $R(s)/\sigma(s)$ estimates at the same lag not to be separated from each other by more than $2\mu(s)$.

Clearly an important feature in the computation of the estimate $R(s)/\sigma(s)$ is the choice of the integers t_i. A possible method, which is used in the examples discussed in this book, is one in which the integers t_i are selected as in the following equation. Let $t_i = t_0 + (i - 1)s$, $i = 1, 2, \ldots, n$, where n is chosen so that $t_n + s \leq N$. The estimated rescaled range $R(s)/\sigma(s)$, along with the bounds $[R(s)/\sigma(s) \pm \mu(s)]$, will be plotted against s on log-log paper. The bounds $[R(s)/\sigma(s) \pm \mu(s)]$ roughly indicate the accuracy of estimation of the corresponding mean value $R(s)/\sigma(s)$.

Much credence cannot be given to those portions of the $R(s)/\sigma(s)$ graphs where the value of s is nearly equal to N, the total number of available data points. When the variable s is of the same order as N, different choices of t_i may yield different estimates of $R(s)/\sigma(s)$.

Hurst *et al.* (1965), Mandelbrot and Wallis (1969b), and others have analyzed a variety of time series of naturally occurring processes, such as the thickness of tree rings and annual river flow, with the concept of range, which is defined in a slightly different manner. It was found that the estimated $R(s)/\sigma(s)$ values, when plotted against lag s, plot as straight lines, with their slopes varying from 0.55 to 0.95, provided the series did not exhibit any strong cyclical components. If the series has a strong cyclical component such as those in monthly river flow or annual sunspot series, then the R/σ vs. s characteristic is made up of two straight lines, the break occurring at a value of s slightly greater than the appropriate period of oscillations in the series. Hence, a natural question at this stage is the determination of the analytical structure of the process which yields such R/σ characteristics, i.e., which obey the rule

$$R_4(s)/\sigma_4(s) \approx ks^{\beta}, \qquad 0.5 < \beta < 1.0. \tag{2i.1.11}$$

If the process y obeys a weak stationary ARMA equation, then asymptotically we have

$$R_4(s)/\sigma_4(s) \sim ks^{0.5}. \tag{2i.1.12}$$

One might perhaps conclude from the relation (2i.1.12) that the ARMA processes cannot explain the observed R/σ characteristic. However, such a strong conclusion is premature for two reasons. First, we know only that (2i.1.12) is valid asymptotically, but we do not know how large s should be so that (2i.1.12) is valid. Second, if we simulate a weak stationary ARMA process on a computer and plot the R/σ characteristics of the output of the model, then its R/σ characteristics are also straight lines with slopes between 0.5 and 0.9. This leads us to believe that the expected value of the estimate $R(s)/\sigma(s)$ considered here may not be $R_4(s)/\sigma_4(s)$, i.e., the estimate may be biased. The analysis of the estimate $R(s)/\sigma(s)$ under the assumption that y obeys an ARMA process is extremely difficult.

There have been attempts to consider entirely different classes of models to explain the observed $R(s)/\sigma(s)$ characteristics. For instance, if the process y obeys a so-called fractional noise process (to be discussed in the next section), then the relation (2i.1.11) is true. Hence some investigators (Mandelbrot and Wallis, 1969a) have suggested that all naturally occurring processes such as rainfall, obey the fractional noise models and not the finite difference equation models. A detailed comparison of the fractional noise models with the constant coefficient AR models for river flows performed in Chapter X reveals that the best fitted fractional noise (FN) model is considerably inferior to the best fitted AR model from the point of forecasting, hypothesis tests, etc. The validity of the FN models on purely physical grounds has been questioned by Scheidegger (1970) and Klemes (1974).

2j. Fractional Noise Models

2j.1. Description

The correlation function $R(k)$ of lag k associated with a weak stationary ARMA equation is either an exponentially decaying function of k or a sum of terms every one of which decays exponentially with k. In modeling processes such as atmospheric turbulence, we need to consider models in which the correlation function $R(k)$ decays at a rate slower than the exponential rate. We will discuss in detail some processes whose correlation function $R(k)$ decays at a rate $1/k^\alpha$, $\alpha > 0$; i.e., the rate is considerably slower than the exponential rate. Our interest in this class of processes arises because of the suggestion that many of the naturally occurring processes such as river flows are better handled by such models rather than ARMA models. We will make direct comparison of such a model with the usual ARMA model in representing river flow processes in Chapter X. Now, we will outline the theory of such processes (Yaglom, 1967; Mandelbrot and Van Ness, 1968; Mandelbrot and Wallis, 1969a).

Consider a so-called fractional noise process y that possesses the infinite moving average representation

$$y(t) = \sum_{k=1}^{\infty} w(t-k)/k^{1.5-h}, \tag{2j.1.1}$$

where $0.5 < h < 1.0$ and $\{w(\cdot)\}$ is a Gaussian zero mean white noise; i.e., it is an IID sequence with normal distribution $N(0, \rho)$. Note that the sequence of gains $\{1/k^{1.5-h}\}$ is a divergent sequence, but the sequence $\{(1/k^{1.5-h})^2\}$ is convergent. Hence it is easy to show that a process $y(\cdot)$ as in (2j.1.1) exists in the mean square sense. It is easy to find its mean square value and other related expressions:

$$E[y^2(t)] = \rho \sum_{k=1}^{\infty} (1/k^{3-2h}) < \infty \tag{2j.1.2}$$

$$E[y(t)y(t-k)] \triangleq R(k) = \rho \sum_{j=k+1}^{\infty} 1/(j(j-k))^{1.5-h}. \tag{2j.1.3}$$

It is difficult to obtain explicit algebraic expressions for the sums on the right-hand side of (2j.1.2) and (2j.1.3), but it is easy to see that the function $R(k)$ in (2j.1.3) asymptotically decays at a rate considerably slower than a^k for any positive number a.

When we want to simulate the fractional noise model on a computer, the infinite series is replaced by a finite series so that the function $R(k)$ of the corresponding model also decays at a rate slower than a^k, $a > 0$.

We will give a brief explanation of the term fractional noise. Consider the partial sum $z(t) = \sum_{j=1}^{t} y(j)$. If $y(\cdot)$ obeys the fractional noise model in (2j.1.1), then $z(\cdot)$ asymptotically obeys the following equation as shown in Appendix 2.1:

$$E[(z(t+s) - z(t))^2] = cs^{2h}, \qquad h > 0. \tag{2j.1.4}$$

If in (2j.1.4), the parameter $h = 0.5$, then $z(\cdot)$ would have been the usual Brownian motion process and the process y derived from it by first differencing (the derivative in the continuous case) is a discrete white noise process. If in (2j.1.4), $0.5 < h < 1$, then the corresponding z process is called a fractional Brownian process, since it is like a Brownian motion process characterized by a fraction h different from 0.5. The process $y(\cdot)$ derived from $z(\cdot)$ is then called the fractional white noise process, since it is characterized by a fraction h other than 0.5.

Now let us consider the R/σ characteristics of the fractional noise model. With considerable manipulation, one can argue that $R_4(s)/\sigma_4(s)$, the expected value of the rescaled range, behaves as s^h for large h. Hence, the corresponding slope of the $R_4(s)/\sigma_4(s)$ vs. s characteristic on a log–log paper will be h, which is different from 0.5. Hence, if we find that the slope h of the empirical R/σ characteristic is different from 0.5, there is a possibility that it could be a fractional noise process with parameter h. However, the final decision of whether the process obeys a FN model or an ARMA model has to be based on a comparison of the two fitted models. Such a comparison is described in Chapter X while discussing the validity of a FN model for river flow data.

When we are dealing with a relatively small observation set, we can usually construct a satisfactory stochastic difference equation model driven by a white noise process. The possibility of the choice of a FN model arises only when we are dealing with processes with a relatively large observation history.

2j.2. Prediction in an FN Process

A one-step-ahead predictor of $y(t)$ based on all information until time $(t-1)$ is denoted $\hat{y}(t|t-1)$, and is defined by

$$\hat{y}(t|t-1) = f(y(t-1), y(t-2), \ldots), \tag{2j.2.1}$$

where $f(\cdot)$ is a deterministic function. It is customary to choose the function f so that the mean square prediction error $E[(y(t) - y(t|t-1))^2]$ is minimized. The optimal predictor that minimizes the mean square prediction error is called the least squares predictor of $y(t)$, given $y(t-j)$, $j \geq 1$, and is denoted $y^*(t|t-1)$. Since the FN process $y(\cdot)$ is a stationary moving average process, the theoretical minimum mean square prediction error is the variance ρ of noise w in (2j.1.1). To obtain an explicit formula for the least squares predictor $y^*(t|t-1)$, we proceed as follows. Since the FN process is normal, the least squares predictor $y^*(t|t-1)$ is a *linear* combination of the observations $y(t-j)$, $j = 1, 2, 3, \ldots$. A formula for $\hat{y}(t|t-1)$, which is an approximation to $y^*(t|t-1)$, can be derived by discretizing the least squares prediction formula for continuous time fractional noise processes (Yaglom, 1967):

$$\hat{y}(t+1|t) = \sum_{j=1}^{\infty} \beta_j(x(t-j) - x(t)). \tag{2j.2.2}$$

where

$$\beta_j = \frac{\sin \pi(h+1.5)}{\pi} \frac{1}{j^{h+1.5}(j+1)}, \qquad x(i) = \sum_{j=-\infty}^{i} y(j). \tag{2j.2.3}$$

We can simplify formula (2j.2.2) to display the explicit dependence of $\hat{y}(t+1|t)$ on the previous observations $y(t), y(t-1), \ldots,$

$$\begin{aligned}\hat{y}(t+1|t) &= \sum_{j=1}^{\infty} \beta_j \sum_{k=0}^{j-1} y(t-k) = \sum_{k=0}^{\infty} \left(\sum_{j=k+1}^{\infty} \beta_j \right) y(t-k) \\ &= \sum_{k=0}^{\infty} c_k y(t-k),\end{aligned} \tag{2j.2.4}$$

where

$$c_k = \sum_{j=k+1}^{\infty} \beta_j. \tag{2j.2.5}$$

Formula (2j.2.2) is not very useful since knowledge of the semi-infinite history of $y(j)$, $-\infty \leq j \leq i$, is needed to use it, whereas only a finite number of observations are available in practice. Still, formula (2j.2.4) shows that the dependence of $\hat{y}(t+1|t)$ on an observation in the remote past is small. The truth of this statement can be ascertained, using (2j.2.3) and (2j.2.5), by establishing

$$c_k/c_0 \leq 0(1/k^2). \tag{2j.2.6}$$

Equation (2j.2.6) suggests that it may be possible to find predictors that operate on a finite number of past observations and yield a mean square prediction error that is only slightly greater than that of $y^*(t|t-1)$. Hence, the following linear predictor $\bar{y}(t+1|t)$ is considered, where r is an integer that must be appropriately selected:

$$\bar{y}(t+1|t) = \sum_{j=0}^{r-1} d_{j+1,r} y(t-j).$$

The coefficients $d_{j,r}$ are determined by minimizing the mean square error $J(\mathbf{d}_r)$:

$$J(\mathbf{d}_r) = E[(y(t+1) - \bar{y}(t+1|t))^2], \qquad \mathbf{d}_r = (d_{1,r}\, d_{2,r}, \ldots, d_{r,r})^{\mathrm{T}};$$

$J(\mathbf{d}_r)$ can be minimized with respect to $\mathbf{d}$ since $J(\mathbf{d})$ is a quadratic form in $\mathbf{d}$. The minimizing value of $\mathbf{d}_r$ is represented by

$$\mathbf{d}_r{}^* = \mathbf{B}_r{}^{-1}\boldsymbol{\delta} = (d^*_{1,r}, d^*_{2,r}, \ldots, d^*_{r,r})^{\mathrm{T}},$$
$$\boldsymbol{\delta} = (R(1), \ldots, R(r))^{\mathrm{T}}; \qquad \mathbf{B}_r \text{ is an } r \times r \text{ matrix}, \tag{2j.2.7}$$
$$(B_r)_{ij} = R(|i - j|), \qquad i, j = 1, 2, \ldots, r.$$

The correlation coefficients $R(k)$ are computed as in Eq. (2j.1.3). The minimum value of the mean square error $J(\mathbf{d})$ is $J(\mathbf{d}_r{}^*) = E[y^2(j)](1 - \boldsymbol{\delta}^{\mathrm{T}}\mathbf{B}_r^{-1}\boldsymbol{\delta})$.

Let us rewrite the final predictor as

$$\bar{y}(t + 1|t) = \sum_{j=0}^{r-1} d^*_{j+1,r} y(i - j). \tag{2j.2.8}$$

In order to choose r, a number of different values of r such as 2, 3, 4, ..., 20, may be considered, and the $\mathbf{d}_r{}^*$ vector computed in each case. The predicted values $\bar{y}_r(t|t - 1)$ are also computed for $t = 21, \ldots, N$ for each case, from (2j.2.8). Let

$$\hat{J}_r = \frac{1}{N - 20} \sum_{j=21}^{N} (y(t) - y_r(t|t - 1))^2.$$

The empirical mean square prediction errors $\hat{J}_1, \ldots, \hat{J}_{20}$ are computed for each case. The value of r that yields the smallest value of $\hat{J}$ is then selected. Since the estimate $\hat{J}_r$ is computed from $(N - 20)$ observations, the standard deviation of the estimate $\hat{J}_r$ is given by $\{[2/(N - 20)]\hat{J}_r\}^{1/2}$.

2k. Conclusions

We have considered a number of topics dealing mainly with prediction in stochastic difference equation models, of both finite and infinite orders. A number of related topics such as spectral representation, prediction, and rescaled range–lag analysis have also been discussed.

Appendix 2.1. Characteristics of Fractional Noise Models

Let $y(t) = \sum_{k=1}^{\infty} w(t - k)/k^{1.5-h}$ and

$$x(t, s) = \sum_{j=0}^{s} y(t - j) = \sum_{j=0}^{s} \sum_{k=1}^{\infty} w(t - j - k)/k^{1.5-h}$$
$$= \sum_{l=1}^{\infty} w(t - l) \sum_{j=0}^{\min[l-1,s]} \frac{1}{(l - j)^{1.5-h}}.$$

To show $E[x^2(t, s)] \approx cs^{2h}$:

$$\begin{aligned}
E[x^2(t, s)] &= \sum_{l=1}^{\infty} \rho\left(\sum_{j=0}^{\min[l-1,s]} \frac{1}{(l-j)^{1.5-h}}\right)^2 \\
&\approx \rho \int_0^{\infty} dl\left(\int_0^{\min[l,s]} \frac{da}{(l-a)^{1.5-h}}\right)^2 \\
&= \rho \int_0^{s} dl\left(\int_0^{l} \frac{da}{(l-a)^{1.5-h}}\right)^2 + \rho \int_s^{\infty} dl\left(\int_0^{s} \frac{da}{(l-a)^{1.5-h}}\right)^2 \\
&= \mathrm{I} + \mathrm{II}.
\end{aligned}$$

$$\begin{aligned}
\mathrm{I} &= \rho \int_0^{s} dl\left\{\frac{(l-a)^{h-1.5+1}}{h-0.5}\bigg|_0^l\right\}^2 \\
&= \frac{\rho}{h-0.5}\int_0^{s} l^{2h-1}\, dl = \frac{s^{2h}}{2h(h-0.5)}.
\end{aligned}$$

Similarly, we can show that $\mathrm{II} = c_1 s^{2h}$.

Problems

1. Let $\mathbf{x}(t) = \mathbf{A}\mathbf{x}(t-1) + \mathbf{G}\mathbf{w}(t)$, where $\mathbf{A}$ is an $n \times n$ matrix with all eigenvalues numerically less than one, and $\mathbf{w}(\cdot)$ is an m-vector with covariance matrix $\boldsymbol{\rho}$ obeying (A1). Let $\mathbf{P}(t_1, t_2) = \mathrm{cov}[\mathbf{x}(t_1), \mathbf{x}(t_2)]$ and $\mathbf{P}_\infty = \lim_{t\to\infty} \mathbf{P}(t, t)$. (a) Obtain the difference equation for $\mathbf{P}(t, t)$ and the algebraic equation for $\mathbf{P}_\infty$. (b) What are the conditions on $\mathbf{P}(0, 0)$ so that the sequence $\{\mathbf{x}(\cdot)\}$ is weak stationary for all $t \geq 0$?

2. Prove the assertion regarding the weak stationarity of $\mathbf{y}(\cdot)$ in Section 2b.

3. Consider the noninvertible system

$$\mathbf{y}(t) = \mathbf{A}_1\mathbf{y}(t-1) + \mathbf{B}_0\mathbf{w}(t) + \mathbf{B}_1\mathbf{w}(t-1),$$

where $\mathbf{w}(\cdot)$ obeys (A1) with $\boldsymbol{\rho} = \mathbf{I}$,

$$\mathbf{A}_1 = \begin{bmatrix} 0.5 & 0 \\ 0 & 0.8 \end{bmatrix}, \quad \mathbf{B}_0 = \begin{bmatrix} 0 & 0 \\ 0 & 1 \end{bmatrix}, \quad \mathbf{B}_1 = \begin{bmatrix} 2 & 0 \\ 1 & 3 \end{bmatrix}.$$

Find the one-step-ahead forecasts of $y_1(t)$ and $y_2(t)$ based on the entire past history of y_1 and y_2.

4. Consider the following system where $w(\cdot)$ and $\eta(\cdot)$ both obey (A1) with variances ρ_1 and ρ_2 and are mutually uncorrelated:

$$y(t) = \theta_1 y(t-1) + \theta_2 y(t-2) + w(t), \qquad z(t) = y(t) + \eta(t).$$

Obtain the predictor of $y(t)$ based only on $z(t-j), j > 1$, using both Kalman theory and the theory of Section 2h.

5. Show that the spectral density of $y(\cdot)$ obeying $y(t) = T(D)w(t)$ has the forms

(i) $T(D) = \dfrac{1}{1 - \theta_1 D - \theta_2 D^2}$,

$$S_{yy}(\omega) = \frac{\rho}{(1 + \theta_2^2)[1 + (\theta_1^2/4\theta_2)] - 4\theta_2(\cos\omega - \cos\omega_1)^2}$$

where

$$\cos\omega_1 = \frac{\theta_1(\theta_2 - 1)}{4\theta_2}, \qquad -1 < \theta_2 < 0, \quad \theta_1^2 + 4\theta_2 < 0;$$

(ii) $T(D) = \dfrac{1 + \theta_2 D}{1 - \theta_1 D}$, $\qquad S_{yy}(\omega) = \dfrac{\rho(1 + \theta_2^2 + 2\theta_2\cos\omega)}{1 + \theta_1^2 - 2\theta_1\cos\omega}$.

6. Find the spectral density matrix of the following process $\mathbf{y}$ and hence find the cross correlation coefficients $\text{cov}[y_1(t), y_2(t-j)], j = 0, 1, 2, \ldots$:

$$(1 - \theta_1 D)y_1(t) = (1 + \beta_1 D)w(t) \qquad (1 - \theta_2 D)y_2(t) = w(t),$$

where $w(\cdot)$ obeys (A1) with variance ρ.

7. Consider the multiplicative process $y(t) = (y(t-1))^{0.8}\eta(t)$, where $\{\eta(\cdot)\}$ is IID with log-normal distribution $\Lambda(0, \rho)$. Obtain the best linear one-step-ahead predictor of $y(t)$ of the form $\hat{y}(t|t-1) = \alpha_1 y(t-1) + \alpha_2 y(t-2)$ according to the quadratic loss function. Compare its mean square error (MSE) with that of the best nonlinear predictor according to the quadratic criterion.

8. Consider the fractional noise process in Eq. (2j.1.1) and truncate it to 300 terms with $h = 0.8$ and $\rho = 1.0$. Assuming $y(\cdot)$ to be stationary, compute $E[y(t)y(t-j)]$ for $j = 1, \ldots, 20$. Construct a linear second-order linear predictor and compare its MSE with that of the best linear predictor of infinite memory. Repeat the problem with the order of linear predictor as 3, 4, 5 and $h = 0.2$.

Chapter III | *Structure of Univariate Models*

Introduction

In this chapter, we shall consider the structure of univariate stochastic difference equations. We will first consider difference equations without exogenous inputs. A stochastic difference equation can have a variety of terms such as autoregressive terms, moving average terms, and deterministic trend functions like sinusoids or polynomials, and the coefficients could be either constants or time varying. Furthermore, the equation could be written in terms of the output variable y or in terms of the first differenced variable $\nabla y = y(t) - y(t-1)$ or the Tth difference $\nabla_T y = y(t) - y(t-T)$, leading to the class of autoregressive integrated moving average (ARIMA) models. Further, the difference equation can be written in terms of a nonlinear transform of y such as $\log y$. In other words, we can make the difference equation weak stationary, covariance stationary, or nonstationary by an appropriate choice of the terms of the difference equation. The main question of this chapter is the following: "Given an empirical time series, what are the classes of models that are most appropriate for modeling the given time series?" Since the question is rather general, an illustration is suitable. Consider a time series $S1$ as in Fig. 3a.1.1, which represents the monthly sales of a certain company X (Chatfield and Prothero, 1973). Clearly, the series has a strong growth component modulated by an approximate periodic component. We shall demonstrate later that there are at least two important classes of models which have time series possessing such behavior. The first one is the class of covariance stationary processes involving a difference equation with autoregressive terms and deterministic trend terms such as t, $\sin \omega t$, $\cos \omega t$, and $t \cos \omega t$. The second class is the class of seasonal autoregressive integrated moving average models. A member in each of the two classes can give rise to a time series such as the one given in Fig. 3a.1.1. We should consider all the relevant classes of models before choosing the appropriate class because the performance of the best model in each class may be different and the computational effort in constructing the best model in each class also differs considerably from class to class. For instance, the parameter estimation and the prediction problems in the seasonal ARIMA models are considerably more involved than the corresponding problems with covariance stationary models. We should first consider the simplest possible class of models and move to another class if the "best" model in the first class is unsatisfactory. We will

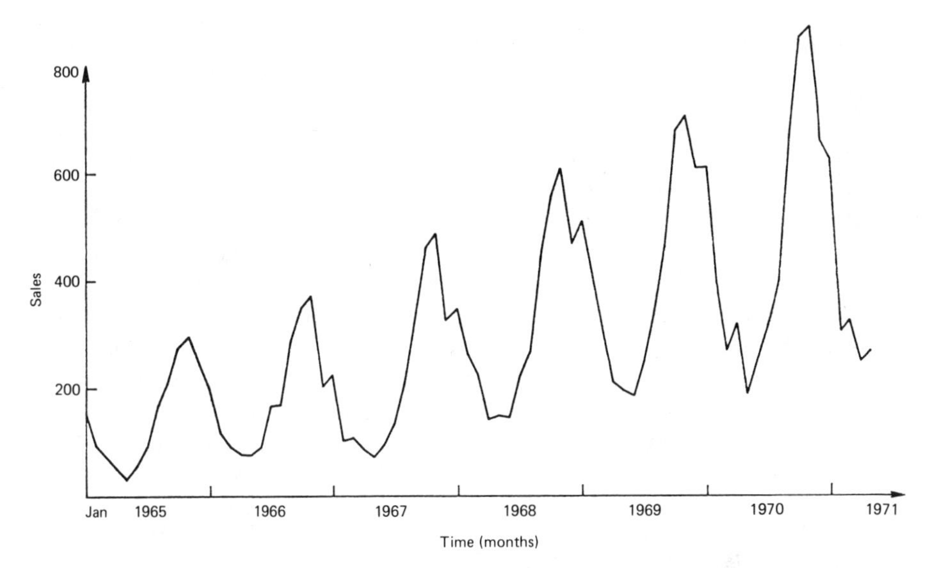

FIG. 3a.1.1. Observed monthly sales of company X (series *S*1).

also consider nonstationary models in which the coefficients may vary with time.

The final choice of the class of models for the given time series is considered in Chapter VIII. We will currently restrict ourselves to the investigation of the dominant features of the time series having different types of terms and the determination of the different classes of models which may yield time series with similar characteristics. Some general discussion on modeling of empirical time series can be found in the books by Box and Jenkins (1970), Kendall (1973), Parzen (1974), and others.

The choice of a suitable time scale for modeling is considered next. Suppose we have hourly data for a process and are interested in a model whose principal purpose is to yield 12-hour-ahead predictions. The appropriate unit of time for modeling, whether it is 1 or 12 hours, is to be decided. Alternatively, the question may be posed as follows: "Should a model be developed for the hourly data $[y(t)]$ or for the 12-hour aggregated data $[y_1(t)]$?" This problem is also related to the effects of aggregation. Whether an aggregated model, e.g., the 12-hour model, yields better predictions than the unaggregated (hourly) model is also discussed. If the best fitted model for the hourly data belongs to a certain class, say autoregressive moving average (ARMA) models, then the best fitting model for the 12-hour data may belong to an entirely different class, such as the class of covariance stationary models. Hence the problem posed here is not completely separate from the problem posed earlier.

In the following discussion we refer to the concept of a "best" model, which is derived with a set of criteria as discussed in some detail in Chapter VIII.

3a. Types of Dynamic Stochastic Models

3a.1. Deterministic Models

Historically, it was common to model time series by deterministic functions of time such as combinations of polynomials in time t or sinusoidal or other functions of time. Such models are usually very uneconomical in the sense that they may involve quite a large number of sinusoidal or polynomial terms. The more important criticism of such models is that their predictive ability is usually very small. For example, if we have a time series having 100 points and if we use the first 50 points to fit an nth-order polynomial to the data, and then use the resulting model to predict the values of the 51st observation, 52nd observation, and so on up to the 100th observation, then usually the prediction error so obtained will be extremely high. This indicates the low quality of the model. The reason for such poor performance is that the probabilistic nature of the time series is completely ignored in models involving only deterministic functions.

An illustration regarding the preceding discussion may be revealing. Let us consider the problem of modeling a biological population, such as the annual population of the United States. The corresponding time series $P1$ is given in Fig. 3a.1.2 for these data. A purely polynomial fit cannot be satisfactory in view of the exponential growth of the population. It is customary to represent the population time series by means of the deterministic function of time,

$$y(t) = K/(1 + Be^{-Ct}), \tag{3a.1.1}$$

the so-called logistic function, involving three positive parameters K, C, and B,

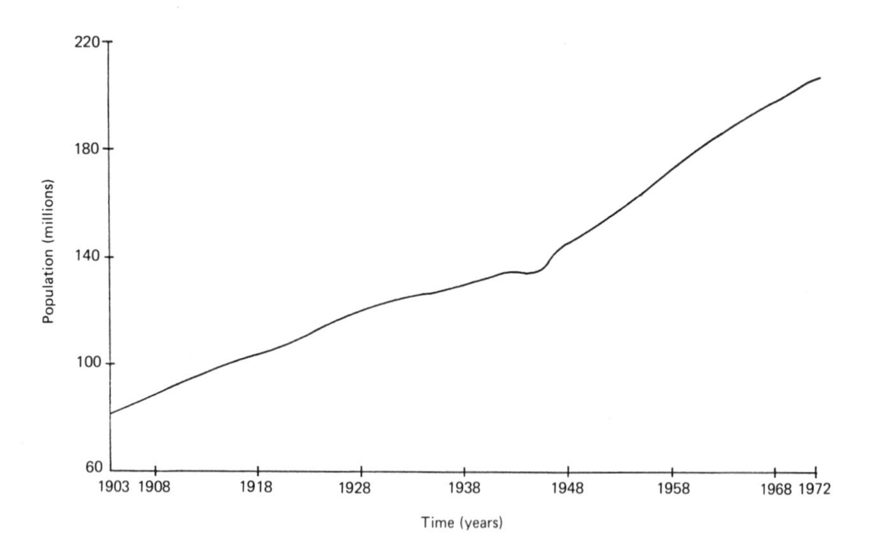

FIG. 3a.1.2. Annual population of the United States (series $P1$).

where y is the population and t represents time in years. Expression (3a.1.1) is a solution to the so-called Volterra differential equation

$$\frac{dy}{dt} = Cy\left(1 - \frac{y}{K}\right), \tag{3a.1.2}$$

which states that the rate of population growth is a maximum if the value $y = K/2$ and becomes negative when y is greater than K. Some other similar models are discussed in Chapter X.

By using the total population of the United States in the years 1840, 1900, and 1960, the constants K, B, and C in Eq. (3a.1.1) can be evaluated and the resulting equation (Montroll, 1968) is

$$y(t) = \frac{246.5}{1 + 2.243 \exp[-0.02984(t - 1900)]}, \tag{3a.1.3}$$

where $y(t)$ is population in millions and t is the calendar year after 1900. As can be seen from Fig. 3a.1.3, the fit provided by Eq. (3a.1.3) up to 1960 is satisfactory. To measure the effectiveness of the model, we use the model to predict the population in the years 1961–1972 and compare these predictions with the observed values. The predicted values are shown in Fig. 3a.1.4 and they are not very good. The prediction error can be reduced to a fraction of these values by using the stochastic difference equation model discussed in Chapter XI. This example dramatically illustrates the limitations of pure deterministic modeling of any time series, even with the aid of complicated functions such as the logistic function. As such, we will not pursue this topic any further because this class of deterministic models will be subsumed by the class of covariance stationary models discussed in Section 3a.3.

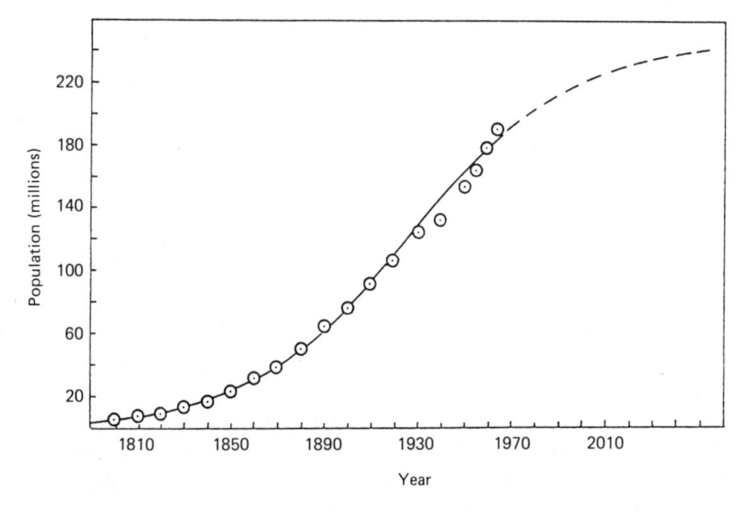

FIG. 3a.1.3. Population of the United States. Logistic curve fitted so that observed points at 1840, 1900, and 1960 are exact. Points represent census data. (After Montroll, 1968.)

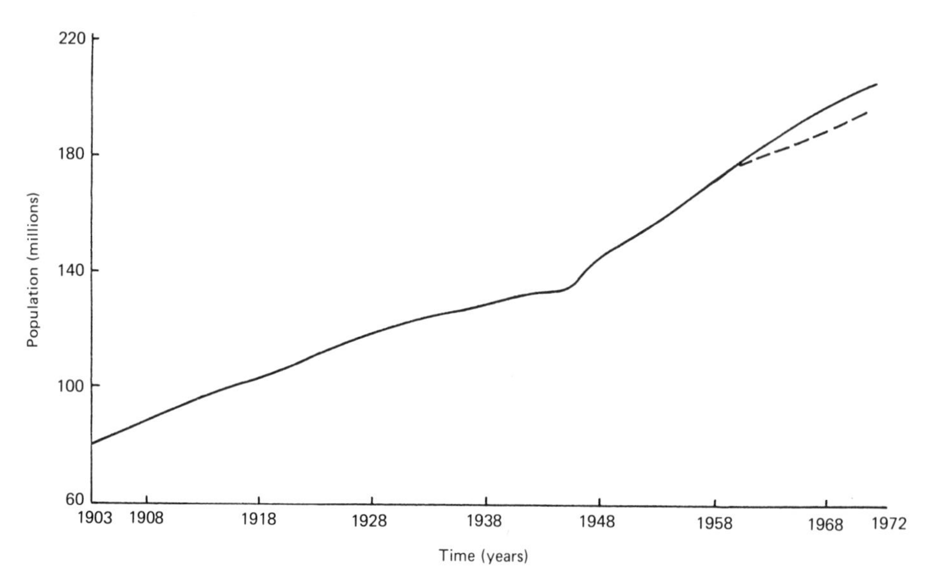

FIG. 3a.1.4. Observed annual population of the United States (solid line) and its 1-year-ahead forecast by Eq. (3a.1.3) during 1961–1971 (broken line). (After Montroll, 1968.)

3a.2. Weakly Stationary Difference Equation Models

3a.2.1. The Autoregressive (AR) Models

This is an important class of stochastic dynamic models; it is also, computationally, the simplest class. We can represent the given process y as

$$A(D)y(t) = w(t) + F_0. \tag{3a.2.1}$$

All zeros of the polynomial $A(D)$ should be outside the unit circle. The constant term F_0 is inserted to account for the nonzero mean of the process. Sometimes, a process y may not obey an AR process, but its transform, say $\ln y(t)$, may. Further, the AR processes can be generalized to include equations such as

$$y(t) = \sum_{j=1}^{m_1} A_j y(t-j) + G_1 \ln y(t-1) + w(t) \tag{3a.2.2}$$

provided the equations are asymptotically stable and the coefficients of the various terms occur linearly. We will refer to equations such as (3a.2.2) as generalized AR processes.

As indicated in Chapter II, a second-order AR process with suitable coefficients can have a sinusoidally damped correlogram. Thus, a second-order AR process [or any AR process with the polynomial $A(D)$ having a pair of complex zeros] is a possible candidate for representing empirical time series with an approximate cyclical behavior.

Another subclass of AR models used in modeling "cyclical" time series has terms such as $y(t - T)$, where T is the approximate period of the time series or an integer near the period. For instance, consider

$$y(t) = A_0 + A_1 y(t-1) + A_2 y(t-2) + A_3 y(t-10) + w(t). \quad (3a.2.3)$$

In (3a.2.3) there are no AR terms corresponding to $y(t-3), y(t-4), \ldots,$ and hence this equation is said to have noncontiguous AR terms. Some time series, such as the well-known annual sunspot index series, are best modeled by a member of the family given in Eq. (3a.2.3). Further details are given in Chapter XI.

3a.2.2. Autoregressive Moving Average (ARMA) Models

The class of ARMA models is the natural generalization of the AR models. The notation ARMA(m_1, m_2) represents an ARMA model with m_1 consecutive AR terms $y(t-1), \ldots, y(t-m_1)$ and m_2 consecutive MA terms $w(t-1), \ldots,$ $w(t-m_2)$. The AR and MA parts of the equation should obey assumptions (A5) and (A6) so that the process y obeying the ARMA model is weak stationary and invertible.

Accurate estimation of parameters in a system involving moving average terms is considerably more difficult than the estimation problem in a system without moving average terms. The difficulty is compounded if the parameters are to be estimated in real time. Consequently, in modeling empirical time series, the possibility of developing a model free of moving average terms should be explored so that the model satisfactorily represents the data. Typically, if a process obeys an ARMA model such as

$$A(D)y(t) = B(D)w(t), \quad (3a.2.4)$$

it can be equivalently represented as an infinite AR process

$$w(t) = [B(D)]^{-1}A(D)y(t) \triangleq y(t) + \sum_{j=1}^{\infty} A_j' y(t-j). \quad (3a.2.5)$$

One can consider the truncation of the infinite series in (3a.2.5). In such a case the adequacy of such a truncated process to represent the given ARMA process must be investigated.

When any zero of the polynomial $B(D)$ in (3a.2.4) is near unity, the corresponding truncated autoregressive model may involve a large number of terms for obtaining a good fit to the original ARMA model. These coefficients have to be estimated from the given data, and the accuracy of the estimates may not be high since the accuracy of estimation is inversely proportional to the number of simultaneously estimated terms. This feature may considerably degrade the performance of the AR model. In such cases, it may be better to use a model with MA terms.

Even otherwise, when any zero of $B(D)$ is not near the unit circle, it is

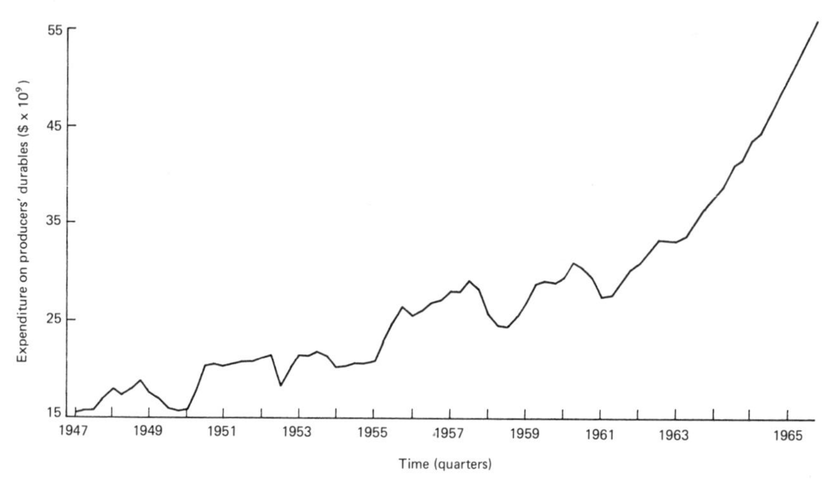

FIG. 3a.2.1. Expenditures on producers' durables (EPD) (series $E1$). (After Nelson, 1973.)

difficult to say anything definitive about the effect of truncation on the performance of the model in designing optimal controls for the process. However, regarding the use of truncated autoregressive models for prediction, there are two conflicting schools of thought. One point of view is that the predictive ability of the truncated model could be made approximately equal to that of the original ARMA process by choosing a sufficiently large number of terms in the truncated AR model. The second viewpoint is that it is not possible to

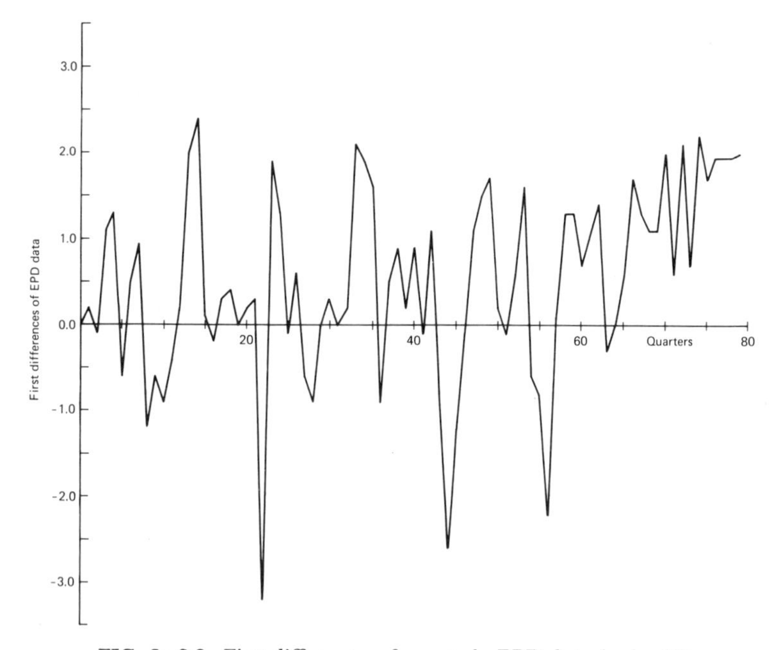

FIG. 3a.2.2. First differences of quarterly EPD data (series $E2$).

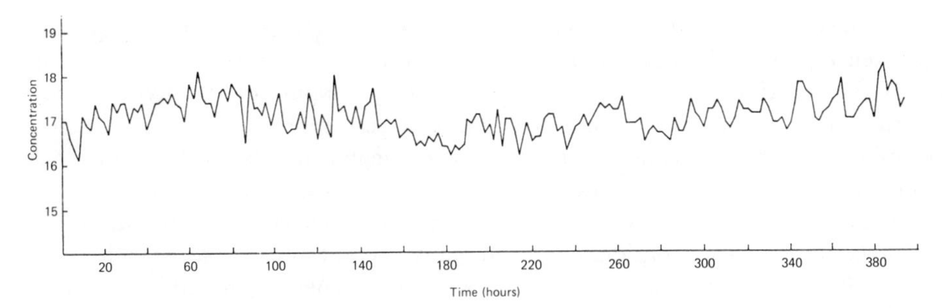

FIG. 3a.2.3. Two-hour chemical process concentration readings (series *C*1). (After Box and Jenkins, 1970.)

achieve the same level of predictive ability in the truncated AR model as in the original ARMA model in every problem because increasing the length of the truncated series does not necessarily result in corresponding increases in the predictive ability of the model.

To illustrate the second view, consider the series *E*1 of the observed quarterly data of expenditure on producer's durables (EPD) shown in Fig. 3a.2.1 and the series of first differences of the quarterly EPD data in billions of current dollars (Nelson, 1973) (Fig. 3a.2.2). The series has 80 points in it. We shall use the first 56 points to estimate the coefficients in the model that is to be fitted. The predictive ability of the fitted model will be estimated by computing the one-step-ahead predictions of the remaining 24 points (which were not used to estimate the coefficients in the model) and the mean square error (MSE) of these predictions. The best fitting model in the class of moving average models was a MA(1) model and the corresponding MSE was 1.05. The best fitting model in the class of AR models was an AR(3) model and the corresponding MSE was 1.36. Further increase in the order of the AR(3) model only resulted in an increase in the MSE value. This example clearly shows the limitation of AR models in adequately representing a time series which truly obeys a moving average or ARMA process. The details of this discussion are found in Section 11d. Similarly, examples of empirical time series supporting the first point of view can be given.

3a.3. Covariance Stationary Series

The difference equations corresponding to covariance stationary series are the usual AR or ARMA equations considered earlier with the addition of deterministic trend functions. These functions may be t, $\cos \omega t$, $\sin \omega t$, $t \cos \omega t$, $t \sin \omega t$, to account for linear growth, sinusoidal behavior, etc. Thus, such processes can be considered to be a sum of a weakly stationary process and a deterministic function of time. The class of purely deterministic models or the class of models with a deterministic signal and additive noise is subsumed by the class of covariance stationary models.

The deterministic trend functions of time occurring in the difference equation often represent the deterministic part of the exogenous inputs to the system. If the contribution of the individual exogenous inputs to the given process cannot be evaluated or monitored, the next best approach is to insert deterministic terms to account for the various exogenous inputs into the difference equation. For instance, consider the monthly flows in a river. The dominant contribution to river flow may be from precipitation. However, the measured precipitation varies considerably from point to point in river basins and it may be difficult to assess the precipitation over a large river basin and its contribution to the river flow. In such cases, as a first step, it is advisable to omit the rainfall terms from the equation representing the monthly flows and insert a sinusoidal function of time to represent the deterministic part of the rainfall contributions at various points in the watershed. Models for monthly river flows with sinusoidal trend terms in them perform much better than those without the sinusoidal trend terms. Another example is the times series of the annual sales of company X, which was discussed earlier. The growth in sales is partly a reflection of the general growth in the economy. It is difficult to identify the corresponding exogenous variables. Instead, we can represent the corresponding expected value of the exogenous variables by a combination of terms such as $\alpha_1 t$, $\alpha_2 t \cos \omega t$, and $\alpha_3 t \sin \omega t$.

Covariance stationary processes are also used in modeling systems for analyzing the causal relationships between variables. For instance, in environmental studies it is important to know whether the rainfall in a region is affected by urbanization or whether the presence of large man-made lakes alters the rainfall characteristics of a region. These questions must be answered with the aid of a dynamic model for the rainfall process since the successive rainfall values may be strongly correlated. If $y(t)$ denotes the rainfall at instant t, then one can consider the following model for determining a possible causal relationship:

$$y(t) = \theta_0 + \theta_1 y(t-1) + \theta_2 \psi_1(t) + w(t), \tag{3a.3.1}$$

$$\psi_1(t) = \begin{cases} 0 & \text{if } t \le t_1 \\ 1 & \text{if } t > t_1, \end{cases}$$

and t_1 is the time when exogenous effects such as urbanization can be considered to be substantial and thus possibly affect rainfall characteristics thereafter.

The trend function ψ_1 is a discontinuous trend, but it can be handled in the same way as other continuous functions in parameter estimation.

3a.4. Nonstationary Processes

3a.4.1. Nonseasonal IAR or ARIMA Models

Let $\nabla_d y(t) = y(t) - y(t-d)$, where d is a positive integer. When $d = 1$ we will denote $\nabla_1 y(t)$ simply by $\nabla y(t)$. If $\nabla_d y(t)$ obeys the AR(m_1) process

$$A_1(D)\,\nabla_d y(t) \triangleq A(D)y(t) = w(t), \tag{3a.4.1}$$

for some integer d, then $y(\cdot)$ is said to obey an integrated autoregressive model, IAR(d, m_1). By definition, $A(D)$ has at least one zero on the unit circle, showing the nonstationarity of y. If we add moving average terms to (3a.4.1), then we obtain the autoregressive integrated moving average model ARIMA(d, m_1, m_2), m_2 being the order of the MA part, d and m_1 having the same meaning as before. The seasonal and nonseasonal IAR and ARIMA processes were popularized by Box and Jenkins (1970). They are special cases of random processes with stationary increments discussed by Gladeshev (1961) and Yaglom (1958). Such processes have been used in modeling some turbulent flows (Tatarski, 1961).

The simplest type of IAR process is given by

$$\nabla y(t) = w(t), \tag{3a.4.2}$$

where $w(\cdot)$ is the usual zero mean IID process with variance ρ, i.e.,

$$y(t) - y(0) = \sum_{j=1}^{t} w(j). \tag{3a.4.3}$$

The process $y(t)$ is a sum of t independent noise variables. This is the classical model for Brownian motion, which is the random motion of microscopic particles such as pollen in water or dust in air caused by the collision of particles with molecules of the surrounding medium. For a process obeying (3a.4.2),

$$E[y(t) - y(0)] = 0, \tag{3a.4.4}$$

$$\operatorname{var}[y(t) - y(0)] = t\rho. \tag{3a.4.5}$$

The absolute value of $[y(t) - y(0)]$ has a growth rate proportional to $\sqrt{t}$, growth being nonmonotonic. This growth behavior is true for all ARIMA processes, expressed as an ARMA equation in terms of the first difference $\nabla y(t)$ without involving a constant term. We will illustrate this behavior by an example.

Example 3a.1. Consider the first-order IAR process

$$\nabla y(t) = A\, \nabla y(t-1) + w(t), \tag{3a.4.6}$$

which yields

$$\begin{aligned} E[y(t) - y(0)] &= 0, \qquad \text{asymptotically}, \\ \operatorname{var}[y(t) - y(0)] &= E[y(t) - y(0)]^2, \qquad \text{asymptotically} \\ &= E\left[\left(\sum_{j=1}^{t} \nabla y(j)\right)^2\right] = P(t), \end{aligned}$$

or

$$P(t) = E\left[\left(\sum_{j=1}^{t-1} \nabla y(j)\right)^2 + (\nabla y(t))^2 + 2(\nabla y(t))\left(\sum_{j=1}^{t-1} \nabla y(j)\right)\right]. \tag{3a.4.7}$$

We may recall from Chapter II the asymptotic expression for correlation for a process obeying (3a.4.6):

$$E[\nabla y(t)\, \nabla y(t-j)] \approx A^j \rho/(1 - A^2), \qquad \text{for large } j. \tag{3a.4.8}$$

Substitution of (3a.4.8) in (3a.4.7) yields the required expression

$$P(t) = P(t-1) + \left(1 + 2\sum_{j=1}^{t-1} A^j\right)\rho/(1 - A^2). \tag{3a.4.9}$$

Clearly, an asymptotic solution for $P(t)$ is of the form tc where c is a constant depending on ρ and A.

We should emphasize that when we actually look at the realization of the $y(\cdot)$ sequence obeying (3a.4.2) or (3a.4.6), the growth may not be apparent. From the foregoing theory we know that $P(t)$ behaves as ct as t tends to infinity, but no information is given by the theory about the sign of $y(t)$ for large t. It may be positive or negative. To be specific, consider Eq. (3a.4.2). By the law of iterated logarithms (Feller, 1966) one can show that

$$\frac{y(t) - y(0)}{(2t \ln \ln t)^{1/2}} \to \pm 1 \qquad \text{with probability 1.}$$

Therefore, $y(t)$ may tend to $\pm(2t \ln \ln t)^{1/2}$ for large t. Moreover, as t increases, the sign of $y(t) - y(0)$ may not be the same. It can be positive for several intervals of time and negative over other intervals of time. In each case it is difficult to say whether it is tending to $+1$ or -1. As an illustration we recall a famous experiment involving the simulation of (3a.4.2) with $w(\cdot)$ being drawn from a sequence of IID variables which can assume only one of two values ± 1 with equal probability. Let $y(0) = 0$. The graph of $y(t)$ when t is large is shown in Fig. 3a.4.1 (Feller, 1966). By inspecting Fig. 3a.4.1 it is not obvious whether the series is regularly growing or decaying, but the variance of the process is clearly growing with t.

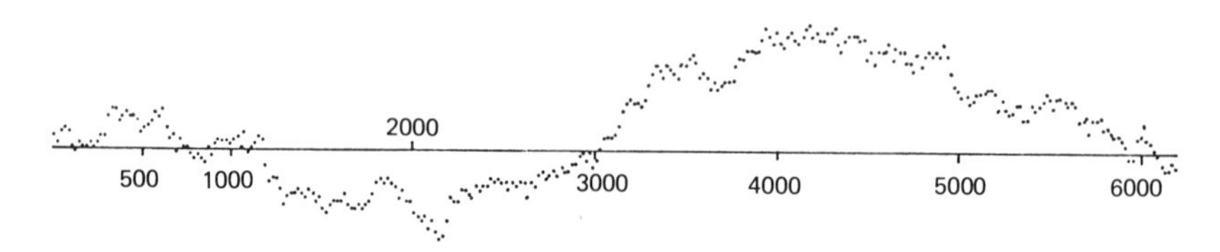

FIG. 3a.4.1. A portion of the record of 10,000 tosses of an ideal coin. (After Feller, 1966.)

The IAR or ARIMA Equation with a Constant Term. Consider the behavior of a process y obeying an ARIMA equation with a constant term in it. For example, consider the first-order IAR process

$$\nabla y(t) = F + A\,\nabla y(t-1) + w(t). \tag{3a.4.10}$$

The process y obeying (3a.4.10) can be represented as

$$y(t) = Ft/(1 - A) + y_1(t) \tag{3a.4.11}$$

where y_1 obeys the first-order ARIMA process without the constant term.

$$\nabla y_1(t) = A\,\nabla y_1(t-1) + w(t) \tag{3a.4.12}$$

Equations (3a.4.11) and (3a.4.12) lead to (3a.4.10). In order to see this apply the difference operation ∇ throughout Eq. (3a.4.11),

$$\nabla y(t) = F/(1 - A) + \nabla y_1(t). \tag{3a.4.13}$$

Express $\nabla y_1(t)$ in terms of $\nabla y(t)$ by using (3a.4.12).

$$\nabla y(t) - F/(1 - A) = A[\nabla y(t - 1) - F/(1 - A)] + w(t),$$

or

$$\nabla y(t) = F + A \nabla y(t - 1) + w(t).$$

The growth behavior of $y(\cdot)$ can be represented as $C_1 t \pm C_2\sqrt{t}$ for large t by using (3a.4.11) since the variance of $y_1(t)$ is proportional to t. Thus, the effect of the constant term in the ARIMA equation is to introduce a *linear* trend into the process.

Similarly, consider a process y where the second difference $\nabla_2 y$ obeys an ARMA equation, with an added constant; i.e., y obeys an ARIMA equation with a constant term. Then y can be represented as the sum of a deterministic quadratic function of time and a signal y_1 that obeys an ARIMA model without a constant term.

ARIMA models with constant terms in them satisfactorily fit a wide variety of empirical time series which display growth. For instance, consider the annual U.S. population series (Fig. 3a.1.2). This example was mentioned in Section 3a.1, where the ineffectiveness of a purely deterministic model was suggested. Let $y(t)$ be the total population in the tth year. We can represent $\ln y(t)$ as

$$\ln y(t) = A_0 t + \ln y'(t), \tag{3a.4.14}$$

where $\ln y'(t)$ obeys an IAR process without the constant term, say

$$A(D) \nabla \ln y'(t) = w(t), \qquad A(D) = 1 + A_1 D + \cdots + A_m D^m. \tag{3a.4.15}$$

Applying the difference operator on (3a.4.14) and using (3a.4.15) we get the following IAR model with constant term for $\ln y(t)$:

$$A(D) \ln y(t) = A_0' + w(t), \tag{3a.4.16}$$

where $A_0' = (1 + \sum_{i=1}^{m} A_i)A_0$. For the U.S. population, model (3a.4.16) with $m = 1$ is appropriate. Note that both the mean and variance of $\ln y(t)$ increase linearly with t.

Second Difference IAR Equation. Consider

$$\nabla^2 y(t) = w(t), \tag{3a.4.17}$$

where $\nabla^2 y(t) = \nabla(\nabla y(t)) = y(t) - 2y(t - 1) + y(t - 2)$. Solving (3a.4.17) for $w(t)$, we obtain

$$y(t) - y(0) = \sum_{j=1}^{t} \nabla y(j) = \sum_{j=1}^{t} \sum_{k=1}^{j} w(k)$$

$$= tw(1) + (t - 1)w(2) + \cdots + w(t), \tag{3a.4.18}$$

$$E[(y(t) - y(0))^2] = kt^3 + t^2 O(1). \tag{3a.4.19}$$

Process y may grow or decay at the rate of $t^{3/2}$ in the sense of Eq. (3a.4.19). As before, the series can oscillate (not, of course, even approximately periodically). It increases for some time, then decreases, and so on.

3a.4.2. Seasonal IAR or ARIMA Models

Seasonal models are suggested if the time series has periodicity in it, i.e., if an approximate periodic oscillation is present in the time series. For illustration let us consider monthly data with an annual cycle, when $T = 12$. Then, in seasonal ARIMA models, the value of the variable $y(t)$ is assumed to depend on the values $y(t-12)$, $y(t-24), \ldots$. Consequently, the difference $y(t) - y(t-12)$ is assumed to obey the difference equation

$$A_1(D^{12})\ \nabla_{12}y(t) = B_1(D^{12})\zeta(t), \tag{3a.4.20}$$

where $\nabla_{12} = 1 - D^{12}$, and $\zeta(t)$ is the disturbance, which may not be white. Equation (3a.4.20) is an ordinary ARMA-like equation with 12 as the unit of time instead of unity. Consequently, the operator D in (3a.4.20) appears in A_1 and B_1 only as multiples of D^{12}. The disturbance $\zeta(\cdot)$ is assumed to obey the ARIMA process

$$A_2(D)\ \nabla_d\zeta(t) = B_2(D)w(t). \tag{3a.4.21}$$

where w is a white noise process.

The process is allowed to have an ARIMA representation instead of ARMA representation to account for the "growth" in the series referred to earlier. Combining (3a.4.20) and (3a.4.21) we get the requisite seasonal ARIMA equation, which can be written as (Box and Jenkins, 1970)

$$A_2(D)A_1(D^{12})\ \nabla_d\ \nabla_{12}y(t) = B_1(D^{12})B_2(D)w(t). \tag{3a.4.22}$$

A fundamental criticism of the seasonal ARIMA model in (3a.4.22) is its counterintuitive nature. It is difficult to imagine that $w(t)$ should be explicitly influenced by the disturbances $w(t-12)$, $w(t-24)$ which occurred 12 or 24 months earlier. Delay terms such as $w(t-12)$ and $w(t-24)$ can be explained only if we postulate some storage mechanism that stores the information and releases it later. It is difficult to imagine that the noise variable w, which is the "residual" variable left over after fitting the known parts, should have such a sophisticated storage mechanism associated with it.

The second criticism of the model is its complexity. The parameter estimation problem is difficult in view of the presence of moving average terms in the system equation. In conclusion, we should consider the class of seasonal ARIMA models only if we find that the best fitted models from other classes of models such as the covariance stationary class are unsatisfactory.

In introducing seasonal ARIMA models, Box and Jenkins point out the disadvantages of modeling a time series (such as those discussed above) purely with sinusoidal terms. But avoiding a purely deterministic model does not imply that we have to dispense with deterministic trend terms altogether, nor does it imply that we have to resort to a seasonal ARIMA model Summing up, there is no a priori reason for considering only seasonal ARIMA models.

3a.4.3. Models with Time-Varying Coefficients

Here we will consider AR or ARMA processes in which the coefficients are not constants. The time sequence of values assumed by some (or all) of the coefficients is itself random, obeying a dynamical system (Kalman, 1963). For instance, consider the process $y(\cdot)$ obeying the first-order time-varying AR model

$$y(t) = a_0(t) + a_1(t)y(t-1) + w(t), \qquad (3a.4.23)$$

where $\{a_0(t), t = 1, 2, \ldots\}$ and $\{a_1(t), t = 1, 2, \ldots\}$ are sequences obeying the constant coefficient AR models

$$\begin{aligned} a_i(t) &= b_i(t) + c_i, \\ b_i(t) &= \gamma_i b_i(t-1) + \eta_i(t), \qquad i = 0, 1; \quad t = 1, 2, \ldots, \end{aligned} \qquad (3a.4.24)$$

where c_i, γ_i, $i = 0, 1$, are constants and $\{\eta_i(t), t = 1, 2, \ldots\}$ are the usual zero mean IID random sequences, independent of each other as well as of $w(\cdot)$.

The constants γ_i are either less than one or equal to one. The case $\gamma_i = 1$ is used in processes involving strong growth, as in some economic time series. Equations (3a.4.23) and (3a.4.24) can be generalized to handle higher order AR models. Very little work has been done on the analysis of the process $y(\cdot)$ in (3a.4.23) and (3a.4.24) for properties such as stability and stationarity.

Regarding the need for time-varying models, there are two extreme views. One view (Mandelbrot and Wallis, 1968) is that the use of time-varying parameters "is rather pointless because the usefulness of a statistical model lies in its large sample predictions." This view can hardly be defended. First of all there is the confusion of the term "constant coefficient." Even though the linear Equation (3a.4.23) has time-varying coefficients, still in the parameter estimation problem posed by (3a.4.23) and (3a.4.24) we need to estimate only the unknown constants γ_i, ρ_i, ... In this sense all modeling problems deal with "constant coefficients." Furthermore, the large-scale predictions of the model in Eqs. (3a.4.23) and (3a.4.24) are quite different from that of the model which obeys (3a.4.23) with $a_0(t)$ and $a_1(t)$ constant for all t. The second view is that all natural processes are highly complicated and nonlinear and naturally need models with time-varying coefficients. In our opinion, this view misinterprets the role played by models. Models are meant for specific purposes and are not photographic representations of reality.

Since we are interested in choosing the simplest model possible, we should consider a time-varying model only if we find that the given observation sequence cannot be adequately fitted by any model of the family in Eq. (1b.1.1) with constant coefficients. One common way in which the inadequacy of the constant coefficient model is exhibited is as follows: Let $\hat{a}_i(t)$ be the least squares estimate of the unknown coefficient a_i based on observations until time t. If the model is inadequate then the variation in the sequence of estimates $\hat{a}_i(n_1), \ldots, \hat{a}_i(N)$ is very much greater than the corresponding theoretical standard deviation of the estimates as given by the model. This aspect is discussed in Chapter X.

The next question is whether or not any process is intrinsically stationary. To be specific, suppose we have N observations of some climatic process taken over a calendar time interval T, say T years. If T is very large, it is often claimed that a stationary model or a constant coefficient model is inappropriate. We think that such claims are inappropriate. The relevant variable for discussion is N, the number of observations, and not T, the calendar time interval. Usually when the number of observations N is very small, a constant coefficient model is satisfactory, i.e., it passes all the validation tests of Chapter VIII. For instance, if we have 20 observations from an environmental process such as atmospheric temperature, at intervals of 50 years each, i.e., if $N = 20$ and $T = 1000$ years, a constant coefficient model is usually sufficient to explain all the variations. In other words, any discrepancy between the characteristics of the process obeying the model and those of the observed process will be within one standard deviation of the corresponding estimates. But if N is large for the same process, then a constant coefficient model is usually not completely satisfactory. If we have 10,000 observations of a process, we usually need a time-varying model regardless of whether the observations are taken hourly, monthly, or yearly. In other words, "stationarity" is in the eyes of the beholder. If we observe any process often enough, a stationary model may not be appropriate. Typically, with any environmental process such as rainfall and temperature, a constant coefficient model is satisfactory as long as N is 100 or even several hundred. While dealing with some economic time series, a constant coefficient model is not appropriate even if $N = 100$.

For alternative approaches for handling linear problems with randomly varying coefficients, see Swamy (1971).

3b. Types of Empirical Time Series

It is difficult to categorize all the observed time series. Hence we will restrict ourselves to a few classes of time series.

3b.1. Pseudoperiodic Series with No Visible "Growth"

Consider empirical series such as the annual sunspot numbers (Fig. 3b.1.1), the annual catch of Canadian lynx (Fig. 3b.1.2), and monthly river flows (Fig. 3b.1.3), which exhibit strong cyclical behavior of approximate periods T but which do not have any indication of "growth." The periodicity and the corresponding period can also be ascertained by plotting the periodogram of the data. The periodogram should display a sharp peak at the frequency $f = 1/T$ cycles

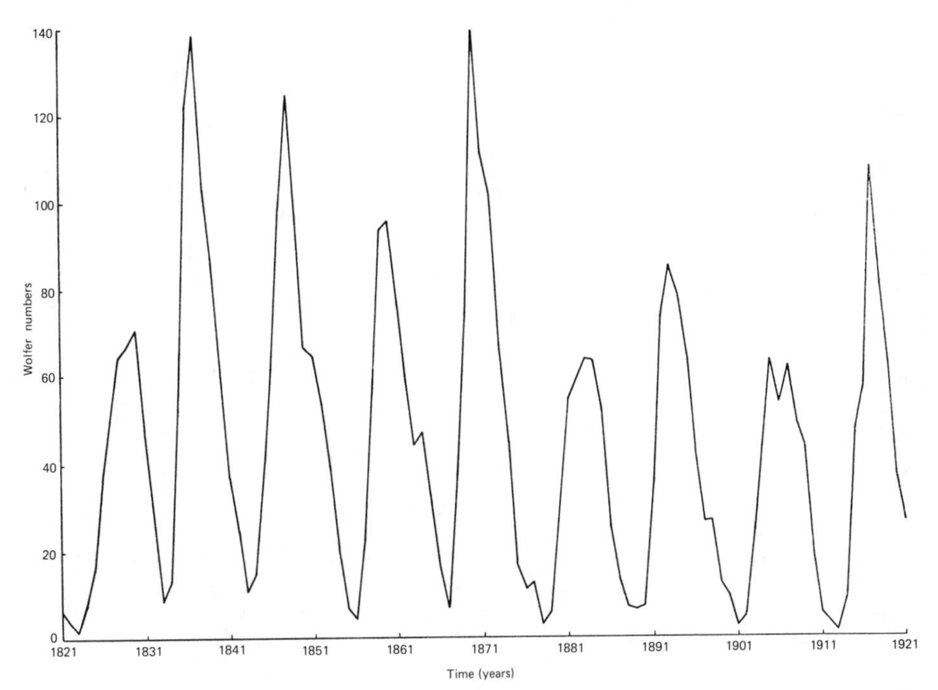

FIG. 3b.1.1. Annual sunspot activity (Wolfer's numbers; series $S2$).

per unit time. Such series could have come from any one of the following classes of models.

Class 1. Weak stationary processes obeying AR or ARMA models with contiguous or noncontiguous AR terms so that the polynomial $A(D)$ in every one of them has a pair of complex zeros with frequency ω_1, the observed frequency.

Class 2. Covariance stationary processes obeying AR or ARMA models with deterministic sinusoidal trend terms of frequency ω_1 and its multiples.

Class 3. $y = y_1 + A \cos \omega_1 t + B \sin \omega_1 t$, where y_1 obeys an ARIMA (d, m_1, m_2) model without the constant term.

Class 4. Seasonal ARIMA model with period T.

Class 1 processes are also known as Slutzky–Yule processes because they were the first investigators to demonstrate that stochastic sequences can have systematic oscillations without explicitly having sinusoidal terms.

We have explicitly excluded the class of deterministic models involving combinations of sinusoids because of the reasons given in Section 3a. However, this statement does not imply that the models of Class 2, i.e., the covariance stationary models, are always poor in comparison with the members of Class 1. We cannot exclude any class of models purely on ideological grounds. We have

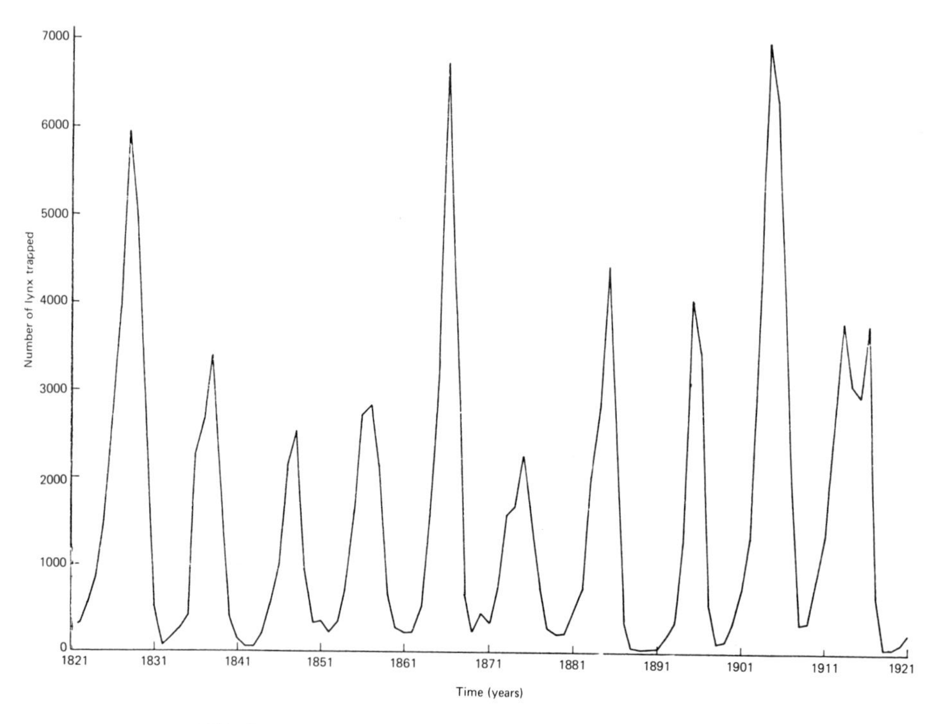

FIG. 3b.1.2. Annual catch of Canadian lynx (series *P*3).

to construct best models from each class and compare them before arriving at definitive conclusions.

We have taken the annual sunspot series, the annual catch of Canadian lynx, and monthly river flow (Krishna River at Vijayawada, India) as illustrations. Parts of these time series are shown in Figs. 3b.1.1, 3b.1.3, respectively. Note that there are no marked visual characteristics to distinguish them.

It is found that the best model for the monthly river flow data is from Class 2 i.e., AR equations with sinusoidal terms of corresponding frequency. The details

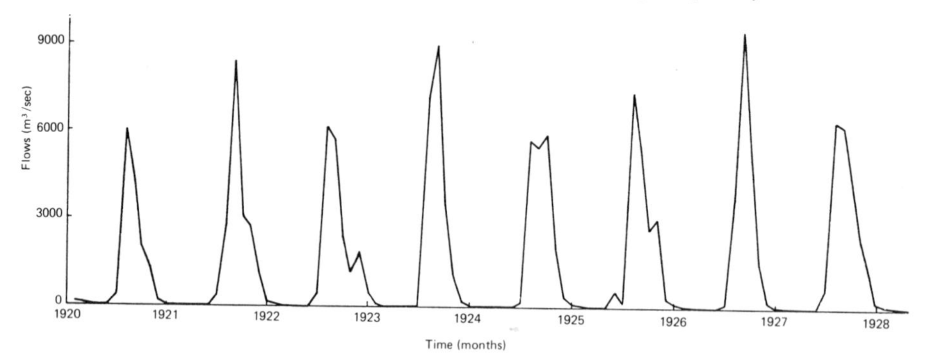

FIG. 3b.1.3. Monthly flows of Krishna River at Vijayawada, India (m^3/sec) (series *F*1).

of the model are given in Chapter X. The annual sunspot series is best fitted by a model from Class 1. Specifically, for the sunspot series, the best model is of the form

$$y(t) = A_0 + A_1 y(t-1) + A_2 y(t-2) + A_3 y(t-T) + w(t).$$

The best value of T is found to be 9, even though the periodicity of the data is about 11.6 years. This model gives a better (in a sense to be described in Chapter XI) fit to the data than the usual AR(2) model widely mentioned in the literature, which was developed by Yule (1927). The AR(2) model was a tremendous improvement over the deterministic model having cosine and sine terms of frequency ω_1 and its multiples.

For Canadian lynx series, a model in Class 1 also gives acceptable results provided we add a term such as $y(t-8)$ to it as shown in Chapter XI.

3b.2. Approximately Periodic Series with Growth

Consider a series such as the monthly sales of company X given in Fig. 3a.1.1. The series indicates a strong growth component and an approximate cyclical character with a period of 12 months. There are two candidate classes for modeling such a series.

Class 1. Covariance stationary processes obeying AR or ARMA equations with deterministic trend functions such as t, $\cos \omega t$, $\sin \omega t$, $t \cos \omega t$, $t \sin \omega t$ added to them.

Class 2. Seasonal ARIMA models.

The sales data are better fitted by a model from Class 1. The best fitting model from Class 2 is not very good from the point of view of prediction as discussed in Chapter XI. However, based on such a small number of examples, we cannot generalize about the relative advantages of seasonal ARIMA models versus covariance stationary models.

3b.3. Time Series with Growth but with No Periodic (or Almost Periodic) Components

Consider time series which exhibit growth (or decay), but which do not exhibit periodic (or approximately periodic) oscillations. Some examples are series $S3$ of the IBM stock (daily closing) price series in Fig. 3b.3.1 (Box and

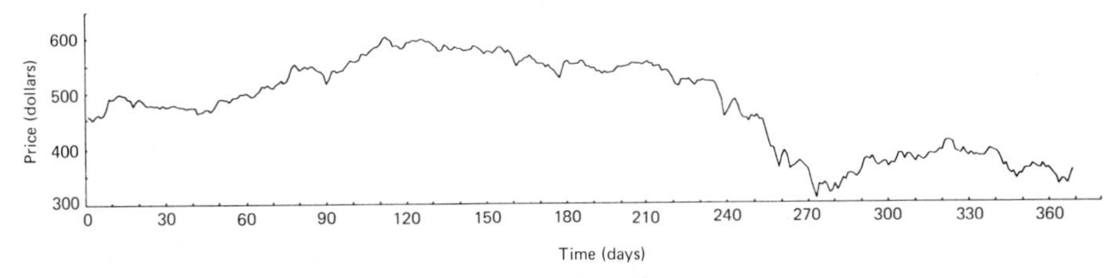

FIG. 3b.3.1. Daily closing IBM stock prices from May 17, 1961 to November 2, 1962 (series $S3$).

Jenkins, 1970), the U.S. annual population series ($P1$) in Fig. 3a.1.2, and the EPD data series in Fig. 3a.2.1. The possible classes of models which may contain such processes are the following:

Class 1. The IAR, IMA, or ARIMA equations without a constant term in them.

Class 2. The IAR, IMA, or ARIMA equations with a constant term in them.

Class 3. AR, MA, or ARMA equations with deterministic trend functions such as $\sqrt{t}$, t, t^2, ...

Class 1 models are useful in representing series where the growth is not very strong, as in the IBM stock data. Otherwise, one has to use Class 2 or 3 models. We may recall that if a model obeys Class 2, then it can be represented as a sum of ct or a suitable polynomial in t and another process y' which obeys an ARIMA equation without a constant term. Thus, a linear deterministic trend term is effectively present in Class 2 models. In Class 3, we allow for a larger class of trend terms.

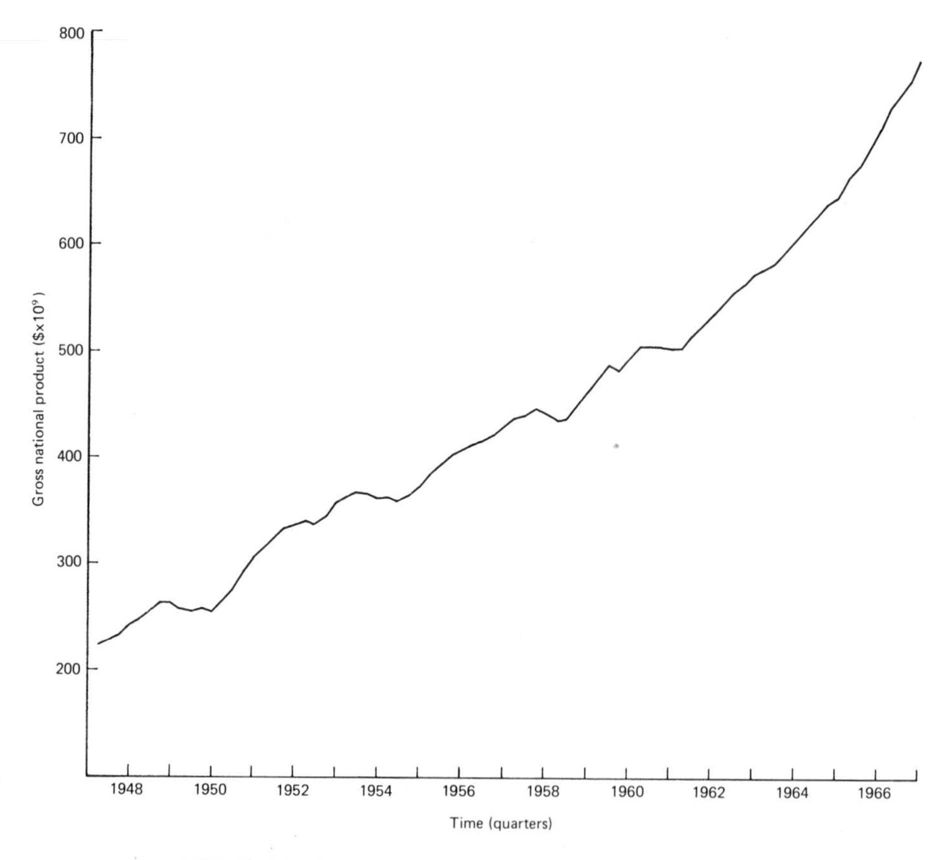

FIG. 3b.3.2. Quarterly GNP of the United States (series $G1$).

Typical series that can be satisfactorily modeled by Class 2 models are the annual U.S. gross national product series ($G1$) involving a first-order IAR equation and the EPD series involving a first-order IMA equation. The gross national product ($G1$) series is given in Fig. 3b.3.2.

Some series, such as the U.S. population and whooping crane population series, exhibit exponential growth. One could try to model them from a member of a Class 3 family with exponential trends. Often (though not always), such fitted models are not satisfactory. In such cases, one could perform a logarithmic transformation on the data and fit a Class 2 model to the transformed data such as (3a.4.16). The details are discussed in Chapter XI.

3b.4. Series without Periodicity or Growth

Consider time series such as series $T1$ (Fig. 3b.4.1), which is the series of chemical process temperature readings observed every minute, or the speech waveform series $S4$ (Fig. 4b.4.2), which do not exhibit any sustained growth, decay, or approximately periodic oscillations. In time series $T1$, which is taken from Box and Jenkins (1970), there is growth for some time and decay for some time, whereas in the related series $C1$ (Fig. 3a.2.3) there are irregular oscillations. The speech waveform series $S4$ could be smooth or rough, depending on the phoneme used.

We need to consider only two classes of models to represent such series:

Class 1. The class of AR, MA, or ARMA models.

Class 2. The class of IAR, IMA, or ARIMA models without any constant term.

By an inspection of the speech waveform series it is difficult to decide the appropriate class for modeling it. The usual practice is to use Class 1 first. If the best fitting model in Class 1 is such that the corresponding polynomial $A(D)$ has a zero on or near the unit circle, then one should investigate the Class 2 models also. Otherwise, Class 1 is enough.

If the process does obey an IAR equation, but we try to approximate it by an AR model, the accuracy of the estimates in the AR equation may be relatively low. This fact is established in Chapter VII. For instance, suppose the process y obeys a first-order IAR process

$$\nabla y(t) = a\,\nabla y(t-1) + w(t), \tag{3b.4.1}$$

which can be rewritten as

$$\begin{aligned} y(t) &= (1+a)y(t-1) - ay(t-2) + w(t) \\ &= a_1 y(t-1) + a_2 y(t-2) + w(t), \end{aligned} \tag{3b.4.2}$$

where

$$a_1 + a_2 = 1. \tag{3b.4.3}$$

Any estimate of the parameters of the AR equation (3b.4.2) obtained without explicitly using the constraint (3b.4.3) is rather inaccurate as compared to the estimate of the single parameter a in (3b.4.1).

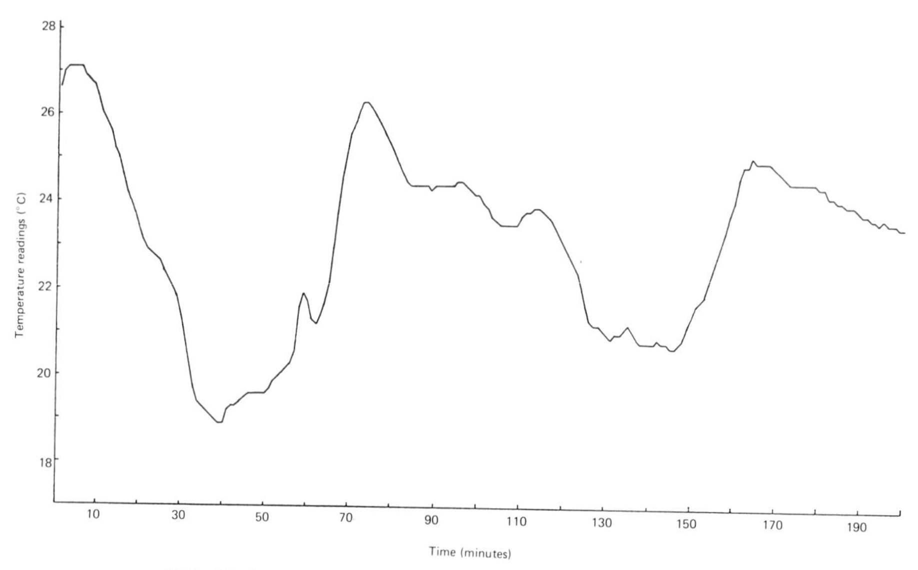

FIG. 3b.4.1. Chemical process temperatures data (series $T1$).

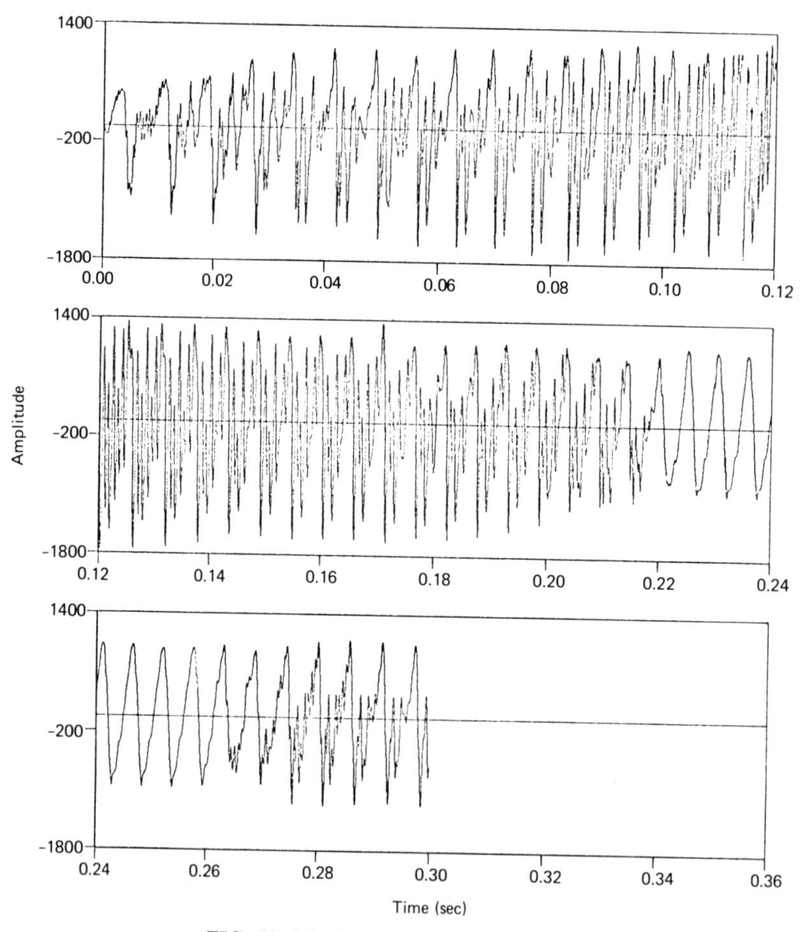

FIG. 3b.4.2. Speech waveform (series $S4$).

Next, let us consider the relative one-step-ahead prediction capability of the two classes of models. Sometimes the best fitting model in Class 1 gives the best fit for the process among all classes. At other times the best predictive model may be in Class 2, and finally, the best fitting model from both classes may give the same performance. For instance, for series $T1$, the best fitting models in Classes 1 and 2 give more or less the same performance. For the speech waveform series $S4$, the best fitting model is an ordinary AR model.

3c. Causality

In the preceding discussion, we omitted the so-called exogenous inputs from the system. Of course, if we are dealing with an engineering system such as aircraft or process control, the exogenous variables are unambiguously known and omitting them from the model can only result in degraded model performance. We are not concerned with such systems here. We would like to pose the following question: "What are the criteria for determining whether a variable is a relevant (exogenous) input for the given output?"

It is easy to state these criteria loosely. If we think that a certain input variable is relevant, then the best fitting model which contains that input must be superior to the best fitting model which does not contain the input. Before proceeding further, we have to clarify the notion of "superior." In a deterministic environment, the test for causality is relatively easy. We hold all input variables except one constant and vary that particular variable in question. The corresponding change in the output is a measure of the degree of the causality of the particular variable with respect to the output variable. Such simple tests are clearly inoperable in a stochastic environment. We can consider the following operational definition of causality in stochastic systems. Suppose we want to test whether the input variable $x(t)$ is a causal variable as far as the output variable $y(t)$ is concerned. If it is so, then one would expect that a term such as $x(t-1)$ or $x(t-2)$ should be present in the equation for $y(t)$, in addition to terms such as $y(t-1), y(t-2), \ldots$. If we omit the term $x(t-1)$ [or $x(t-2)$] from the equation, the ability of the model to represent the data is reduced and this loss can be detected by checking the prediction behavior. As an illustration, we can divide the given observation set into two parts, say equal in length. We can use the first part to construct the two best fitting models for $y(t)$, the first one having only the output terms and the second one output terms plus input terms $x(t-1), x(t-2), \ldots$. We use both fitted models to obtain one-step-ahead predictions of the observations in the second part mentioned earlier and compute the corresponding mean square prediction errors (MSE). The model that gives the smaller MSE is considered to be the superior one. The experiment may be repeated by dividing the observation history into two parts in some other proportion, to make sure that the results are not sensitive to partition of the data. Other methods for determining the contribution of an exogenous term to the predictive ability of a model are discussed in Chapters VIII and IX.

One of the important inferences from such an investigation is that a (best fitting) model with a large number of input terms does not necessarily give any better prediction than a (best fitting) model with a few or no input variables. As a matter of fact, addition of the "irrelevant" input variables may result in a sharp reduction in the predictive ability of the model. This result is not unexpected. It has been observed even in ordinary autoregressive systems. A best fitted AR(4) model does not necessarily give better prediction than a best fitting AR(2) model.

We recall that trend terms may be introduced to account for a number of exogenous variables. In a manner similar to that discussed above, one can determine the suitability of introducing trend terms into the model or the corresponding exogenous terms.

There have been attempts at defining the notion of causality in a stochastic environment in terms of the empirical (or observed) correlation coefficient. If we are dealing with two time series $x(\cdot)$ and $y(\cdot)$ and if the estimated correlation coefficient between them, say between $x(t)$ and $y(t)$, is very near unity, then x and y are said to be causally related to one another. Such a definition is unsatisfactory for several reasons. First, the two series x and y may be strongly correlated with each other even though they are entirely unrelated to each other. An example is given by

$$y(t) = at + w_1(t), \qquad x(t) = bt + w_2(t) \tag{3c.1.1}$$

where x and y are statistically independent because of the independence of w_1 and w_2, and yet x and y have a high empirical correlation caused by the similar trend terms. Another reason is that the correlation coefficient is symmetrical in the two variables x and y. But causality, as it is usually understood, is not a symmetric property. x may be a causal input to y, but y may not be a causal input to x. For instance, x may be rainfall and y can be groundwater levels. It is clear that rainfall affects groundwater levels, but not vice versa. There are other reasons for not considering the correlation coefficients. Other methods of handling causality are discussed in Chapter IX.

3d. Choice of Time Scale for Modeling

An important problem in modeling is the choice of time scale. Specifically, suppose we have monthly data about a process, each data point representing the average value of the variable in a month, and we want to develop a model for obtaining *yearly* forecasts, which involves forecasts up to 12 months ahead. Then a decision must be made as to whether the monthly data are to be used to construct a monthly model where the interval of discretization $\Delta = 1$ or construct a model with aggregated data. The model with aggregated data may have a discretization rate Δ, where Δ may vary from 2 to 12. A simple-minded decision is to construct the monthly model (model with $\Delta = 1$) since the models with other levels of discretization can be obtained from this model. For instance, let the given data be $y(t)$ $(t = 1, 2, \ldots)$, where Δ corresponds to a month. Let the model constructed with $\Delta = 1$ have spectral density $s_{yy}(e^{-i\omega})$. Consider the aggregated process for which $\Delta = 7$:

$$x(t) = y(t) + \cdots + y(t-6).$$

Then the spectral density of $x = s_{xx}(e^{-i\omega})$ is given by

$$s_{xx}(e^{-i\omega}) = \|1 + e^{-i\omega} + \cdots + e^{-6i\omega}\|^2 s_{yy}(e^{-i\omega}),$$

which can be factorized to obtain the difference equation for x. But the order of the difference equation so obtained may be high. If the order of the difference equation is to be reduced, there will be an approximation error. Consequently, the equation for the y process should give at least as good a prediction as the equation for $x(\cdot)$, and more often the predictions obtained from the model for the y process should be better than the predictions obtained from the model of the x process because of the approximation error mentioned above.

Analysis such as that outlined above is too simplistic, and would be correct if we knew that the given monthly data were exactly representable as a linear difference equation with $\Delta = 1$. In practice, seldom do we have such knowledge. The constructed model is only an approximation to the given process. Moreover, the model itself, whether it is constructed for y or x places restrictions on the information which is used to estimate the unknown coefficients in it. For instance, if we fit an AR(1) model to the given process $y(\cdot)$, then the estimate of the AR coefficient will involve only the zero- and first-order properties $(1/N) \sum_{t=1}^{N} y(t)y(t-j)$, $j = 0$ and 1. Such a model will not absorb any other information, such as $(1/N) \sum_{t=1}^{N} y(t)y(t-j), j > 1$, which is available in the data. Now if a model for the aggregated process $x(i)$ is fitted by using the data directly, the model for $x(\cdot)$ will involve the estimated autocovariances of the $x(\cdot)$ process which will be computed from the covariances

$$(1/j) \sum_{t} y(t)y(t-j), \qquad j = 0, 1, \ldots, 8,$$

and as such there is a possibility of the aggregated model for x giving better predictions than the original model for $y(\cdot)$ because of its additional information content. In view of the foregoing, it may be concluded that different models with various levels of discretization have to be constructed so that the correct model for the given task can be selected from these different models.

Some discussion of the parameter estimation from aggregated data can be found in the papers by Engle and Liu (1972) and Lee *et al.* (1970).

3e. Conclusions

The main purpose of this chapter was to illustrate the behavior of different types of difference equations with various types of terms such as AR, MA, trends, and time-varying coefficients. We also suggested the different classes of models that should be considered in modeling empirical series with such prominent properties as growth and systematic oscillations.

Notes

The sales data of company X in Fig. (3a.1.1) are obtained from Chatfield and Prothero (1973). The annual U.S. population data are from Statistical Abstracts (1974). The GNP data in Fig. 3b.3.2 and the EPD data in Fig. 3a.2.1 are from Nelson (1973). The IBM stock price data in Fig. 3b.3.1, the chemical process concentrations in Fig. 3a.2.3, and the temperature data in Fig. 3b.4.1, are from Box and Jenkins (1970). The annual sunspots data in Fig. 3b.1.1 are from Waldemeir (1961); the lynx data in Fig. 3b.1.2 are from Moran (1953); and the flow data of the Krishna river are from UNESCO (1971).

Problems

1. Consider processes y_1 and y_2, where w_1 and w_2 have variances $\rho_1 = 1$ and ρ_2:

$$y_1(t) = 0.5y_1(t-1) + t + w_1(t)$$
$$\nabla y_2(t) = \theta_1 \nabla y_2(t-1) + \theta_2 + w_2(t).$$

(a) Choose θ_1, θ_2, and ρ_2 so that $\nabla y_1(t)$ and $\nabla y_2(t)$ have asymptotically the same expectation, variance, and first correlation coefficient.
(b) Simulate the processes y_1 and y_2 on a computer and discuss whether it is possible to distinguish between the two types of processes by inspection alone.

2. Consider the following three pseudoperiodic series having the same period of 6 where ρ_i is the variance of w_i. Let $\rho_1 = 1.0$.

(i) $y_1(t) = 1.33y_1(t-1) - 0.5y_1(t-2) + w_1(t)$,
(ii) $y_2(t) = \theta_1 y_2(t-1) + \theta_2 \cos 2\pi t/6 + \theta_3 \sin(2\pi t/6) + w_2(t)$,
(iii) $(1 - \theta_4 D)(1 - \theta_5 D^6)[y_3(t) - y_3(t-6)] = w_3(t)$.

Normalize process $y_2(\cdot)$ by choosing θ_1, θ_2, θ_3, and ρ_2 so that y_2 and y_1 have the same time averaged correlation coefficients c_i, $i = 0, 1, 2$. Normalize $y_3(\cdot)$ by choosing θ_4, θ_5, and ρ_3 so that $y_1(t) - y_1(t-6)$ and $y_3(t) - y_3(t-6)$ have the same time-averaged correlation coefficients c_i, $i = 0, 1, 2$. Simulate all three processes on a digital computer and verify that they cannot be distinguished by inspection alone.

3. Consider process y_1, of Problem 2. Let

$$\hat{\gamma}_1(\omega) = \frac{2}{N}\sum_{t=1}^{N} y_1(t) \cos \omega t, \qquad \hat{\gamma}_2(\omega) = \frac{2}{N}\sum_{t=1}^{N} y_1(t) \sin \omega t.$$

Find $E[\hat{\gamma}_i(\omega)^j]$, $i = 1, 2, j = 1, 2$, and verify that when $\omega = 2\pi/6$, $E[(\hat{\gamma}_1(\omega)^2 + \hat{\gamma}_2(\omega)^2]$ could be relatively large for N as large as 50.

4. Discuss the advantages and disadvantages of summarizing the empirical data of a time series in the form of empirical correlation coefficients if (i) N is small (say <100), (ii) N is large (say >1000).

Chapter IV | Estimability in Single Output Systems

Introduction

In Chapter III, we considered various forms of single output stochastic difference equations that are useful for modeling different types of time series. In this chapter we investigate the conditions under which the unknown coefficients in the difference equation can be recovered with the aid of the observation histories of the processes y and $\mathbf{u}$; i.e., we will determine the conditions under which the second-order properties of the process y are uniquely characterized by the coefficients in the difference equation, the trend vector $\boldsymbol{\psi}$, and the properties of the input $\mathbf{u}$. This is the first step in the estimation of the unknown coefficients from a finite observation history. We may mention that not all the unknowns in every equation can be recovered even with a semi-infinite observation history in view of the possible multiplicity of the process, i.e., the possibility of more than one difference equation yielding the same process. The estimability conditions here ensure that there is at most one difference equation in the given family of difference equations for the given empirical process. We shall restrict ourselves to single output processes in this chapter [or $m = 1$ in (2a.1.1)].

We first derive a necessary and sufficient condition on the difference equation (2a.1.1) with $m = 1$ or the scalar equation, so that the input $\mathbf{u}$, the trends $\boldsymbol{\psi}$, and the set of coefficients $\boldsymbol{\theta}$ in it uniquely characterize the process. Since the validity of the necessary and sufficient conditions cannot be ascertained easily, a set of sufficiency conditions, which are easily verifiable, is developed to test the uniqueness of characterization. The characterization problems of two related classes of systems, namely systems with autoregressive disturbances and systems in which the output can be observed only after it is corrupted by additive noise, are also considered. Finally, we will discuss the accuracy of estimation of the parameters and the limits that can be placed on the achievable accuracy of estimation with a given set of observations.

Astrom *et al.* (1965), Caines and Rissanen (1974), Bohlin (1971), and others have considered the estimability of single difference equations with observable random input terms obeying the independence assumption (A3). One of the important aspects of the theory developed in this chapter is the possibility of relaxing the independence assumption (A3). (A7) and (A7′) are identical when y is a scalar and imply that (4a.1.1) cannot be further simplified.

4a. Estimability of Systems in Standard Form

4a.1. Necessary and Sufficient Conditions

Consider the difference equation

$$A(D)y(t) = \mathbf{G}(D)\mathbf{u}(t-1) + B(D)w(t) + \mathbf{F}\boldsymbol{\psi}(t-1), \qquad (4a.1.1)$$

where $\mathbf{G}$ and $\mathbf{F}$ are row vectors. Let $\boldsymbol{\theta}$ denote the set of all the coefficients of the terms y, $\mathbf{u}$, $\boldsymbol{\psi}$, and w in (4a.1.1), $\boldsymbol{\theta}$ being a vector of dimension $n = m_1 + m_2 + l_1 m_3 + l_2$. When $\boldsymbol{\theta}$ is unknown, it is our intention to estimate it from the observed time histories of the processes y and $\mathbf{u}$. The coefficient vector $\boldsymbol{\theta}$ is said to be estimable† if its true value can be recovered with the aid of the semi-infinite observation histories $\{y(t), \mathbf{u}(t), 1 \le t \le \infty\}$.

Since we are dealing with the coefficients in a linear system, a knowledge of the second-order properties of the processes y and $\mathbf{u}$, i.e., their mean and covariances of all orders, is sufficient for their estimation.

Hence, we can equivalently define estimability as follows. A vector $\boldsymbol{\theta}$ is said to be estimable (or more precisely, second-order estimable) if the parameter vector, in conjunction with the second-order properties of the $\mathbf{u}$ process, uniquely characterizes the second-order properties of the y process.

We will determine some necessary and sufficient conditions for the estimability of the elements of $\boldsymbol{\theta}$ and investigate the adequacy of some of the assumptions among (A1)–(A7) to satisfy these conditions.

Let us rewrite (4a.1.1) as

$$\begin{aligned} y(t) &= \boldsymbol{\theta}^{\mathrm{T}}\mathbf{x}(t-1) + w(t), \qquad (4a.1.2)\\ \mathbf{x}(t-1) &= (y(t-1), \ldots, y(t-m_1), \mathbf{u}^{\mathrm{T}}(t-1), \ldots, \mathbf{u}^{\mathrm{T}}(t-m_3),\\ &\quad \boldsymbol{\psi}^{\mathrm{T}}(t-1), w(t-1), \ldots, w(t-m_2))^{\mathrm{T}}. \end{aligned}$$

An obvious sufficient condition for the estimability of $\boldsymbol{\theta}$ is that all the components of $\mathbf{x}(t)$ be linearly independent for all t. However, such a condition is not necessary for estimability. We need only rule out the possibility of asymptotic linear dependency, the possibility of a linear combination of $\mathbf{x}(t)$ going to zero at a sufficiently fast rate, among the elements of $\mathbf{x}(t)$. The requisite condition is stated as (B1).

Condition B1. There *does not* exist a constant nonzero n-vector $\boldsymbol{\phi}$ so as to satisfy

$$\lim_{N\to\infty} \sum_{t=1}^{N} g^2(t) < \infty, \qquad (4a.1.3)$$

where $g(t) \triangleq \boldsymbol{\phi}^{\mathrm{T}}\mathbf{x}(t)$.

Theorem 4a.1. Condition B1 is necessary and sufficient for the estimability of the coefficient vector $\boldsymbol{\theta}$ in (4a.1.1) which obeys assumptions (A1), (A2), (A5), and (A6).

† The choice of the word "estimable" is from Rao (1965). Synonyms for the word in current usage are "identifiable" and "recoverable."

Proof. The sufficiency part of the proposition is established in Appendix 4.1. Only the necessity part is established here. Suppose condition (B1) is not satisfied, i.e., there exists a nonzero vector $\boldsymbol{\phi}$ obeying (4a.1.3). Consider a related scalar process y' defined as

$$\begin{aligned} y'(t) &= y(t) + \boldsymbol{\phi}^{\mathrm{T}}\mathbf{x}(t-1) = (\boldsymbol{\theta} + \boldsymbol{\phi})^{\mathrm{T}}\mathbf{x}(t-1) + w(t) \\ &= (\boldsymbol{\theta}')^{\mathrm{T}}\mathbf{x}(t-1) + w(t), \end{aligned} \tag{4a.1.4}$$

where $\boldsymbol{\theta}' = \boldsymbol{\theta} + \boldsymbol{\phi}$.

We will establish that the processes y and y' have the same second-order properties asymptotically, which makes us unable to distinguish between $\boldsymbol{\theta}$ and $\boldsymbol{\theta}'$ for the characterization of the process y:

$$\lim_{N\to\infty} \frac{1}{N}\sum_{t=1}^{N} (y'(t))^2 = \lim_{N\to\infty} \frac{1}{N}\sum_{t=1}^{N} [y(t) + \boldsymbol{\phi}^{\mathrm{T}}\mathbf{x}(t-1)]^2 \tag{4a.1.5}$$

$$= \lim_{N\to\infty} \left[\frac{1}{N}\sum_{t=1}^{N} y^2(t) + \frac{1}{N}\sum_{t=1}^{N} (\boldsymbol{\phi}^{\mathrm{T}}\mathbf{x}(t-1))^2 + \frac{2}{N}\sum_{t=1}^{N} y(t)\boldsymbol{\phi}^{\mathrm{T}}\mathbf{x}(t-1)\right]. \tag{4a.1.6}$$

As N tends to infinity, the second term on the right-hand side of (4a.1.6) tends to zero in view of (4a.1.3). By the Schwarz inequality, the absolute value of the third term of (4a.1.6) is less than or equal to

$$\frac{2}{N}\left[\left(\frac{1}{N}\sum_{t=1}^{N} y^2(t)\right)\sum_{t=1}^{N} (\boldsymbol{\phi}^{\mathrm{T}}\mathbf{x}(t-1))^2\right]^{1/2},$$

which by (4a.1.3), goes to zero asymptotically. Consequently, the mean square values of the processes y and y' are asymptotically identical. Similarly, we can establish the asymptotic equality of all the autocorrelations of processes y and y'.

Comment 1. Some examples of systems in which (B1) is not satisfied are in order. Since it is trivial to produce examples in which $g(t)$ [defined in (4a.1.3)] is zero, only those examples where $g(t)$ is not identically zero are given. It should be noted that $g(t)$ can be nonzero even if there are trend terms in (4a.1.1).

Example 4a.1. Consider system (4a.1.1) with $\psi(t) = \psi_1(t) = 1/(t+1)$. We can find a nonzero $\boldsymbol{\phi}$ so that

$$\boldsymbol{\phi}^{\mathrm{T}}\mathbf{x}(t-1) \triangleq \psi_1(t-1) \triangleq g(t).$$

By definition of ψ_1, $\sum_{t=1}^{\infty} g^2(t) < \infty$. Hence the corresponding set of coefficients in (4a.1.1) is not estimable.

Example 4a.2 (Malinvaud, 1970). Consider system (4a.1.1) with

$$\boldsymbol{\psi}(t) = (\psi_1(t), \psi_2(t))^{\mathrm{T}}, \qquad \psi_1(t) = 1, \qquad \psi_2(t) = \left(1 - \frac{1}{t+1}\right).$$

Again, we can find a nonzero $\boldsymbol{\phi}$ such that

$$\boldsymbol{\phi}^{\mathrm{T}}\mathbf{x}(t-1) = \psi_1(t-1) - \psi_2(t-1) \triangleq g(t).$$

By definition of ψ_1 and ψ_2, $\sum_{t=1}^{\infty} g^2(t) < \infty$. Again all the coefficients in a system with a trend vector are not estimable.

Comment 2. Several interesting consequences of Theorem 4a.1 may now be pointed out. Theorem 4a.1 does not involve total knowledge of the probability distribution of the disturbances. The disturbances need only be independent with a common finite variance as specified in (A1). Further, the disturbances need not be identically distributed. However, if the probability distribution of the disturbances is known, the accuracy of the estimates can be determined and the estimates of the highest possible accuracy can be obtained. If interest is in obtaining consistent estimates of the unknown parameters, then partial knowledge of parameters governing the probability distribution such as zero mean, symmetry, finite variance, etc., is sufficient.

For estimability of parameters, the observed input $\mathbf{u}$ need not be independent of the disturbance w. All we need is assumption (A1), which only implies the independence of $w(t)$ from $\mathbf{u}(t-j), j \geq 1$. This aspect is further discussed in Section 4a.2.

Furthermore, polynomial or exponentially increasing trends can be present in the stochastic difference equation (SDE) and the corresponding coefficients can be consistently estimated. The process y need only be covariance stationary and not necessarily weakly stationary or asymptotically weakly stationary.

The following example deals with a system that does not obey (B1). It is a system that also does not obey (A7).

Example 4a.3. Consider the system

$$(1 - A_1D)(1 - A_2D)y(t) = (1 - A_1D)(1 - G_1D)u(t-1) + (1 - A_1D)w(t). \tag{4a.1.7}$$

The corresponding $\mathbf{x}$-vector is

$$\mathbf{x}(t-1) = (y(t-1), y(t-2), u(t-1), u(t-2), u(t-3), w(t-1))^{\mathrm{T}}.$$

We will show that $\mathbf{x}(t)$ is linearly dependent. In order to see this, note the common zero $(1 - A_1D)$ in all terms of (4a.1.7) and obtain a difference equation by canceling this factor:

$$(1 - A_2D)y(t) = (1 - G_1D)u(t-1) + w(t), \tag{4a.1.8}$$

or

$$y(t) - A_2y(t-1) - u(t-1) + G_1u(t-2) - w(t) = 0. \tag{4a.1.9}$$

Clearly, Eq. (4a.1.8) describes the same process as (4a.1.7) since both of them lead to the same transfer function. Equation (4a.1.9) represents a linear combination of $\mathbf{x}(t)$ which is identically zero for all t.

As stated earlier, it is not easy to check whether condition (B1) is satisfied. Consequently, we will translate condition (B1) into explicit conditions on the polynomials A, B, and G, and the inputs $\mathbf{u}$ and $\boldsymbol{\psi}$. It is relatively easy to check whether or not these conditions on the polynomials are satisfied.

Theorem 4a.2. Consider a process y obeying Eq. (4a.1.1) with the trend vector $\boldsymbol{\psi}$ omitted. Let the coefficients and the inputs in it obey assumptions (A1)–(A3), (A5), and (A6). Then assumption (A7) is necessary and sufficient for the estimability of all the coefficients in (4a.1.1).

Proof. The necessity of (A7) is clear from Example 4a.3. To show its sufficiency we need to show that assumption (A7) implies condition (B1). If condition (B1) is not satisfied, then a linear combination of $\mathbf{x}(t)$ is either zero or equal to $g(t)$, obeying (4a.1.3). The latter case can be ruled out in view of the absence of the trend terms $\boldsymbol{\psi}$.

A linear combination of $\mathbf{x}(t-1)$ can be zero in only one of the two following ways, which are mutually exclusive: (i) A linear combination of $\mathbf{u}(t-1), \ldots, \mathbf{u}(t-m_3)$ and $w(t-1), \ldots, w(t-m_2)$ is zero for all t, and (ii) $y(t)$ is equal to a linear combination of $y(t-1), \ldots, y(t-m_1+1), \mathbf{u}(t), \ldots, \mathbf{u}(t-m_3+1)$, and $w(t), \ldots, w(t-m_2+1)$.

Case (i) cannot occur in view of assumptions (A2) and (A3). The possibility of the occurrence of case (ii) arises if the corresponding polynomials A, B, and G_j, $j = 1, \ldots, l_1$, have a common factor. This common factor can be factored out of (4a.1.1) and the reduced equation is a linear combination of $\mathbf{x}(t-1)$ equaling zero as stated in case (ii). This contingency is ruled out by assumption (A7).

Comment 1. To emphasize the fact that Theorem 4a.2 is not true when trend terms are present [i.e., (A7) is not necessary for estimability, when trends are present], the following example is given.

Example 4a.4. Let system $y(\cdot)$ obey

$$(1 - A_1 D)y(t) = (1 - A_1 D)w(t) + A_2 \cos t, \tag{4a.1.10}$$

where $|A_1| < 1$, $A_2 \neq 0$. Clearly, (4a.1.10) does not satisfy (A7) in view of the common factor $(1 - A_1 D)$. However, we will show that (B1) is still satisfied and consequently A_1 and A_2 are both recoverable.

Define two additional parameters A_3 and A_4 by the relation

$$A_2 \cos t = (1 - A_1 D)(A_3 \cos t + A_4 \sin t). \tag{4a.1.11}$$

Solving (4a.1.11) for A_3 and A_4, we obtain

$$A_3 = A_2(1 - A_1 \cos 1)/(1 + A_1^2 - 2A_1 \cos 1) \tag{4a.1.12}$$

$$A_4 = A_1 A_2 (\sin 1)/(1 + A_1^2 - 2A_1 \cos 1). \tag{4a.1.13}$$

Substituting (4a.1.11) into (4a.1.10), we obtain

$$(1 - A_1 D)\{y(t) - w(t) - A_3 \cos t - A_4 \sin t\} = 0. \tag{4a.1.14}$$

Removing the common factor in (4a.1.14), we get the following reduced expression for $y(t)$:

$$y(t) = w(t) + A_3 \cos t + A_4 \sin t. \tag{4a.1.15}$$

Since a linear combination of cos t and sin t cannot be identically zero for all t, condition (B1) is automatically satisfied. Hence, A_1 and A_2 are estimable since it is easy to recover them from A_3 and A_4 using (4a.1.12) and (4a.1.13).

To consider the estimability of systems that have trend terms, let us consider assumptions (A4) and (A4′) on the trend vector $\boldsymbol{\psi}$. In both (A4) and (A4′), the components of $\boldsymbol{\psi}(t)$ must be linearly independent. Assumption (A4) is sufficient for most purposes. But an important class of trend functions, namely $\boldsymbol{\psi}(t) = t$, is *not* obeyed by (A4). Since this trend is required in many problems, we have to investigate the variant (A4′). In addition to ensuring that the components of $\boldsymbol{\psi}$ are linearly independent, assumption (A4′) eliminates trends such as $\psi(t) = 1/t + 1$ which cause inestimability. Assumption (A4) implies (A4′), but (A4′) does not imply (A4). Sinusoidal terms such as cos t and sin t, both obey (A4) and (A4′). Trends such as $\psi(t) = t, 1/\sqrt{t}$ obey (A4′), but not (A4).

We will modify Theorem 4a.2 so that the corresponding equation could have trend terms in it.

Theorem 4a.3. Consider a process y obeying (4a.1.1) along with assumptions (A1)–(A3), (A4′), (A5), and (A6). Then (A7) is sufficient (but not necessary) for the estimability of various coefficients in (4a.1.1).

This theorem can be proved by showing that assumptions (A1)–(A3), (A4′), and (A5)–(A7) imply condition (B1), which implies the estimability of all coefficients in the equation in view of Theorem 4a.1.

In order to understand the reason for (A7) not being necessary for estimability, we consider two examples. The system equation in Example 4a.4. does not obey (A7). Still, the coefficients in it are estimable. On the other hand, (A7) is not satisfied by the system equation in the following example, and the system is not estimable.

Example 4a.5. Let the system equation be

$$(1 - A_1 D)y(t) = (1 - A_1 D)w(t) + F_1 \cos t + F_2 \sin t. \tag{4a.1.16}$$

The trend vector $\boldsymbol{\psi} = (\cos t, \sin t)$ obeys (A4′), but (A7) is not satisfied and consequently all the coefficients are not estimable. To investigate further we will divide (4a.1.16) throughout by $(1 - A_1 D)$:

$$\begin{aligned} y(t) &= w(t) + (1 - A_1 D)^{-1}(F_1 \cos t - F_2 \sin t) \\ &= w(t) + F_3 \cos t + F_4 \sin t. \end{aligned} \tag{4a.1.17}$$

In order to express F_3 and F_4 in terms of A_1, F_1, and F_2, we use the identity

$$\begin{aligned} F_1 \cos t + F_2 \sin t &= (1 - A_1 D)(F_3 \cos t + F_4 \sin t) \\ &= F_3 \cos t + F_4 \sin t - A_1[F_3 \cos (t - 1) + F_4 \sin (t - 1)]. \end{aligned} \tag{4a.1.18}$$

From (4a.1.17) we can recover F_3 and F_4. But we cannot recover the three parameters F_1, F_2, and A_1 from F_3 and F_4.

The difference in the estimability behavior in Examples 4a.4 and 4a.5 can be traced to the fact that the components of the trend vector in Example 4a.4 are not "complete," i.e.,

$$(1 - A_1 D)\boldsymbol{\psi}(t) = \mathbf{C}_1 \boldsymbol{\psi}(t) \tag{4a.1.19}$$

is not satisfied for any $\mathbf{C}_1$. But, in Example 4a.5, the two trends $\psi_1 = \cos t$ and $\psi_2 = \sin t$ are complete, i.e., there exist constants c_{ij} such that

$$(1 - A_1 D)\begin{pmatrix} \psi_1(t) \\ \psi_2(t) \end{pmatrix} = \begin{pmatrix} c_{11} & c_{12} \\ c_{21} & c_{22} \end{pmatrix}\begin{pmatrix} \psi_1(t) \\ \psi_2(t) \end{pmatrix} \tag{4a.1.20}$$

is satisfied.

In Example (4a.4), the system Equation (4a.1.7) still preserves the information about A_1 after factoring out $(1 - A_1 D)$, whereas in Example 4a.5 the removal of the factor $(1 - A_1 D)$ does not yield any information about A_1, and hence the system is not estimable.

These two examples suggest that condition (A7) in Theorem 4a.3 can become necessary as well if assumption (A4′) in the theorem is replaced by the stronger assumption (A4″). Assumption (4a.1.21) in (A4″) is a generalization of the assumption on the trends mentioned in (4a.1.20):

A4″. The trend vector $\boldsymbol{\psi}$ obeys (A4′) and

$$A(D)\psi_i(t) = \mathbf{h}_i^{\mathrm{T}} \boldsymbol{\psi}(t) + g_i(t), \qquad i = 1, \ldots, l_2, \tag{4a.1.21}$$

where $\mathbf{h}_i$, $i = 1, \ldots, l_2$, are constant l_2-vectors and the functions $g_i(t)$ obey

$$\sum_{t=1}^{\infty} g_i^2(t) < \infty. \tag{4a.1.22}$$

The system in Example 4a.5 satisfies (A4″), whereas the system in Example 4a.4 does not.

Finally, consider the inestimability caused by the trend terms not obeying assumption (A4′). Then the estimability can be restored by regrouping the trend terms or by omitting certain inconsequential trends and *reducing* the number of coefficients of the trend terms. For instance, in Example 4a.2, the system parameters can be made estimable by having only one trend term in the system equation, namely $\psi_1 = 1$.

4a.2. Inestimability Caused by Feedback

The use of assumption (A3), namely the independence of $\mathbf{u}$ and w in Theorems 4a.2 and 4a.3, is highly restrictive. In particular, systems of linear feedback, i.e., systems in which $\mathbf{u}(t)$ is a linear function of $y(t), y(t-1), \ldots$. are outside the purview of Theorem 4a.2. Hence, we will discuss them separately.

Consider a first-order single input–single output system

$$y(t) = \sum_{j=1}^{m_1} A_j y(t-j) + G_1 u(t-1) + w(t), \tag{4a.2.1}$$

where the control u obeys the feedback law

$$u(t) = K_1 y(t). \tag{4a.2.2}$$

Substitute for $u(\cdot)$ in (4a.2.1) from (4a.2.2), to obtain

$$y(t) = (A_1 + G_1 K_1) y(t-1) + \sum_{j=2}^{m_1} A_j y(t-j) + w(t), \tag{4a.2.3}$$

from which the coefficients $(A_1 + G_1 K_1)$ and A_j, $j = 2, \ldots, m_1$, are estimable. By using the histories of y and u, and Eq. (4a.2.2), K_1 is recoverable. But the coefficients A_1 and G_1 cannot be separately recovered. We clearly see how linear feedback destroys estimability.

Estimability can be restored in such feedback systems in two ways. First, suppose there is a small amount of noise in the control, i.e., the feedback law is

$$u(t) = K_1 y(t) + \eta(t), \tag{4a.2.4}$$

where η is a zero mean white noise, independent of w. The system equation is (4a.2.1) as before. Such a system comes under the purview of Theorem 4a.1 since (B1) is satisfied. Consequently both parameters A_1 and G_1 are estimable by Theorem 4a.1. The variance of η can have any value as long as it is greater than zero. This example is very instructive in the role played by small amounts of noise in restoring estimability. Exact recovery of the parameters is possible with noisy control, but impossible with a pure deterministic control.

The second method of restoring estimability is to have the control $u(t)$ depend not only on $y(t)$ but also on $y(t-1), \ldots, y(t-m_4)$, where $m_4 \geq (m_1 + 1)$, m_1 being the order of the autoregressive part of Eq. (4a.2.1):

$$\begin{aligned} u(t) = {} & K_1 y(t) + K_2 y(t-1) + \cdots + K_{m_1} y(t - m_1 + 1) \\ & + \cdots + K_{m_4} y(t - m_4 + 1). \end{aligned} \tag{4a.2.5}$$

Substituting for $u(t)$ in (4a.2.1) from (4a.2.5), we obtain

$$y(t) = \sum_{j=1}^{m_1} (A_j + G_1 K_j) y(t-j) + \sum_{j=m_1+1}^{m_4} G_1 K_j y(t-j) + w(t). \tag{4a.2.6}$$

From (4a.2.5), $K_1, \ldots, K_{m_4}$ are estimable. From (4a.2.6), $(A_j + G_1 K_j)$, $j = 1, \ldots, m_1$, and $(G_1 K_j)$, $j = m_1 + 1, \ldots, m_4$, are estimable. Since the K_j are known, G_1 is known from $G_1 K_{m_1+1}$. Hence, we can recover A_j, $j = 1, \ldots, m_1$, from $(A_j + K_j G_1)$, $j = 1, \ldots, m_1$. Thus, complete estimability is possible even with linear feedback.

For a more detailed treatment of estimability in feedback systems, see Ljung *et al.* (1974), Box and McGregor (1972), and Caines and Chan (1974).

4b. Estimability in Systems with Noisy Observations

Let the output y obey the difference equation

$$A(D)y(t) = B(D)w(t), \qquad A(0) = B(0) = 1. \tag{4b.1.1}$$

Suppose that $y(\cdot)$ is not directly observable. Let the actual observable variable be $x(\cdot)$ defined by

$$x(t) = y(t) + \eta(t), \tag{4b.1.2}$$

where $\eta(\cdot)$ is an additive zero mean white noise with covariance R_η. R_w is the covariance of w. We are interested in determining the conditions under which the coefficients of the polynomials $A(D)$ and $B(D)$ and the covariances R_η and R_w can be estimated from the semi-infinite observation history of $x(\cdot)$ only. The basic idea behind the technique used here is that the process $x(\cdot)$ obeys an ARMA process similar to (4b.1.1). This idea has already been used in Section 2h.

The problem of noisy observations is of some importance in modeling and monitoring of environmental processes such as pollution levels and river flows. Often, information about the observation errors involved in the published record of the time series is inadequate. From the observed series, we have to decide whether to choose a model as in (4b.1.1) and (4b.1.2), explicitly allowing for the observation noise, or to choose the simpler ARMA model in (4b.1.1), without explicitly allowing for the observation noise. In the latter case, the disturbance w subsumes the observation errors as well. We have already alluded to the fact that a more complicated model does not necessarily lead to better performance. In some cases, the simpler ARMA model may be superior to the more complex signal-plus-noise model. Example 4b.1 illustrates the method of choice in such problems. We will present the following theorem (Kashyap, 1970a) for the estimability of systems (4b.1.1) and (4b.1.2).

Theorem 4b.1. Consider the signal process y and the observation process x in (4b.1.1) and (4b.1.2) where w and η are two independent zero mean processes with variances R_w and R_η. In addition, $\eta(\cdot)$ is independent of both $w(\cdot)$ and $y(\cdot)$. Let system (4b.1.1) obey (A1) and (A5)–(A7). Then the condition $m_1 \geq m_2 + 1$ is sufficient for the estimability of the coefficients of the polynomials $A(D)$ and $B(D)$ and the covariances R_w and R_η, where

$$A(D) = 1 + A_1'D + \cdots + A'_{m_1}D^{m_1}, \qquad A_j = -A_j',$$
$$B(D) = 1 + B_1D + \cdots + B_{m_2}D^{m_2}.$$

Proof. As shown in Section 2h, process x obeys the ARMA process

$$A(D)x(t) = B'(D)\zeta(t), \tag{4b.1.3}$$

where

$$B'(D)\zeta(t) = A(D)\eta(t) + B(D)w(t). \tag{4b.1.4}$$

$\{\zeta(\cdot)\}$ is a zero mean white noise sequence with covariance R_ζ. As shown in Section 2h, $\zeta(t) = x(t) - \hat{x}(t|t-1)$, where $\hat{x}(t|t-1)$ is the least squares one-step-ahead predictor of $x(t)$ based on $x(t-j), j \geq 1, B'(D)$ is a polynomial of degree m_1, and its coefficients can be found by equating the $0, 1, \ldots, m_1$ order correlations of the left- and right-hand sides of Eq. (4b.1.4).

We will first show that the polynomials $A(D)$ and $B'(D)$ in (4b.1.3) are estimable by using the observation history $x(\cdot)$, i.e., assumption (A7) for the polynomial pair $(A(D), B(D))$ implies the validity of assumption (A7) for the polynomial pair $(A(D), B'(D))$. We will prove this statement by assuming the contrary and demonstrating a contradiction. Suppose $(A(D), B'(D))$ does not obey (A7). Then $A(D)$ and $B'(D)$ have a common left polynomial factor, say $\phi(D)$. Then, by taking out the factor $\phi(D)$ throughout, Eq. (4b.1.3) can be simplified to

$$A'(D)x(t) = B''(D)\zeta(t), \tag{4b.1.5}$$

where $A(D) = \phi A'(D)$ and $B'(D) = \phi B''(D)$. But combining (4b.1.3) and (4b.1.4) and left multiplying by $(\phi(D))^{-1}$ yields

$$A'(D)x(t) = A'(D)\eta(t) + [\phi(D)]^{-1}B(D)w(t). \tag{4b.1.6}$$

The degree of polynomial $B'(D)$ is m_1 and hence the degree of $B''(D)$ is at most $(m_1 - 1)$. Hence all the autocorrelations of the variable representing the right-hand side of (4b.1.5) of order greater than $(m_1 - 1)$ are identically zero. But the variable representing the right-hand side of (4b.1.6) possesses nonzero correlation of all orders since the polynomial $B(D)$ is not divisible by $\phi(D)$ in view of (A7). This is a contradiction. Hence $(A(D), B'(D))$ obeys (A7), implying the estimability of the coefficients of A and B'.

We will next show that the coefficients B' and R_η can be recovered from $A(D)$, $B'(D)$, and R_ζ, the covariance of the noise $\zeta(\cdot)$. We shall use Eqs. (2h.2.8), rewritten here as

$$\begin{aligned}
A'_{m_1}R_\eta &= B'_{m_1}R_\zeta \\
A'_{m_1-1}R_\eta + A'_{m_1}R_\eta A_1{}' + B_{m_1-1}R_w &= B'_{m_1-1}R_\zeta + B'_{m_1}R_\zeta B_1{}' \\
&\;\vdots \\
\sum_{j=0}^{m_1} (A_j{}')R_\eta A_j{}' + \sum_{j=0}^{m_1-1} B_j R_w B_j{}^{\mathrm{T}} &= \sum_{j=0}^{m_1} B_j{}'R_\zeta B_j{}'.
\end{aligned} \tag{4b.1.7}$$

From the first equation in this set we can determine R_η, and from the remaining we can obtain $B_1, \ldots, B_{m_1-1}$.

To illustrate the choice between a signal-plus-noise model and a model without the observation noise, we will give the following example.

Example 4b.1. Suppose we fit a stochastic difference equation for a scalar process x and find that ARMA(1, 1) process

$$x(t) = A_1x(t-1) + \zeta(t) + B_1\zeta(t-1) \tag{4b.1.8}$$

is quite satisfactory for it. We have two models:

Model 1. $y(t) = A_1y(t-1) + w(t), x(t) = y(t) + \eta(t)$.

Model 2. Equation (4b.1.8) and the identity $y(t) = x(t)$.

We want to find out if the coefficients A_1 and B_1 throw any light on the selection between the two models. Let us consider model 1. By using this model, we should be able to derive an ARMA equation (4b.1.8) for x, and the corresponding coefficients should satisfy

$$(1 + B_1^2)R_\zeta = (1 + A_1^2)R_\eta + R_w, \qquad B_1 R_\zeta = -A_1 R_\eta, \tag{4b.1.9}$$

which is only a specialization of (4b.1.7). Let us simplify (4b.1.9). Solve for R_ζ using the second equation in (4b.1.9), substitute it in the first equation of (4b.1.9), and divide throughout by R_η:

$$-(1 + B_1^2)(A_1/B_1) - (1 + A_1^2) = R_w/R_\eta > 0. \tag{4b.1.10}$$

Simplifying (4b.1.10), we obtain

$$\left(1 + \frac{A_1}{B_1}\right)(1 + A_1 B_1) < 0. \tag{4b.1.11}$$

Since $|A_1| < 1$ and $|B_1| < 1$ by (A5) and (A6), (4b.1.11) implies that

$$A_1/B_1 < -1. \tag{4b.1.12}$$

Hence, if the coefficients A_1 and B_1 in (4b.1.8) obey (4b.1.12), then model 1 is applicable. Even though model 2 is also applicable, model 1 is preferred because it is more realistic. If (4b.1.12) is not satisfied, then model 1 is *not* applicable and model 2 is the only possibility.

Condition (4b.1.12) is satisfied, for example, by the model for monthly sunspot series. Hence, the observation process of the monthly sunspot series can be regarded as a sum of first-order AR process and noise. The presence of observation noise in such a process is very realistic.

4c. Estimability in Systems with AR Disturbances

While developing the basic model of the book in Section 1b, we started with a difference equation for y having AR terms, trends, input u, and a disturbance ζ, which was assigned a pure moving average structure. In this section, we investigate the consequences of assigning an AR structure to the disturbance ζ. Such models are used in econometrics.

Let the given process $y(\cdot)$ obey

$$A(D)y(t) = \zeta(t), \tag{4c.1.1}$$

where $\zeta(\cdot)$ is an AR process obeying

$$H(D)\zeta(t) = w(t). \tag{4c.1.2}$$

The polynomials A and H obey (A5) and $w(\cdot)$ obeys (A1). Substituting (4c.1.2) in (4c.1.1) the AR process

$$\theta(D)y(t) = w(t) \tag{4c.1.3}$$

is obtained for y, where

$$\theta(D) = H(D)A(D). \tag{4c.1.4}$$

Clearly the polynomial $\theta(D)$ is estimable. However, it does not follow that the factors A and H can be uniquely recovered from θ without additional information. As a matter of fact, consider any other factorization of θ as

$$\theta(D) = H'(D)A'(D), \tag{4c.1.5}$$

where H' and A' are also polynomials. Then, another model for $y(\cdot)$ is given in

$$A'(D)y(t) = \zeta'(t), \qquad H'(D)\zeta'(t) = w(t), \tag{4c.1.6}$$

where $w(\cdot)$ is the noise sequence obeying (A1). We cannot choose the "true" model among the various possible models such as Eqs. (4c.1.6), (4c.1.1), and (4c.1.2). The ambiguity can be resolved only if we have additional information about the process $\zeta(\cdot)$ in (4c.1.2). This ambiguity has been emphasized by Kendall (1971), who also gave the following example.

Example 4c.1. Let the given process y be

$$y(t) = Ay(t-1) + \zeta(t) \tag{4c.1.7}$$
$$\zeta(t) = H\zeta(t-1) + w(t). \tag{4c.1.8}$$

Substituting (4c.1.8) into (4c.1.7) we obtain the AR process

$$y(t) = (A + H)y(t-1) + AHy(t-2) + w(t). \tag{4c.1.9}$$

Clearly, $(A + H)$ and AH are estimable. Using these, two possible solution pairs (A^0, H^0) and (A^1, H^1) may be obtained for the unknowns (A, H). We cannot identify the so-called correct solution pair without additional information.

4d. The Estimation Accuracy

In the proof of Theorem 4a.1, an estimate of the unknown vector $\boldsymbol{\theta}^0$ based on a finite observation history was introduced. The construction of this estimate does not involve the probability distribution of the disturbances. This estimate is labeled the *quasi-maximum likelihood* (QML) estimate since it can be interpreted as a maximum likelihood estimate (to be defined later) when the distribution of the disturbances is Gaussian. In Chapter VI this estimator is discussed in some detail. At present, we will discuss the mean square error matrix associated with the estimate. The conditions under which the estimate yields the smallest value of mean square error in the class of asymptotically unbiased estimates will be investigated by using the Cramér–Rao inequality (Rao, 1965).

4d.1. Mean Square Error (MSE) of the Quasi-Maximum Likelihood (QML) Estimate

We will redefine the QML estimate. Let the vector of the unknown coefficients in

$$A^0(D)y(t) = \mathbf{G}^0(D)\mathbf{u}(t-1) + \mathbf{F}^0\boldsymbol{\psi}(t-1) + B^0(D)w^0(t) \tag{4d.1.1}$$

be labeled $\boldsymbol{\theta}^0$, and the covariance matrix of the disturbance w^0 be ρ^0. Let $\boldsymbol{\theta}$ be a dummy vector of the same dimension as $\boldsymbol{\theta}^0$ and let $A(D)$, $B(D)$, $\mathbf{G}(D)$, and $\mathbf{F}$ be constructed from $\boldsymbol{\theta}$ just as A^0, B^0, $\mathbf{G}^0$, and $\mathbf{F}^0$ are constructed from $\boldsymbol{\theta}^0$. Let $w(t, \boldsymbol{\theta})$ be an estimate of $w^0(t)$ constructed recursively from

$$B(D)w(t, \boldsymbol{\theta}) = A(D)y(t) - \mathbf{G}(D)\mathbf{u}(t-1) - \mathbf{F}\boldsymbol{\psi}(t-1) \qquad (4d.1.2)$$

based on $\boldsymbol{\theta}$ and the observation history prior to time t.

Consider the loss function

$$J(\boldsymbol{\theta}) = \sum_{t=1}^{N} w^2(t, \boldsymbol{\theta}).$$

The QML estimate of $\boldsymbol{\theta}^0$ based on the observation history of $y(t)$, $\mathbf{u}(t-1)$ $\{t = 1, \ldots, N\}$, is the value of $\boldsymbol{\theta}$ that minimizes $J(\boldsymbol{\theta})$ and is denoted by $\hat{\boldsymbol{\theta}}(N)$.

The estimate $\hat{\boldsymbol{\theta}}(N)$ has the following MSE matrix for large N as shown in Section B of Appendix 4.1:

$$E[(\hat{\boldsymbol{\theta}}(N) - \boldsymbol{\theta}^0)(\hat{\boldsymbol{\theta}}(N) - \boldsymbol{\theta}^0)^{\mathrm{T}}] \approx \left[E\left\{\sum_{t=1}^{N} \nabla_{\boldsymbol{\theta}} w(t, \boldsymbol{\theta}^0)(\nabla_{\boldsymbol{\theta}} w(t, \boldsymbol{\theta}^0)^{\mathrm{T}}\right\}\right]^{-1} \rho^0. \qquad (4d.1.3)$$

Assumption (B1) or its equivalent assures the existence of the inverse in (4d.1.3). The matrix in (4d.1.3) is the celebrated Fisher information matrix used extensively in statistical literature.

Since Eq. (4d.1.3) or its equivalent involves the unknown $\boldsymbol{\theta}^0$, this equation can be used only in making qualitative assessments of the accuracy of the estimate $\hat{\boldsymbol{\theta}}$. An approximate expression for the mean square error matrix of the estimate $\boldsymbol{\theta}(N)$ of $\boldsymbol{\theta}^0$ can be obtained by replacing $\boldsymbol{\theta}^0$ in (4d.1.3) by its estimate $\hat{\boldsymbol{\theta}}$. Expression (4d.1.3) can be evaluated by algebraic methods in relatively simple problems such as in Example 4d.2. The evaluation of expression (4d.1.3) is considered in some detail in Chapter VII.

4d.2. Convergence Rates of the QML Estimate

(i) If assumption (B1) is satisfied, the MSE matrix will tend to zero as N tends to infinity. Otherwise, one or more elements of the matrix may not approach zero as N tends to infinity. One such instance is given in the following example.

Example 4d.1. Let $y(\cdot)$ obey

$$y(t) = (\theta_1{}^0/t) + \theta_2{}^0 y(t-1) + w^0(t), \qquad (4d.2.1)$$

which does not obey (B1). The MSE matrix of the QML estimates of $\theta_1{}^0$ and $\theta_2{}^0$ is denoted by

$$\bar{\mathbf{S}}(N) = \begin{bmatrix} E\sum_{t=1}^{N}(1/t)^2 & E\sum_{t=1}^{N}(y(t-1)/t) \\ E\sum_{t=1}^{N} y(t-1)/t & E\sum_{t=1}^{N} y^2(t-1) \end{bmatrix}^{-1} \rho^0 = \begin{bmatrix} a_{11} & a_{12} \\ a_{12} & a_{22} \end{bmatrix}^{-1}.$$

By inspection $a_{11} \approx \sum_{t=1}^{\infty} (1/t^2) = \text{const } c < \infty$. By the Schwarz inequality $a_{11}a_{22} \geq a_{12}^2$,

$$\bar{\mathbf{S}}_{11}(N) = a_{22}/(a_{11}a_{22} - a_{12}^2) \geq a_{22}/(a_{11}a_{22}) = 1/a_{11} > 0 \qquad \text{for all} \quad N.$$

Hence $\hat{\theta}_1$ will not tend to $\theta_1{}^0$ even if N tends to infinity.

(ii) If the difference equation does not have any trend terms and if the input $\mathbf{u}(\cdot)$ is weak stationary (with zero or constant mean), then the MSE of the estimate of every coefficient in the equation is of the order $O(1/N)$, by a version of the law of large numbers. This follows from the fact that the gradient vector sequence $\{\nabla_{\boldsymbol{\theta}} w(t, \boldsymbol{\theta}^0)\}$ is weak stationary:

$$E[(\hat{\boldsymbol{\theta}}(N) - \boldsymbol{\theta}^0)(\hat{\boldsymbol{\theta}}(N) - \boldsymbol{\theta}^0)^{\mathrm{T}}] = \frac{1}{N} E[\{\nabla_{\boldsymbol{\theta}} w(t, \boldsymbol{\theta}^0)\}\{\nabla_{\boldsymbol{\theta}} w(t, \boldsymbol{\theta}^0)\}^{\mathrm{T}}]^{-1}\rho^0.$$

The method of evaluation of the expectation term in the above expression is given in Chapter VII. Now we will give an example to discuss the implications.

Example 4d.2.

$$\begin{aligned} y(t) &= A_1{}^0 y(t-1) + B_1{}^0 w(t-1) + w(t), \\ \boldsymbol{\theta}^0 &= (A_1{}^0, B_1{}^0)^{\mathrm{T}}, \qquad \boldsymbol{\theta} = (A_1, B_1)^{\mathrm{T}}. \end{aligned} \tag{4d.2.2}$$

In Chapter VII we will show that the expression for the mean square error matrix is

$$E[(\hat{\boldsymbol{\theta}}(N) - \boldsymbol{\theta}^0)(\hat{\boldsymbol{\theta}}(N) - \boldsymbol{\theta}^0)^{\mathrm{T}} | \boldsymbol{\theta}^0, \rho^0]$$

$$= \frac{1}{N} \begin{bmatrix} \dfrac{(1 - (A_1{}^0)^2)(1 + A_1{}^0 B_1{}^0)^2}{(A_1{}^0 + B_1{}^0)^2} & \dfrac{-(1 - (A_1{}^0)^2)(1 - (B_1{}^0)^2)(1 + A_1{}^0 B_1{}^0)}{(A_1{}^0 + B_1{}^0)^2} \\ \dfrac{-(1 - (A_1{}^0)^2)(1 - (B_1{}^0)^2)(1 + A_1{}^0 B_1{}^0)}{(A_1{}^0 + B_1{}^0)^2} & \dfrac{(1 - (B_1{}^0)^2)(1 + A_1{}^0 B_1{}^0)}{(A_1{}^0 + B_1{}^0)^2} \end{bmatrix}$$

$$+ o(1/N). \tag{4d.2.3}$$

Notice that the covariance matrix is independent of ρ^0! The expression in (4d.2.3) becomes infinite if $A_1{}^0 + B_1{}^0 = 0$, i.e., if (4d.2.3) does not satisfy assumption (A7). We should be careful about the case when (A7) is strictly satisfied, but when $A_1{}^0 + B_1{}^0$ is nearly zero. In such a case there is a pole and zero pair near one another. Such cases lead to excessive errors in their estimates. The appropriate action in such cases is to use a model of reduced order.

(iii) When the difference equation has trend terms, the coefficients of these terms possess estimates whose MSE may decay at a rate other than $O(1/N)$, depending on the nature of the trend. For instance, consider the trend in the following example.

Example 4d.3. $y(t) = (\theta^0/\sqrt{t}) + w(t)$. Let $\hat{\theta}$ be an estimate of θ^0:

$$E[(\hat{\theta} - \theta^0)^2] \approx \rho^0 \Big/ \left(\sum_{t=1}^{N} \frac{1}{t} \right) = O(1/\log N).$$

Example 4d.4. $y(t) = \theta^0 t + w(t)$.

$$E[(\hat{\theta} - \theta^0)^2] \approx \rho^0\left[\sum_{t=1}^{N} t^2\right]^{-1} = O(1/N^3).$$

These examples may give the false impression that the mean square error of a coefficient of a trend term $\psi_i(t)$ is of the order $O(1/\sum_{t=1}^{\infty}\psi_i^2(t))$. In order to allay this notion, the following example may be considered.

Example 4d.5.

$$y(t) = \theta_1^0 t + \theta_2^0\left(t + \frac{1}{\sqrt{t}}\right) + w(t). \quad (4d.2.4)$$

Here

$$\psi_1(t) = t, \qquad \psi_2(t) = t + 1/\sqrt{t}, \qquad \sum_{t=1}^{N}\psi_1^2(t) = O(N^3).$$

$$E[(\hat{\boldsymbol{\theta}} - \boldsymbol{\theta}^0)(\hat{\boldsymbol{\theta}} - \boldsymbol{\theta}^0)^T] = \begin{bmatrix} \sum_{t=1}^{N} t^2 & \sum_{t=1}^{N}(t^2 + \sqrt{t}) \\ \sum_{t=1}^{N}(t^2 + \sqrt{t}) & \sum_{t=1}^{N}(t + 1/\sqrt{t})^2 \end{bmatrix}^{-1} \rho^0. \quad (4d.2.5)$$

The determinant of the matrix on the right-hand side of (4d.2.5) is of the order $O(1/(N^3 \log N))$. Consequently,

$$E[(\hat{\theta}_1 - \theta_1^0)^2] = O(1/\log N), \qquad E[(\hat{\theta}_2 - \theta_2^0)^2] = O(1/\log N). \quad (4d.2.6)$$

This example shows the effect of improper grouping of the trend terms also. By properly grouping the trend terms, the accuracy of the parameter estimates can be considerably increased. To illustrate this feature we rewrite Eq. (4d.2.4) as

$$y(t) = \theta_3^0 t + \theta_2^0/\sqrt{t} + w(t), \qquad \theta_3^0 = (\theta_1^0 + \theta_2^0). \quad (4d.2.7)$$

If $\hat{\theta}_3$ and $\hat{\theta}_2$ are the QML estimates of θ_3^0 and θ_2^0, then their accuracy is given by

$$E[(\hat{\theta}_3 - \theta_3^0)^2] = O(1/N^3), \qquad E[(\hat{\theta}_2 - \theta_2^0)^2] = O(1/\log N). \quad (4d.2.8)$$

Note the vast difference between the orders of the mean square errors of the estimates of θ_1^0 and θ_3^0, as given in (4d.2.6) and (4d.2.8). By properly grouping the terms, the accuracy of estimation can be increased considerably.

(iv) When the equation has autoregressive, moving average, and trend terms, then the estimates of the coefficients of the AR and MA terms will have mean square errors of order $O(1/N)$. However, the estimates of the coefficients of the trend terms can have mean square errors whose order is quite different from the corresponding values mentioned in item (ii). The following example will illustrate this statement.

Example 4d.6. $y(t) = \theta_1{}^0 t + \theta_2{}^0 y(t-1) + w(t)$.

$$E[(\hat{\boldsymbol{\theta}} - \boldsymbol{\theta}^0)(\hat{\boldsymbol{\theta}} - \boldsymbol{\theta}^0)^{\mathrm{T}}] \approx \mathbf{S}^{-1}(N)\rho^0.$$

$$\mathbf{S}(N) = \begin{bmatrix} E\sum_{t=1}^{N} t^2 & E\sum_{t=1}^{N} ty(t-1) \\ E\sum_{t=1}^{N} ty(t-1) & E\sum_{t=1}^{N} y^2(t-1) \end{bmatrix} = \begin{bmatrix} a_{11} & a_{12} \\ a_{12} & a_{22} \end{bmatrix}.$$

We can simplify the expression above by splitting $y(t)$ into a weak stationary part $y_1(t)$ and a deterministic trend βt:

$$y(t) = y_1(t) + \beta t, \qquad \beta = \theta_1{}^0/(1 - \theta_2{}^0).$$

Then,

$$a_{22} = NE[y_1{}^2(t)] + \beta^2 \sum_{t=1}^{N} t^2 + \beta E\left[\sum_{t=1}^{N} (t-1)y_1(t-1)\right] = O(N^3)$$

$$a_{11} = \sum_{t=1}^{N} t^2 = O(N^3), \qquad a_{12} = E\left[\sum_{t=1}^{N} ty_1(t-1)\right] + \beta \sum_{t=1}^{N} t(t-1)$$

$$a_{11}a_{22} - a_{12}^2 = NE\{y_1{}^2(t)\}\left[\sum_{t=1}^{N} t^2\right] - \left[E\sum_{t=1}^{N} (t-1)y_1(t-1)\right]^2 = O(N^4)$$

$$S_{11}^{-1}(N) = a_{22}/(a_{11}a_{22} - a_{12}^2) = O(1/N)$$

$$S_{22}^{-1}(N) = a_{11}/(a_{11}a_{22} - a_{12}^2) = O(1/N).$$

The mean square error of $\hat{\theta}_1$ is $O(1/N)$, whereas it would have been $O(1/N^3)$ if $\theta_2{}^0$ were zero, as in Example 4d.4. The presence of the autoregressive term can drastically reduce the accuracy of the estimate of the coefficient of the trend term.

One could consider the possibility of developing a scheme to obtain a better estimate of $\theta_1{}^0$ in the example given above. We will demonstrate in Section 4d.4 that if we restrict ourselves to asymptotically unbiased estimation schemes that yield strong consistent estimates of both parameters, then the scheme given here is asymptotically optimal, provided the disturbance $w(\cdot)$ is Gaussian; i.e., there does not exist another asymptotically unbiased estimator that yields smaller errors than the estimator given above.

4d.3. Parsimony and the Accuracy of Estimation

If a number of models explain the given data adequately, prudence suggests the adoption of the simplest model in the set. Before making this statement operational, we have to specify the concept of the simplicity of a model. A measure of model simplicity is the number of parameters in the model that have to be estimated. This measure is appropriate in dealing with a family of models wherein the computational complexity is a monotonically increasing function of the number of parameters. This is the case with the family of autoregressive models. The computation involved in fitting an AR(5) model is greater than that involved in fitting an AR(4) model. In such cases, among all the competing models we should prefer the one having the least number of parameters. This rule is called the principle of parsimony.

The fact that estimation complexity is not always related monotonically to the number of unknown parameters must be emphasized. For instance, the computation involved in fitting an AR(3) model having three unknowns is very much smaller than fitting a MA(1) model involving one unknown. For such cases, the use of the principle of parsimony cannot be justified solely on computational grounds.

We will now argue that the simpler model is preferable to a complicated model from the point of view of accuracy of the parameter estimates as well. The estimates of all the parameters in an overspecified model will usually be less accurate than the estimates of the corresponding parameters of the simpler model. For instance, if we attempt to fit an AR process of order $(n + 1)$ to the observations obtained from a process that obeys an AR(n) process, the estimate of every one of the parameters in the $(n + 1)$ parameter model will usually have a larger mean square error than that of the corresponding parameters of the nth order model. This aspect is illustrated in the example given below.

Example 4d.7. Consider the process y:

$$y(t) = \theta_1{}^0 y(t-1) + w(t).$$

Suppose we try to fit a first-order AR model to a set of N observations of the y process. Then the mean square error of the estimate $\hat{\theta}_1$ of the first AR coefficient is

$$E[(\hat{\theta}_1 - \theta_1{}^0)^2] = (\rho^0/N)r_0; \qquad r_0 = E(y^2).$$

We may fit a second-order AR model to the same N observations of the y process and let $\bar{\theta}_1$, $\bar{\theta}_2$ represent the estimates of $\theta_1{}^0$ and $\theta_2{}^0$. Note that $\theta_2{}^0 = 0$, which is not obviously utilized in the estimation process. The two estimates have the following mean square error matrix:

$$E\left[\begin{pmatrix}\bar{\theta}_1 - \theta_1{}^0 \\ \bar{\theta}_2 - \theta_2{}^0\end{pmatrix}(\bar{\theta}_1 - \theta_1{}^0, \bar{\theta}_2 - \theta_2{}^0)\right] = \begin{bmatrix} r_0 & r_1 \\ r_1 & r_0 \end{bmatrix}^{-1}\left(\frac{\rho^0}{N}\right)$$

$$= \frac{\rho}{N(r_0{}^2 - r_1{}^2)}\begin{bmatrix} r_0 & -r_1 \\ -r_1 & r_0 \end{bmatrix},$$

where $r_j = E[y(i)y(i-j)]$. Since $1/r_0 < r_0/(r_0{}^2 - r_1{}^2)$, $E[(\hat{\theta}_1 - \theta_1{}^0)^2] < E[(\bar{\theta}_1 - \theta_1{}^0)^2]$. Therefore, the estimate of the first AR coefficient in the AR(1) model is more accurate than the estimate of the same coefficient in the AR(2) model.

4d.4. Lower Bounds on Estimation Accuracy (Cramér–Rao Bound)

In this section we will investigate the conditions under which the QML estimate mentioned earlier yields the smallest value of the mean square error in the class of unbiased estimates with the aid of the Cramér–Rao lower bound on the accuracy of any estimate. This bound depends on the probability distribution of the disturbances.

Suppose we are given N observation pairs $(y(t), \mathbf{u}(t-1))$, $t = 1, \ldots, N$, obeying the difference equation (4a.1.1). Let $\boldsymbol{\theta}^0$ denote the vector of unknown coefficients in the equation, and ρ^0, the unknown variance. Let $\boldsymbol{\phi}^0 = (\boldsymbol{\theta}^0, \rho^0)$, and let $\hat{\boldsymbol{\phi}}$ be any asymptotically *unbiased estimate* of $\boldsymbol{\phi}^0$ based on the history $\{y(t), \mathbf{u}(t-1), t = 1, \ldots, N\}$. Then the covariance matrix of $\hat{\boldsymbol{\phi}}$ satisfies the following Cramér–Rao inequality where $\mathbf{A} \geq \mathbf{B}$ means $(\mathbf{A} - \mathbf{B})$ is nonnegative-definite:

$$E[(\hat{\boldsymbol{\phi}} - \boldsymbol{\phi}^0)(\hat{\boldsymbol{\phi}} - \boldsymbol{\phi}^0)^{\mathrm{T}} | \boldsymbol{\phi}^0] \geq E[-\boldsymbol{\nabla}^2_{\boldsymbol{\phi\phi}} \ln p(y(N), \ldots, y(1), \boldsymbol{\xi}_1(N); \boldsymbol{\phi}^0)] \tag{4d.4.1}$$

where $\boldsymbol{\xi}_1(N) \triangleq \{y(0), \ldots, y(-m_1), \mathbf{u}(N-1), \ldots, \mathbf{u}(-m_3)\}$, and $p(\cdot)$ denotes the joint probability density of the various observations indicated in the argument. Note that p should *not* be interpreted as a joint density of $y(\cdot)$ and $\boldsymbol{\phi}^0$. The presence of $\boldsymbol{\phi}^0$ only indicates the appropriate functional relation.

We will simplify expression (4d.4.1)

$$\ln p(y(N), \ldots, y(1), \boldsymbol{\xi}_1(N); \boldsymbol{\phi}^0) = \ln p(y(N), \ldots, y(1) | \boldsymbol{\xi}_1(N); \boldsymbol{\phi}^0) + \ln p(\boldsymbol{\xi}_1(N); \boldsymbol{\phi}^0) \tag{4d.4.2}$$

$$= L_N(\boldsymbol{\phi}^0) + \text{a function asymptotically independent of } \boldsymbol{\phi}^0 \quad \text{[in view of assumptions (A3) and (A5)]}, \tag{4d.4.3}$$

where

$$L_N(\boldsymbol{\phi}^0) = \ln p(y(N), \ldots, y(1) | \boldsymbol{\xi}_1(N); \boldsymbol{\phi}^0).$$

Equations (4d.4.1) and (4d.4.3) yield

$$E[(\hat{\boldsymbol{\phi}} - \boldsymbol{\phi}^0)(\hat{\boldsymbol{\phi}} - \boldsymbol{\phi}^0)^{\mathrm{T}} | \boldsymbol{\phi}^0] \geq E[-\boldsymbol{\nabla}^2_{\boldsymbol{\phi\phi}} L_N(\boldsymbol{\phi}) | \boldsymbol{\phi} = \boldsymbol{\phi}^0]. \tag{4d.4.4}$$

We can construct the function $L_N(\boldsymbol{\phi})$ in terms of the residuals $w(t, \boldsymbol{\theta})$ recursively generated as in (4d.1.2). Assuming the normal distribution of the variables $w(\cdot)$, we have the following expression for L_N:

$$L_N(\boldsymbol{\phi}^0) = -(N/2) \ln 2\pi\rho^0 - \tfrac{1}{2} \sum_{t=1}^{N} w^2(t, \boldsymbol{\theta}^0)/\rho^0. \tag{4d.4.5}$$

Substituting (4d.4.5) into (4d.4.4) and carrying out the computations yields

$$E[(\hat{\boldsymbol{\theta}} - \boldsymbol{\theta}^0)(\hat{\boldsymbol{\theta}} - \boldsymbol{\theta}^0)^{\mathrm{T}} | \boldsymbol{\theta}^0, \rho^0] \geq E\left[\sum_{t=1}^{N} (\boldsymbol{\nabla}_{\boldsymbol{\theta}} w(t, \boldsymbol{\theta}^0))(\boldsymbol{\nabla}_{\boldsymbol{\theta}} w(t, \boldsymbol{\theta}^0))^{\mathrm{T}} | \boldsymbol{\theta}^0, \rho^0\right]^{-1} \rho^0 \tag{4d.4.6}$$

$$E[(\hat{\rho} - \rho^0)^2 | \rho^0] \geq 2(\rho^0)^2/N + o(1/N). \tag{4d.4.7}$$

The details are given in Appendix 4.2.

Thus the covariance matrix of the QML estimate $\hat{\boldsymbol{\theta}}$ in Eq. (4d.1.3) is identical to the right-hand side (4d.4.6). Therefore, in the class of asymptotically unbiased estimates, the QML estimate has the least variance, provided the distribution of w^0 is Gaussian.

It should be noted that if we are prepared to accept a biased estimate, then we can obtain, in the Gaussian case, an estimate that has smaller MSE than the QML estimate. The QML estimate yields the least MSE only in the class of unbiased estimates.

4d.5. Robust Estimators

The Cramér–Rao bound evaluated above is crucially dependent on the normal distribution. When the distribution of w^0 is not normal, the bound in (4d.4.6) may not be valid and the QML estimate does not necessarily yield the minimum MSE even in the class of unbiased estimates.

When the distribution of w^0 is not normal, we can consider two possibilities. If the distribution is not normal but known, then we can construct the maximum likelihood (ML) estimate obtained by maximizing the likelihood function based on the particular probability density of the noise. In such a case the variance of the estimate again equals the Cramér–Rao lower bound.

The second case is more interesting. Suppose we do not know the exact probability distribution but know that the distribution is approximately normal, i.e., it is made up of, say, 95% of a normal distribution and 5% of some other unknown distribution. Contrary to intuition, in such a case the MSE of the QML estimate is very sensitive to small changes in the probability distribution, even though the Cramér–Rao lower bound of the MSE is not so sensitive to small deviations from normality. Thus, in dealing with mixture distributions, the QML estimate is not satisfactory, and we have to resort to a new class of estimators, the so-called robust estimators having the following properties.

When the distribution of the noise is exactly normal, then the mean square error of the robust estimator is slightly greater than the MSE of the QML estimator. However, the MSE of the robust estimator does not vary very much if the noise has a mixture distribution, whereas that of the QML estimator may increase considerably. Thus, the MSE of the robust estimator is considerably less than that of the QML estimator with mixture distribution.

Robust estimation of parameters using IID observations is extensively discussed by Andrews *et al.* (1972). However, the application of the theory of robust estimation to the time series problems is not well developed. Preliminary work (Nasburg and Kashyap, 1975) indicates that the computation of the robust estimates of the parameters in an ARMA process is as easy as the computation of the QML estimates of the same parameters.

4e. Conclusions

We have considered various aspects of the estimability of the coefficients in the stochastic difference equation for a single output such as the role of feed-back, the role of additive noise in the observations, and the presence of deterministic trend terms in the difference equation. In addition, we discussed the accuracy of the estimates of the various coefficients in the equation. The QML estimates are used because their consistency does not involve any restrictive assumptions such as (A3), i.e., the independence of the input u and the disturbance w. The manner in which the presence of trend terms affects the accuracy of estimates, the conditions under which QML estimates are satisfactory, and the need for robust estimates are also discussed.

Appendix 4.1

A. Proof of Theorem 4a.1

Since the necessity part of the proof has already been given in the text, we will concern ourselves only with the *sufficiency* part. The proof is rather lengthy, which is caused by the fact that assumption (A3) is not used. Moreover, condition (B1) implies only condition (A4′) on the trends and not the stronger condition (A4). We can give a relatively simple proof of estimability under conditions (A1)–(A7). We will not prove this result here since a multivariate version of it appears as Theorem 5b.2. The proof of Theorem 5b.2 is relatively easy compared to the present proof.

Let the given equation be

$$\begin{aligned} y(t) &= (\boldsymbol{\theta}^0)^{\mathrm{T}}\mathbf{x}^0(t-1) + w^0(t) \\ \mathbf{x}^0(t-1) &= [y(t-1),\ldots,y(t-m_1),\mathbf{u}^{\mathrm{T}}(t-1),\ldots,\mathbf{u}^{\mathrm{T}}(t-m_3), \\ &\qquad \boldsymbol{\psi}^{\mathrm{T}}(t-1), w^0(t-1),\ldots,w^0(t-m_2)]^{\mathrm{T}}, \\ \text{dimension of } \mathbf{x}^0 &= n = m_1 + l_1 m_3 + l_2 + m_2, \end{aligned} \tag{1}$$

where y is a scalar and $\boldsymbol{\theta}^0$ is the unknown vector parameter to be estimated. By using the given observation history $\{y(t), \mathbf{u}(t-1), t = 1,\ldots,N\}$ and an estimate $\boldsymbol{\theta}$ of $\boldsymbol{\theta}^0$, we can recursively obtain an estimate $w(t, \boldsymbol{\theta})$ of $w^0(t)$:

$$\begin{aligned} w(t,\boldsymbol{\theta}) &= y(t) - \boldsymbol{\theta}^{\mathrm{T}}\mathbf{x}(t-1,\boldsymbol{\theta}) \\ \mathbf{x}(t-1,\boldsymbol{\theta}) &= [y(t-1),\ldots,y(t-m_1),\mathbf{u}^{\mathrm{T}}(t-1),\ldots,\mathbf{u}^{\mathrm{T}}(t-m_3),\boldsymbol{\psi}^{\mathrm{T}}(t-1), \\ &\qquad w(t-1,\boldsymbol{\theta}),\ldots,w(t-m_2,\boldsymbol{\theta})]^{\mathrm{T}}. \end{aligned} \tag{2}$$

Note that by (A6), $w(t, \boldsymbol{\theta}^0) = w^0(t)$ for large t. Consider the criterion function

$$J(\boldsymbol{\theta}) = \tfrac{1}{2}\sum_{t=1}^{N} w^2(t,\boldsymbol{\theta}). \tag{3}$$

An estimate $\hat{\boldsymbol{\theta}}(N)$ of $\boldsymbol{\theta}^0$ is obtained by minimizing $J(\boldsymbol{\theta})$. An approximate expression for $\hat{\boldsymbol{\theta}}(N)$ is obtained by considering the quadratic approximation to $J_N(\boldsymbol{\theta})$ given by its Taylor series expansion,

$$\begin{aligned} J(\boldsymbol{\theta}) &= J(\boldsymbol{\theta}^0) + (\nabla_{\boldsymbol{\theta}} J(\boldsymbol{\theta}^0))^{\mathrm{T}}(\boldsymbol{\theta}-\boldsymbol{\theta}^0) + \tfrac{1}{2}(\boldsymbol{\theta}-\boldsymbol{\theta}^0)^{\mathrm{T}}(\nabla^2_{\boldsymbol{\theta}\boldsymbol{\theta}} J(\boldsymbol{\theta}^0))(\boldsymbol{\theta}-\boldsymbol{\theta}^0) \\ &\quad + O(|\boldsymbol{\theta}-\boldsymbol{\theta}^0|^3), \end{aligned}$$

where

$$\nabla_{\boldsymbol{\theta}} J(\boldsymbol{\theta}^0) \triangleq \left.\frac{\partial J(\theta)}{\partial \boldsymbol{\theta}}\right|_{\boldsymbol{\theta}=\boldsymbol{\theta}^0}, \qquad \nabla^2_{\boldsymbol{\theta}\boldsymbol{\theta}} J(\boldsymbol{\theta}^0) \triangleq \left.\frac{\partial^2 J(\boldsymbol{\theta})}{\partial \boldsymbol{\theta}^2}\right|_{\boldsymbol{\theta}=\boldsymbol{\theta}^0}.$$

Minimizing the quadratic part of the above expression, the following equation is obtained for $\bar{\boldsymbol{\theta}}(N)$, which is an approximation to $\hat{\boldsymbol{\theta}}(N)$:

$$\nabla^2_{\boldsymbol{\theta}\boldsymbol{\theta}} J(\boldsymbol{\theta}^0)(\bar{\boldsymbol{\theta}}(N) - \boldsymbol{\theta}^0) = \nabla_{\boldsymbol{\theta}} J(\boldsymbol{\theta}^0). \tag{4}$$

By direct differentiation of $J(\boldsymbol{\theta})$, we obtain

$$\nabla_{\boldsymbol{\theta}} J(\boldsymbol{\theta}) = \sum_{t=1}^{N} w(t, \boldsymbol{\theta})\, \nabla_{\boldsymbol{\theta}} w(t, \boldsymbol{\theta}), \tag{5}$$

$$\nabla^2_{\boldsymbol{\theta}\boldsymbol{\theta}} J(\theta) = \sum_{t=1}^{N} w(t, \boldsymbol{\theta})\, \nabla_{\boldsymbol{\theta}}{}^2 w(t, \boldsymbol{\theta}) + \sum_{t=1}^{N} \nabla_{\boldsymbol{\theta}} w(t, \boldsymbol{\theta})(\nabla_{\boldsymbol{\theta}} w(t, \boldsymbol{\theta}))^{\mathrm{T}}, \tag{6}$$

where

$$\nabla_{\boldsymbol{\theta}} w(t, \boldsymbol{\theta}) = \partial w(t, \boldsymbol{\theta})/\partial \boldsymbol{\theta} = [\nabla_{\theta_1} w(t, \boldsymbol{\theta}), \ldots, \nabla_{\theta_n} w(t, \boldsymbol{\theta})]^{\mathrm{T}},$$

$$\nabla^2_{\boldsymbol{\theta}\boldsymbol{\theta}} w(t, \boldsymbol{\theta}) = \left\{\frac{\partial^2 w(t, \boldsymbol{\theta})}{\partial \theta_i\, \partial \theta_j}\right\}, \qquad i, j = 1, \ldots, n.$$

Let

$$\mathbf{P}_N = \sum_{t=1}^{N} \{\nabla_{\boldsymbol{\theta}} w(t, \boldsymbol{\theta}^0)\}\{\nabla_{\boldsymbol{\theta}} w(t, \boldsymbol{\theta}^0)\}^{\mathrm{T}},$$

and let $F_1(N), \ldots, F_n(N)$ be the eigenvalues of $\mathbf{P}_N$, and the matrix $\mathbf{M}$ be defined as shown below, $\mathbf{M}$ being unitary:

$$\mathbf{M}\mathbf{P}_N\mathbf{M}^{\mathrm{T}} = \mathrm{diag.}(F_1(N), \ldots, F_n(N)) \triangleq \mathbf{F}(N).$$

In Lemma 1 (below) we show that $\mathbf{F}(N)$ is asymptotically nonsingular. By left multiplying both sides of (4) by $\mathbf{F}^{-1}(N)\mathbf{M}$ and substituting (6) into (4), we obtain

$$[\mathbf{G}(N) + \mathbf{Q}(N)]\mathbf{F}(N)\mathbf{M}(\bar{\boldsymbol{\theta}}(N) - \boldsymbol{\theta}^0) = \mathbf{F}^{-1}(N)\mathbf{M}\, \nabla_{\boldsymbol{\theta}} J(\boldsymbol{\theta}^0), \tag{7}$$

where

$$\mathbf{G}(N) = \mathbf{F}^{-1}(N)\mathbf{M}\mathbf{P}_N\mathbf{M}^{\mathrm{T}}\mathbf{F}^{-1}(N),$$

$$\mathbf{Q}(N) = \mathbf{F}^{-1}(N)\mathbf{M} \sum_{t=1}^{N} w(t, \boldsymbol{\theta}^0)\, \nabla^2_{\boldsymbol{\theta}\boldsymbol{\theta}} w(t, \boldsymbol{\theta}^0)\mathbf{M}^{\mathrm{T}}\mathbf{F}^{-1}(N).$$

At this stage we need the following lemmas, which are proved later:

Lemma 1. The eigenvalues $F_i(N)$, $i = 1, \ldots, N$, are monotonically increasing functions of N, and $\mathbf{G}(N)$ tends to a constant positive-definite matrix.

Lemma 2. The right-hand side of (7), $\mathbf{F}^{-1}(N)\mathbf{M}\, \nabla_{\boldsymbol{\theta}} J(\boldsymbol{\theta}^0)$, has finite mean square value.

Lemma 3. The matrix $(\mathbf{Q}N)$ tends to zero asymptotically.

We can solve (7) for $\mathbf{F}(N)\mathbf{M}(\bar{\boldsymbol{\theta}}(N) - \boldsymbol{\theta}^0)$, and this solution will exist and have finite mean square value by Lemmas 1–3. Since $\mathbf{M}$ is nonsingular and $F_i(N)$ are monotonically increasing functions of N for all i tending to infinity as $N \to \infty$, $(\bar{\boldsymbol{\theta}}(N) - \boldsymbol{\theta}^0)$ should go to zero with probability 1, in view of the finiteness of $\mathbf{F}(N)\mathbf{M}(\bar{\boldsymbol{\theta}}(N) - \boldsymbol{\theta}^0)$.

Proof of Lemma 1. A difference equation for $\nabla_{\theta_i} w(t, \boldsymbol{\theta})$ can be obtained by differentiating Eq. (2) with respect to θ_i:

$$\nabla_{\boldsymbol{\theta}} w(t, \boldsymbol{\theta}) + \sum_{j=1}^{m_2} \theta_{n-m_2+j} \nabla_{\boldsymbol{\theta}} w(t-j, \boldsymbol{\theta}) = -\mathbf{x}(t-1, \boldsymbol{\theta}). \tag{8}$$

Assume $\mathbf{x}(t, \boldsymbol{\theta}) = 0$, $\forall t < 0$. We can express $\nabla_{\boldsymbol{\theta}} w(t, \boldsymbol{\theta})$ in terms of $\mathbf{x}(t-j, \boldsymbol{\theta})$, $j = 1, 2, \ldots,$

$$\nabla_{\boldsymbol{\theta}} w(t, \boldsymbol{\theta}) = -[\mathbf{x}(t-1, \boldsymbol{\theta}) + k_1 \mathbf{x}(t-2, \boldsymbol{\theta}) + \cdots + k_{t-1}\mathbf{x}(0, \boldsymbol{\theta})].$$

Let

$$\mathbf{X}_N = [\mathbf{x}(0, \boldsymbol{\theta}), \ldots, \mathbf{x}(N-1, \boldsymbol{\theta})]$$

and

$$\mathbf{K}_N = \begin{bmatrix} 1 & k_1 & k_2 & \cdots & k_{N-1} \\ & 1 & k_1 & & k_{N-2} \\ & \vdots & & \vdots & \\ & 0 & & \cdots & 1 \end{bmatrix}$$

$$\mathbf{P}_N \triangleq \sum_{t=1}^{N} \{\nabla_{\boldsymbol{\theta}} w(t, \boldsymbol{\theta})\}\{\nabla_{\boldsymbol{\theta}} w(t, \boldsymbol{\theta})\}^{\mathrm{T}} = \mathbf{X}_N \mathbf{K}_N \mathbf{K}_N^{\mathrm{T}} \mathbf{X}_N^{\mathrm{T}}. \tag{9}$$

Consider any n-vector $\mathbf{h}$:

$$\mathbf{h}^{\mathrm{T}} \sum_{t=1}^{N} \{\nabla_{\boldsymbol{\theta}} w(t, \boldsymbol{\theta}^0)\}\{\nabla_{\boldsymbol{\theta}} w(t, \boldsymbol{\theta}^0)\}^{\mathrm{T}} \mathbf{h} = \mathbf{h}^{\mathrm{T}} \mathbf{X}_N \mathbf{K}_N \mathbf{K}_N^{\mathrm{T}} \mathbf{X}_N^{\mathrm{T}} \mathbf{h} \geq c \|\mathbf{h}^{\mathrm{T}} \mathbf{X}_N\|^2,$$

since $\mathbf{K}_N \mathbf{K}_N^{\mathrm{T}}$ is a strictly positive-definite matrix and $c > 0$. Thus,

$$\begin{aligned} \mathbf{h}^{\mathrm{T}} \mathbf{P}_N \mathbf{h} &\geq c \sum_{t=1}^{N} \{\mathbf{h}^{\mathrm{T}} \mathbf{x}(t-1)\}^2 \\ &\to \infty \quad \text{as} \quad N \to \infty \quad \text{by (B1).} \end{aligned} \tag{10}$$

Since $\mathbf{h}$ in (10) can be chosen to be any eigenvector of $\mathbf{P}_N$, $F_i(N)$, $i = 1, \ldots, m$, are all monotonically increasing functions of N, approaching infinity. Consequently, $\mathbf{P}_N$ is also an asymptotically positive-definite matrix. By definition of $\mathbf{M}$ and $F_i(N)$, $\mathbf{G}(N)$ tends to a constant positive-definite matrix as N tends to infinity.

Proof of Lemma 2. We will show that the vector $F^{-1}(N)\mathbf{M}\,\nabla_{\boldsymbol{\theta}} J(\boldsymbol{\theta}^0)$ tends to a finite order as N tends to infinity. To do this, multiply this vector by its own transpose, substitute for $\nabla_{\boldsymbol{\theta}} J$ from (5), and take expectation. Then split the double summation $\sum_{t_1=1}^{N} \sum_{t_2=1}^{N}$ into three groups, namely those with $t_1 = t_2$, those with $t_1 > t_2$, and those with $t_1 < t_2$:

$$
\begin{aligned}
&E[\mathbf{F}^{-1}(N)\mathbf{M}\,\nabla_{\boldsymbol{\theta}}J(\boldsymbol{\theta}^0)\{\nabla_{\boldsymbol{\theta}}J(\boldsymbol{\theta}^0)\}^{\mathrm{T}}\mathbf{M}^{\mathrm{T}}\mathbf{F}^{-1}] \\
&\quad = E\Bigg[\mathbf{F}^{-1}(N)\mathbf{M} \sum_{\substack{t_1=1 \\ t_1=t_2}}^{N} \sum_{t_2=1}^{N} w(t_1, \boldsymbol{\theta}^0)\,\nabla_{\boldsymbol{\theta}} w(t_1, \boldsymbol{\theta}^0) \\
&\qquad \times \sum_{t_2=1}^{N} w(t_2, \boldsymbol{\theta}^0)(\nabla_{\boldsymbol{\theta}} w(t_2, \boldsymbol{\theta}^0))^{\mathrm{T}}\mathbf{M}^{\mathrm{T}}(\mathbf{F}^{-1}(N))^{\mathrm{T}}\Bigg] \\
&\qquad + E\Bigg[\mathbf{F}^{-1}(N)\mathbf{M}\Bigg\{\sum_{\substack{t_1=1 \\ t_1>t_2}}^{N} \sum_{t_2=1}^{N} w(t_1, \boldsymbol{\theta}^0)w(t_2, \boldsymbol{\theta}^0) \\
&\qquad \times \nabla_{\boldsymbol{\theta}} w(t_1, \boldsymbol{\theta}^0)(\nabla_{\boldsymbol{\theta}} w(t_2, \boldsymbol{\theta}^0))^{\mathrm{T}}\Bigg\}\mathbf{M}^{\mathrm{T}}\mathbf{F}^{-1}(N)\Bigg] \\
&\qquad + E\Bigg[\mathbf{F}^{-1}(N)\mathbf{M}\Bigg\{\sum_{\substack{t_1=1 \\ t_2>t_1}}^{N} \sum_{t_2=1}^{N} w(t_2, \boldsymbol{\theta}^0) \\
&\qquad \times w(t_1, \boldsymbol{\theta}^0)\,\nabla_{\boldsymbol{\theta}} w(t_1, \boldsymbol{\theta}^0)(\nabla_{\boldsymbol{\theta}} w(t_2, \boldsymbol{\theta}^0))^{\mathrm{T}}\Bigg\}\mathbf{M}^{\mathrm{T}}\mathbf{F}^{-1}(N)\Bigg].
\end{aligned}
\tag{11}
$$

We will show that the second and third terms in (11) are zero. Let us denote the expression within the braces of the second term of (11) by $\mathbf{H}(N)$:

$$
\mathbf{H}(N) = \sum_{\substack{t_1=1 \\ t_1>t_2}}^{N} \sum_{t_2=1}^{N} w(t_1, \boldsymbol{\theta}^0)w(t_2, \boldsymbol{\theta}^0)\,\nabla_{\boldsymbol{\theta}} w(t_1, \boldsymbol{\theta}^0)(\nabla_{\boldsymbol{\theta}} w(t_2, \boldsymbol{\theta}^0))^{\mathrm{T}}
$$

$$
\begin{aligned}
\text{second term in (11)} &= E[\mathbf{F}^{-1}(N)\mathbf{M}\mathbf{H}(N)\mathbf{M}^{\mathrm{T}}\mathbf{F}^{-1}(N)] \\
&= E[\mathbf{F}^{-1}(N)\mathbf{M}E\{\mathbf{H}(N)|\mathbf{P}(N)\}\mathbf{M}^{\mathrm{T}}\mathbf{F}^{-1}(N)]
\end{aligned}
$$

$$
\begin{aligned}
E\{\mathbf{H}(N)|\mathbf{P}(N)\} &= \sum_{\substack{t_1=1 \\ t_1>t_2}}^{N} \sum_{t_2=1}^{N} E[w(t_1, \boldsymbol{\theta}^0)|\mathbf{P}(N)]E[w_2(t_2, \boldsymbol{\theta}^0)\,\nabla_{\boldsymbol{\theta}} w(t_1, \boldsymbol{\theta}^0) \\
&\quad \times (\nabla_{\boldsymbol{\theta}} w(t_2, \boldsymbol{\theta}^0))^{\mathrm{T}}|\mathbf{P}(N)]
\end{aligned}
$$

by using assumption (A1), i.e., $w(t_1, \boldsymbol{\theta}^0)$ is independent of the part $\mathbf{x}(t_1 - j, \boldsymbol{\theta}^0)$ for all $j \geq 1$

$= 0$, by (A1).

Hence, the second term in (11) is equal to zero. Similarly, the third term in (11) can be shown to be equal to zero. Hence, the left-hand side of expression (11) simplifies as follows:

left-hand side of (11)

$$
= E(w^2(t, \boldsymbol{\theta}^0))\mathbf{F}^{-1}(N) \sum_{t=1}^{N} \mathbf{M}E\{\nabla_{\boldsymbol{\theta}} w(t, \boldsymbol{\theta}^0)(\nabla_{\boldsymbol{\theta}} w(t, \boldsymbol{\theta}^0))^{\mathrm{T}}\}\mathbf{M}^{\mathrm{T}}\mathbf{F}^{-1}(N), \tag{12}
$$

which is finite by Lemma 1.

Proof of Lemma 3. Recall that

$$\mathbf{Q}(N) = (\bar{\mathbf{F}}(N))^{-1}\bar{\mathbf{Q}}(N)(\bar{\mathbf{F}}(N))^{-1}, \tag{13}$$

where

$$\mathbf{F}(N) = (\bar{\mathbf{F}}(N))^2 \tag{14}$$

and

$$\bar{\mathbf{Q}}(N) = (\bar{\mathbf{F}}(N))^{-1}\mathbf{M} \sum_{t=1}^{N} w(t, \boldsymbol{\theta}^0)\, \nabla_{\boldsymbol{\theta}}{}^2 w(t, \boldsymbol{\theta}^0)\mathbf{M}^{\mathrm{T}}(\bar{\mathbf{F}}(N))^{-1}. \tag{15}$$

By direct differentiation of the difference equation for $w(t, \boldsymbol{\theta})$, we can see that $\nabla_{\boldsymbol{\theta}}{}^2(t, \boldsymbol{\theta}^0)$ is a function of $\mathbf{x}(t-j, \boldsymbol{\theta}^0)$, $j > 1$, and consequently $w(t, \boldsymbol{\theta}^0)$ is independent of $\nabla_{\boldsymbol{\theta}}{}^2(t, \boldsymbol{\theta}^0)$ in view of (A1). Utilizing this fact, we can show, as in the proof of Lemma 2, that the quantity $\bar{\mathbf{Q}}(N)$ has finite mean square value, i.e., $\|E\bar{\mathbf{Q}}(N)(\bar{\mathbf{Q}}(N))^{\mathrm{T}}\|$ is finite. Consequently, the expression $\mathbf{Q}(N)$ in (13) goes to zero asymptotically in the mean square sense in view of the additional factor $(\bar{\mathbf{F}}(N))^{-1}$ in it.

B. The Mean Square Error of the QML Estimate for Large N

We will begin with Eq. (7) derived above:

$$\mathbf{G}(N)\mathbf{F}(N)\mathbf{M}(\bar{\boldsymbol{\theta}}(N) - \boldsymbol{\theta}^0) = \mathbf{F}^{-1}(N)\mathbf{M}\, \nabla_{\boldsymbol{\theta}} J(\boldsymbol{\theta}^0), \tag{16}$$

where $\mathbf{Q}(N)$ has been omitted because of Lemma 3. For large N, we can replace $\mathbf{G}(N)$ in (16) by its limit $\mathbf{G}(\infty)$, which is a constant positive-definite matrix by Lemma 1. By using these changes we get, from (16),

$$\begin{aligned}\mathbf{F}(N)\mathbf{M}(\bar{\boldsymbol{\theta}}(N) - \boldsymbol{\theta}^0) &= \mathbf{G}^{-1}(\infty)\mathbf{F}^{-1}(N)\mathbf{M}\, \nabla_{\boldsymbol{\theta}} J(\boldsymbol{\theta}^0)\\ &= \mathbf{G}^{-1}(\infty)\mathbf{F}^{-1}(N)\mathbf{M} \sum_{t=1}^{N} w(t, \boldsymbol{\theta}^0)\, \nabla_{\boldsymbol{\theta}} w(t, \boldsymbol{\theta}^0). \end{aligned} \tag{17}$$

Multiplying (17) by its own transpose and taking expectation on both sides, we obtain

$$\begin{aligned}&\mathbf{F}(N)\mathbf{M}E[(\bar{\boldsymbol{\theta}}(N) - \boldsymbol{\theta}^0)(\bar{\boldsymbol{\theta}}(N) - \boldsymbol{\theta}^0)^{\mathrm{T}}]\mathbf{M}^{\mathrm{T}}\mathbf{F}(N)\\ &\quad = \mathbf{G}^{-1}(\infty)\mathbf{F}^{-1}(N)\mathbf{M}E\left[\left\{\sum_{t=1}^{N} w(t, \boldsymbol{\theta}^0)\, \nabla_{\boldsymbol{\theta}} w(t, \boldsymbol{\theta}^0)\right\}\right.\\ &\qquad \times \left.\left\{\sum_{t=1}^{N} w(t, \boldsymbol{\theta}^0)\, \nabla_{\boldsymbol{\theta}} w(t, \boldsymbol{\theta}^0)\right\}^{\mathrm{T}}\right]\mathbf{M}^{\mathrm{T}}\mathbf{F}^{-1}(N)\mathbf{G}^{-1}(\infty). \end{aligned} \tag{18}$$

Recalling that $w(t, \boldsymbol{\theta}^0)$ is independent of $\nabla_{\boldsymbol{\theta}} w(t, \boldsymbol{\theta}^0)$, we can simplify the expectation term on the right-hand side of (18):

$$\begin{aligned}\text{left-hand side of (18)} &= \mathbf{G}^{-1}(\infty)\mathbf{F}^{-1}(N)\mathbf{M}E[w^2(t, \boldsymbol{\theta}^0)]E\left[\sum_{t=1}^{N} \{\nabla_{\boldsymbol{\theta}} w(t, \boldsymbol{\theta}^0)\}\right.\\ &\qquad \times \left.\{\nabla_{\boldsymbol{\theta}} w(t, \boldsymbol{\theta}^0)\}^{\mathrm{T}}\right]\mathbf{M}^{\mathrm{T}}\mathbf{F}^{-1}(N)\mathbf{G}^{-1}(\infty)\\ &= \rho^0\mathbf{G}^{-1}(\infty)\mathbf{G}(\infty)\mathbf{G}^{-1}(\infty) = \rho^0\mathbf{G}^{-1}(\infty). \end{aligned} \tag{19}$$

In (19), we can replace **G** by its equivalent given in the line below (7) and note that **M** is unitary. Removing the left and right factors of $\mathbf{F}(N)\mathbf{M}$ and $\mathbf{M}^T\mathbf{F}(N)$ from (19) we obtain the following expression for the mean square error matrix:

$$E[(\bar{\boldsymbol{\theta}}(N) - \boldsymbol{\theta}^0)(\bar{\boldsymbol{\theta}}(N) - \boldsymbol{\theta}^0)^T] \approx \rho^0\left[E\sum_{t=1}^{N} \nabla_{\boldsymbol{\theta}} w(t, \boldsymbol{\theta}^0)(\nabla_{\boldsymbol{\theta}} w(t, \boldsymbol{\theta}^0))^T\right]^{-1}. \quad (20)$$

Appendix 4.2. Evaluation of the Cramér–Rao Matrix Lower Bound in Single Output Systems

Recall that

$$L(\boldsymbol{\theta}, \rho) = -\frac{N}{2}\ln 2\pi\rho - \frac{1}{2\rho}\sum_{t=1}^{N} w^2(t, \boldsymbol{\theta}). \quad (21)$$

We will obtain the expressions for the various derivatives

$$\nabla_{\boldsymbol{\theta}} L(\boldsymbol{\theta}, \rho) = -\frac{1}{\rho}\sum_{t=1}^{N} w(t, \boldsymbol{\theta})\, \nabla_{\boldsymbol{\theta}} w(t, \boldsymbol{\theta}) \quad (22)$$

$$\nabla^2_{\boldsymbol{\theta\theta}} L(\boldsymbol{\theta}, \rho) = -\frac{1}{\rho}\sum_{t=1}^{N} \{\nabla_{\boldsymbol{\theta}} w(t, \boldsymbol{\theta})(\nabla_{\boldsymbol{\theta}} w(t, \boldsymbol{\theta}))^T + w(t, \boldsymbol{\theta})\, \nabla^2_{\boldsymbol{\theta\theta}} w(t, \boldsymbol{\theta})\} \quad (23)$$

$$\nabla_{\boldsymbol{\rho}} L(\boldsymbol{\theta}, \rho) = -\frac{N}{2\rho} + \frac{1}{2\rho^2}\sum_{t=1}^{N} w^2(t, \boldsymbol{\theta}) \quad (24)$$

$$\nabla^2_{\boldsymbol{\theta\rho}} L(\boldsymbol{\theta}, \rho) = \frac{1}{\rho^2}\sum_{t=1}^{N} w(t, \boldsymbol{\theta})\, \nabla_{\boldsymbol{\theta}} w(t, \boldsymbol{\theta}) \quad (25)$$

$$\nabla^2_{\boldsymbol{\rho\rho}} L(\boldsymbol{\theta}, \rho) = -\frac{1}{\rho^3}\sum_{t=1}^{N} w^2(t, \boldsymbol{\theta}). \quad (26)$$

Consider the expectation of the second term on the right-hand side of (23) at $\boldsymbol{\theta} - \boldsymbol{\theta}^0$:

$$E[w(t, \boldsymbol{\theta}^0)\, \nabla^2_{\boldsymbol{\theta\theta}} w(t, \boldsymbol{\theta}^0)] = 0, \qquad \text{in view of (A1)} \quad (27)$$

since $\nabla^2_{\boldsymbol{\theta\theta}} w(t, \boldsymbol{\theta}^0)$ is a function of the history $\boldsymbol{\xi}(t-1)$. Similarly, the expectation of (25) is also zero. Using (26)

$$E[\nabla^2_{\rho\rho} L(\boldsymbol{\theta}, \rho) | \boldsymbol{\theta} = \boldsymbol{\theta}^0, \rho = \rho^0] = -\frac{N}{(\rho^0)^2}$$

Hence,

$$E[\nabla^2_{\boldsymbol{\phi\phi}} L(\boldsymbol{\phi}^0) | \boldsymbol{\phi}^0] = -\frac{1}{\rho^0}\begin{bmatrix} E\sum_{t=1}^{N} \nabla_{\boldsymbol{\theta}} w(t, \boldsymbol{\theta}^0)(\nabla_{\boldsymbol{\theta}} w(t, \boldsymbol{\theta}^0))^T & 0 \\ 0 & \dfrac{N}{2\rho^0} \end{bmatrix}.$$

using (4d.4.1)

$$E[(\boldsymbol{\theta}^* - \boldsymbol{\theta}^0)(\boldsymbol{\theta}^* - \boldsymbol{\theta}^0)^{\mathrm{T}}|\boldsymbol{\phi}^0] \geqslant \rho^0\left[E\left\{\sum_{t=1}^{N} \nabla_{\boldsymbol{\theta}} w(t, \boldsymbol{\theta}^0)(\nabla_{\boldsymbol{\theta}} w(t, \boldsymbol{\theta}^0))^{\mathrm{T}}\right\}\right]^{-1}$$

$$E[(\rho^* - \rho^0)^2|\rho^0] \geqslant \frac{2(\rho^0)^2}{N}.$$

Problems

1. Consider the following process $y(\cdot)$ generated by the nonwhite disturbances ξ and w obeying (A1): $A(D)y(t) = \xi(t)$, $H(D)\xi(t) = w(t)$. Let the degrees of the polynomials A and H be 2 and 1, respectively. Obtain a set of sufficient conditions on A and H so that they can be uniquely identified by means of an infinite observation sequence $\{y(\cdot)\}$.

2. Let θ_1 and θ_2 be two parameters whose unbiased estimates $\hat{\theta}_1$ and $\hat{\theta}_2$ have a known matrix lower bound on their variance. Suppose there is another parameter $\theta_4 = f(\theta_1, \theta_2)$, where f is a known function. Find the Cramer–Rao bound on the variance of an unbiased estimate of θ_4.

3. Consider an IAR(1) system that can be written as an AR(2) system with coefficients $\theta_1{}^0$ and $\theta_2{}^0$, where $\theta_1{}^0 + \theta_2{}^0 = 1$. What happens to the mean square error matrix of the unbiased estimates of $\theta_1{}^0$ and $\theta_2{}^0$? What is the correct way of estimating the Cramer–Rao bound on the variance of the estimates of the coefficient of an IAR(1) process?

4. Consider the bivariate process $\mathbf{y}(\cdot)$ obeying

$$\mathbf{y}(t) = \mathbf{A}\mathbf{y}(t-1) + \mathbf{g}u(t) + \mathbf{w}(t),$$

where $u(t) = \mathbf{k}^{\mathrm{T}}\mathbf{y}(t-1)$, u being a scalar and $\mathbf{k}$ a vector. Obtain a set of necessary and sufficient conditions on $\mathbf{A}$ and $\mathbf{g}$ so that their true values can be recovered from the observations of $\mathbf{y}$ and u.

5. For the system in Eq. (4d.2.7), establish the bounds in (4d.2.8).

6. Establish the equality

$$E\left[\frac{\partial^2 \ln p(\mathbf{x}; \boldsymbol{\theta})}{\partial \boldsymbol{\theta}^2}\right] = -E\left[\frac{\partial \ln p(\mathbf{x}, \boldsymbol{\theta})}{\partial \boldsymbol{\theta}}\left(\frac{\partial \ln p(\mathbf{x}, \boldsymbol{\theta})}{\partial \boldsymbol{\theta}}\right)^{\mathrm{T}}\right].$$

where $p(x; \boldsymbol{\theta})$ is the probability density of $\mathbf{x}$ indexed by vector parameter $\boldsymbol{\theta}$. Establish (4d.1.3) using the above equality.

Chapter V | Structure and Estimability in Multivariate Systems

Introduction

In the preceding chapters, we considered the structure of univariate systems and associated problems such as estimability. In this chapter we discuss similar topics in multivariate systems. Every one of the problems and many of the properties considered in the univariate case have their multivariate counterpart. In addition, there are new problems in the estimability of multivariate systems which are absent in the univariate case. There is an extensive discussion of identifiability in multivariate systems arising in econometrics (Dhrymes, 1972; Chow, 1974; Fisher, 1966).

A notable feature of the multivariate system is that the same process $\mathbf{y}$ can be described by an infinite number of difference equations. This statement is true even if we impose condition (A7) and (A7′), which remove all common left factors. Recall that assumption (A7), with the other usual assumptions, was sufficient in univariate systems to ensure the uniqueness of the difference equation in describing the process y. Hence our first task is to develop a method for generating all the possible difference equations (i.e., the polynomial matrices $\mathbf{A}$, $\mathbf{B}$, $\mathbf{G}$ and the covariance matrix $\boldsymbol{\rho}$), the outputs of all of which have identical first- and second-order statistical properties such as the spectral densities $(\mathbf{S}_{yy}, \mathbf{S}_{yu})$ and the mean.

One can show that all the 4-tuples $(\mathbf{A}, \mathbf{B}, \mathbf{G}, \boldsymbol{\rho})$ corresponding to the process having the same spectral pair $(\mathbf{S}_{yy}, \mathbf{S}_{yu})$ can be derived from one 4-tuple, say $(\mathbf{A}^*, \mathbf{B}^*, \mathbf{G}^*, \boldsymbol{\rho}^*)$, by means of unimodular transformations. Hence we can call the 4-tuple $(\mathbf{A}^*, \mathbf{B}^*, \mathbf{G}^*, \boldsymbol{\rho}^*)$ a canonical tuple and the difference equation involving this 4-tuple a canonical difference equation corresponding to the given process. However, a canonical form is not unique in $\mathscr{P}$, the set of all 4-tuples $\{\mathbf{A}, \mathbf{B}, \mathbf{G}, \boldsymbol{\rho}\}$ that obey (A5)–(A6); i.e., there exists more than one 4-tuple in $\mathscr{P}$ which possesses the above-mentioned property. But for each canonical form, say the ith, we can find a subset $\mathscr{P}_i$ of the set $\mathscr{P}$ such that there is only one canonical tuple in this subset corresponding to a spectral pair $(\mathbf{S}_{yy}, \mathbf{S}_{yu})$. Using this canonical tuple, we can generate all the 4-tuples in $\mathscr{P}$ which possess the same spectral pair.

The utilitarian justification of the canonical form is the following. Consider the problem of estimation of the coefficients of a 4-tuple $\{\mathbf{A}, \mathbf{B}, \mathbf{G}, \boldsymbol{\rho}\}$

corresponding to the given process $\mathbf{y}$ based on its observation history. We will search for a best fitting model or its 4-tuple for the given observation set in the class $\mathscr{P}$. But, since $\mathscr{P}$ has a countably infinite number of members which lead to the same process (i.e., the same spectral density), no computational method can yield even one solution. All the algorithms will usually oscillate between the various solutions or diverge. However, corresponding to the given process, the subclass $\mathscr{P}_i$ of $\mathscr{P}$ of the 4-tuples has only one member in it, namely the ith canonical form. Hence, any standard search algorithm can search for the required canonical form by searching in the subset $\mathscr{P}_i$. Once we recover one canonical form, we can recover all the other equivalent forms, canonical and noncanonical.

The topic of canonical forms dealing with multiple input–output deterministic systems in state variable representation has been extensively discussed in the last 10 years (Popov, 1969; Wolovich, 1974; Rosenbrock, 1970). But these results can be used only when one has some idea of the order of the system. This information is rarely available with empirical time series. Hence, we have limited ourselves to the discussion of canonical forms that will be of use in analyzing time series with little prior knowledge.

Among the large number of canonical forms, we will consider three, because of their connection to the estimation problem.

Canonical form I: $\mathbf{A}(D)$ is lower triangular, $\mathbf{A}(0) = \mathbf{I}$,
$\mathbf{B}$, $\mathbf{G}$, and $\boldsymbol{\rho}$ are arbitrary.

Canonical form II: $\mathbf{B}(D)$ is diagonal, $\mathbf{A}(0) = \mathbf{I}$,
$\mathbf{A}(D)$, $\mathbf{G}$, $\boldsymbol{\rho}$ are arbitrary.

Canonical form III: $\mathbf{B}(D)$, $\boldsymbol{\rho}$ are diagonal; $\mathbf{A}(0)$ is triangular,
$\mathbf{A}(D)$, $\mathbf{G}$ are arbitrary.

The canonical forms with $\mathbf{B}$ diagonal are important because they permit the estimation of the unknown parameters in the individual equations separately. The importance of this feature in a system of high dimension cannot be underestimated. The canonical form with $\mathbf{A}$ triangular is important because of its parsimonious structure; i.e., if we count the total number of scalar coefficients in the polynomial matrices $\mathbf{A}$, $\mathbf{G}$, and $\mathbf{B}$, the number associated with a triangular canonical form (or the variants of the triangular form) is often the smallest. The property of parsimony is strongly related to the question of accuracy of estimation.

There are canonical forms in which $\mathbf{A}$ is diagonal. Such forms only complicate the estimation problem instead of simplifying it, since they vastly increase the importance of the moving average part of the system, which always causes problems in estimation. Hence, we will not discuss such forms. However, such a form may be useful in system synthesis. But whenever the canonical form in which $\mathbf{A}$ is diagonal is needed, it can be generated from any one of the three canonical forms previously discussed by means of a unimodular transformation.

In addition to the canonical forms, we will also introduce the concept of pseudocanonical forms. This concept is also useful in estimation problems. Each pseudocanonical form is unique in a subset $\hat{\mathscr{P}}$ of $\mathscr{P}$ just like a canonical form. But the difference between the pseudocanonical form and the canonical form lies in the fact that every process may not possess a pseudocanonical form of every type having the same spectral density as the process, whereas every process obeying a difference equation must have a unique canonical form of each type associated with it.

Let us briefly mention the principal similarities and differences in the univariate and multivariate systems. We have already mentioned the multiplicity of the difference equations for the same process even under assumption (A7) and the concept of different canonical forms being unique to multiple output systems. But the conditions of the estimability of the parameters in each of the canonical forms and the accuracy of estimation are quite similar to the corresponding discussion of the univariate system. This is especially so in the case of canonical forms with $\mathbf{B}$ diagonal since the estimation problem in the multiple output system can be decomposed into m single equation estimation problems. Similarly, the problems due to observation noise can be handled in conjunction with a canonical form in a manner analogous to that in single output systems.

5a. Characterization

Consider the basic equation where the trend term $\boldsymbol{\psi}$ has been omitted for the sake of simplicity:

$$\mathbf{A}(D)\mathbf{y}(t) = \mathbf{G}(D)\mathbf{u}(t-1) + \mathbf{B}(D)\mathbf{w}(t). \tag{5a.1.1}$$

The difference equation is characterized by the 4-tuple $\{\mathbf{A}(D), \mathbf{B}(D), \mathbf{G}(D), \boldsymbol{\rho}\}$, $\boldsymbol{\rho}$ being the positive-definite covariance matrix of $\mathbf{w}$ and $\mathbf{A}(D)$, $\mathbf{B}(D)$, and $\mathbf{G}(D)$ being the polynomial matrices of appropriate dimensions. The symbol D in $\mathbf{A}(D), \mathbf{B}(D), \ldots$ will be omitted if there is no cause for confusion. Let the inputs $\mathbf{u}$ and $\mathbf{w}$ in (5a.1.1) obey (A1)–(A3), and the triple $\{\mathbf{A}, \mathbf{B}, \mathbf{G}\}$ obey (A5)–(A6).

A brief comment on the need for the strong assumption (A3) is in order. Since we are interested in determining all the difference equations with the same second-order characteristics obeyed by the process, we need to obtain an explicit expression for the spectral density of the process obeying the difference equation. To do this we need to know the precise relationship between $\mathbf{u}$ and $\mathbf{w}$. One such relationship is (A3). However, if we are interested only in determining the estimability of coefficients in the difference equation, then (A3) is not necessary. This feature is discussed in Section 5c.

Assumption (A7) states that if $\mathbf{U}$ is a nonsingular common left factor matrix as in

$$\mathbf{A} = \mathbf{U}\hat{\mathbf{A}}, \qquad \mathbf{B} = \mathbf{U}\hat{\mathbf{B}}, \qquad \mathbf{G} = \mathbf{U}\hat{\mathbf{G}}, \tag{5a.1.2}$$

then $\mathbf{U}$ will be unimodular; i.e., the determinant of $\mathbf{U}(D)$ is a nonzero constant independent of D. One can check whether $(\mathbf{A}, \mathbf{B}, \mathbf{G})$ obeys (A7) by finding the

Smith form of the triple and determining whether it has the form $\{\mathbf{I} \vdots \mathbf{0}\}$ where $\mathbf{I}$ is an $m \times m$ identity matrix. Standard algorithms are available for finding the Smith form of a matrix.

If a triple $\{\mathbf{A}, \mathbf{B}, \mathbf{G}\}$ does not obey (A7), we can repeatedly remove common nonmodular left factors to obtain a set $\{\hat{\mathbf{A}}, \hat{\mathbf{B}}, \hat{\mathbf{G}}\}$ that does obey (A7). Removal of such common factors in the multivariate case is similar to the removal of the common poles and zeros in the univariate case. It is a first step in reducing the redundancy in the system equations.

On the other hand, if $\{\mathbf{A}, \mathbf{B}, \mathbf{G}\}$ obeys (A7′), then everyone of the m scalar difference equations in (5a.1.1) cannot be individually simplified further. It is easy to show that (A7) implies (A7′) but not vice versa. See problem 5 for an illustration.

Throughout this chapter we will assume that the weakly stationary process $\mathbf{u}$ has a spectral density $\mathbf{S}_{uu}$. Since $\mathbf{y}$ is also weakly stationary, all second-order information about the process $\mathbf{y}$ is contained in the spectral pair $(\mathbf{S}_{yy}, \mathbf{S}_{yu})$. The relation between the spectral pair $\{\mathbf{S}_{yy}, \mathbf{S}_{yu}\}$ and the 4-tuple $\{\mathbf{A}, \mathbf{B}, \mathbf{G}, \boldsymbol{\rho}\}$ is given by

$$\mathbf{S}_{yy} = \mathbf{T}_1(e^{i\omega})\boldsymbol{\rho}\mathbf{T}_1{}^*(e^{i\omega}) + \mathbf{T}_2(e^{i\omega})\mathbf{S}_{uu}\mathbf{T}_2{}^*(e^{i\omega}), \tag{5a.1.3}$$

$$\mathbf{S}_{yu} = \mathbf{T}_2(e^{i\omega})\mathbf{S}_{uu}, \tag{5a.1.4}$$

which are obtained by using assumption (A3), and where the asterisk stands for the conjugate transpose. The transfer matrices $\mathbf{T}_1$, $\mathbf{T}_2$ mentioned above have the form

$$\mathbf{T}_1(D) = [\mathbf{A}(D)]^{-1}\mathbf{B}(D), \qquad \mathbf{T}_2(D) = [\mathbf{A}(D)]^{-1}\mathbf{G}(D). \tag{5a.1.5}$$

The inverse in (5a.1.5) exists everywhere except at a finite number of points in view of assumption (A5).

When the processes $\mathbf{y}$ and $\mathbf{u}$ are zero mean Gaussian, all the information contained in $\mathbf{y}$ is contained in the pair $\{\mathbf{S}_{yy}, \mathbf{S}_{yu}\}$. In such a case, Eqs. (5a.1.3) and (5a.1.4) represent all the available information in characterizing the 4-tuple $\{\mathbf{A}, \mathbf{B}, \mathbf{G}, \boldsymbol{\rho}\}$ in terms of the properties of the processes $\mathbf{y}$ and $\mathbf{u}$. However, when $\mathbf{y}$ and $\mathbf{u}$ are non-Gaussian, additional relations can be obtained from (5a.1.1) connecting the higher order statistics of $\mathbf{y}$ to the matrices $\mathbf{A}$, $\mathbf{B}$, $\mathbf{G}$, and $\boldsymbol{\rho}$, besides (5a.1.3) and (5a.1.4). Still, we will restrict ourselves to using only the second-order properties of $\mathbf{y}$ and $\mathbf{u}$ and characterize $\mathbf{A}$, $\mathbf{B}$, $\mathbf{G}$, and $\boldsymbol{\rho}$ correspondingly.

Our task is to characterize the class of processes $\mathbf{y}$ represented by 4-tuples $[\mathbf{A}, \mathbf{B}, \mathbf{G}, \boldsymbol{\rho}]$ which lead to the same spectral pair $\{\mathbf{S}_{yy}, \mathbf{S}_{yu}\}$. We will introduce the definition

$$\begin{aligned}\mathscr{P} = \{(\mathbf{A}, \mathbf{B}, \mathbf{G}, \boldsymbol{\rho}) \mid &\text{(i) } \mathbf{A}, \mathbf{B}, \mathbf{G} \text{ obey (A5)–(A6), (ii) } \boldsymbol{\rho} > \mathbf{0}, \\ &\text{(iii) } \mathbf{A}, \mathbf{B}, \boldsymbol{\rho} \text{ are } m \times m \text{ matrices, and } \mathbf{G} \text{ is an } m \times l_1 \text{ matrix}\}.\end{aligned}$$

Definition. Two members H_1 and H_2 belonging to $\mathscr{P}$ are said to obey the relation $\mathscr{E}$, i.e., $H_1 \mathscr{E} H_2$ if the two 4-tuples H_1 and H_2 lead to the same spectral density pair $(\mathbf{S}_{yy}, \mathbf{S}_{yu})$.

By inspection, we can see that the relation $\mathscr{E}$ is an equivalence relation since it possesses the following properties:

(i) $\mathscr{E}$ is reflexive $(H_1 \mathscr{E} H_1)$,

(ii) $\mathscr{E}$ is symmetric $(H_1 \mathscr{E} H_2 \rightarrow H_2 \mathscr{E} H_1)$,

(iii) $\mathscr{E}$ is transitive $(H_1 \mathscr{E} H_2$ and $H_2 \mathscr{E} H_3 \rightarrow H_1 \mathscr{E} H_3)$.

An equivalence relation $\mathscr{E}$ divides the set $\mathscr{P}$ into equivalence classes, say $\mathscr{P}_1, \mathscr{P}_2, \ldots$, which are mutually exclusive and totally inclusive subsets of $\mathscr{P}$. Any two members in the same equivalence class obey the relation $\mathscr{E}$; i.e., all the 4-tuples in an equivalence class correspond to the same spectral density pair $\{\mathbf{S}_{yy}, \mathbf{S}_{yu}\}$. Thus, each equivalence class is characterized uniquely by a spectral density pair $(\mathbf{S}_{yy}, \mathbf{S}_{yu})$, and vice versa. Theorem 5a.1 gives a method for generating all possible 4-tuples contained in an equivalence class from the knowledge of any one element of the equivalence class. Thus, Theorem 5a.1 gives the structure of these equivalence classes.

Theorem 5a.1. Consider any two 4-tuples H_1, H_2 belonging to $\mathscr{P}$ where $H_1 = (\mathbf{A}_1, \mathbf{B}_1, \mathbf{G}_1, \boldsymbol{\rho}_1)$, $H_2 = (\mathbf{A}_2, \mathbf{B}_2, \mathbf{G}_2, \boldsymbol{\rho}_2)$.

(i) If $H_1 \mathscr{E} H_2$, there exists a unimodular matrix $\mathbf{U} \in \mathscr{R}^{m \times m}(D)$ and a nonsingular matrix $\mathbf{R} \in \mathscr{R}^{m \times m}$ satisfying

$$\mathbf{A}_2 = \mathbf{U}\mathbf{A}_1, \qquad \mathbf{B}_2 = \mathbf{U}\mathbf{B}_1\mathbf{R}^{-1}, \qquad \mathbf{G}_2 = \mathbf{U}\mathbf{G}_1, \qquad \boldsymbol{\rho}_2 = \mathbf{R}\boldsymbol{\rho}_1\mathbf{R}^{\mathrm{T}}. \quad (5\text{a}.1.6)$$

(ii) The converse of (i) is also true; i.e., if there exists a pair of 4-tuples H_1 and H_2 belonging to $\mathscr{P}$ and obeying (5a.1.6), then $H_1 \mathscr{E} H_2$.

The proof of the theorem is in Appendix 5.1.

Comment 1. The importance of Theorem 5a.1 lies in the fact that it offers a constructive procedure for generating *all* members in an equivalence class starting from one member. Different pairs $(\mathbf{U}, \mathbf{R})$ defined in the theorem yield different members. The question is how to generate all the possible pairs $(\mathbf{U}, \mathbf{R})$. The next observation clears this question.

Suppose the operation in (5a.1.6) between H_1, H_2, and $(\mathbf{U}, \mathbf{R})$ is denoted by the symbol ϕ. We may rewrite (5a.1.6) as

$$H_2 = H_1 \phi(\mathbf{U}, \mathbf{R}). \qquad (5\text{a}.1.7)$$

Consider the set K:

$$K = \{(\mathbf{U}, \mathbf{R}) \mid H_2 = H_1 \phi(\mathbf{U}, \mathbf{R}) \text{ for any } H_1, H_2 \text{ such that } H_1 \mathscr{E} H_2 \text{ and } (\mathbf{U}, \mathbf{R}) \text{ obeys the conditions in Theorem (5a.1)}\}.$$

Theorem 5a.1 shows that the set K is a group. Then, successive application of all members of K to any one member in an equivalence class as indicated by (5a.1.6) and (5a.1.7) yields *all* the members of that equivalence class.

Comment 2. Even though all the 4-tuples $(\mathbf{A}, \mathbf{B}, \mathbf{G}, \boldsymbol{\rho})$ in an equivalence class obey assumption (A7), the degrees of the corresponding polynomials **A, B, G** may not be the same for all members. This is illustrated by the following example.

Example 5a.1. Consider the 4-tuple $H_1 = (\mathbf{A}_1, \mathbf{B}_1, \mathbf{G}_1, \boldsymbol{\rho})$

$$\mathbf{A}_1 = \begin{bmatrix} 1 + aD & d_1 D^2 \\ 0 & 1 + bD \end{bmatrix}, \qquad \mathbf{B}_1 = \begin{bmatrix} 1 + Dd_2 & Dd_3 \\ 0 & 1 \end{bmatrix},$$

$\mathbf{G}_1 = 0$.

Construct the pair $(\mathbf{A}_2, \mathbf{B}_2)$ from $(\mathbf{A}_1, \mathbf{B}_1)$ as shown below using the unimodular matrix $\mathbf{U}$:

$$\mathbf{A}_2 = \mathbf{U}\mathbf{A}_1 = \begin{bmatrix} 1 + aD & -d_1 D/b \\ 0 & 1 + bD \end{bmatrix},$$

$$\mathbf{B}_2 = \mathbf{U}\mathbf{B}_1 = \begin{bmatrix} 1 + Dd_2 & (d_3 - d_1/b)D \\ 0 & 1 \end{bmatrix},$$

where

$$\mathbf{U} = \begin{bmatrix} 1 & -d_1 D/b \\ 0 & 1 \end{bmatrix}.$$

Let $H_2 = (\mathbf{A}_2, \mathbf{B}_2, \mathbf{0}, \boldsymbol{\rho})$. By definition $H_1 \mathscr{E} H_2$.

Since the maximum degree of a polynomial in H_1 is greater than that in H_2, the order of difference equation obtained from H_1 is greater than that obtained from H_2 even though both difference equations describe the same process $\mathbf{y}$ and, in addition, obey (A7). Such an occurrence is impossible in the case of a scalar difference equation system.

Our next step would be to identify a member in each equivalence class having some special property. For instance, we may desire a member that often leads to a difference equation with the smallest order among all the difference equations corresponding to some equivalence class. This particular member can be called a canonical form since it possesses some special property and all other members in the equivalence class can be generated from this member. One can also define a subset $\mathscr{P}^*$ of $\mathscr{P}$ such that $\mathscr{P}^*$ has exactly one member from each equivalence class, the member being the corresponding canonical form. This subset is called a canonical subset.

The concept of canonical subset is very important in the estimation of parameters of difference equations. If we try to search for a best fit in the class $\mathscr{P}$ for the given observation set using any standard parameter estimation algorithm, the algorithm rarely converges (or it may even diverge) because the set $\mathscr{P}$ has many members that are equally good fits for the given process and the algorithm oscillates between these values. On the other hand, if we search for a best fit in a canonical subset, there is no problem of convergence since it has only one member corresponding to an equivalence class.

We can also think of defining canonical forms and the corresponding subsets which simplify the parameter estimation problem considerably. These aspects are considered in the next two sections.

5b. The Triangular Canonical Forms

5b.1. The Basic Form

The equivalence relation $\mathscr{E}$ has been shown to partition the set of 4-tuples $(\mathbf{A}, \mathbf{B}, \mathbf{G}, \boldsymbol{\rho})$ into equivalence classes so that all the tuples corresponding to the same equivalence class have the same spectral pair $(\mathbf{S}_{yy}, \mathbf{S}_{yu})$. Our intention here is to construct a subset $\mathscr{P}_1^*$ of $\mathscr{P}$, the so-called canonical subset, such that $\mathscr{P}_1^*$ has exactly one member from every distinct equivalence class generated by the relation $\mathscr{E}$. Given any arbitrary member $H \in \mathscr{P}$, H_1^* is said to be a canonical 4-tuple or the canonical form corresponding to H (or the associated spectral density) if $H_1^* \in \mathscr{P}_1^*$ and $H_1^* \mathscr{E} H$. The difference equation obtained from H_1^* is called a canonical difference equation. The canonical set $\mathscr{P}_1^*$ has a one-to-one correspondence with the quotient set $[\mathscr{P}/\mathscr{E}]$ induced by the relation $\mathscr{E}$. Stated differently, there exists a one-to-one map between the spectral density pairs $(\mathbf{S}_{yy}, \mathbf{S}_{yu})$ and $\mathscr{P}_1^*$.

The set $\mathscr{P}_1^*$ is not unique. We can construct different canonical sets $\mathscr{P}_2^*$, $\mathscr{P}_3^*, \ldots$, where $\mathscr{P}_i^* \subseteq \mathscr{P}$, $i = 2, 3, \ldots$. To see this, suppose we find a canonical set $\mathscr{P}_1^*$ and choose nonsingular matrices $\mathbf{W}, \mathbf{R}$ such that $\mathbf{W}$ is unimodular and $\mathbf{R}$ is any matrix over the reals. Then we define $\mathscr{P}_2^*$ as

$$\mathscr{P}_2^* = \{H_2 | H_2 = H_1 \phi(\mathbf{W}, \mathbf{R}) \qquad \text{and} \qquad H_1 \in \mathscr{P}_1^*\}.$$

Then $\mathscr{P}_2^*$ is also a canonical set. By selecting different matrices $(\mathbf{W}_i, \mathbf{R}_i)$ we can generate other canonical sets $\mathscr{P}_i^*$ by the relationships above. All possible canonical sets can be generated in this way, in view of the group property of set K.

We will give a number of theorems for the generation of the various canonical forms. In this connection we need the following assumption:

A9. The matrix polynomial $\mathbf{A}$ obeys the following conditions: (i) $\mathbf{A}$ is lower triangular; (ii) $\mathbf{A}(0) = \mathbf{I}$; (iii) degree of $A_{ij} \leq$ degree of A_{ii}, $\forall j < i$.

$$\mathbf{A} = \begin{bmatrix} A_{11} & & & & 0 \\ & \ddots & & & \\ & & A_{ii} & & \\ & & & \ddots & \\ A_{m1} & & A_{mi} & & A_{mm} \end{bmatrix}.$$

Define the subset $\mathscr{P}_1^* \subseteq \mathscr{P}$ as follows:

$$\mathscr{P}_1^* = \{(\mathbf{A}, \mathbf{B}, \mathbf{G}, \boldsymbol{\rho}) | \mathbf{A} \text{ obeys (A9)}; (\mathbf{A}, \mathbf{B}, \mathbf{G}) \text{ obeys } \mathbf{A}\ (7); (\mathbf{A}, \mathbf{B}, \mathbf{G}, \boldsymbol{\rho}) \in \mathscr{P}\}.$$

Theorem 5b.1 (**Canonical Form I**) (Hannan, 1971). (i) If H_1 and H_2 belong to $\mathscr{P}_1^*$ and $H_1 \mathscr{E} H_2$, then $H_1 = H_2$. (ii) Given any $H = (\mathbf{A}, \mathbf{B}, \mathbf{G}, \boldsymbol{\rho}) \in \mathscr{P}$, there exists a 4-tuple $H^1 = (\mathbf{A}^1, \mathbf{B}^1, \mathbf{G}^1, \boldsymbol{\rho}^1) \in \mathscr{P}_1^*$ such that $H^1 \mathscr{E} H$.

The proof of part (i) of the theorem is in Appendix 5.1. The proof of part (ii) follows from the algorithm given below. A 4-tuple H is said to be in canonical form i if $H \in \mathscr{P}_1^*$.

Algorithm for Constructing Canonical Form I. Given any tuple $(\mathbf{A}, \mathbf{B}, \mathbf{G}, \boldsymbol{\rho})$, we will construct a 4-tuple $(\mathbf{A}', \mathbf{B}', \mathbf{G}', \boldsymbol{\rho})$ in canonical form i. It is a well-known result that any nonsingular matrix $\mathbf{A} \in \mathscr{R}^{m \times m}(D)$ can be reduced to a lower triangular form in the class $\mathscr{R}^{m \times m}(D)$ by means of elementary operation on the rows only. This statement is equivalent to stating the existence of a unimodular matrix $\mathbf{U}$ satisfying

$$\mathbf{A}' = \mathbf{UA}.$$

Applying the transformation $\mathbf{U}$ on $\mathbf{B}$ and $\mathbf{G}$ gives us the remaining members $\mathbf{B}'$ and $\mathbf{G}'$ of the 4-tuple $H' = (\mathbf{A}', \mathbf{B}', \mathbf{G}', \boldsymbol{\rho}) \in \mathscr{P}_1^*$:

$$\mathbf{B}' = \mathbf{UB}, \qquad \mathbf{G}' = \mathbf{UG}.$$

Comment 1. Part (i) of the theorem states that all members in $\mathscr{P}_1^*$ possess distinct spectral density pairs. Part (ii) states that given any $H = (\mathbf{A}, \mathbf{B}, \mathbf{G}, \boldsymbol{\rho})$, we can find a unique $H^1 \in \mathscr{P}_1^*$ such that $H \mathscr{E} H^1$. The required $H^1 = \{\mathbf{A}^1, \mathbf{B}^1, \mathbf{G}^1, \boldsymbol{\rho}\}$ is given by

$$\mathbf{A}^1 = \mathbf{UA}, \qquad \mathbf{B}^1 = \mathbf{UB}, \qquad \mathbf{G}^1 = \mathbf{UG}, \tag{5b.1.1}$$

where $\mathbf{U}$ is a unimodular matrix. For a given $\mathbf{A}$, $\mathbf{U}$ is unique since $\mathbf{U} = (\mathbf{A}^1)^{-1}\mathbf{A}$, and $\mathbf{A}^1$ is unique.

Comment 2. The canonical difference equation corresponding to canonical form I has the form

$$\mathbf{A}(D)\mathbf{y}(t) = \mathbf{G}(D)\mathbf{u}(t-1) + \mathbf{B}(D)\mathbf{w}(t),$$

where $\mathbf{A}(D) = \mathbf{I} - \mathbf{A}_1 D - \cdots - \mathbf{A}_{m_1} D^{m_1}$. All the matrices $\mathbf{A}_j$ are lower triangular. If $(A_i)_{kk} = 0$, $\forall i > j$, and $(A_j)_{kk} \neq 0$ for any k, then by (A9) all elements of the kth row of $(\mathbf{A}_i)$ must be zero for all $i > j$.

Comment 3. Theorem 5b.1 is valid if the matrix $\mathbf{A}$ in $\mathscr{P}_1^*$ is defined in terms of upper triangular matrices instead of lower triangular matrices.

Comment 4. Theorem 5b.1 is valid if assumption (A9) is replaced by (A9′) in defining $\mathscr{P}_1^*$ and the corresponding canonical form is labeled I′.

A9′. $\mathbf{A}(D)$ obeys the following conditions:

(i) $\mathbf{A}(D)$ is lower triangular,
(ii) $\mathbf{A}(0)_{ii} = \mathbf{I}$, $\forall i = 1, \ldots, m$,
(iii) degree of A_{ji} < degree of A_{jj}, $\forall i < j$,

i.e., $\mathbf{A}(0)$ is not necessarily diagonal, but is lower triangular.

The canonical form I′ is of interest from the point of view of parameter complexity. Parameter complexity could be measured in one of two ways:

It could be measured by the number of distinct coefficients in the polynomials **A**, **B**, **G**, and the matrix $\boldsymbol{\rho}$, i.e., the number of coefficients that have to be estimated from the data if the 4-tuple (**A**, **B**, **G**, and $\boldsymbol{\rho}$) was unknown. This is labeled complexity measure I. Alternatively, it could be measured by the sums of the degrees of the polynomials **A**, **B**, **G**, i.e., by $(m_1 + m_2 + m_3)$. This measure is labeled complexity measure II. In either case the parameter complexity measures defined above are only indirectly related to the computational complexity of the estimation of the coefficients in the system. For instance, it is much easier to estimate the parameters in an AR system than those in an ARMA system even though the number of equations in the former may be larger than that in the latter.

The advantage of canonical form I′ is that its complexity measure (I or II) is usually lower than that of form I, which obeys (A9).

Comment 5. Theorem 5a.1 is valid if **B** rather than **A** obeys (A9) or (A9′). The corresponding canonical forms are labeled I″ and I‴. There are occasions when canonical forms I″ and I‴ may have lower complexity than form I. An example of this kind is the autoregressive system with 4-tuple (**A**, **I**, **G**, $\boldsymbol{\rho}$). This is, by definition, in form I″. But if we try to express it in form I, it will lead to a 4-tuple (**A**′, **B**′, **G**′, $\boldsymbol{\rho}$) where **A**′ may be of higher degree than **A**; thus, form I has greater complexity than form I″.

Comment 6. Canonical form I may not always have minimal complexity among all the different forms corresponding to the same spectral pair. But one of the forms among I, I′, I″, and I‴ usually has minimal complexity. The complexity of form I is often, but not always, less than that of the corresponding forms II or III. Example 5c.1 illustrates this feature very well.

5b.2. Estimability

The uniqueness of the canonical 4-tuple in the class $\mathscr{P}_1^*$ corresponding to a particular process **y** does not automatically ensure the estimability of all the coefficients of the 4-tuple though it is reasonable to make such a guess. In the next theorem, we will prove the estimability of the coefficients in (**A**, **B**, **G**) in the canonical difference equation.

Theorem 5b.2. Consider the difference equation (5b.2.1) where $(\mathbf{A}^0, \mathbf{B}^0, \mathbf{G}^0, \boldsymbol{\rho}^0) \in \mathscr{P}_1^*$:

$$\mathbf{y}(t) = \sum_{j=1}^{m_1} \mathbf{A}_j{}^0\mathbf{y}(t-j) + \sum_{j=1}^{m_3} \mathbf{G}_j{}^0\mathbf{u}(t-j) + \mathbf{w}(t) + \sum_{j=1}^{m_2} \mathbf{B}_j{}^0\mathbf{w}(t-j). \quad (5b.2.1)$$

Let $\boldsymbol{\theta}^0$ denote the vector of coefficients in the matrices $\{\mathbf{A}_j{}^0, \mathbf{B}_j{}^0, \mathbf{G}_j{}^0\}$. Then the vector $\boldsymbol{\theta}^0$ is estimable.

The proof of Theorem 5b.2 is in Appendix 5.1.

Theorem 5b.2 appears to be true even if we add a vector of trend terms, $\boldsymbol{\psi}$, into Eq. (5b.2.1), provided $\boldsymbol{\psi}$ obeys (A4). But the restriction (A3) appears to be basic. There is every reason to believe that the estimability results do not need as strong an assumption as (A3). But proving Theorem 5b.2 without (A3) appears to be difficult.

5c. Diagonal Canonical Forms

5c.1. The Basic Forms with $\boldsymbol{\rho}$ Arbitrary

We will develop a canonical form for an arbitrary element of $\mathscr{P}$ such that the matrix $\mathbf{B}$ in the canonical tuple is diagonal. In this form we set $\mathbf{A}(0) = \mathbf{I}$. As before, we set up a canonical set $\mathscr{P}_2{}^*$ using *assumption (A10)*:

$$\mathscr{P}_2{}^* = \{(\mathbf{A}, \mathbf{B}, \mathbf{G}, \boldsymbol{\rho}) \,|\, (\mathbf{A}, \mathbf{B}) \text{ obeys (A10)}, (\mathbf{A}, \mathbf{B}, \mathbf{G},) \text{ obeys (A7}'), (\mathbf{A}, \mathbf{B}, \mathbf{G}, \boldsymbol{\rho}) \in \mathscr{P}\}.$$

A10. $\mathbf{B}(D)$ is diagonal with $\mathbf{B}(0) = \mathbf{I}$ and $\mathbf{A}(0) = \mathbf{I}$.

Theorem 5c.1 (Canonical Form II) (Kashyap and Nasburg, 1974). (i) Consider two 4-tuples $H' = (\mathbf{A}', \mathbf{B}', \mathbf{G}', \boldsymbol{\rho})$ and $H'' = (\mathbf{A}'', \mathbf{B}'', \mathbf{G}'', \boldsymbol{\rho})$ belonging to $\mathscr{P}_2{}^*$ with $H' \mathscr{E} H''$. Then $H' = H''$. (ii) Given any arbitrary $H = (\mathbf{A}, \mathbf{B}, \mathbf{G}, \boldsymbol{\rho})$ belonging to $\mathscr{P}$, there exists an $H' = (\mathbf{A}', \mathbf{B}', \mathbf{G}', \boldsymbol{\rho}) \in \mathscr{P}_2{}^*$ such that $H \mathscr{E} H'$.

The proof of part (i) is in Appendix 5.1, and that of part (ii) follows from the algorithm given below.

Algorithm for Construction of Canonical Form II. Starting with any 4-tuple $(\mathbf{A}, \mathbf{B}, \mathbf{0}, \boldsymbol{\rho})$, the following algorithm gives the corresponding 4-tuple $(\mathbf{A}', \mathbf{B}', \mathbf{0}, \boldsymbol{\rho})$ in canonical form II. Let $\mathbf{T}(D) = (\mathbf{A}(D))^{-1}\mathbf{B}(D)$; $(T^{-1})_{ij} = p_{ij}/f_{ij}$, $i, j = 1, \ldots, m$, where p_{ij} and f_{ij} are relatively prime; and define $b_{ii}(D) = \gamma_i \operatorname{LCM}(f_{i1}, \ldots, f_{im})$, $i = 1, 2, \ldots, m$, where γ_i is chosen so that $b_{ii}(0) = 1$ and LCM means least common multiple. Let $\mathbf{B}'(D) = \operatorname{diag.}(b'_{11}(D), \ldots, b'_{mn}(D))$ and $\mathbf{A}'(D) = (\mathbf{T}(D))^{-1}\mathbf{B}'(D)$. Then $(\mathbf{A}', \mathbf{B}', \mathbf{0}, \boldsymbol{\rho})$ is the required 4-tuple.

Comment. Consider the difference equation from a 4-tuple in canonical form II:

$$\begin{aligned} y_i(t) &= \sum_{j=1}^{m} \sum_{k=1}^{m_1} A_{ijk} y_j(t-k) + w_i(t) + \sum_{k=1}^{m_2} b_{ik} w_i(t-k) \\ &\quad + \sum_{j=1}^{l_1} \sum_{k=1}^{m_3} G_{ijk} u_j(t-k), \qquad i = 1, \ldots, m. \end{aligned} \tag{5c.1.1}$$

Note that in the equation for y_i only the noise variable w_i appears. Hence, the equation for the output y_i is an ordinary scalar autoregressive moving average equation with w_i as the noise input, $y_j, j \neq i$ and u_j, being the observable inputs.

Thus, the unknown coefficients in the ith equation can be estimated using the observation history of $\mathbf{y}$ and $\mathbf{u}$ without reference to the equation for $y_j, j \neq i$. Therefore, for purposes of estimation, the multivariate equation can be decomposed into separate univariate equations. This decomposition is one of the important advantages of this canonical form. This advantage is not present in a canonical form (Mayne, 1968) in which $\mathbf{A}$ is diagonal instead of $\mathbf{B}$.

Another advantage of the canonical form II is that it involves only (A7′) and not (A7); i.e., if the m individual equations in the corresponding equation (5a.1.1) are in a simplified form, then the entire set of equations cannot be simplified further. Such a decomposition is not valid in canonical form I.

However, there is a drawback to this decomposition. First, in separately estimating the unknowns in the individual equations, we ignore the fact that the disturbances $w_1, \ldots, w_m$ in the m equations are correlated according to the covariance matrix $\boldsymbol{\rho}$. The inability to use this knowledge reduces the accuracy of the estimates. The estimates of coefficients obtained by the decomposition method are less accurate than those obtained by simultaneously estimating all the unknowns in the entire multivariate system by a method such as maximum likelihood, considering the fact that $\boldsymbol{\rho}$ may be nondiagonal.

The second reason for the possible inefficiency of the estimates obtained with the decomposition method is the possible redundancy of parameters in canonical form II. Often, the complexity measure associated with canonical form II is greater than the corresponding value for canonical form I.

If we consider a system in canonical form I involving n_1 coefficients to be estimated, then the same system in canonical form II involves the estimation of n_2 coefficients where $n_2 \geq n_1$. The redundancy in vector $\boldsymbol{\theta}_2$, the vector of coefficients in canonical form II, can be represented as a set of simultaneous algebraic equations in the vector $\boldsymbol{\theta}_2$:

$$\theta_i = f_i(\theta_{l_3+1}, \theta_{l_3+2}, \ldots, \theta_{n_2}), \qquad i = 1, \ldots, l_3.$$

This relation is ignored while separately estimating the unknowns in the m equations, which causes the loss in estimation accuracy. We will illustrate the parameter redundancy (or the additional complexity) in canonical form II by the following example.

Example 5c.1. Consider the following system in canonical form I, where $m = 2$:

$$\mathbf{A}_1(D)\mathbf{y}(t) = \mathbf{B}_1(D)\mathbf{w}(t) \tag{5c.1.2}$$

$$\mathbf{A}_1 = \begin{bmatrix} 1 - a_1 D & 0 \\ -a_2 D & 1 - a_3 D \end{bmatrix}, \qquad \mathbf{B}_1 = \begin{bmatrix} 1 & b_1 D \\ b_2 D & 1 + b_3 D \end{bmatrix}.$$

Here $\boldsymbol{\theta}_1$ is the vector of coefficients in (5c.1.2), $(a_1, a_2, a_3, b_1, b_2, b_3)^{\mathrm{T}}$, that is, $n_1 = 6$. Let $\bar{\mathbf{S}}_1$ denote the maximum likelihood estimate of $\boldsymbol{\theta}_1$. We will use the algorithm given in the proof of Theorem 5c.1 to obtain canonical form II for the system in (5c.1.2). The difference equation in canonical form II is given by

$$\mathbf{A}_2\mathbf{y}(t) = \mathbf{B}_2\mathbf{w}(t) \tag{5c.1.3}$$

where

$$\mathbf{A}_2 = \begin{bmatrix} 1 - \theta_1 D - \theta_2 D^2 & -\theta_3 D - \theta_4 D^2 \\ -\theta_5 D - \theta_6 D^2 & 1 - \theta_7 D \end{bmatrix}$$

$$\mathbf{B}_2 = \begin{bmatrix} 1 + \theta_8 D + \theta_9 D^2 & 0 \\ 0 & 1 + \theta_{10} D + \theta_{11} D^2 \end{bmatrix}$$

and

$$\begin{aligned} &\theta_1 = a_1 - b_3, \qquad \theta_2 = b_3 a_3 - b_1 a_2 \\ &\theta_3 = b_1, \qquad \theta_4 = -b_1 a_3, \qquad \theta_5 = b_2 + a_2, \qquad \theta_6 = -b_2 a_1, \\ &\theta_8 = \theta_{10} = b_3, \qquad \theta_9 = \theta_{11} = -b_1 b_2. \end{aligned} \tag{5c.1.4}$$

Let

$$\begin{aligned} \boldsymbol{\theta}_{21} &= (\theta_1, \theta_2, \theta_3, \theta_4, \theta_8, \theta_9)^{\mathrm{T}} \\ \boldsymbol{\theta}_{22} &= (\theta_5, \theta_6, \theta_7, \theta_{10}, \theta_{11})^{\mathrm{T}} \\ \boldsymbol{\theta}_2 &= \begin{bmatrix} \boldsymbol{\theta}_{21} \\ \boldsymbol{\theta}_{22} \end{bmatrix}. \end{aligned} \tag{5c.1.5}$$

We can estimate $\boldsymbol{\theta}_2$ in two ways. In the first method, replace the variables $a_1, a_2, \ldots,$ in (5c.1.4) by the corresponding estimate from $\bar{\mathbf{S}}_1$. The corresponding estimate of $\boldsymbol{\theta}_2$ from (5c.1.5) will be denoted by $\bar{\mathbf{S}}_2$. In the second method, obtain an estimate $\hat{\boldsymbol{\theta}}_{21}$ of $\boldsymbol{\theta}_{21}$ from the first scalar equation in (5c.1.3) and an estimate $\hat{\boldsymbol{\theta}}_{22}$ of $\boldsymbol{\theta}_{22}$ from the second equation in (5c.1.3).

Let $\hat{\boldsymbol{\theta}}_2 = ((\hat{\boldsymbol{\theta}}_{21})^{\mathrm{T}}, (\hat{\boldsymbol{\theta}}_{22})^{\mathrm{T}})^{\mathrm{T}}$. In the second method of estimation, we have to estimate a total of 11 parameters, ignoring the knowledge contained in (5c.1.4) such as $\theta_8 = \theta_{10}$, $\theta_9 = \theta_{11}$, etc., whereas we estimate only six unknowns in computing $\bar{\mathbf{S}}_1$. Naturally, the estimate $\hat{\boldsymbol{\theta}}_2$, which utilizes less knowledge and involves the estimation of more parameters, is less accurate than the estimate $\bar{\mathbf{S}}_2$ derived from $\hat{\boldsymbol{\theta}}_1$.

In addition, the redundancy in canonical form II may cause problems when we want to obtain the minimal state variable realization needed in the work on optimal control.

Even with these disadvantages, canonical form II is very useful because all the methods of estimation in which all the unknowns are simultaneously estimated, do not work even for small values of m such as 4 or 5. For instance, with $m = l_1 = 4$, $m_1 = m_2 = m_3 = 2$, we have to estimate at least 96 unknowns simultaneously if we use canonical form I. In such cases the decomposition method of estimation is the only reasonable way to proceed. This is possible only if canonical form II is used. Furthermore, once we construct a model in canonical form II, we can determine the corresponding transfer functions $\mathbf{T}_1$ and $\mathbf{T}_2$ and thus obtain the canonical form I from these transfer functions after removing from them all pole zeros that cancel approximately.

5c.2. The Basic Form with $\boldsymbol{\rho}$ Diagonal

Next we consider a canonical form $\{\mathbf{A}, \mathbf{B}, \mathbf{G}, \boldsymbol{\rho}\}$ in which $\boldsymbol{\rho}$ is diagonal in addition to $\mathbf{B}$ (Kashyap and Nasburg, 1974). In such cases $\mathbf{A}(0)$ cannot always be an identity matrix. Instead it can only be a triangular matrix, with the values on the main diagonal being unity. We will define the corresponding canonical set $\mathscr{P}_3^*$ as follows:

$$\mathscr{P}_3^* = \{(\mathbf{A}, \mathbf{B}, \mathbf{G}, \boldsymbol{\rho}) \mid (\mathbf{A}, \mathbf{B}, \mathbf{G}, \boldsymbol{\rho}) \in \mathscr{P}, (\mathbf{A}, \mathbf{B}, \mathbf{G}) \text{ obeys (A7')} \text{ and } (\mathbf{A}, \mathbf{B}, \boldsymbol{\rho}) \text{ obeys assumption (A11)}\}.$$

A11. $(\mathbf{A}, \mathbf{B}, \boldsymbol{\rho})$ has the following properties: (i) $\mathbf{B}(D)$ is diagonal with $\mathbf{B}(0) = \mathbf{I}$; (ii) $\mathbf{A}(0)$ is lower triangular with $(A(0))_{ii} = 1$, $\forall i = 1, \ldots, m$; (iii) $\boldsymbol{\rho}$ is diagonal.

Theorem 5c.2 corresponding to canonical form III is identical to Theorem 5c.1 except that canonical set $\mathscr{P}_2^*$ in Theorem 5c.1 is replaced by $\mathscr{P}_3^*$ in Theorem 5c.2.

Theorem 5c.2 (Canonical Form III). (i) Given two 4-tuples $H' = (\mathbf{A}', \mathbf{B}', \mathbf{G}', \boldsymbol{\rho}')$ and $H'' = (\mathbf{A}'', \mathbf{B}'', \mathbf{G}'', \boldsymbol{\rho}'')$ belonging to $\mathscr{P}_3^*$ with $H' \mathscr{E} H''$, we have $H' = H''$. (ii) Given any arbitrary $H = (\mathbf{A}, \mathbf{B}, \mathbf{G}, \boldsymbol{\rho}) \in \mathscr{P}$, there exists an $H'' = (\mathbf{A}'', \mathbf{B}'', \mathbf{G}'', \boldsymbol{\rho}'') \in \mathscr{P}_3^*$ such that $H \mathscr{E} H''$.

The proof of this theorem is in Appendix 5.1.

Algorithm for the Construction of Canonical Form III from Any Arbitrary $H = (\mathbf{A}, \mathbf{B}, \mathbf{0}, \boldsymbol{\rho})$. Using any technique, factor the given covariance matrix $\boldsymbol{\rho}$ as

$$\boldsymbol{\rho} = \mathbf{A}_0^{-1} \boldsymbol{\rho}'' (\mathbf{A}_0^{-1})^{\mathrm{T}}, \tag{5c.2.1}$$

where $\boldsymbol{\rho}''$ is diagonal and $\mathbf{A}_0$ is lower triangular with $(\mathbf{A}_0)_{ii} = 1$. Let $\mathbf{T}'(D) = (\mathbf{A}(D))^{-1}(\mathbf{B}(D))\mathbf{A}_0^{-1}$ and $(\mathbf{T}')_{ij}^{-1} = \rho'_{ij}/f'_{ij}$, $i, j = i, \ldots, m$, where ρ'_{ij}, f'_{ij} are relatively prime.

Let $b'_{ii}(D) = \gamma_i \operatorname{LCM}(f_{ij}, \ldots, f_{im})$, $i = 1, \ldots, m$, where γ_i is chosen so that $b'_{ii}(0) = 1$, and let $\mathbf{B}''(D) = \operatorname{diag.}(b'_{11}(D), \ldots, b'_{mm}(D))$, $\mathbf{A}''(D) = (\mathbf{T}'(D))^{-1} \times \mathbf{B}''(D)$. Then the 4-tuple $(\mathbf{A}'', \mathbf{B}'', \mathbf{0}, \boldsymbol{\rho}'')$ is in canonical form III possessing the same spectral density as $(\mathbf{A}, \mathbf{B}, \mathbf{0}, \boldsymbol{\rho})$.

The validity of the algorithm is evident from the proof of Theorem 5c.2 in Appendix 5.1.

Canonical form III retains the principal advantage of canonical form II, namely the possibility of decomposition of the multivariate system into a number of single equation systems. Since $\boldsymbol{\rho}$ is diagonal in canonical form III, one of the reasons for the inefficiency of the corresponding estimate of the unknown coefficient vector in the case of canonical form II is absent in canonical form III; i.e., since the noise components $w_1, w_2, \ldots$, above are uncorrelated, nothing is lost on this count by treating the m difference equations separately.

However, the parameter redundancy in canonical form III is often slightly more than that in canonical form II and the inefficiency of the estimate caused by the parameter redundancy still remains. We will give an example to illustrate the increased redundancy with canonical form III.

Example 5c.2. Consider Eq. (5c.1.2), the same system of Example 5c.1 with the covariance matrix $\boldsymbol{\rho}_1$ given below. Use the algorithm given above to obtain canonical form III ($\mathbf{A}_3$, $\mathbf{B}_3$, $\mathbf{0}$, $\boldsymbol{\rho}_3$). We indicate the various steps. Let

$$\boldsymbol{\rho}_1 = \begin{bmatrix} 1.0 & 0.5 \\ 0.5 & 2.25 \end{bmatrix} \tag{5c.2.2}$$

be the noise covariance matrix in canonical form I. By factorization we get

$$\boldsymbol{\rho}_1 = \mathbf{A}_0^{-1}\boldsymbol{\rho}_3(\mathbf{A}_0^{\mathrm{T}})^{-1},$$

where

$$\mathbf{A}_0^{-1} = \begin{bmatrix} 1 & 0 \\ 0.5 & 1 \end{bmatrix}, \qquad \boldsymbol{\rho}_3 = \begin{bmatrix} 1 & 0 \\ 0 & 2 \end{bmatrix},$$

and $\boldsymbol{\rho}_3$ is the noise covariance matrix in canonical form III. We will construct canonical form III in two steps. Rewrite the system equation (5c.1.2) so that the noise $\mathbf{w}'$ has covariance $\boldsymbol{\rho}_3$:

$$\mathbf{A}_0\mathbf{A}_1(D)\mathbf{y} = [\mathbf{A}_0\mathbf{B}_1(D)(\mathbf{A}_0^{-1})]\mathbf{A}_0\mathbf{w}$$

or

$$\mathbf{A}_1'(D)\mathbf{y} = \mathbf{B}_1'(D)\mathbf{w}', \tag{5c.2.3}$$

where $\mathbf{A}_1'(D) = \mathbf{A}_0\mathbf{A}_1(D)$, $\mathbf{B}_1'(D) = \mathbf{A}_0\mathbf{B}_1(D)\mathbf{A}_0^{-1}$, and $\mathbf{w}' = \mathbf{A}_0\mathbf{w}$. Transform (5c.2.3) so that the corresponding $\mathbf{B}(D)$ is diagonal:

$$\begin{aligned}
(\mathbf{B}_1'(D))^{-1}\mathbf{A}_1'(D) &= (\mathbf{A}_0\mathbf{B}_1(D)\mathbf{A}_0^{-1})^{-1}\mathbf{A}_0\mathbf{A}_1(D) \\
&= \mathbf{A}_0\mathbf{B}_1^{-1}(D)\mathbf{A}_1(D) \\
&= \mathbf{A}_0\mathbf{B}_2^{-1}(D)\mathbf{A}_2(D), && \text{using (5c.1.2) and (5c.1.3)} \\
&= \mathbf{B}_2^{-1}(D)\mathbf{A}_0\mathbf{A}_2(D), && \text{since } \mathbf{B}_2(D) \text{ is a diagonal matrix} \\
& && \text{with identical diagonal elements} \\
&= \mathbf{B}_3^{-1}(D)\mathbf{A}_3(D),
\end{aligned}$$

where

$$\mathbf{B}_3(D) \triangleq \mathbf{B}_2(D) = \begin{bmatrix} 1 + \theta_{10}'D + \theta_{11}'D & 0 \\ 0 & 1 + \theta_{12}' + \theta_{13}'D^2 \end{bmatrix}$$

$$\mathbf{A}_3(D) = \mathbf{A}_0\mathbf{A}_2(D) = \begin{bmatrix} 1 - \theta_1'D - \theta_2'D^2 & -\theta_3'D - \theta_4'D^2 \\ -\theta_6'D - \theta_7'D^2 & 1 - \theta_8'D - \theta_9'D^2 \end{bmatrix}$$

$$\theta_{10}' = \theta_{12}' = \theta_8, \qquad \theta_{11}' = \theta_{13}' = \theta_9$$

$$\theta_1' = a_1 - b_3, \qquad \theta_2' = b_3a_3 - b_1a_2$$

$$\theta_3' = b_1, \qquad \theta_4' = -b_1a_3$$

$$\theta_5' = 0.5, \qquad \theta_6' = b_2 + a_2 - 0.5b_3 + 0.5a_1$$

$$\theta_7' = -b_2a_1 - 0.5b_3a_3 + 0.5b_1a_2$$

$$\theta_8' = a_3 - 0.5b_1, \qquad \theta_9' = 0.5b_1a_3.$$

Canonical form III has 13 coefficients $\theta_1', \ldots, \theta_{13}'$, in contrast to canonical form II, which has 11 coefficients, and canonical form I, which has six coefficients. This example is quite typical regarding the relative redundancies of the three canonical forms.

5c.3. Estimability

The questions of estimability in difference equations with the 4-tuple $(\mathbf{A}, \mathbf{B}, \mathbf{G}, \boldsymbol{\rho})$ in canonical form II or III can be handled directly in terms of the estimability theory of single equation systems. Consider the difference equations in canonical forms II or III. We can regard the equation for y_i as an autoregressive moving average system having observable inputs such as u_k and y_k, $k \neq i$. Since $(\mathbf{A}, \mathbf{B}, \mathbf{G})$ in (5a.1.1) obeys (A7′), each of the individual equations in (5a.1.1) obey (A7). Hence the estimability of all the coefficients in each of the equations follows trivially, under the assumptions associated with the canonical form.

As a matter of fact, we can relax some of the assumptions if we are interested only in estimability and not in the question of uniqueness of the canonical forms. Clearly, (A3) can be relaxed since we are dealing with the estimability theory of single equations. Similarly, we can also insert trend terms obeying (A4) in the difference equation, and still preserve estimability.

5d. Pseudocanonical Forms

In defining the three canonical forms, we started with the corresponding canonical set $\mathscr{P}_i^* \subseteq \mathscr{P}$ having the following two properties:

(i) The spectral density pairs corresponding to any two members of $\mathscr{P}_i^*$ are distinct; i.e., if $H_1 \mathscr{E} H_2$ and $H_1, H_2 \in \mathscr{P}_i^*$, then $H_1 = H_2$.

(ii) For any $H \in \mathscr{P}$, there exists one and only one member $H' \in \mathscr{P}_i^*$ such that $H \mathscr{E} H'$; i.e., H and H' have the same spectral density.

In some instances, we can construct a subset $\hat{\mathscr{P}}_i \subseteq \mathscr{P}$ such that $\hat{\mathscr{P}}_i$ possesses only property (i) and not property (ii). One such subset $\hat{\mathscr{P}}_1$ is the set of all tuples corresponding to autoregressive processes, i.e.,

$$\hat{\mathscr{P}}_1 = \{(\mathbf{A}, \mathbf{B}, \mathbf{G}, \boldsymbol{\rho});\, (\mathbf{A}, \mathbf{B}, \mathbf{G}, \boldsymbol{\rho}) \in \mathscr{P}, (\mathbf{A}, \mathbf{B}, \mathbf{G}) \text{ obeys (A7)}, \mathbf{B}(D) = \mathbf{I}, \mathbf{A}(0) = \mathbf{I}\}.$$

In estimation problems, property (i) is crucial for any subset of $\mathscr{P}$ in which the search is made for the best fitting model. If property (i) is lacking, the search algorithm may never converge. On the other hand, property (ii) is not crucial. For instance, if we try to search the class $\hat{\mathscr{P}}_1$ for a suitable model for a process known to have an AR representation, then we obtain a best fitting autoregressive fit.

Another set $\hat{\mathscr{P}}_2$ which includes $\hat{\mathscr{P}}_1$ is given below, $\hat{\mathscr{P}}_1 \subseteq \hat{\mathscr{P}}_2$:

$$\hat{\mathscr{P}}_2 = \{(\mathbf{A}, \mathbf{B}, \mathbf{G}, \boldsymbol{\rho}) \mid (\mathbf{A}, \mathbf{B}, \mathbf{G}) \text{ obeys (A8)}; \\ (\mathbf{A}, \mathbf{B}, \mathbf{G}) \text{ obeys (A7)}, \mathbf{A}(0) = \mathbf{B}(0) = \mathbf{I} \text{ and } (\mathbf{A}, \mathbf{B}, \mathbf{G}, \boldsymbol{\rho}) \in \mathscr{P}\}$$

where

A8. $\operatorname{rank}[\mathbf{A}_{m_1} \quad \mathbf{B}_{m_2} \quad \mathbf{G}_{m_3}] = m.$

The 4-tuples in $\hat{\mathscr{P}}_2$ correspond to difference equations having both autoregressive and moving average terms.

Theorem 5d.1 (Hannan, 1969). All members in $\hat{\mathscr{P}}_2$ have different spectral density pairs; i.e., if there are two members H_1, H_2 in $\hat{\mathscr{P}}_2$ such that $H_1 \mathscr{E} H_2$, then $H_1 = H_2$. A similar result is true for $\hat{\mathscr{P}}_1$.

The proofs are in Appendix 5.1. Hannan (1971) has given an example to illustrate the incompleteness of the set $\hat{\mathscr{P}}_2$.

If the given 4-tuple H is such that there exists an equivalent 4-tuple H' for it in $\hat{\mathscr{P}}_2$, then H' is called the pseudocanonical form II of H. If H' exists, then it is often less redundant than even canonical form I or its variants. But as mentioned earlier, the estimation problems in the difference equations derived from this pseudocanonical form II and canonical form I involve the simultaneous estimation of all the unknowns, and this is often unwieldy and may be impractical.

5e. Discussion of the Three Canonical Forms

We will discuss only the highlights of the three forms since they have been discussed in detail earlier. The three canonical forms illustrate very well the tradeoff involved between the simplicity of estimation and parameter redundancy. When we are trying to build models for multivariate data without much knowledge of the system, we have to use many trial and error methods to determine the appropriate variables which influence each variable. Then canonical form II is appropriate because it allows us to handle the equations separately. Form I should be used when we desire the most accurate parameter estimate. Form III has been introduced to illustrate models in which the equation for $y_i(t)$ has terms $y_j(t)$, $j \neq i$. Such models are called simultaneous equation models in econometric literature. Note that the triangular nature of $\mathbf{A}_0$ has been obtained at the cost of making $\boldsymbol{\rho}$ diagonal. An important point in form III is that we should not attach too much importance to the fact that there may be two variables y_1, y_2 such that a variable, say $y_2(t)$, occurs on the right-hand side of the equation for $y_1(t)$, whereas $y_1(t)$ does not occur in the equation for y_2. By rearranging the variables, it is possible to obtain another canonical set in which the reverse is true. Therefore, we cannot draw any causal conclusions from form III. For discussions of causality, form II is the appropriate form.

5f. Estimation Accuracy

In the proof of Theorem 5b.2, an estimate of the unknown vector $\boldsymbol{\theta}$ is introduced, the so-called *quasi-maximum likelihood estimate.* We shall discuss the mean square error associated with this estimate. The conditions under which this estimate yields the smallest value of the mean square error in the class of asymptotically unbiased estimator will be discussed by using the Cramer–Rao inequality.

Let the vector of coefficients of the set of matrices $\{\mathbf{A}_j^0, \mathbf{B}_j^0, \mathbf{G}_j^0\}$ in Eq. (5b.2.1) be denoted by $\boldsymbol{\theta}^0$. Let $\boldsymbol{\theta}$ be a dummy vector of the same dimensions as $\boldsymbol{\theta}^0$ and let the individual matrices in the set $\{\mathbf{A}_j, \mathbf{B}_j, \mathbf{G}_j\}$ be constructed from $\boldsymbol{\theta}$ just as the elements of $\{\mathbf{A}_j^0, \mathbf{B}_j^0, \mathbf{G}_j^0\}$ can be constructed from $\boldsymbol{\theta}^0$. Let $\mathbf{w}(t, \boldsymbol{\theta})$ be an estimate of $\mathbf{w}^0(t)$ constructed recursively as

$$\mathbf{B}(D)\mathbf{w}(t, \boldsymbol{\theta}) = \mathbf{A}(D)\mathbf{y}(t) - \mathbf{G}(D)\mathbf{u}(t-1) \tag{5f.1.1}$$

based on $\boldsymbol{\theta}$ and the observation history prior to time t. Consider the following loss function:

$$J(\boldsymbol{\theta}) = \frac{1}{2N}\sum_{t=1}^{N}\sum_{i,j=1}^{m}(\rho^{-1})_{ij}w_i(t, \boldsymbol{\theta})w_j(t, \boldsymbol{\theta}).$$

The QML estimate of $\boldsymbol{\theta}^0$ based on the observation history of $\{\mathbf{y}(t), \mathbf{u}(t-1), t = 1, \ldots, N\}$ is the value of $\boldsymbol{\theta}$ that minimizes $J(\boldsymbol{\theta})$ and is denoted by $\hat{\boldsymbol{\theta}}(N)$.

5f.1. The Mean Square Error of the Multivariate QML Estimate

For large N, the mean square error matrix of the QML estimate has the form

$$\begin{aligned} &E[\hat{\boldsymbol{\theta}}(N) - \boldsymbol{\theta}^0)(\hat{\boldsymbol{\theta}}(N) - \boldsymbol{\theta}^0)^{\mathrm{T}}] \\ &\quad = \mathscr{I}_N^{-1}(\boldsymbol{\theta}^0, \boldsymbol{\rho}^0) + o(1/N) \\ &\quad = \left[E\left\{\sum_{t=1}^{N}\sum_{i,j=1}^{m}((\boldsymbol{\rho}^0)^{-1})_{ij}(\nabla_{\boldsymbol{\theta}}w_i(t, \boldsymbol{\theta}^0)) \times (\nabla_{\boldsymbol{\theta}}w_j(t, \boldsymbol{\theta}^0))^{\mathrm{T}}\right\}\right]^{-1} + o(1/N). \end{aligned} \tag{5f.1.2}$$

The assumptions made in connection with the canonical forms assure the existence of the inverse in this expression. The proof of (5f.1.2) is established in Appendix 5.1. While proving (5f.1.2), we restrict ourselves to such systems as (5b.2.1) which do not have trend terms. But expression (5f.1.2) is valid even if there is a trend vector $\boldsymbol{\psi}$ in (5b.2.1) as long as (5b.2.1) satisfies (A4′).

Since Eq. (5f.1.2) involves the unknowns $\boldsymbol{\theta}^0$ and $\boldsymbol{\rho}^0$, this equation can be used only in making quantitative assessment of the accuracy of the estimate. An approximate expression for the MSE matrix of the QML estimate can be obtained from (5f.1.2) by ignoring the expectation sign and replacing $\boldsymbol{\theta}^0$ by $\hat{\boldsymbol{\theta}}$, and the derivatives $\nabla_{\boldsymbol{\theta}}\mathbf{w}(t, \boldsymbol{\theta}^0)$ by their numerical values $\nabla_{\boldsymbol{\theta}}w(t, \hat{\boldsymbol{\theta}})$, $t = 1$,

2, ... In relatively simple problems, expression (5f.1.2) can be given as an explicit function of $\boldsymbol{\theta}^0$, as indicated in Chapter VII.

5f.2. Lower Bounds on the Estimation Accuracy

This section parallels the corresponding discussion in single equation systems in Section 4d.4.

Suppose we are given N observation pairs $\{\mathbf{y}(t), \mathbf{u}(t-1), t = 1, \ldots, N\}$ obeying (5b.2.1). Let $\boldsymbol{\theta}^0$ denote the vector of the values of the coefficients in (5b.2.1), and let $\boldsymbol{\theta}^*$ denote an estimate of $\boldsymbol{\theta}^0$ based on the observation history until time N.

The variance of $\boldsymbol{\theta}^*$ is clearly dependent on whether $\boldsymbol{\theta}^0$ is the only unknown or both $\boldsymbol{\theta}^0$ and $\boldsymbol{\rho}^0$ are unknown. Clearly, the variance of the estimate $\boldsymbol{\theta}^*$ in the latter case is larger than that in the former. However, we have already seen in Chapter IV that when $m = 1$ (ρ is scalar), knowledge of ρ does not make a drastic difference in the lower bound on the variance of $\boldsymbol{\theta}^*$. The contribution to the lower bound for the variance of $\boldsymbol{\theta}^*$ by the ignorance of ρ is $o(1/N)$. A similar statement is true for multivariate systems also. Hence we will assume that $\boldsymbol{\rho}^0$ is known while computing the lower bound on the variance of $\boldsymbol{\theta}^*$, an estimate of $\boldsymbol{\theta}^0$.

Let

$$E[\boldsymbol{\theta}^*|\boldsymbol{\theta}^0] = \mathbf{h}(\boldsymbol{\theta}^0) = (h_1(\boldsymbol{\theta}^0), \ldots, h_n(\boldsymbol{\theta}^0))^{\mathrm{T}}, \tag{5f.2.1}$$

i.e., $h_i(\boldsymbol{\theta}^0)$ is the bias of the estimate θ_i^*.

The variance of the component θ_i^* of $\boldsymbol{\theta}^*$ obeys the inequality

$$E[(\theta_i^* - h_i(\boldsymbol{\theta}^0))^2|\boldsymbol{\theta}^0, \boldsymbol{\rho}^0] \geq (\nabla_\theta h_i(\boldsymbol{\theta}^0))^{\mathrm{T}}[\mathscr{I}_N(\boldsymbol{\theta}^0, \boldsymbol{\rho}^0)]^{-1} \nabla_{\boldsymbol{\theta}} h_i(\boldsymbol{\theta}^0) \tag{5f.2.2}$$

where

$$\mathscr{I}_N(\boldsymbol{\theta}^0, \boldsymbol{\rho}^0) = -E[\nabla^2_{\boldsymbol{\theta}\boldsymbol{\theta}} L_N(\boldsymbol{\theta}^0)], \tag{5f.2.3}$$

is the Fisher information matrix, and $L_N(\boldsymbol{\theta})$ is the log likelihood function of the observation until time N.

When the estimate θ_j^* is unbiased (or asymptotically unbiased) i.e., when $h_j(\boldsymbol{\theta}^0) = (\boldsymbol{\theta}^0)_j$, inequality (5f.2.2) simplifies to

$$E[(\theta_j^* - \theta_j^0)^2] \geq [\mathscr{I}_N^{-1}(\boldsymbol{\theta}^0, \boldsymbol{\rho}^0)]_{jj}. \tag{5f.2.4}$$

For the unbiased estimators, a Cramér–Rao inequality can also be given for the mean square error matrix as follows:

$$E[(\boldsymbol{\theta}^* - \boldsymbol{\theta}^0)(\boldsymbol{\theta}^* - \boldsymbol{\theta}^0)^{\mathrm{T}}|\boldsymbol{\theta}^0, \boldsymbol{\rho}] \geq \mathscr{I}_N^{-1}(\boldsymbol{\theta}^0, \boldsymbol{\rho}^0), \tag{5f.2.5}$$

where $\geq$ indicates positive semidefiniteness.

It should be clear that knowledge of the probability distribution of the disturbances is necessary for the evaluation of the lower bounds. We will first evaluate the bound when $\mathbf{w}^0(\cdot)$ is Gaussian. For this case the log likelihood function has the form

$$L_N(\boldsymbol{\theta}^0) = -\tfrac{1}{2} \sum_{t=1}^{N} \sum_{i,j=1}^{m} w_i(t, \boldsymbol{\theta}^0)(\rho)_{ij}^{-1} w_j(t, \boldsymbol{\theta}^0) - \frac{Nm}{2} \ln 2\pi - \frac{N}{2} \ln|\det \boldsymbol{\rho}|.$$

$\mathbf{w}(t, \boldsymbol{\theta})$, $t = 1, 2, \ldots$, are generated recursively as in (5f.1.1) using the dummy $\boldsymbol{\theta}$ and observation history until time t. By successive differentiation of L_N we obtain

$$\nabla_{\boldsymbol{\theta}} L_N(\boldsymbol{\theta}) = -\sum_{t=1}^{N} \sum_{i,j=1}^{m} (\rho)_{ij}^{-1} w_i(t, \boldsymbol{\theta})\, \nabla_{\boldsymbol{\theta}} w_j(t, \boldsymbol{\theta}) \tag{5f.2.6}$$

$$\begin{aligned}\nabla_{\boldsymbol{\theta}\boldsymbol{\theta}}^2 L_N(\boldsymbol{\theta}) = -\sum_{t=1}^{N} \sum_{i,j=1}^{m} (\rho)_{ij}^{-1} \{ & w_i(t, \boldsymbol{\theta})\, \nabla_{\boldsymbol{\theta}\boldsymbol{\theta}}^2 w_j(t, \boldsymbol{\theta}) \\ & + (\nabla_{\boldsymbol{\theta}} w_i(t, \boldsymbol{\theta}))(\nabla_{\boldsymbol{\theta}} w_j(t, \boldsymbol{\theta}))^{\mathrm{T}} \}.\end{aligned} \tag{5f.2.7}$$

When $\boldsymbol{\theta} = \boldsymbol{\theta}^0$, the expectation of the first term on the right-hand side of (5f.2.7) is zero since $w_i^0(t)$ is independent of the past history $\boldsymbol{\xi}(t-1)$. Consequently, the Fisher information matrix has the expression

$$\mathscr{I}_N(\boldsymbol{\theta}^0, \boldsymbol{\rho}^0) = E\left[\sum_{t=1}^{N} \sum_{i,j=1}^{m} (\rho^0)_{ij}^{-1} (\nabla_{\boldsymbol{\theta}} w_i(t, \boldsymbol{\theta}^0)(\nabla_{\boldsymbol{\theta}} w_j(t, \boldsymbol{\theta}^0))^{\mathrm{T}}\right] \tag{5f.2.8}$$

or

$$E[(\theta_j^* - \theta_j^0)^2 | \boldsymbol{\theta}^0, \boldsymbol{\rho}^0] \geq (\mathscr{I}_N^{-1}(\boldsymbol{\theta}^0, \boldsymbol{\rho}^0))_{jj} + o(1/N). \tag{5f.2.9}$$

Let us consider the quasi-maximum likelihood estimate of $\boldsymbol{\theta}^*$. We have seen that it is asymptotically unbiased and that its covariance matrix is equal to $\mathscr{I}_N^{-1}(\boldsymbol{\theta}^0, \boldsymbol{\rho}^0)$. By definition, the QML estimate yields the smallest value of MSE among all asymptotically unbiased estimates, provided $\mathbf{w}$ is Gaussian.

The limitations of the QML estimators with mixture distributions pointed out in Chapter IV are valid here also. Consequently, we have to use the robust estimators when we have reason to believe that the noise obeys a mixture distribution.

We may emphasize that the lower bound given above has been obtained by considering the class of asymptotically unbiased estimators. It is interesting to know that there exist biased estimators which yield lower mean square error values than those given here. The consequences of using such estimators are not clear and, therefore, have not been discussed here.

5g. Conclusions

We have dealt mainly with those structural aspects of the difference equations of multiple output systems which do not have a counterpart with those of single output systems. The various canonical forms and pseudocanonical forms of the difference equations are discussed at some length, emphasizing the different roles played by the different forms. We have stressed canonical forms II and III, which considerably simplify the estimation of parameters. Algorithms for converting the forms into each other, and the estimability of the parameters in the various canonical forms and the accuracy of the corresponding quasi-maximum likelihood estimates are also discussed.

Appendix 5.1. Proofs of Theorems

A. Proof of Theorem 5a.1

We need the following lemma, which will be proved later:

Lemma 1. Given two positive-definite matrices $\boldsymbol{\rho}_1$ and $\boldsymbol{\rho}_2$, there exists a nonsingular matrix $\mathbf{R}$ obeying

$$\boldsymbol{\rho}_2 = \mathbf{R}\boldsymbol{\rho}_1\mathbf{R}^{\mathrm{T}} \qquad \text{and} \qquad \mathbf{R}^{\mathrm{T}} = \mathbf{R}^{-1}. \tag{1}$$

Proof of Part (i) of Theorem 5a.1. (A) We will show by construction, the existence of matrices $\mathbf{U}$ and $\mathbf{R}$ satisfying Eq. (5a.1.6). Define $\mathbf{R}$ as in (1). Let

$$\mathbf{U} = \mathbf{A}_2\mathbf{A}_1^{-1} \qquad \text{or} \quad \mathbf{A}_2 = \mathbf{U}\mathbf{A}_1. \tag{2}$$

Equating the spectral densities $\mathbf{S}_{yu}$ from the two 4-tuples $(\mathbf{A}_1, \mathbf{B}_1, \mathbf{C}_1, \boldsymbol{\rho}_1)$ and $(\mathbf{A}_2, \mathbf{B}_2, \mathbf{G}_2, \boldsymbol{\rho}_2)$, we obtain

$$\mathbf{A}_1^{-1}\mathbf{G}_1 = \mathbf{A}_2^{-1}\mathbf{G}_2. \tag{3}$$

Substituting for $\mathbf{A}_1$ in (3) from (2), we obtain

$$\mathbf{G}_2 = \mathbf{U}\mathbf{G}_1. \tag{4}$$

Similarly, equating the spectral densities $\mathbf{S}_{yy}$ of the above-mentioned two 4-tuples, and using (3), we get

$$(\mathbf{A}_1^{-1}\mathbf{B}_1)\boldsymbol{\rho}_1(\mathbf{A}_1^{-1}\mathbf{B}_1)^* = (\mathbf{A}_2^{-1}\mathbf{B}_2)\boldsymbol{\rho}_2(\mathbf{A}_2^{-1}\mathbf{B}_2)^*. \tag{5}$$

Substitute for $\boldsymbol{\rho}_1$ in the left-hand side of (5) from (1), to obtain

$$\mathbf{A}_1^{-1}\mathbf{B}_1\mathbf{R}^{-1}\boldsymbol{\rho}_2(\mathbf{A}_1^{-1}\mathbf{B}_1\mathbf{R}^{-1})^* = (\mathbf{A}_2^{-1}\mathbf{B}_2)\boldsymbol{\rho}_2(\mathbf{A}_2^{-1}\mathbf{B}_2)^*, \tag{6}$$

and since $\boldsymbol{\rho}_2$ is nonsingular, this implies

$$\mathbf{A}_1^{-1}\mathbf{B}_1\mathbf{R}^{-1} = \mathbf{A}_2^{-1}\mathbf{B}_2. \tag{7}$$

Substituting for $\mathbf{A}_1^{-1}$ from (2) in (7) and simplifying, we obtain

$$\mathbf{B}_2 = \mathbf{U}\mathbf{B}_1\mathbf{R}^{-1}. \tag{8}$$

Equations (2), (4), and (8) consitute Eq. (5a.1.6).

(B) We will show that the matrix $\mathbf{U}$ constructed above is unimodular, i.e., det $\mathbf{U}$ is a nonzero constant.

Any matrix in $\mathscr{R}^{m\times m}(D)$ can be reduced to a diagonal matrix in $\mathscr{R}^{m\times m}(D)$ by elementary operations involving only elements of $\mathscr{R}(D)$; i.e., there exist matrices $\mathbf{V}, \mathbf{W} \in \mathscr{R}^{m\times m}(D)$ so that

$$\mathbf{U} = \mathbf{V}\mathbf{C}\mathbf{W} \tag{9}$$

is true, where $\mathbf{C} \in \mathscr{R}^{m \times m}(D)$ is diagonal and $\mathbf{V}$ and $\mathbf{W}$ are unimodular. Since $\mathbf{C}$ is diagonal, it can be represented as

$$\mathbf{C} = \mathbf{P}^{-1}\mathbf{Q}, \tag{10}$$

where $\mathbf{P}$ and $\mathbf{Q}$ are diagonal and $\mathbf{P}, \mathbf{Q} \in \mathscr{R}^{m \times m}(D)$ such that P_{ii} and Q_{ii} are relatively prime. Consider Eq. (2) and, from (9), substitute for $\mathbf{U}$ in it:

$$\mathbf{A}_2 = \mathbf{VCWA}_1 = \mathbf{VP}^{-1}\mathbf{QWA}_1, \qquad \text{by using (10).} \tag{11}$$

Simplifying (11) we obtain

$$\mathbf{PV}^{-1}\mathbf{A}_2 = \mathbf{QWA}_1. \tag{12}$$

Similarly, (4) and (8) yield

$$\mathbf{PV}^{-1}\mathbf{G}_2 = \mathbf{QWG}_1 \tag{13}$$

$$\mathbf{PV}^{-1}\mathbf{B}_2 = \mathbf{QWB}_1\mathbf{R}^{-1}, \tag{14}$$

respectively. Consider Eqs. (12)–(14). $\mathbf{PV}^{-1}$ is a nonsingular greatest common left divisor of $\{\mathbf{PV}^{-1}\mathbf{A}_2, \mathbf{PV}^{-1}\mathbf{G}_2$, and $\mathbf{PV}^{-1}\mathbf{B}_2\}$ since $(\mathbf{A}_2, \mathbf{G}_2, \mathbf{B}_2)$ is relatively left prime by (A7). But $\mathbf{QW}$ is also a nonsingular greatest common left divisor of $\{\mathbf{PV}^{-1}\mathbf{A}_2, \mathbf{PV}^{-1}\mathbf{G}_2$, and $\mathbf{PV}^{-1}\mathbf{B}_1\}$ by (12)–(14) since the triple $\{\mathbf{A}_1, \mathbf{G}_1, \mathbf{B}_1\mathbf{R}^{-1}\}$ is relatively left prime by (A7). Any two greatest common divisors of the same triple are related to each other by a unimodular matrix, say $\mathbf{S}$ here. Hence,

$$\mathbf{PV}^{-1}\mathbf{S} = \mathbf{QW},$$
$$\mathbf{S} = (\mathbf{PV}^{-1})^{-1}\mathbf{QW} = \mathbf{VP}^{-1}\mathbf{QW} = \mathbf{U}, \qquad \text{by (9).}$$

Hence, $\mathbf{U}$ is unimodular.

Proof of Part (ii) of Theorem 5a.1. It is easy to verify that if the two 4-tuples $(\mathbf{A}_1, \mathbf{B}_1, \mathbf{G}_1, \boldsymbol{\rho}_1)$ and $(\mathbf{A}_2, \mathbf{B}_2, \mathbf{G}_2, \boldsymbol{\rho}_2)$ obey (5a.1.6), then they have the same spectral density $(\mathbf{S}_{yy}, \mathbf{S}_{yu})$ for a given $\mathbf{S}_{uu}$, and hence the two 4-tuples obey the equivalence relation $\mathscr{E}$. Q.E.D.

Proof of Lemma 1. Since $\boldsymbol{\rho}_1$ and $\boldsymbol{\rho}_2$ are symmetric and nonsingular, there exist nonsingular matrices $\mathbf{F}_1$ and $\mathbf{F}_2$ which diagonalize them as

$$\mathbf{F}_2\boldsymbol{\rho}_2\mathbf{F}_2^{\mathrm{T}} = \boldsymbol{\rho}_2, \qquad \mathbf{F}_1\boldsymbol{\rho}_1\mathbf{F}_1^{\mathrm{T}} = \boldsymbol{\rho}_1, \tag{15}$$

where $\boldsymbol{\rho}_1$ and $\boldsymbol{\rho}_2$ are diagonal. There exists a nonsingular symmetric matrix $\mathbf{C}$ obeying

$$\boldsymbol{\rho}_2' = \mathbf{C}\boldsymbol{\rho}_1'\mathbf{C}^{\mathrm{T}} \tag{16}$$

since $\boldsymbol{\rho}_1'$ and $\boldsymbol{\rho}_2'$ are diagonal. Substitute for $\boldsymbol{\rho}_2'$ and $\boldsymbol{\rho}_1'$ in (16) from (15):

$$\mathbf{F}_2\boldsymbol{\rho}_2\mathbf{F}_2^{\mathrm{T}} = \mathbf{CF}_1\boldsymbol{\rho}_1\mathbf{F}_1^{\mathrm{T}}\mathbf{C}^{\mathrm{T}}. \tag{17}$$

Simplifying (17), we get

$$\boldsymbol{\rho}_2 = (\mathbf{F}_2)^{-1}\mathbf{CF}_1\boldsymbol{\rho}_1(\mathbf{CF}_1)^{\mathrm{T}}(\mathbf{F}_2^{-1})^{\mathrm{T}}. \tag{18}$$

Equation (18) is identical to (1) if we define $\mathbf{R} = \mathbf{F}_2^{-1}\mathbf{CF}_1$.

B. Proof of Theorem 5b.1

Part (*i*). Let $H, H' \in \mathscr{P}_1^*$, where

$$H = (\mathbf{A}, \mathbf{B}, \mathbf{G}, \boldsymbol{\rho}) \tag{19}$$

$$H' = (\mathbf{A}', \mathbf{B}', \mathbf{G}', \boldsymbol{\rho}'). \tag{20}$$

Let $H \mathscr{E} H'$. Then there exists a unimodular matrix $\mathbf{U}(D)$ such that

$$\mathbf{A}' = \mathbf{U}\mathbf{A} \tag{21}$$

is true. For simplicity, let us restrict $\mathbf{U}(D)$ to the form

$$\mathbf{U} = \mathbf{U}_0 + \mathbf{U}_1 D. \tag{22}$$

Since $\mathbf{A}'$ and $\mathbf{A}$ are lower triangular, $\mathbf{U}(D)$ must also be lower triangular. Since $\mathbf{A}(0) = \mathbf{I}$ and $\mathbf{A}'(0) = \mathbf{I}$, (21) implies

$$\mathbf{U}(0) = \mathbf{I}. \tag{23}$$

Since $\mathbf{U}(D)$ is lower triangular,

$$\det \mathbf{U}(D) = \prod_{i=1}^{m} U_{ii}(D), \tag{24}$$

$$= \text{const}, \qquad \text{since } \mathbf{U} \text{ is unimodular.} \tag{25}$$

Combining (23)–(25), we obtain

$$U_{ii}(D) = 1, \qquad \forall i = 1, \ldots, m, \tag{26}$$

i.e., $(U_1)_{ii} = 0$, $i = 1, \ldots, m$, by (22). Let

$$\mathbf{A}(D) = \mathbf{I} + \sum_{i=1}^{m_1} \mathbf{A}_j D^j, \qquad \mathbf{A}'(D) = \mathbf{I} + \sum_{j=1}^{m_1'} \mathbf{A}_j' D^j.$$

$$\mathbf{A}'(D) = (\mathbf{I} + \mathbf{U}_1 D)\left(\mathbf{I} + \sum_{j=1}^{m_1} \mathbf{A}_j D^j\right) \qquad \text{by (21).} \tag{27}$$

The coefficient of the term D^{m_1+1}, the highest possible degree of the terms on the left-hand side of (27), is $\mathbf{U}_1 \mathbf{A}_{m_1}$:

$$\begin{aligned}(\mathbf{U}_1 \mathbf{A}_{m_1})_{ii} &= \sum_{j=1}^{m} (\mathbf{U}_1)_{ij} (\mathbf{A}_{m_1})_{ji} \\ &= \sum_{j=1}^{i-1} (U_1)_{ij} (A_{m_1})_{ji}, \qquad \text{since } \mathbf{U}_1 \text{ is triangular and obeys (26)} \\ &= 0, \qquad \text{since } \mathbf{A} \text{ is triangular.}\end{aligned} \tag{28}$$

Since $\mathbf{A}'$ obeys (A9), (27) and (28) imply

$$\mathbf{U}_1 \mathbf{A}_{m_1} = 0. \tag{29}$$

Since (29) must be true for any $\mathbf{A}_{m_1}$, we have the equality

$$\mathbf{U}_1 = \mathbf{0} \Rightarrow \mathbf{A}(D) = \mathbf{A}'(D). \qquad \text{Q.E.D.}$$

C. Proof of Theorem 5b.2

Consider the criterion function $J(\boldsymbol{\theta})$:

$$J(\boldsymbol{\theta}) = \frac{1}{2N} \sum_{t=1}^{N} \mathbf{w}^{\mathrm{T}}(t, \boldsymbol{\theta}) \boldsymbol{\rho}^{-1} \mathbf{w}(t, \boldsymbol{\theta}), \tag{30}$$

where $\boldsymbol{\theta}$ is a dummy vector similar in structure to $\boldsymbol{\theta}^0$ constructed from $\mathbf{A}_j$, $\mathbf{B}_j$, $\mathbf{G}_j, \ldots$, and $\mathbf{w}(t, \boldsymbol{\theta})$ is generated with the aid of $\boldsymbol{\theta}$ and the observation history of $\mathbf{y}$ and $\mathbf{u}$ using

$$\mathbf{w}(t, \boldsymbol{\theta}) = -\sum_{j=1}^{m_2} \mathbf{B}_j \mathbf{w}(t-j, \boldsymbol{\theta}) + \mathbf{y}(t) - \sum_{j=1}^{m_1} \mathbf{A}_j \mathbf{y}(t-j) - \sum_{j=1}^{m_3} \mathbf{G}_j \mathbf{u}(t-j). \tag{31}$$

An estimate $\hat{\boldsymbol{\theta}}(N)$ of $\boldsymbol{\theta}^0$ is obtained by minimizing $J(\boldsymbol{\theta})$. An expression for $\hat{\boldsymbol{\theta}}(N)$ is obtained by considering the quadratic approximation to $J(\boldsymbol{\theta})$ given by its Taylor series expansion

$$\begin{aligned} J(\boldsymbol{\theta}) &= J(\boldsymbol{\theta}^0) + (\nabla_{\boldsymbol{\theta}} J(\boldsymbol{\theta}^0))^{\mathrm{T}}(\boldsymbol{\theta} - \boldsymbol{\theta}^0) + \tfrac{1}{2}(\boldsymbol{\theta} - \boldsymbol{\theta}^0)^{\mathrm{T}} \nabla^2_{\boldsymbol{\theta}\boldsymbol{\theta}} J(\boldsymbol{\theta}^0)(\boldsymbol{\theta} - \boldsymbol{\theta}^0) \\ &\quad + O(|\boldsymbol{\theta} - \boldsymbol{\theta}^0|^3), \ldots. \end{aligned} \tag{32}$$

Minimizing the quadratic part of (32) with respect to $\boldsymbol{\theta}$, the following expression is obtained for $\bar{\mathbf{S}}(N)$, which is an approximation to $\hat{\boldsymbol{\theta}}(N)$:

$$\nabla^2_{\boldsymbol{\theta}\boldsymbol{\theta}} J(\boldsymbol{\theta}^0)[\bar{\mathbf{S}}(N) - \boldsymbol{\theta}^0] = \nabla_{\boldsymbol{\theta}} J(\boldsymbol{\theta}^0). \tag{33}$$

By direct differentiation of $J(\boldsymbol{\theta})$, we obtain

$$\nabla_{\boldsymbol{\theta}} J(\boldsymbol{\theta}) = \frac{1}{N} \sum_{t=1}^{N} \sum_{i,j=1}^{m} w(t, \boldsymbol{\theta})\, \nabla_{\boldsymbol{\theta}} w_j(t, \boldsymbol{\theta}) (\boldsymbol{\rho}^{-1})_{ij}, \tag{34}$$

$$\begin{aligned} \nabla^2_{\boldsymbol{\theta}\boldsymbol{\theta}} J(\boldsymbol{\theta}) &= \frac{1}{N} \sum_{t=1}^{N} \sum_{i,j=1}^{m} \{\nabla_{\boldsymbol{\theta}} w_i(t, \boldsymbol{\theta}) (\nabla_{\boldsymbol{\theta}} w_j(t, \boldsymbol{\theta}))^{\mathrm{T}} (\boldsymbol{\rho}^{-1})_{ij} \\ &\quad + w_i(t, \boldsymbol{\theta})\, \nabla_{\boldsymbol{\theta}}{}^2 w_j(t, \boldsymbol{\theta}) (\boldsymbol{\rho}^{-1})_{ij}\}, \end{aligned} \tag{35}$$

$$\nabla_{\boldsymbol{\theta}} w_i(t, \boldsymbol{\theta}) = (\nabla_{\theta_1} w_i(t, \boldsymbol{\theta}), \ldots, \nabla_{\theta_{n_0}} w_i(t, \boldsymbol{\theta}))^{\mathrm{T}}.$$

We will subsequently establish

$$\nabla_{\boldsymbol{\theta}} J(\boldsymbol{\theta}^0) \to 0 \qquad \text{with probability one (WP1)} \tag{36}$$

$$\nabla^2_{\boldsymbol{\theta}\boldsymbol{\theta}} J(\boldsymbol{\theta}^0) \text{ is positive definite WP1 for all } N > N_1. \tag{37}$$

In view of (37), we can solve (33) for $(\bar{\mathbf{S}}(N) - \boldsymbol{\theta}^0)$:

$$(\bar{\mathbf{S}}(N) - \boldsymbol{\theta}^0) = [\nabla^2_{\boldsymbol{\theta}\boldsymbol{\theta}} J(\boldsymbol{\theta}^0)]^{-1}\, \nabla_{\boldsymbol{\theta}} J(\boldsymbol{\theta}^0).$$

This expression implies that $(\bar{\mathbf{S}}(N) - \boldsymbol{\theta}^0)$ will go to zero WP1 by (36) and (37).

Proof of *(36)*. Note that the sequences $w_i(t, \boldsymbol{\theta}^0)$ and $\nabla_{\boldsymbol{\theta}} w_i(t, \boldsymbol{\theta}^0)$ are ergodic. Hence, as N tends to infinity, we have

$$\begin{aligned} \lim_{N \to \infty} \nabla_{\boldsymbol{\theta}} J(\boldsymbol{\theta}^0) &= E\left[\sum_{i,j=1}^{m} w_i(t, \boldsymbol{\theta}^0)\, \nabla_{\boldsymbol{\theta}} w_j(t, \boldsymbol{\theta}^0)((\boldsymbol{\rho}^0)^{-1})_{ij}\right] \qquad \text{WP1} \\ &= 0 \qquad \text{by (A1).} \end{aligned}$$

Proof of (37). The second term in (35) tends to zero WP1 for the same reason as $\nabla_{\boldsymbol{\theta}} J(\boldsymbol{\theta}^0)$. Hence

$$\nabla^2_{\boldsymbol{\theta\theta}} J(\boldsymbol{\theta}^0) = \frac{1}{N} \sum_{t=1}^{N} \sum_{i,j=1}^{m} \nabla_{\boldsymbol{\theta}} w_i(t, \boldsymbol{\theta}^0) (\nabla_{\boldsymbol{\theta}} w_j(t, \boldsymbol{\theta}^0))^{\mathrm{T}} (\rho^{-1})_{ij}. \tag{38}$$

Let $\mathbf{h}$ be any n_0-vector

$$\mathbf{h}^{\mathrm{T}} \nabla^2_{\boldsymbol{\theta\theta}} J(\boldsymbol{\theta}^0)\mathbf{h} = \frac{1}{N} \sum_{t=1}^{N} \sum_{i,j=1}^{m} (\mathbf{h}^{\mathrm{T}} \nabla_{\boldsymbol{\theta}} w_i(t, \boldsymbol{\theta}^0))(\mathbf{h}^{\mathrm{T}} \nabla_{\boldsymbol{\theta}} w_j(t, \boldsymbol{\theta}^0))(\rho^{-1})_{ij} \tag{39}$$

and

$$\mathbf{h}^{\mathrm{T}} \nabla_{\boldsymbol{\theta}} w_i(t, \boldsymbol{\theta}^0) \triangleq v_i(t), \qquad \mathbf{v}(t) = (v_1(t), \ldots, v_m(t))^{\mathrm{T}}.$$

By differentiating the difference equation for w_i in (31) with respect to $\boldsymbol{\theta}$ and multiplying the entire equation by a vector $\mathbf{h}^{\mathrm{T}}$ we obtain the difference equation

$$v_i(t) = f_i(\mathbf{v}(t-1), \ldots, \mathbf{v}(t-m_2)) + \mathbf{x}_i^{\mathrm{T}}\mathbf{h}, \qquad i = 1, \ldots, m, \tag{40}$$

where the f_i are scalar linear functions and $\mathbf{x}_i^{\mathrm{T}}$ is made up of components such as $\mathbf{w}(t-j)$, $\mathbf{u}(t-j)$, $\mathbf{y}(t-j)$ for various j, or

$$\mathbf{v}(t) = \mathbf{f}(\mathbf{v}(t-1), \ldots, \mathbf{v}(t-m_2)) + \mathbf{X}(t)\mathbf{h}, \tag{41}$$

where

$$\mathbf{f} = (f_1, \ldots, f_m)^{\mathrm{T}}, \qquad \mathbf{X}(t) = \begin{bmatrix} \mathbf{x}_1^{\mathrm{T}}(t) \\ \vdots \\ \mathbf{x}_m^{\mathrm{T}}(t) \end{bmatrix}.$$

Since Eq. (41) is asymptotically stable and the forcing function $\mathbf{X}(t)\mathbf{h}$ has finite mean square value in view of (5a.1.1) and associated assumptions, we have

$$\frac{1}{N} \sum_{t=1}^{N} \|\mathbf{v}(t)\|^2 \geq C_h' > 0, \qquad \forall N \geq N_1. \tag{42}$$

Now let us simplify (39). Let μ be the minimum eigenvalue of $\boldsymbol{\rho}^{-1}$. Clearly, $\mu > 0$.

$$\begin{aligned} \mathbf{h}^{\mathrm{T}} \nabla^2_{\boldsymbol{\theta\theta}} J(\boldsymbol{\theta}^0)\mathbf{h} &\geq \sum_{i,j=1}^{m} \frac{1}{N} \sum_{t=1}^{N} (\mathbf{h}^{\mathrm{T}} \nabla_{\boldsymbol{\theta}} w_i(t, \boldsymbol{\theta}^0))(\mathbf{h}^{\mathrm{T}} \nabla_{\boldsymbol{\theta}} w_j(t, \boldsymbol{\theta}^0)) \\ &\geq \sum_{i,j=1}^{m} \left(\frac{1}{N} \sum_{t=1}^{N} \|\mathbf{v}(t)\|^2 \right) \mu \geq \mu C_h \\ &> 0, \qquad \forall N \geq N_1 \quad \text{and} \quad \forall \mathbf{h} \neq 0. \end{aligned} \tag{43}$$

Thus, (37) is established.

D. Mean Square Error of the QML Estimate in the Multivariate Case

As shown in the proof of Theorem 5b.2, the second term in (35) goes to zero with probability 1 as N tends to infinity. In view of the ergodicity of the sequence $\{\nabla_{\theta_i} w_j(t, \boldsymbol{\theta}^0), t = 1, 2, \ldots\}$ for all i and j and (35), we have the expression

$$\nabla^2_{\boldsymbol{\theta}\boldsymbol{\theta}} J(\boldsymbol{\theta}^0) \to \frac{1}{N} E\left[\sum_{t=1}^{N} \sum_{i,j=1}^{m} (\nabla_{\boldsymbol{\theta}} w_i(t, \boldsymbol{\theta}^0))(\nabla_{\boldsymbol{\theta}} w_j(t, \boldsymbol{\theta}^0))^{\mathrm{T}}\right] \quad \text{with probability 1} \tag{44}$$

$$\triangleq \frac{1}{N} \mathscr{I}_N(\boldsymbol{\theta}^0, \boldsymbol{\rho}^0).$$

Hence, (33) can be written asymptotically as

$$\frac{1}{N} \mathscr{I}_N(\boldsymbol{\theta}^0, \boldsymbol{\rho}^0)(\bar{\mathbf{S}}(N) - \boldsymbol{\theta}^0) = \nabla_{\boldsymbol{\theta}} J(\boldsymbol{\theta}^0). \tag{45}$$

Since the matrix $\mathscr{I}_N$ has an inverse as established in the proof of Theorem 5b.2, Eq. (45) can be solved for $(\bar{\mathbf{S}}(N) - \boldsymbol{\theta}^0)$:

$$(\bar{\mathbf{S}}(N) - \boldsymbol{\theta}^0) = N \mathscr{I}_N^{-1}(\boldsymbol{\theta}^0, \boldsymbol{\rho}^0)\, \nabla_{\boldsymbol{\theta}} J(\boldsymbol{\theta}^0). \tag{46}$$

Multiplying (46) by its own transpose and taking expectation throughout, we obtain

$$E[(\bar{\mathbf{S}}(N) - \boldsymbol{\theta}^0)(\bar{\mathbf{S}}(N) - \boldsymbol{\theta}^0)^{\mathrm{T}}] = N^2 \mathscr{I}_N^{-1}(\boldsymbol{\theta}^0, \boldsymbol{\rho}^0) E[(\nabla_{\boldsymbol{\theta}} J(\boldsymbol{\theta}^0))(\nabla_{\boldsymbol{\theta}} J(\boldsymbol{\theta}^0))^{\mathrm{T}} | \boldsymbol{\theta}^0, \boldsymbol{\rho}^0] \mathscr{I}_N^{-1}(\boldsymbol{\theta}^0, \boldsymbol{\rho}^0). \tag{47}$$

We will evaluate the expectation term on the right-hand side of (47). Recall from (34) that

$$\nabla_{\boldsymbol{\theta}} J(\boldsymbol{\theta}^0) = \frac{1}{N} \sum_{t=1}^{N} \sum_{i,j=1}^{m} w_i(t, \boldsymbol{\theta})\, \nabla_{\boldsymbol{\theta}} w_j(t, \boldsymbol{\theta}) (\rho^0)^{-1}_{ij}. \tag{48}$$

Multiply (48) by its own transpose and take expectation:

$$E[(\nabla_{\boldsymbol{\theta}} J(\boldsymbol{\theta}^0))(\nabla_{\boldsymbol{\theta}} J(\boldsymbol{\theta}^0))^{\mathrm{T}} | \boldsymbol{\theta}^0, \boldsymbol{\rho}^0] = \frac{1}{N^2} \sum_{t_1, t_2 = 1}^{N} \sum_{i_1, i_2, j_1, j_2 = 1}^{m} (\rho^0)^{-1}_{i_1 j_1} (\rho^0)^{-1}_{i_2 j_2} \times E[w_{i_1}(t_1, \boldsymbol{\theta}^0) w_{i_2}(t_2, \boldsymbol{\theta}^0)(\nabla_{\boldsymbol{\theta}} w_{j_1}(t_1, \boldsymbol{\theta}^0)) \times (\nabla_{\boldsymbol{\theta}} w_{j_2}(t_2, \boldsymbol{\theta}^0))^{\mathrm{T}}]. \tag{49}$$

Consider the summation of those terms in (49) for which $t_1 > t_2$. Then $w_{i_1}(t_1, \boldsymbol{\theta}^0)$ is independent of all the remaining terms within the bracket of the right-hand side of (49) since they are all functions of the (past) observation history until time t_2, $t_1 > t_2$. Hence for $t_1 > t_2$,

$$\begin{aligned} &E[w_{i_1}(t_1, \boldsymbol{\theta}^0) w_{i_2}(t_2, \boldsymbol{\theta}^0)(\nabla_{\boldsymbol{\theta}} w_{j_1}(t_1, \boldsymbol{\theta}^0))(\nabla_{\boldsymbol{\theta}} w_{j_2}(t_2, \boldsymbol{\theta}^0))^{\mathrm{T}}] \\ &= E[w_{i_1}(t_1, \boldsymbol{\theta}^0)] E[w_{i_2}(t_2, \boldsymbol{\theta}^0)(\nabla_{\boldsymbol{\theta}} w_{j_1}(t_1, \boldsymbol{\theta}^0))(\nabla_{\boldsymbol{\theta}} w_{j_2}(t_1, \boldsymbol{\theta}^0))^{\mathrm{T}}] \\ &= 0, \end{aligned} \tag{50}$$

since the first term is zero. Similarly, (50) is also true for $t_2 > t_1$. Next, consider the left-hand side of (50) when $t_1 = t_2$:

$$E[w_{i_1}(t_1, \boldsymbol{\theta}^0) w_{i_2}(t_1, \boldsymbol{\theta}^0)\ \nabla_\theta w_{j_1}(t_1, \boldsymbol{\theta}^0)(\nabla_{\boldsymbol{\theta}} w_{j_2}(t_1, \boldsymbol{\theta}^0))^{\mathrm{T}}]$$
$$= E[w_{i_1}(t_1, \boldsymbol{\theta}^0) w_{i_2}(t_1, \boldsymbol{\theta}^0)] E[\nabla_{\boldsymbol{\theta}} w_{j_1}(t_1, \boldsymbol{\theta}^0)\ \nabla_{\boldsymbol{\theta}} w_{j_2}(t_1, \boldsymbol{\theta}^0)^{\mathrm{T}}]$$

since $w_i(t, \boldsymbol{\theta}^0)$ is independent of the past history $\mathbf{y}(j), j \leq t$ and $\mathbf{u}(j), j \leq t$ and hence is independent of $\nabla_{\boldsymbol{\theta}} w_j(t_1, \boldsymbol{\theta})$,

$$= \rho^0_{i_1, i_2} E[\nabla_{\boldsymbol{\theta}} w_{j_1}(t_1, \boldsymbol{\theta}^0)(\nabla_{\boldsymbol{\theta}} w_{j_2}(t_1, \boldsymbol{\theta}^0))^{\mathrm{T}}]. \tag{51}$$

Substitute (50) and (51) into (49) and regroup the terms:

$$\text{left-hand side of (49)} = \frac{1}{N^2} \sum_{t_1=1}^{N} \sum_{j_1, j_2=1}^{m} E[\nabla_{\boldsymbol{\theta}} w_{j_1}(t_1, \boldsymbol{\theta}^0)(\nabla_{\boldsymbol{\theta}} w_{j_2}(t_1, \boldsymbol{\theta}^0))^{\mathrm{T}}]$$
$$\times \left\{ \sum_{i_1, i_2=1}^{m} (\rho^0)^{-1}_{j_1 i_1} (\rho^0)_{i_1 i_2} (\rho^0)^{-1}_{i_2 j_2} \right\}$$
$$\times \frac{1}{N} \cdot \frac{1}{N} \sum_{t=1}^{N} \sum_{j_1, j_2=1}^{m} E[\nabla_{\boldsymbol{\theta}} w_{j_1}(t, \boldsymbol{\theta}^0)(\nabla_{\boldsymbol{\theta}} w_{j_2}(t, \boldsymbol{\theta}^0))^{\mathrm{T}}] (\rho^0)^{-1}_{j_1 j_2}$$
$$\to \frac{1}{N^2} \mathscr{I}_N(\boldsymbol{\theta}^0, \boldsymbol{\rho}^0). \tag{52}$$

Substituting (52) in (47), we get

$$E[\bar{\mathbf{S}}(N) - \boldsymbol{\theta}^0)(\bar{\mathbf{S}}(N) - \boldsymbol{\theta}^0)^{\mathrm{T}} | \boldsymbol{\theta}^0, \boldsymbol{\rho}^0]$$
$$\approx E\left[\sum_{t=1}^{N} \sum_{i,j=1}^{m} (\nabla_{\boldsymbol{\theta}} w_i(t, \boldsymbol{\theta}^0))(\nabla_{\boldsymbol{\theta}} w_j(t, \boldsymbol{\theta}^0))^{\mathrm{T}} (\rho^0)^{-1}_{ij}\right]^{-1}$$
$$\triangleq \mathscr{I}_N^{-1}(\boldsymbol{\theta}^0, \boldsymbol{\rho}^0). \tag{53}$$

Q.E.D.

E. Proof of Theorem 5c.1

To simplify the discussion we will assume $\mathbf{G} = 0$ since the matrix triple $(\mathbf{A}, \mathbf{B}, \boldsymbol{\rho})$ represents all the important properties. The generalization to the case of $\mathbf{G} \neq 0$ is relatively easy.

Part (i). Suppose part (i) is not true, i.e., $(\mathbf{A}', \mathbf{B}')$, $(\mathbf{A}'', \mathbf{B}'')$ are distinct and lead to the same spectral density $\mathbf{S}_{yy}$. It is relatively easy to show that there must exist two diagonal nonsingular polynomial matrices $\mathbf{H}$ and $\mathbf{K}$ such that

$$\mathbf{HA}' = \mathbf{KA}'', \qquad \mathbf{HB}' = \mathbf{KB}'' \tag{54}$$

is true.

Since $\mathbf{A}'$ and $\mathbf{B}'$ obey (A7′), $\mathbf{H}$ is a diagonal GCLD (greatest common left divisor) of $\mathbf{HA}'$ and $\mathbf{HB}'$. But by (54), $\mathbf{K}$ is a diagonal GCLD of $\mathbf{HA}'$ and $\mathbf{HB}'$ since $\mathbf{A}''$ and $\mathbf{B}''$ obey (A7′). Since $\mathbf{H}$ and $\mathbf{K}$ are GCLD's, they are related by a unimodular matrix $\mathbf{U}$ as in

$$\mathbf{H} = \mathbf{KU}. \tag{55}$$

(54) can be rewritten using (55):

$$\mathbf{KUB}' = \mathbf{KB}'', \qquad \text{or} \quad \mathbf{UB}' = \mathbf{B}'', \tag{56}$$

since $\mathbf{K}$ is nonsingular.

Since $\mathbf{U}$ is diagonal by Eq. (55) and unimodular, it must be composed of constant elements. This, combined with (56) and the fact that $\mathbf{B}'(0) = \mathbf{B}''(0) = \mathbf{I}$, implies that $\mathbf{U} = \mathbf{I}$ or $\mathbf{K} = \mathbf{H}$. Since $\mathbf{K}$ and $\mathbf{H}$ are nonsingular, we must have

$$\mathbf{A}' = \mathbf{A}'', \qquad \mathbf{B}' = \mathbf{B}''.$$

F. Proof of Theorem 5c.2

The proof of part (i) is identical to that of part (i) of Theorem 5c.1.

Part (ii). Consider the system difference equation

$$\mathbf{y}(t) = \mathbf{T}(D)\mathbf{w}(t), \qquad \mathbf{T} = \mathbf{A}^{-1}\mathbf{B}, \qquad \mathbf{T}(0) = \mathbf{I}. \tag{57}$$

Let $\mathbf{w}'(t) = \mathbf{A}_0\mathbf{w}(t)$, where $\mathbf{A}_0$ is defined in (5c.2.1). Then $\mathbf{w}'(t)$ has a covariance matrix $\boldsymbol{\rho}'$. Hence rewrite Eq. (57) as

$$\mathbf{y}(t) = \mathbf{T}'(D)\mathbf{w}'(t), \tag{58}$$

where $\mathbf{T}'(D) = (\mathbf{A}(D))^{-1}\mathbf{B}(D)\mathbf{A}_0^{-1}$. We will apply the algorithm in Section 5c.1 to obtain the required form from $\mathbf{T}'(D)$. Let $(T')_{ij}^{-1} = e_{ij}'/f_{ij}'$, where e_{ij}', f_{ij}' are relatively prime. Define $b_{ii}''(D) = \gamma_i \,\text{LCM}(f_{i1}, \ldots, f_{im})$, where γ_i is chosen so that $b_{ii}''(0) = 1$. Then $\mathbf{B}''(D) = \text{diag}(b_{11}''(D), \ldots, b_{mm}''(D))$ and $\mathbf{A}''(D) = (\mathbf{T}'(D))^{-1}\mathbf{B}''(D)$. One can see by inspection that the 4-tuple $(\mathbf{A}'', \mathbf{B}'', \mathbf{0}, \boldsymbol{\rho}'')$ leads to the same spectral density as $(\mathbf{A}, \mathbf{B}, \mathbf{0}, \boldsymbol{\rho})$.

G. Proof of Theorem 5d.1

As before we will set $\mathbf{G} = 0$. Suppose both $H = (\mathbf{A}, \mathbf{B}, \mathbf{0}, \boldsymbol{\rho})$ and $H^1 = (\mathbf{A}^1, \mathbf{B}^1, \mathbf{0}, \boldsymbol{\rho}^1) \in \mathscr{P}_2$ and $H \mathscr{E} H^1$. Then the 2-tuples should be connected by a unimodular matrix $\mathbf{U}(D)$:

$$\mathbf{A}^1(D) = \mathbf{U}(D)\mathbf{A}(D), \qquad \mathbf{B}^1(D) = \mathbf{U}(D)\mathbf{B}(D). \tag{59}$$

Let

$$\mathbf{A}(D) = \mathbf{I} + \mathbf{A}_1 D + \cdots + \mathbf{A}_{m_1} D^{m_1}, \qquad \mathbf{B}(D) = \mathbf{I} + \cdots + \mathbf{B}_{m_2} D^{m_2}.$$

The primed matrices are similarly defined. Let $\mathbf{U}(D) = \mathbf{C}_0 + \mathbf{C}_1 D + \cdots + \mathbf{C}_r D^r$. Since $H, H^1 \in \mathscr{P}_2$, (59) yields

$$\mathbf{A}^1(0) = \mathbf{U}(0)\mathbf{A}(0) = \mathbf{I} \Rightarrow \mathbf{U}(0) = \mathbf{I} \Rightarrow \mathbf{C}_0 = \mathbf{I}. \tag{60}$$

The unimodularity of $\mathbf{U}(D)$ implies that $\det \mathbf{C}_r = 0$. Suppose $\mathbf{C}_r \neq 0$, and the coefficients of the highest degree polynomials in $\mathbf{A}^1(D)$ and $\mathbf{B}^1(D)$ are $\mathbf{C}_r\mathbf{A}_{m_1}$

and $\mathbf{C}_r\mathbf{B}_{m_1}$. But $[\mathbf{C}_r\mathbf{A}_{m_1} \vdots \mathbf{C}_r\mathbf{B}_{m_1}]$ cannot have maximal rank and hence cannot satisfy (A8), since $\mathbf{C}_r$ is singular. This contradictis the fact that H and H^1 belong to $\hat{\mathscr{P}}_2$. Hence $\mathbf{C}_r = \mathbf{0}$. Similarly, we can show that $\mathbf{C}_j = \mathbf{0}$, $j = 1, \ldots, r-1$. Hence $\mathbf{A} = \mathbf{A}^1$, $\mathbf{B} = \mathbf{B}^1$, and $H = H^1$.

Problems

1. Consider the scalar process y obeying

$$\mathbf{x}(t+1) = \mathbf{A}\mathbf{x}(t) + \mathbf{g}w(t), \qquad y(t) = \mathbf{h}^{\mathrm{T}}\mathbf{x}(t),$$

where the scalar $w(\cdot)$ obeys (A1) and $\mathbf{x}$, an n-vector. Let $\mathscr{P}$ be the set of all triples $(\mathbf{A}, \mathbf{h}, \mathbf{g})$. Let $\mathscr{P}_1 = \{(\mathbf{A}, \mathbf{h}, \mathbf{g}) | (\mathbf{A}, \mathbf{h}, \mathbf{g}) \in \mathscr{P}, \mathbf{A}$ is in companion form and $\mathbf{h} = (1, 0, \ldots, 0)^{\mathrm{T}}\}$. Let two members of $\mathscr{P}$ be considered equivalent if they lead to the same spectral density for y. Show that $\mathscr{P}_1$ is a canonical subset of $\mathscr{P}$; i.e., Theorem 5b.1 is satisfied by $\mathscr{P}_1$.

2. Let

$$\mathbf{y}(t) = \begin{bmatrix} \dfrac{(D+2)}{(D+0.5)^2} & \dfrac{D+3}{D+0.5} \\ \dfrac{D+4}{D+0.6} & \dfrac{D+1}{D+0.8} \end{bmatrix} \mathbf{w}(t).$$

Find the difference equations for $\mathbf{y}$ in canonical forms I, II, and III and the pseudocanonical form II (if possible), and the canonical form in which $\mathbf{A}(D)$ is diagonal, $\mathbf{A}(0) = \mathrm{I}$ and $\mathbf{B}(0) = \mathrm{I}$. Assume $\boldsymbol{\rho} = \operatorname{cov} \mathbf{w} = \left[\begin{smallmatrix} 2 & 1 \\ 1 & 4 \end{smallmatrix}\right]$.

3. Suppose we slightly perturb the pole of the (1, 1) element of the transfer matrix in Problem 2. Find the linearized equation for the deviation between $\mathbf{y}(t)$ and $\mathbf{y}'(t)$, $\mathbf{y}'(t)$ being the perturbed output.

4. Consider the following stationary sequence $\mathbf{y}(\cdot)$ obeying

$$y_1(t) = \theta_1 y_1(t-1) + \theta_2 y_2(t-1) + w_1(t),$$

where $w_1(\cdot)$ obeys (A1) with variance P_1 and $w_1(t)$ is independent of $\mathbf{y}(t-j)$, $j \geq 1$, and $\mathbf{y}(t) = (y_1(t), y_2(t))^{\mathrm{T}}$. Find the greatest possible lower bounds on the variance of the unbiased estimates of θ_1 and θ_2 in the following three cases given $E[y_1{}^2(t)]$ and $E[y_1(t)y_2(t)]$: (i) y_2 and w_1 are mutually independent processes; (ii) $y_2(\cdot)$ obeys

$$y_2(t) = \theta_3 y_1(t-1) + \theta_4 y_2(t-1) + w_2(t),$$

where $w_2(\cdot)$ obeys (A1), $w_2(t)$ is independent of $\mathbf{y}(t-1)$ for all t, and $E[w_1(t)w_2{}^{\mathrm{T}}(j)] = \rho_{12}\delta_{tj}$, where ρ_{12} is nonzero; (iii) same as case (ii) except that ρ_{12} is zero.

5. Consider the following triple $(\mathbf{A}, \mathbf{B}, \mathbf{G})$ with $\mathbf{G} = 0$

$$\mathbf{A} = \begin{bmatrix} 1 + 0.5D & 0 \\ -D^2 - 0.5D^3 & 1 - 0.5D - 0.25D^2 + 0.125D^3 \end{bmatrix},$$

$$\mathbf{B} = \begin{bmatrix} 1 - 0.5D & 0 \\ 0 & 1 - 0.5D \end{bmatrix}$$

Verify that $(\mathbf{A}, \mathbf{B}, \mathbf{G})$ obeys (A7′) but not (A7), i.e., the Smith form of $(\mathbf{A}, \mathbf{B})$ is not of the form $[\mathbf{1} \quad \mathbf{0}]$. Next consider any other polynomial triple $(\hat{\mathbf{A}}, \hat{\mathbf{B}}, \hat{\mathbf{G}})$ and a polynomial $\mathbf{U}$ related to $(\mathbf{A}, \mathbf{B}, \mathbf{G})$ via (5a.1.2). Show that if we restrict $\mathbf{U}$ to diagonal matrices, then $\mathbf{U}$ is unimodular, but $\mathbf{U}$ is not necessarily unimodular if no restriction is imposed on $\mathbf{U}$.

Chapter VI | Estimation in Autoregressive Processes

Introduction

In the preceding chapters, we discussed the structure of the multivariate and univariate systems and the conditions under which the parameters can be recovered from the given observation history.

In this chapter we consider the estimation of the coefficients in the difference equations of the given process, based only on the observations of the given process and related processes. Even though the coefficient vector is unknown, it is known to belong to a certain set. In particular, the coefficient vector can be either a constant or obey a known random process characterized by some unknown parameters. The given process $\mathbf{y}$ can be described entirely by a single equation or a system of equations. A restriction on the difference equation is that it should not have any moving average terms. This restriction will be relaxed in subsequent chapters. Also, the difference equation should be linear in the unknown coefficients, but it could be nonlinear in other variables.

We will also consider the problems of predicting the value of the process one-step-ahead by using all the available observations at the current time instant. The solution to the prediction problem of Chapter II was based on the knowledge of the coefficients in the difference equation obeyed by the process. We will develop a prediction technique that does not require knowledge of the parameters in the difference equation.

There are three principal approaches for parameter estimation, namely the maximum likelihood (CML), Bayesian, and limited information approaches. In the likelihood and Bayesian approaches, the joint probability distribution of both the input $\mathbf{u}$ and the disturbance $\mathbf{w}$ must be known. Between the two, the Bayesian approach has the ability to absorb any available prior knowledge on the parameters. In the limited information estimates, only the probability density of the disturbance $\mathbf{w}$ must be known. Moreover, in multiple output systems, the limited information estimates of the unknowns in each equation can be determined separately, unlike the Bayesian and maximum likelihood methods, and this aspect is of considerable computational importance.

There are three types of computational methods for evaluating the estimates. The conditional maximum likelihood estimate of the coefficients in a multivariate system in which $\boldsymbol{\rho}$, the covariance matrix of the disturbance, is unknown can only be evaluated by an iterative procedure. This iterative procedure is not usually implementable in real time. In other words, if $\hat{\boldsymbol{\theta}}(N)$ denotes the estimate of the unknown parameter based on the observation until time N, then $\hat{\boldsymbol{\theta}}(N)$ cannot be updated into $\hat{\boldsymbol{\theta}}(N+1)$ in a unit of time as N tends to infinity by any finite computer. However, the CML estimates of the parameters of a single output system or the limited information estimates of the multivariate system can be computed in real time. Next we consider the problem of fitting a number of different AR models of different orders to the same empirical data.

Finally, we consider the problem of forecasting with a system whose difference equation involves unknown parameters. We inquire whether the solution to this problem can be expressed in two parts, namely (i) the estimation of the unknown parameters and (ii) forecasting the process by replacing the unknown parameters in it by their estimates.

6a. Maximum Likelihood Estimators

6a.1. Statement of the Problem

6a.1.1. Single Output Process

Consider the difference equation $\mathscr{s}$ in terms of the scalar variable y:

$$\mathscr{s}: f_0[y(t)] = \mathbf{z}^{\mathrm{T}}(t-1)\boldsymbol{\theta} + w(t) \tag{6a.1.1}$$

$$\mathbf{z}(t-1) = [f_1(y(t-1)), \ldots, f_{m_1}(y(t-m_1)), \mathbf{u}^{\mathrm{T}}(t-1), \ldots, \mathbf{u}^{\mathrm{T}}(t-m_3), \psi_1(t-1), \ldots, \psi_{l_2}(t-1)]^{\mathrm{T}}$$

$$\boldsymbol{\theta} = n\text{-vector}, \qquad n = m_1 + l_1 m_3 + l_2.$$

Often we use the notation $f_0(t)$ for $f_0(y(t))$. In (6a.1.1), f_i, $i = 0, \ldots, m_1$, are known functions, $\mathbf{u}(\cdot)$ is the observable vector input of dimension l_1, the $\psi_i(\cdot)$ are the deterministic scalar trends, and $w(\cdot)$ is a zero mean Gaussian disturbance. A class of univariate dynamic models is a triplet $[\mathscr{s}, \mathscr{H}, \Omega]$ where $\mathscr{s}$ is the scalar difference equation (6a.1.1) involving the parameter $\boldsymbol{\theta}$, $\mathscr{H}$ is the set of all the values that can be assumed by $\boldsymbol{\theta}$, and Ω is the set of values $[0, \infty]$ that can be assumed by ρ, the covariance of the noise w in $\mathscr{s}$.

We will impose the usual assumptions (A1)–(A4) on w, $\mathbf{u}$, and $\boldsymbol{\psi}$. In addition we assume that the distribution of w is normal. We are not assuming the normality of y.

Finally, the difference equation will satisfy (A5′) for every $\boldsymbol{\theta} \in \mathscr{H}$.

A5′. The homogeneous part of the difference equation (6a.1.1) is asymptotically stable.

Under these assumptions it is easy to show that there is at most one model in the class $[\mathscr{s}, \mathscr{H}, \Omega]$ for any given process. We will indicate methods of estimation in which the assumption of independence of $\mathbf{u}$ and w is not needed. Assumption (A4) excludes linear trends. However, many of the methods of estimation are valid even if linear trend terms are present in the equation, i.e., (A4) can be replaced by (A4′) or (A4″). When all the functions f_i in (6a.1.1) are identical for all i, then (A5′) reduces to assumption (A5), which states that all zeros of the corresponding polynomial are outside the unit circle. Otherwise, we have to check each equation separately to ensure the validity of (A5′).

Suppose at any instant N, we have accumulated the following set of observations of the processes y and $\mathbf{u}$, y belonging to the class $[\mathscr{s}, \mathscr{H}, \Omega]$:

$$\xi(N) = \{y(N), \ldots, y(1), y(0), \ldots, y(-m_1), \mathbf{u}(N-1), \ldots, \mathbf{u}(-m_3)\} \in \mathscr{R}^{N'}.$$

Let $\boldsymbol{\theta}^0$ be the parameter vector characterizing the particular model in the class $[\mathscr{s}, \mathscr{H}, \Omega]$ obeyed by the observed process y, and let ρ^0 be the corresponding covariance of the noise. By the estimators $\hat{\boldsymbol{\theta}}$ of $\boldsymbol{\theta}^0$ and $\hat{\rho}$ of ρ^0, based on $\xi(N)$, we mean the maps

$$\hat{\boldsymbol{\theta}}: \mathscr{R}^{N'} \to \mathscr{H}, \qquad \hat{\rho}: \mathscr{R}^{N'} \to \Omega.$$

We will use the word estimator to mean the map or function and the word estimate to represent a particular value of the function. Our intention is to find estimators $\hat{\boldsymbol{\theta}}$ and $\hat{\rho}$ according to some suitable criterion.

6a.1.2. Vector Processes

Let $\mathscr{M}$ stand for a system of difference equations for the m-vector $\mathbf{y}$:

$$\mathbf{y}(t) = [y_1(t), \ldots, y_m(t)]^{\mathrm{T}}, \qquad \mathbf{w}(t) = [w_1(t), \ldots, w_m(t)]^{\mathrm{T}}$$

$$\mathscr{M}: f_{0i}[y_i(t)] = \mathbf{z}_i^{\mathrm{T}}(t-1)\boldsymbol{\theta}_i + w_i(t), \qquad i = 1, 2, \ldots, m, \tag{6a.1.2}$$

or

$$\mathbf{f}_0[\mathbf{y}(t)] = \mathbf{Z}(t-1)\boldsymbol{\theta} + \mathbf{w}(t), \tag{6a.1.3}$$

where

$$\mathbf{f}_0[\mathbf{y}(t)] = [f_{01}(y_1(t)), \ldots, f_{0m}(y_m(t))]^{\mathrm{T}} \triangleq \mathbf{f}_0(t).$$

f_{0i}, $i = 1, \ldots, m$, are known functions of time, $\boldsymbol{\theta}_i$ is a vector of dimension n_i, $i = 1, \ldots, m$, $\boldsymbol{\theta} = [\boldsymbol{\theta}_1^{\mathrm{T}}, \ldots, \boldsymbol{\theta}_m^{\mathrm{T}}]$ is an n_0-vector, $\boldsymbol{\rho}^0 = E[\mathbf{w}(t)\mathbf{w}^{\mathrm{T}}(t)]$, and $n_0 = \sum_{i=1}^{m} n_i$

$$\mathbf{Z}(t-1) = \begin{bmatrix} \mathbf{z}_1^{\mathrm{T}}(t-1) & & & \\ & \ddots & & \\ & & \mathbf{z}_2^{\mathrm{T}}(t-1) & \\ & & & \ddots \\ & & & & \mathbf{z}_m^{\mathrm{T}}(t-1) \end{bmatrix} = m \times n_0 \text{ matrix.} \tag{6a.1.4}$$

The vector $\mathbf{z}_i(t-1)$ is of dimension n_i and is made up of components of $\mathbf{f}_1(y(t-1)), \ldots, \mathbf{f}_{m_1}(y(t-m)), \mathbf{u}(t-1), \ldots, \mathbf{u}(t-m_3), \psi_1(t-1), \ldots, \psi_{l_2}(t-1)$. A class of multivariate dynamic models is a triplet $(\mathscr{M}, \mathscr{H}, \Omega)$ where $\mathscr{M}$ is the equation in (6a.1.3) involving a vector $\boldsymbol{\theta}$ of dimension n_0, $\mathscr{H}$ is the set of all values that can be assumed by $\boldsymbol{\theta}$, and Ω is the set of all the $(m \times m)$ positive-definite covariance matrices of the noise $\mathbf{w}$ of Gaussian distribution.

Assumptions (A1)–(A4) will be imposed on the inputs $\mathbf{w}$, $\mathbf{u}$, and $\boldsymbol{\psi}$ as stated earlier. On many occasions assumption (A4) can be relaxed into (A4′) or (A4″). In addition, the set $\mathscr{H}$ is defined so that the equation $\mathscr{M}$ with any $\boldsymbol{\theta} \in \mathscr{H}$ obeys (A5′) with (6a.1.3) replacing (6a.1.1). With these assumptions, for any given process $\mathbf{y}$, there can be at most one model in the class $[\mathscr{M}, \mathscr{H}, \Omega]$. The estimators of $\boldsymbol{\theta}^0$ and $\boldsymbol{\rho}^0$, the parameter vector and covariance matrix characterizing the difference equation of the particular process $\mathbf{y}$, can also be defined as in single equation systems.

We will now describe two versions of maximum likelihood estimator and then a third one in Section 6c.

6a.2. Full Information Maximum Likelihood (FIML) Estimators

All the objective knowledge contained in the observation set $\boldsymbol{\xi}(N)$ about the unknowns $\boldsymbol{\theta}^0$ and $\boldsymbol{\rho}^0$ is contained in the joint probability density function of the given observations $\boldsymbol{\xi}(N)$ and the sets $\mathscr{H}$ and Ω. Let

$$p(\boldsymbol{\xi}(N)) = g_1(\boldsymbol{\xi}(N), \boldsymbol{\theta}^0, \boldsymbol{\rho}^0),$$

where g_1 is an explicit function of the arguments. According to the likelihood principle, the values $\boldsymbol{\theta}$ and $\boldsymbol{\rho}$ which maximize the logarithm of the probability density p, the so-called log likelihood function, should be considered as estimates of $\boldsymbol{\theta}^0$ and $\boldsymbol{\rho}^0$:

$$(\boldsymbol{\theta}_1^*(N), \boldsymbol{\rho}_1^*(N)) = \arg\Big[\sup_{\boldsymbol{\theta}\in\mathscr{H},\, \boldsymbol{\rho}\in\Omega} \ln g_1(\boldsymbol{\xi}(N), \boldsymbol{\theta}, \boldsymbol{\rho})\Big]\cdot$$

$\boldsymbol{\theta}_1^*$ and $\boldsymbol{\rho}_1^*$ are the FIML estimates of $\boldsymbol{\theta}^0$ and $\boldsymbol{\rho}^0$.

In general, it is difficult to determine the FIML estimate. For instance, consider the following case when $\mathbf{u}$ is *absent* and y *is a scalar*:

$$\begin{aligned} p(\boldsymbol{\xi}(N)) &= p(y(N), \ldots, y(-m_1)) \\ &= \Big[\prod_{t=1}^{N} p(y(t)\,|\,y(t-1), \ldots, y(-m_1))\Big] p(y(0), \ldots, y(-m_1)) \\ &= \Big[\prod_{t=1}^{N} (2\pi\rho^0)^{-1/2} \exp\{-\tfrac{1}{2}(f_0(y(t)) - \mathbf{z}^{\mathrm{T}}(t-1)\boldsymbol{\theta})^2/\rho_0\}\Big] \\ &\quad \times p(y(0), y(1), \ldots, y(-m_1)). \end{aligned}$$

The probability density $p(y(0), \ldots, y(-m_1))$ is usually a complicated function of $y(0), \ldots, y(-m_1)$, $\boldsymbol{\theta}^0$, and ρ^0 as indicated above; therefore, the determination of FIML estimate is difficult even under the assumption that y is stationary.

If the process y is nonstationary, the precise probability density function of the observations $y(0), \ldots, y(-m_1)$ is often unknown. In addition, when the function f_0 in (6a.1.1) is not linear, it is difficult to find the stationary probability distribution of y and consequently $p(y(0), \ldots, y(-m_1))$ is not known. Furthermore, $\mathbf{u}$ and w may not be independent (unlike the present assumption) and in such cases, the joint probability density of $\mathbf{u}$ and w is usually unknown.

Consequently, the function g_1 is not known in such cases. For these reasons, this FIML estimate is rarely used.

6a.3. Conditional Maximum Likelihood (CML) Estimator in Univariate Systems

6a.3.1. The Form of the CML Estimator

Let $\boldsymbol{\xi}(N) = (\boldsymbol{\xi}_1(N), \boldsymbol{\xi}_2(N))$,

$$\boldsymbol{\xi}_1(N) = (y(N), \ldots, y(1))$$
$$\boldsymbol{\xi}_2(N) = (y(0), \ldots, y(-m_1), \mathbf{u}(N-1), \ldots, \mathbf{u}(-m_3)),$$

Let $p(\boldsymbol{\xi}_1(N)|\boldsymbol{\xi}_2(N)) = g_2(\boldsymbol{\xi}(N), \boldsymbol{\theta}^0, \rho^0)$ and

$$(\boldsymbol{\theta}_2^*, \rho_2^*) = \arg\left[\sup_{\boldsymbol{\theta}\in\mathscr{H},\, \rho\in\Omega} \ln g_2(\boldsymbol{\xi}(N), \boldsymbol{\theta}, \rho)\right]. \tag{6a.3.1}$$

The required maximization in (6a.3.1) can be carried out relatively easily since $\ln g_2$ is a quadratic function in $\boldsymbol{\theta}$:

$$\ln g_2(\boldsymbol{\xi}, \boldsymbol{\theta}, \rho) = -\frac{N}{2}\ln(2\pi\rho) - \tfrac{1}{2}\sum_{t=1}^{N}(f_0(t) - \boldsymbol{\theta}^{\mathrm{T}}\mathbf{z}(t-1))^2/\rho$$

$$\frac{\partial \ln g_2(\boldsymbol{\xi}, \boldsymbol{\theta}, \rho)}{\partial \boldsymbol{\theta}} = 0 \rightarrow \sum_{t=1}^{N} f_0(t)\mathbf{z}(t-1) = \left(\sum_{t=1}^{N}\mathbf{z}(t-1)\mathbf{z}^{\mathrm{T}}(t-1)\right)\boldsymbol{\theta}.$$

Hence,

$$\boldsymbol{\theta}_2^* = \left(\sum_{t=1}^{N}\mathbf{z}(t-1)\mathbf{z}^{\mathrm{T}}(t-1)\right)^{-1}\sum_{t=1}^{N} f_0(t)\mathbf{z}(t-1). \tag{6a.3.2}$$

Replacing $\boldsymbol{\theta}$ in $g_2(\boldsymbol{\xi}, \boldsymbol{\theta}, \boldsymbol{\rho})$ by $\boldsymbol{\theta}_2^*$, we get the so-called *concentrated log likelihood function* $\ln g_3(\boldsymbol{\xi}, \rho)$:

$$\ln g_3(\boldsymbol{\xi}, \rho) \triangleq \ln g_2(\boldsymbol{\xi}, \boldsymbol{\theta}_2^*, \rho) = -\frac{N}{2}\ln 2\pi\rho - \frac{1}{2\rho}\sum_{t=1}^{N}(f_0(t) - (\boldsymbol{\theta}_2^*)^{\mathrm{T}}\mathbf{z}(t-1))^2.$$

We can maximize $\ln g_3(\boldsymbol{\xi}, \rho)$ with respect to ρ by setting its first derivative to zero:

$$\frac{\partial \ln g_3(\boldsymbol{\xi}, \rho)}{\partial \rho} = 0 \rightarrow \frac{N}{\rho} - \frac{1}{\rho^2}\sum_{t=1}^{N}(f_0(t) - (\boldsymbol{\theta}_2{}^*)^{\mathrm{T}}\mathbf{z}(t-1))^2 = 0$$

or

$$\rho_2^* = \frac{1}{N}\sum_{t=1}^{N}(f_0(t) - (\boldsymbol{\theta}_2^*)^{\mathrm{T}}\mathbf{z}(t-1))^2. \tag{6a.3.3}$$

$\boldsymbol{\theta}_2^*$ and ρ_2^* are the CML estimates of $\boldsymbol{\theta}^0$ and ρ^0. When the input $\mathbf{u}$ is absent, the function g_2 is relatively easy to compute, as seen above. If $\mathbf{u}$ is present and is independent of w, the function g_2 has a relatively simple representation in that case also. If the relationship between $\mathbf{u}$ and w is *not* known, g_2 is not known and hence CML estimates cannot be computed.

Let us consider the relationship between FIML and CML estimators:

$$
\begin{aligned}
p(\xi_1(N)|\xi_2(N)) &= \prod_{t=1}^{N} p(y(t)|y(t-1),\ldots,y(1),\xi_2(N)) \\
&= \prod_{t=1}^{N} p(y(t)|y(t-1),\ldots,y(1),\mathbf{u}(N-1),\ldots, \\
&\qquad \mathbf{u}(-m_3),y(0),\ldots,y(-m_1)) \\
&= \prod_{t=1}^{N} p(y(t)|y(t-1),\ldots,y(1),\mathbf{u}(t-1),\ldots, \\
&\qquad \mathbf{u}(-m_3),y(0),\ldots,y(-m_1))
\end{aligned}
$$

by the independence of $\mathbf{u}$ and w. Therefore

$$
p(\xi_1(N)|\xi_2(N)) = \left(\frac{1}{(2\pi\rho^0)^{1/2}}\right)^N \exp\left[-\frac{1}{2\rho^0}\sum_{t=1}^{N}(f_0(y(t)) - (\boldsymbol{\theta}^0)^{\mathrm{T}}\mathbf{z}(t-1))^2\right].
$$

The main difference between the FIML and CML estimates is that the effects of the initial observations $y(0),\ldots,y(-m_1)$ are partially ignored in the latter. Since we are dealing with an asymptotically stable system, this approximation cannot influence the asymptotic behavior and as such FIML and CML estimates have the same asymptotic properties, provided $\mathbf{u}$ is independent of w.

When dealing with small samples, it is tempting to state that the FIML estimator is superior to the CML estimator according to a suitable criterion such as the variance of the estimator. However, the empirical evidence for this statement is not conclusive.

To illustrate the difference in computation of the FIML and CML estimates we will consider the following example (Anderson, 1971).

Example 6a.1. Consider a first-order AR process

$$
\begin{aligned}
y(t) &= \theta y(t-1) + w(t) \\
\mathscr{H} &= \{\theta: |\theta| < 1\}, \qquad \Omega = \{\rho: 0 < \rho < \infty\}.
\end{aligned}
$$

For computing the FIML estimate we assume that $y(\cdot)$ is stationary and that w is Gaussian:

$$
p[y(0)] \sim N[0, \rho^0/(1-\theta^2)]
$$

$$
\begin{aligned}
p[y(N),\ldots,y(1),y(0)] &= \prod_{t=1}^{N} p[y(t)|y(t-1),\ldots,y(0)]p(y(0)) \\
&= \left[\prod_{t=1}^{N}\frac{1}{(2\pi\rho^0)^{1/2}}\exp\left\{-\frac{1}{2}\frac{(y(t)-\theta^0 y(t-1))^2}{\rho^0}\right\}\right] \\
&\quad \times \frac{1}{[2\pi\rho^0/(1-(\theta^0)^2)]^{1/2}}\exp\left\{-\frac{1}{2}\frac{y^2(0)}{\rho^0/(1-(\theta^0)^2)}\right\}
\end{aligned}
$$

$$
\begin{aligned}
\ln g_1(\xi(N),\theta,\rho) &= \frac{-(N+1)}{2}\ln 2\pi - \left(\frac{N+1}{2}\right)\ln\rho + \tfrac{1}{2}\ln(1-\theta^2) \\
&\quad - \frac{1}{2\rho}\sum_{t=1}^{N}(y(t)-\theta y(t-1))^2 - \frac{1}{2\rho}y^2(0)(1-\theta^2).
\end{aligned}
$$

In view of the complexity of the function g_1, it is difficult to obtain an explicit algebraic expression for the maximizing values of θ and ρ even in this simple problem.

The *CML estimate* is, in contrast, easy to determine algebraically:

$$\theta^* = \sum_{t=1}^{N} y(t-1)y(t) \Big/ \sum_{t=1}^{N} y^2(t-1)$$

$$\rho^* = \frac{1}{N}\sum_{t=1}^{N} (y(t) - \theta^* y(t-1))^2.$$

6a.3.2. Properties of the CML Estimator†

(1) **Existence and Consistency.** In Section 4d we proved that the inverse of the matrix in (6a.3.2) exists for sufficiently large N and that the estimator $\boldsymbol{\theta}_2^*(N)$ tends to the value $\boldsymbol{\theta}^0$ in the mean square sense. Without invoking the normality assumption, the consistency of the estimator ρ_2^* can also be proved in a similar manner.

(2) **The Mean and Variance of the Estimator.** It was shown in Chapter IV that the following expression is valid for the asymptotic mean and covariance of $\boldsymbol{\theta}_2^*(N)$ under the same weakened conditions as in property (1) above:

$$E[\boldsymbol{\theta}_2^*(N)|\boldsymbol{\theta}^0, \rho^0] = \boldsymbol{\theta}^0 + O(1/N^{1/2})$$

$$\operatorname{var}[\boldsymbol{\theta}_2^*(N)|\boldsymbol{\theta}^0, \rho^0] \approx \rho^0\left[E\left\{\sum_{t=1}^{N} \mathbf{z}(t-1)\mathbf{z}^{\mathrm{T}}(t-1)\right\}\right]^{-1} + o(1/N).$$

We refer the reader to Section 4d for further details.

(3) **Minimum Variance Property of $\boldsymbol{\theta}_2^*$.** The variance of the estimator $\boldsymbol{\theta}_2^*$ was proved in Section 4d, to be asymptotically equal to the Cramer–Rao lower bound; it was also shown that $\boldsymbol{\theta}_2^*$ has asymptotically minimum variance.

(4) **Probability Distribution of ρ^*.** Recall

$$\rho^*(N) = \frac{1}{N}\sum_{t=1}^{N} w^2(t, \boldsymbol{\theta}^*).$$

Let $\rho_1^* = N\rho^*/\rho^0$. It is well known [Kashyap and Rao, 1975; Anderson, 1971] that the probability density of ρ_1^* is $\chi^2(N-n)$ for large N. Consequently,

$$E[\rho^*(N)] = \rho^0(N-n)/N, \qquad \operatorname{var}[\rho^*(N)] = (\rho^0)^2\left(\frac{2N-2n}{N^2}\right)$$

for large N.

(5) **Probability Distribution of $\boldsymbol{\theta}^*$.** Let

$$\mathbf{S}(N) = \left[\sum_{t=1}^{N} \mathbf{z}(t-1)\mathbf{z}^{\mathrm{T}}(t-1)\right]^{-1}$$

$$p[\boldsymbol{\theta}^*(N)|\mathbf{S}(N)] \approx N(\boldsymbol{\theta}^0, \mathbf{S}(N)\rho^0) \qquad \text{for large } N.$$

† See Mann and Wald (1943).

6a.3.3. Confidence Intervals for the CML Estimator $\boldsymbol{\theta}^*$

We would like to set up an upper bound for the deviation between the CML estimator $\boldsymbol{\theta}^*$ and the value $\boldsymbol{\theta}^0$ in probabilistic terms. Specifically, if we ignore events that occur with a probability of ε or less (ε is usually 0.05) we want to determine a region $C_\varepsilon \subseteq \mathscr{H}$ such that

$$\text{prob}[(\boldsymbol{\theta}^0 - \boldsymbol{\theta}^*) \notin C_\varepsilon] < \varepsilon.$$

We can find such a region as follows. Let

$$r_2 = \frac{\|\boldsymbol{\theta}^0 - \boldsymbol{\theta}^*\|^2_{\mathbf{S}^{-1}(N)}}{\rho^*(N)} \cdot \left(\frac{N-n}{n}\right). \tag{6a.3.4}$$

For large N, the probability density of r_2 is an F-distribution with parameters n and $N - n$.

The probability density of r_2 is independent of $\boldsymbol{\theta}^0$ and ρ^0. By using the tabulated values of the F-distribution, we can easily find a threshold η to satisfy

$$\text{prob}[r_2 \leq \eta] = 0.95. \tag{6a.3.5}$$

Let $\eta_1 = \eta\rho^*(N)(N - n)/n$. Equations (6a.3.4) and (6a.3.5) imply

$$\text{prob}[\|\boldsymbol{\theta}^0 - \boldsymbol{\theta}^*\|^2_{\mathbf{S}^{-1}(N)} \leq \eta_1] = 0.95.$$

If the realizations $\boldsymbol{\theta}(N)$ that occur with probability less than 0.05 are ignored, then every numerical estimate $\boldsymbol{\theta}^*(N)$ is going to be located in the corresponding ellipsoid, $\|\boldsymbol{\theta}^0 - \boldsymbol{\theta}^*\|^2_{\mathbf{S}^{-1}(N)} \leq \eta_1$.

Alternatively, we can also loosely interpret (6a.3.5) as follows. We can approximately say that with probability 0.95, the true value $\boldsymbol{\theta}^0$ is going to be within the ellipsoid $\|\boldsymbol{\theta}^0 - \boldsymbol{\theta}^*\|^2_{\mathbf{S}^{-1}(N)} = \eta_1$ which is centered at $\boldsymbol{\theta}^*(N)$ and whose axes are determined by $\mathbf{S}(N)$. The ellipsoid $\|\boldsymbol{\theta}^0 - \boldsymbol{\theta}^*\|^2_{\mathbf{S}^{-1}(N)} = \eta_1$ centered at $\boldsymbol{\theta}^*$ (with $\boldsymbol{\theta}^0$ as variable) is regarded as a 95% confidence region boundary for the estimate $\boldsymbol{\theta}^*$.

One of the principal uses of this confidence region is the information it gives on the relative accuracies required in computing the various components of the estimate $\boldsymbol{\theta}^*(N)$.

Example 6a.2. Let $y(t) = \theta_1{}^0 y(t-1) + w(t)$; $w(t) \sim N(0, \rho^0)$. In this case $m = 1$ and $n = 1$. The unknown parameters are $\theta_1{}^0$ and ρ^0. Let $N = 100$. Suppose we are given

$$r_{0y} \triangleq \frac{1}{N}\sum_{t=1}^{N} y^2(t) = 1.00, \qquad r_{1y} = \frac{1}{N}\sum_{t=1}^{N} y(t)y(t-1) = 0.5.$$

Then the approximate maximum likelihood estimate of $\theta_1{}^0$ is $\hat{\theta}_1(N)$:

$$\hat{\theta}_1(N) = \frac{r_{1y}}{r_{0y}} = 0.5$$

$$\hat{\rho}(N) = \frac{1}{N}\sum_{t=2}^{N} (y(t) - \hat{\theta}_1(N)y(t-1))^2 = r_{0y} - \hat{\theta}_1(N)r_{1y} = 0.75.$$

Recall that $w(t, \theta_1) = y(t) - \theta_1 y(t-1)$. Hence,

$$\frac{\partial^2[w^2(t, \theta_1)]}{\partial \theta_1^2} = 2y^2(t-1)$$

$$\frac{1}{N} S^{-1}(N) = \sum_{t=1}^{N} y^2(t-1)/N \approx r_{0y} = 1.00$$

$$S(N) = 1/N = 0.01$$

$$r_2 = \frac{(\theta_1^0 - 0.5)^2 \times 100}{0.75} \times \frac{(100-1)}{100}.$$

From the tables of the F-distribution we find prob$[r_2 \leq 3.84] \approx 0.95$, but $r_2 \leq 3.84 \to |\theta_1^0 - 0.5| \leq 0.1705$. Hence, prob$[0.3295 \leq \theta_1^0 \leq 0.6705] = 0.95$.

6a.4. CML Estimators in Multivariate Systems

6a.4.1. Form of the CML Estimator

We recall that the CML estimator can be determined if $\mathbf{u}$ is independent of $\mathbf{w}$ or if $\mathbf{u}$ is absent in $\mathscr{M}$:

$$\begin{aligned} p(\xi_1(N) | \xi_2(N)) &= \prod_{t=1}^{N} p(\mathbf{y}(t) | \mathbf{y}(t-1), \ldots, \mathbf{y}(1), \xi_2(N)) \\ &= \prod_{t=1}^{N} (2\pi)^{-m/2} |\det \boldsymbol{\rho}^0|^{-1/2} \exp[-\tfrac{1}{2} \|\mathbf{f}_0(y(t)) - \mathbf{Z}(t-1)\boldsymbol{\theta}^0\|^2_{(\boldsymbol{\rho}^0)^{-1}}] \\ &= (2\pi)^{-Nm/2} |\det \boldsymbol{\rho}^0|^{-N/2} \exp\left[-\tfrac{1}{2} \sum_{t=1}^{N} \|\mathbf{f}_0(y(t)) - \mathbf{Z}(t-1)\boldsymbol{\theta}^0\|^2_{(\boldsymbol{\rho}^0)^{-1}}\right] \\ &= g_2(\xi(N), \boldsymbol{\theta}^0, \boldsymbol{\rho}^0). \end{aligned} \tag{6a.4.1}$$

In general, it is difficult to obtain explicit solutions for the values of $\boldsymbol{\theta}$ and $\boldsymbol{\rho}$ which maximize the expression $g_2(\xi(N), \boldsymbol{\theta}, \boldsymbol{\rho})$ with respect to $\boldsymbol{\theta}$ and $\boldsymbol{\rho}$. We will present two special cases in which the CML estimators can be expressed as closed form functions of the observations. In other cases, the maximization problem has to be solved by numerical methods.

Case (i). $\boldsymbol{\rho}^0$ is known; i.e., Ω has only one element in it. Maximizing $g_2(\xi(N), \boldsymbol{\theta}, \boldsymbol{\rho}^0)$ with respect to $\boldsymbol{\theta}$ is equivalent to minimizing $J_2(\boldsymbol{\theta})$ with respect to $\boldsymbol{\theta}$:

$$J_2(\boldsymbol{\theta}) = \sum_{t=1}^{N} \|\mathbf{f}_0(\mathbf{y}(t)) - \mathbf{Z}(t-1)\boldsymbol{\theta}\|^2_{(\boldsymbol{\rho}^0)^{-1}}. \tag{6a.4.2}$$

We can obtain the following solution of the above maximization problem by setting the first derivative of $J_2(\boldsymbol{\theta})$ with respect to $\boldsymbol{\theta}$ to zero,

$$\boldsymbol{\theta}^*(N) = \left[\sum_{t=1}^{N} \mathbf{Z}^T(t-1)(\boldsymbol{\rho}^0)^{-1}\mathbf{Z}(t-1)\right]^{-1} \left[\sum_{t=1}^{N} \mathbf{Z}^T(t-1)(\boldsymbol{\rho}^0)^{-1}\mathbf{y}(t)\right]. \tag{6a.4.3}$$

It is relatively easy to demonstrate the existence of the inverse in (6a.4.3) for all sufficiently large N under the assumptions made earlier.

Case (*ii*). The matrix $\boldsymbol{\rho}^0$ is *unknown*, and the equation for $\mathbf{y}$ has the following *special* structure. This structure is chosen when we do not have any knowledge about the system:

$$f_{0i}(\mathbf{y}(t)) = \mathbf{z}^{\mathrm{T}}(t-1)\boldsymbol{\theta}_i + w_i(t), \qquad i = 1, 2, \ldots, m \tag{6a.4.4}$$

or

$$\mathbf{f}_0(t) = \mathbf{Z}(t-1)\boldsymbol{\theta} + \mathbf{w}(t), \tag{6a.4.5}$$

where

$$\mathbf{Z}(t-1) = \begin{bmatrix} \mathbf{z}^{\mathrm{T}}(t-1) & & \\ & \mathbf{z}^{\mathrm{T}}(t-1) & \\ & & \mathbf{z}^{\mathrm{T}}(t-1) \end{bmatrix} \leftarrow m \times n_0 \text{ matrix}$$

$$\mathbf{f}_0(t) = [f_{01}(y_1(t)), \ldots, f_{0m}(y_m(t))], \quad f_{0i}, \ i = 1, 2, \ldots, \text{known functions.}$$

$\boldsymbol{\theta}_i$, $i = 1, 2, \ldots, m$, and $\mathbf{z}$ are all $\bar{n}$-vectors, $\boldsymbol{\theta} = (\boldsymbol{\theta}_1^{\mathrm{T}}, \ldots, \boldsymbol{\theta}_m^{\mathrm{T}})^{\mathrm{T}}$ is an n_0-vector, and $n_0 = m\bar{n}$.

Theorem 6a.1. The parameter $\boldsymbol{\theta}$ in the system above has the following CML estimator $\boldsymbol{\theta}^*$:

$$\boldsymbol{\theta}^*(N) = [(\boldsymbol{\theta}_1^*)^{\mathrm{T}}, \ldots, (\boldsymbol{\theta}_m^*)^{\mathrm{T}}]^{\mathrm{T}}$$
$$\boldsymbol{\theta}_i^* = \left[\sum_{t=1}^{N} \mathbf{z}(t-1)\mathbf{z}^{\mathrm{T}}(t-1)\right]^{-1} \sum_{t=1}^{N} \mathbf{z}(t-1) f_{0i}(\mathbf{y}(t)). \tag{6a.4.6}$$

The corresponding estimator of $\boldsymbol{\rho}^0$ is given by $\boldsymbol{\rho}^*$:

$$\boldsymbol{\rho}^* = \frac{1}{N}\sum_{t=1}^{N} (\mathbf{f}_0(t) - \mathbf{Z}(t-1)\boldsymbol{\theta}^*(N))(\mathbf{f}_0(t) - \mathbf{Z}(t-1)\boldsymbol{\theta}^*(N))^{\mathrm{T}}. \tag{6a.4.7}$$

This theorem is proved in Appendix 6.1.

The result is intriguing because the estimator $\boldsymbol{\theta}_i^*$ is obtained by considering only the ith equation in (6a.4.6) without reference to the other equations, and the collection of such estimators yields the *CML* estimator of $\boldsymbol{\theta}^0$. The result is valid only if the vector $\mathbf{z}(\cdot)$ is the same in all equations of (6a.4.4) for $i = 1, \ldots, m$.

If we had only one equation, such as the equation for y_1, without the knowledge of the equations obeyed by the other variables $y_2, y_3, \ldots$, then the estimator $\boldsymbol{\theta}_1^*$ given by (6a.4.6) would *not* necessarily be CML. It would be only a limited information (LI) estimator. If $y_2, y_3, \ldots, y_n$ also have the structure (6a.4.5), then this LI estimator $\boldsymbol{\theta}_1^*$ is also a CML estimate.

6a.4.2. Properties of the CML Estimators

Let us first consider the special estimator in (6a.4.3) which is derived by assuming knowledge of $\boldsymbol{\rho}^0$:

$$E[\boldsymbol{\theta}^*(N)] = \boldsymbol{\theta}^0 + O(1/N^{1/2}) \tag{6a.4.8}$$

$$E[(\boldsymbol{\theta}^*(N) - \boldsymbol{\theta}^0)(\boldsymbol{\theta}^*(N) - \boldsymbol{\theta}^0)^{\mathrm{T}}] = E\left[\sum_{t=1}^{N} \mathbf{Z}^{\mathrm{T}}(t-1)(\boldsymbol{\rho}^0)^{-1}\mathbf{Z}(t-1)\right]^{-1} + o\left(\frac{1}{N}\right). \tag{6a.4.9}$$

The results in (6a.4.8) and (6a.4.9) are established in Appendix 6.1, B.

In Appendix 6.1, C, expressions (6a.4.8) and (6a.4.9) are shown to be valid even for the general CML estimator, which does not assume knowledge of $\boldsymbol{\rho}^0$. We will show in Appendix 6.4 that the CML estimator given here has asymptotically minimum variance in the sense that the variance given above is the same as the lower bound given by the Cramer–Rao inequality plus a quantity $o(1/N)$.

6a.5. Limited Information (LI) Estimator in Multivariate Systems

Consider the multivariate system in (6a.1.2) or (6a.1.3). Recall that $\boldsymbol{\theta} = (\boldsymbol{\theta}_1^{\mathrm{T}}, \ldots, \boldsymbol{\theta}_m^{\mathrm{T}})^{\mathrm{T}}$, where $\boldsymbol{\theta}_i$ is the n_i-dimensional vector of coefficients in the ith equation and $n_0 = \sum_{i=1}^{m} n_i$ is the dimension of $\boldsymbol{\theta}$. Note that n_i, $i = 1, \ldots, m$, need not be the same.

The limited information estimator $\boldsymbol{\theta}_i^*(N)$ of $\boldsymbol{\theta}_i^0$ is computed by using the ith equation alone and the entire observation history $\boldsymbol{\xi}(N)$. The name *limited information estimate* is used because the unknowns in the ith equation can be estimated without taking into account the information given by the other equations (Kashyap and Nasburg, 1974). Let

$$\boldsymbol{\theta}^*(N) = ((\boldsymbol{\theta}_1^*(N))^{\mathrm{T}}, \ldots, (\boldsymbol{\theta}_m^*(N))^{\mathrm{T}})^{\mathrm{T}},$$

where $\boldsymbol{\theta}_i^*(N)$, $i = 1, \ldots, n$, are defined in (6a.4.6). Then $\boldsymbol{\theta}^*(N)$ is the required limited information estimate of the vector $\boldsymbol{\theta}^0$ of the multivariate system. The corresponding estimator of $\boldsymbol{\rho}^0$ is

$$\boldsymbol{\rho}^*(N) = \frac{1}{N}\sum_{t=1}^{N} (\mathbf{f}_0(\mathbf{y}(t)) - \mathbf{Z}(t-1)\boldsymbol{\theta}^*)(\mathbf{f}_0(\mathbf{y}(t)) - \mathbf{Z}(t-1)\boldsymbol{\theta}^*)^{\mathrm{T}}.$$

Obviously, $\boldsymbol{\theta}^*$ is consistent and the mean square error of the component vectors $\boldsymbol{\theta}_i^*$ can be determined as before.

The important advantage of the LI estimator is that it enables us to determine a consistent estimator of $\boldsymbol{\theta}^0$ *without* the knowledge of the covariance ($\boldsymbol{\rho}$) of the noise. Knowledge of $\boldsymbol{\rho}$ is required in both the likelihood and Bayesian approaches. Moreover, the LI estimate can also be computed in real time unlike the multivariate CML estimator or the Bayesian estimator.

The LI estimate is not always efficient. It is efficient if we represent the system equation in canonical form III.

An important aspect of the limited information estimators is that they allow the decomposition of the estimation problem for the multivariate system into a number of separate estimation problems for the individual equations in the system. This concept is very useful in parameter estimation in systems with moving average terms.

6b. Bayesian Estimators

6b.1. Fundamentals of Bayesian Approach

A Bayesian framework differs from the likelihood framework in two aspects. First of all, in the Bayesian approach, the discrepancy between the estimator and the quantity that is being estimated is quantified in the form of a loss function. The estimator is selected to minimize the expected value of the loss function. Clearly, different choices of the loss function yield different estimators. Still, with each choice of loss function, we have a clear idea of the consequences involved in the use of a particular estimator as reflected by the loss function. Often, a particular Bayesian estimator derived from a loss function suited to the particular occasion is better than the corresponding CML estimator. The most common loss function is the quadratic loss function.

The second important idea in the Bayesian framework is that any prior knowledge available about the parameters $\boldsymbol{\theta}^0$ and $\boldsymbol{\rho}^0$ describing the given process $\mathbf{y}$ can and should be utilized in the estimation process. Such knowledge is ignored in the likelihood approach. Prior knowledge is utilized by regarding the variables $\boldsymbol{\theta}$ and $\boldsymbol{\rho}$ ($\boldsymbol{\theta} \in \mathscr{H}$, $\boldsymbol{\rho} \in \Omega$) as random variables with known probability distribution $F(\boldsymbol{\theta}, \boldsymbol{\rho})$, the so-called prior probability distribution. This distribution will be assumed to have a density $p(\boldsymbol{\theta}, \boldsymbol{\rho})$, the prior probability density. The numerical value $p(\boldsymbol{\theta} = \boldsymbol{\theta}', \boldsymbol{\rho} = \boldsymbol{\rho}')$ reflects our relative degree of belief in the statement that the numerical values $\boldsymbol{\theta}'$ and $\boldsymbol{\rho}'$ characterize parameters of the given process $\mathbf{y}$. This density is called a prior density since such a judgment is made before using the observations about the process $\mathbf{y}$ for estimation of the parameters. Some guidelines for the choice of prior distributions and related topics can be found in the works of Kashyap (1971) and Fine (1973).

Arbitrariness in the choice of the prior distribution is one of the disadvantages of the Bayesian approach. In addition, the Bayesian estimators are difficult to evaluate, especially if there are MA terms in the system. Suppose we have an observation $\boldsymbol{\xi}(N)$. We can define estimators $\hat{\boldsymbol{\theta}}$ and $\hat{\boldsymbol{\rho}}$ based on $\boldsymbol{\xi}$ as maps as in Section 6a.1.1. We will find the optimal estimator $\boldsymbol{\theta}^*$ with respect to the quadratic loss function $L(\boldsymbol{\theta}, \hat{\boldsymbol{\theta}}) = \|\boldsymbol{\theta} - \hat{\boldsymbol{\theta}}\|^2$. The corresponding risk function is

$$J_1(\boldsymbol{\theta}, \hat{\boldsymbol{\theta}}) = E_{\boldsymbol{\theta},\boldsymbol{\rho}}[L(\boldsymbol{\theta}, \hat{\boldsymbol{\theta}}) | \boldsymbol{\theta}, \boldsymbol{\rho}] = \int \|\boldsymbol{\theta} - \hat{\boldsymbol{\theta}}(\boldsymbol{\xi})\|^2 p(\boldsymbol{\xi} | \boldsymbol{\theta}, \boldsymbol{\rho}) \, |d\boldsymbol{\xi}|.$$

It should be noted that $E_{\boldsymbol{\theta}}$ means that the expectation is taken keeping $\boldsymbol{\theta}$ fixed. We will average over $\boldsymbol{\theta}$ and $\boldsymbol{\rho}$ using the prior probability density $p(\boldsymbol{\theta}, \boldsymbol{\rho})$ to obtain the Bayes risk function $J(\hat{\boldsymbol{\theta}})$:

$$J(\hat{\boldsymbol{\theta}}) = E[J_1(\boldsymbol{\theta}, \hat{\boldsymbol{\theta}})] = \int J_1(\boldsymbol{\theta}, \hat{\boldsymbol{\theta}})p(\boldsymbol{\theta}, \boldsymbol{\rho})\,|d\boldsymbol{\theta}|\,|d\boldsymbol{\rho}|.$$

As J_1 is a function of the estimator $\hat{\boldsymbol{\theta}}$, we can choose the estimator $\hat{\boldsymbol{\theta}}$ to minimize the Bayes risk function J. Such an estimator is called a Bayes estimator.

Theorem 6b.1. The Bayes estimator $\boldsymbol{\theta}^*$ that minimizes $J(\boldsymbol{\theta})$ has the form

$$\boldsymbol{\theta}^*(\boldsymbol{\xi}) = \int_{\boldsymbol{\theta}\in\mathscr{H}} \boldsymbol{\theta}p(\boldsymbol{\theta}|\boldsymbol{\xi})\,|d\boldsymbol{\theta}|. \tag{6b.1.1}$$

Proof. Let $\hat{\boldsymbol{\theta}}$ be any other estimator. We will show that $J(\boldsymbol{\theta}^*) \leq J(\boldsymbol{\theta})$:

$$\begin{aligned} J(\hat{\boldsymbol{\theta}}) &= \iint \|\boldsymbol{\theta} - \hat{\boldsymbol{\theta}}\|^2 p(\boldsymbol{\xi}|\boldsymbol{\theta}, \boldsymbol{\rho})p(\boldsymbol{\theta}, \boldsymbol{\rho})\,|d\boldsymbol{\xi}|\,|d\boldsymbol{\theta}|\,|d\boldsymbol{\rho}| \\ &= E[\|\boldsymbol{\theta} - \hat{\boldsymbol{\theta}}\|^2] = E[\|\boldsymbol{\theta} - \boldsymbol{\theta}^* + \boldsymbol{\theta}^* - \hat{\boldsymbol{\theta}}\|^2] \\ &= E[\|\boldsymbol{\theta} - \boldsymbol{\theta}^*\|^2] + E[\|\boldsymbol{\theta}^* - \hat{\boldsymbol{\theta}}\|^2] + 2E[(\boldsymbol{\theta} - \boldsymbol{\theta}^*)^{\mathrm{T}}(\boldsymbol{\theta}^* - \hat{\boldsymbol{\theta}})]. \end{aligned} \tag{6b.1.2}$$

The variable $(\boldsymbol{\theta}^* - \hat{\boldsymbol{\theta}})$ is a function of observation $\boldsymbol{\xi}$ only. We will use the usual rule

$$E[A] = E[E_B(A|B)] \tag{6b.1.3}$$

to simplify (6b.1.2). In (6b.1.3), the inner expectation is evaluated by keeping B fixed; hence, it is a function of B only. The outer expectation is taken over the random variable B. Consider the third term in (6b.1.2):

$$\begin{aligned} E[(\boldsymbol{\theta} - \boldsymbol{\theta}^*)^{\mathrm{T}}(\boldsymbol{\theta}^* - \hat{\boldsymbol{\theta}})] &= E[E\{(\boldsymbol{\theta} - \boldsymbol{\theta}^*)^{\mathrm{T}}(\boldsymbol{\theta}^* - \hat{\boldsymbol{\theta}})|\boldsymbol{\xi}\}] \\ &= E[\{E(\boldsymbol{\theta} - \boldsymbol{\theta}^*)|\boldsymbol{\xi}\}^{\mathrm{T}}(\boldsymbol{\theta}^* - \hat{\boldsymbol{\theta}})], \\ &\quad \text{since } (\boldsymbol{\theta}^* - \hat{\boldsymbol{\theta}}) \text{ is a function of } \boldsymbol{\xi} \text{ only}, \\ &= 0, \end{aligned}$$

by definition of $\boldsymbol{\theta}^*$ in (6b.1.1).

Substituting this expression in (6b.1.2) and simplifying, we obtain

$$J(\hat{\boldsymbol{\theta}}) = J(\boldsymbol{\theta}^*) + E[\|\boldsymbol{\theta}^* - \hat{\boldsymbol{\theta}}\|^2] \geq J(\boldsymbol{\theta}^*). \tag{6b.1.4}$$

The probability density $p(\boldsymbol{\theta}|\boldsymbol{\xi})$ is called the posterior probability density of $\boldsymbol{\theta}$ given $\boldsymbol{\xi}$ and it has all the information contained in $\boldsymbol{\xi}$ about $\boldsymbol{\theta}$. This density is crucial in the Bayes estimation theory. One of the principal difficulties in the application of Bayes theory is the determination of explicit expressions for the posterior density $p(\boldsymbol{\theta}|\boldsymbol{\xi})$. Explicit expressions for the posterior densities can be obtained only with certain prior densities such as the Gaussian probability density function. Otherwise, one has to determine $p(\boldsymbol{\theta}|\boldsymbol{\xi})$ numerically, which may be difficult. In a similar manner we can obtain the Bayes estimator of $\boldsymbol{\rho}^0$. Bayesian analysis has been used in time series by many investigators (Zellner, 1971; Box and Tiao, 1973).

6b.2. Bayesian Estimator in Single Output Systems

The following additional assumptions are needed in addition to the assumptions made at the beginning of Section 6a:

E1. $p(\boldsymbol{\theta}|\boldsymbol{\xi}_2) = p(\boldsymbol{\theta})$, where $\boldsymbol{\xi}_2 = \{y(0), \ldots, y(-m_1); \mathbf{u}(N-1), \ldots, \mathbf{u}(-m_3)\}$.
E2. The variance ρ of the noise w is assumed to be known.
E3. The prior density of $\boldsymbol{\theta}$ is $p(\boldsymbol{\theta}) \sim N[\boldsymbol{\theta}_0^*, \mathbf{S}_0\rho]$, $\boldsymbol{\theta}_0^*$ and $\mathbf{S}_0$ being known.

Assumption (E1) reflects the fact that the components of $\boldsymbol{\theta}$ are coefficients of the difference equation and, as such, no information is gained from either the initial conditions $y(0), \ldots, y(-m_1)$ or the inputs $\mathbf{u}(\cdot)$ alone. Assumption (E2) is given to render the analysis easier, and (E3) assures us that the posterior density is also normal and hence easier to manipulate.

Theorem 6b.2. The posterior density of $\boldsymbol{\theta}$ given $\boldsymbol{\xi}(N)$ has the form

$$p(\boldsymbol{\theta}|\boldsymbol{\xi}(N)) \sim N(\boldsymbol{\theta}^*(N), \mathbf{S}(N)\rho), \tag{6b.2.1}$$

where

$$\boldsymbol{\theta}^*(N) = \mathbf{S}(N)\left[\sum_{t=1}^{N} \mathbf{z}(t-1) f_0(y(t)) + \mathbf{S}_0^{-1}\boldsymbol{\theta}_0\right] \tag{6b.2.2}$$

$$\mathbf{S}(N) = \left(\sum_{t=1}^{N} \mathbf{z}(t-1)\mathbf{z}^{\mathrm{T}}(t-1) + \mathbf{S}_0^{-1}\right)^{-1}. \tag{6b.2.3}$$

By Theorems 6b.1 and 6b.2, $\boldsymbol{\theta}^*(N)$ is the required optimal estimate of $\boldsymbol{\theta}$ under the quadratic loss function. The posterior variance of $\boldsymbol{\theta}(N)$ given the observations until time N, is $\mathbf{S}(N)\rho$. Theorem 6b.2 is proved in part B of Appendix 6.2.

Comment 1. An important aspect of the estimate $\boldsymbol{\theta}^*(N)$ is that it does not involve the variance ρ. It depends on $\mathbf{S}(N)$, which is not dependent on ρ for its evaluation. Thus assumption (E2) is superfluous to a certain extent.

Comment 2. To obtain an idea of the variance of estimate $\boldsymbol{\theta}^*$ when ρ is unknown, we can use the expression $\rho^*(N)$:

$$\rho^*(N) = \frac{1}{N}\sum_{t=1}^{N} (f_0(y(t)) - \mathbf{z}^{\mathrm{T}}(t-1)\boldsymbol{\theta}^*(N))^2.$$

Comment 3. The estimate $\boldsymbol{\theta}^*(N)$ can be regarded as a weighted least squares estimate obtained by minimizing the criterion function $J'(\boldsymbol{\theta})$:

$$J'(\boldsymbol{\theta}) = \sum_{t=1}^{N} [f_0(y(t)) - \mathbf{z}^{\mathrm{T}}(t-1)\boldsymbol{\theta}]^2 + \|\boldsymbol{\theta} - \boldsymbol{\theta}_0\|^2_{\mathbf{S}_0^{-1}}. \tag{6b.2.4}$$

The first factor in $J'(\boldsymbol{\theta})$ represents the lack of fit of the estimated quantity to the observations, whereas the second term represents the lack of fit to the prior estimate of $\boldsymbol{\theta}$.

Comment 4. We want to emphasize the fact that an explicit expression has been determined for $\boldsymbol{\theta}^*(N)$ only because the corresponding difference equation is linear in $\boldsymbol{\theta}$. The function $f_0(y(t))$, however, can be nonlinear and the output y need not be normally distributed, but w must be normal.

6b.3. Bayesian Estimation in Multiple Output Systems

Consider the multivariate system in (6a.1.3) and (6a.1.4). We will make the assumptions.

E1′. $p(\boldsymbol{\theta}|\boldsymbol{\xi}_2) = p(\boldsymbol{\theta})$, where $\boldsymbol{\xi}_2$ is the multivariate version of the corresponding symbol in Section 6b.2.

E2′. The covariance matrix of $\mathbf{w}$ is known and is equal to $\boldsymbol{\rho}$.

E3′. The prior density of $\boldsymbol{\theta}$ is normal: $p(\boldsymbol{\theta}) \sim N(\boldsymbol{\theta}_0, \bar{\mathbf{S}}_0)$, $\boldsymbol{\theta}_0$ and $\bar{\mathbf{S}}_0$ being known.

These assumptions are similar to those made in connection with the univariate case and the comments made therein are valid here also. We will find the expression for the posterior density $p(\boldsymbol{\theta}|\boldsymbol{\xi})$.

Theorem 6b.3. The posterior density of $\boldsymbol{\theta}$ given $\boldsymbol{\xi}(N)$ has the form

$$p(\boldsymbol{\theta}|\boldsymbol{\xi}(N)) \sim N(\boldsymbol{\theta}^*(N), \bar{\mathbf{S}}(N)), \tag{6b.3.1}$$

where

$$\boldsymbol{\theta}^*(N) = \bar{\mathbf{S}}(N)\left[\sum_{t=1}^{N} \mathbf{Z}^{\mathrm{T}}(t-1)\boldsymbol{\rho}^{-1}\mathbf{f}_0(\mathbf{y}(t)) + \bar{\mathbf{S}}_0^{-1}\boldsymbol{\theta}_0\right] \tag{6b.3.2}$$

$$\bar{\mathbf{S}}(N) = \left[\sum_{t=1}^{N} \mathbf{Z}^{\mathrm{T}}(t-1)\boldsymbol{\rho}^{-1}\mathbf{Z}(t-1) + \bar{\mathbf{S}}_0^{-1}\right]^{-1}. \tag{6b.3.3}$$

Theorem 6b.3 is proved in Appendix 6.2. Unlike the scalar case, the estimate $\boldsymbol{\theta}^*(N)$ depends *explicitly* on the covariance matrix $\boldsymbol{\rho}$.

One can regard $\boldsymbol{\rho}$ also as a random variable and find a joint posterior density $\boldsymbol{\theta}$ and $\boldsymbol{\rho}$ given $\boldsymbol{\xi}$, although the resulting final expression is very unwieldy.

6b.4. Comparison of Bayesian and Maximum Likelihood Estimates

The principal advantage of the Bayesian approach is its ability to incorporate the available prior knowledge about the particular process $\mathbf{y}$ into the system. But this particular knowledge can be absorbed only by means of a prior probability distribution. The normal prior distribution has been used here because of its computational ease. The arbitrariness involved in the choice of prior density has been one of the principal criticisms against the Bayesian method. However, the normal prior density of $\boldsymbol{\theta}$ can be justified by information theoretic and other arguments. The maximum likelihood approach is preferred by some investigators because it does not involve any arbitrary assumptions.

The Bayes method has two other advantages. First, the Bayes estimate is always admissible, whereas sometimes the ML estimator may not be admissible with respect to any loss function. Second, the Bayes method can also take into account the cost of observations. Usually one has to pay a price for the observations, and the natural question at each instant of time is whether the additional benefit gained by an additional observation is worth its cost. This tradeoff can be expressed clearly in the Bayes approach. The benefit to be gained by an additional observation at each instant can be quantified in terms of the marginal decrease in the average risk function caused by the additional observation. This benefit can be compared to the cost of the observation.

Apart from these considerations, one can obtain similar results on both the estimators. For instance, we can compute confidence intervals of the estimator in either method, although the confidence intervals are interpreted differently. In the Bayesian approach, $\boldsymbol{\theta}$ is a random variable, whereas in the likelihood approach $\boldsymbol{\theta}^0$ is an unknown constant. Since the two estimates differ only because of the use of the prior distribution in one of them, the flatter the prior density, the closer will be the two estimates. Moreover, the two estimates are usually asymptotically identical to one another since in the large sample case the knowledge of the observations *swamps that of the prior distribution.*

6c. Quasi-Maximum Likelihood (QML) Estimators in Single Output Systems

6c.1. Description

In this section, we consider an estimation scheme that does not involve many of the restrictive assumptions made in Sections 6a and 6b. This estimator was mentioned in Chapter IV. In Sections 6a and 6b, we had to assume the independence of $\mathbf{u}$ and w because we could not get an expression for the joint probability density of the entire observation vector $\boldsymbol{\xi}(N)$ otherwise. In the quasi-maximum likelihood schemes, we consider only a part of the probability density of $\boldsymbol{\xi}(N)$ which can be derived from the knowledge of the probability of the disturbance w alone, and hence the name "quasi-maximum likelihood" (Kashyap and Nasburg, 1974). Under certain conditions, this estimator is consistent and under certain additional conditions this estimator will be shown to have the same asymptotic variance as the FIML estimator.

Consider the estimation problem associated with the scalar process y in the class $[\mathscr{I}, \mathscr{H}, \Omega]$ in Eq. (6a.1.1). We will assume the usual assumptions (A1), (A2), (A4′), and (A5). Note that we are not using (A3), i.e., the independence of $\mathbf{u}$ and w is not assumed. We will use the criterion function

$$J(\boldsymbol{\theta}) = \sum_{t=1}^{N} (f_0(t) - \mathbf{z}^{\mathrm{T}}(t-1)\boldsymbol{\theta})^2.$$

The estimate $\boldsymbol{\theta}^*$ of $\boldsymbol{\theta}$ obtained by minimizing $J(\boldsymbol{\theta})$ is called a quasi-maximum likelihood estimate of $\boldsymbol{\theta}$ based on $\boldsymbol{\xi}(N)$.

$$\boldsymbol{\theta}^*(N) = \left[\sum_{t=1}^{N} \mathbf{z}(t-1)\mathbf{z}^{\mathrm{T}}(t-1)\right]^{-1} \left[\sum_{t=1}^{N} \mathbf{z}(t-1) f_0(t)\right].$$

This estimator is the same as the Bayes estimator in Section 6b.2 if we set $\mathbf{S}_0 = \infty$, i.e., when we do not have any reasonable prior knowledge of the parameter $\boldsymbol{\theta}^0$. The name quasi-maximum likelihood is given because the estimate can be obtained by maximizing a *part* of the full information log likelihood function, as seen below:

$$\begin{aligned} p(\boldsymbol{\xi}(N)) &= p(y(N), \ldots, y(-m_1), \mathbf{u}(N-1), \ldots, \mathbf{u}(-m_3)) \\ &= \mathrm{I} \times \mathrm{II} \times \mathrm{III} \end{aligned}$$

where

$$\begin{aligned} \mathrm{I} &= \prod_{t=1}^{N} p(y(t) | \mathbf{u}(t-1), \ldots, \mathbf{u}(-m_3), y(t-1), \ldots, y(-m_1)), \\ \mathrm{II} &= \prod_{t=1}^{N} p(\mathbf{u}(t-1) | y(t-1), \ldots, y(-m_1), \mathbf{u}(t-2), \ldots, \mathbf{u}(-m_3)), \\ \mathrm{III} &= p(y(0), \ldots, y(-m_1), \mathbf{u}(0), \ldots, \mathbf{u}(-m_3)). \end{aligned}$$

The first term in this product, labeled I, can be expanded as follows:

$$\begin{aligned} \mathrm{I} &= \prod_{t=1}^{N} \frac{1}{(2\pi\rho^0)^{1/2}} \exp\left[-\frac{1}{2\rho^0} (f_0(y(t)) - \mathbf{z}^{\mathrm{T}}(t-1)\boldsymbol{\theta}^0)^2 \right] \\ &\triangleq p(w(1, \boldsymbol{\theta}^0), \ldots, w(N, \boldsymbol{\theta}^0)), \end{aligned}$$

where $w(t, \boldsymbol{\theta}^0) = f_0(y(t)) - \mathbf{z}^{\mathrm{T}}(t-1)\boldsymbol{\theta}^0$. The value of $\boldsymbol{\theta}$ which maximizes the log likelihood function derived from I above is the QML estimate of $\boldsymbol{\theta}^0$.

Clearly the QML and CML estimators are identical when $\mathbf{u}$ and w are independent or if $\mathbf{u}$ is absent. However, under certain conditions, even when $\mathbf{u}$ and w are dependent, the QML estimator can be demonstrated (Section 6c.2) to be identical to the FIML estimator.

6c.2. Relation between CML and QML Estimators

Consider (6a.4.4) with y_1 obeying $y_1(t) = \mathbf{z}^{\mathrm{T}}(t-1)\boldsymbol{\theta}_1{}^0 + w_1(t)$, where

$$\begin{aligned} \mathbf{z}(t-1) = [&y_1(t-1), \ldots, y_1(t-m_1), y_2(t-1), \ldots, \\ &\times y_2(t-m_3), \psi_1(t-1), \ldots, \psi_{l_2}(t-1)]^{\mathrm{T}}. \end{aligned}$$

The variable $y_2(\cdot)$ is observable but it is dependent on w_1, even though $w_1(t)$ is independent of $y_2(t-j)$ for all $j \geq 1$. Let $\boldsymbol{\theta}_1(N)$ denote a QML estimate of $\boldsymbol{\theta}_1$. We will give the conditions under which $\boldsymbol{\theta}_1(N)$ is a CML estimate in the following theorem.

Theorem 6c.1. The QML estimate $\boldsymbol{\theta}_1^*(N)$ of $\boldsymbol{\theta}_1{}^0$ defined above is identical to the CML estimate of $\boldsymbol{\theta}_1{}^0$ if the process y_2 obeys

$$y_2(t) = \mathbf{z}^{\mathrm{T}}(t-1)\boldsymbol{\theta}_2{}^0 + w_2(t),$$

where $w_2(t)$ is a white noise sequence that can be correlated with w_1 and $\boldsymbol{\theta}_2{}^0$ may be unknown.

This theorem is a direct consequence of Theorem 6a.1 in Section 6a.4. The conditions to make a QML estimator correspond with a CML estimator are not very restrictive. However, if we do have some knowledge of $\boldsymbol{\theta}_2$, such as its first component, then the estimator $\boldsymbol{\theta}_1^*$ is not a CML estimator because we are not using all the available information in the estimation process. Still, $\boldsymbol{\theta}_1^*$ is a QML estimator.

6d. Computational Methods

6d.1. Real Time Algorithm for the Single Output System

The Bayes estimate $\boldsymbol{\theta}^*$ given in (6b.2.2) and (6b.2.3) gives the misleading impression that the entire observation history $[y(j), \mathbf{u}(j), j \leq N]$ is to be stored to compute the estimate $\boldsymbol{\theta}^*(N)$. In other words, the memory needed to compute $\boldsymbol{\theta}^*(N)$ appears to increase linearly with time N. However, it is possible to rewrite the estimate $\boldsymbol{\theta}^*(N)$ so that only a finite amount of memory is needed to update the estimate $\boldsymbol{\theta}^*(N-1)$ and $\mathbf{S}(N-1)$ into the new estimates $\boldsymbol{\theta}^*(N)$ and $\mathbf{S}(N)$ by using the new observations $y(N)$ and $\mathbf{u}(N-1)$. *Therefore, the estimate* $\boldsymbol{\theta}^*(N)$ *is real time computable.* The recursive algorithm not only simplifies the computation of estimate, but opens up many interesting possibilities such as real time prediction with unknown systems, the possibility of detecting nonstationarities in a system, etc.

The required algorithm is given below (it is proved in Appendix 6.3). It belongs to the family of Kalman (1963) filter algorithms.

The Algorithm for Estimation

$$\boldsymbol{\theta}^*(t) = \boldsymbol{\theta}^*(t-1) + \mathbf{S}(t)\mathbf{z}(t-1)[f_0(y(t)) - (\boldsymbol{\theta}^*(t-1))^{\mathrm{T}}\mathbf{z}(t-1)]$$

$$\mathbf{S}(t) = \mathbf{S}(t-1) - \frac{\mathbf{S}(t-1)\mathbf{z}(t-1)\mathbf{z}^{\mathrm{T}}(t-1)\mathbf{S}(t-1)}{1 + \mathbf{z}^{\mathrm{T}}(t-1)\mathbf{S}(t-1)\mathbf{z}(t-1)}, \quad t = 1, 2, \ldots . \qquad \textbf{C1}$$

To emphasize the real time computability of algorithm (C1), Table 6d.1.1 shows the quantities in storage at time t, which are computed at time $t = 0, 1, 2,$ etc., for a model with $m_1 = 3$, $m_3 = 2$, $l_2 = 2$.

TABLE 6d.1.1. Information Pattern Used in the Estimator

Time instant t	Quantities in storage at time t	Additional data which arrive at time t	Quantities computed in the time interval between t and $t+1$
4	$y(3), y(2), y(1), \mathbf{u}(2)$ $\boldsymbol{\theta}^*(2), \mathbf{S}(2)$	$y(4), \mathbf{u}(3)$	$\mathbf{z}(3)$ $\boldsymbol{\theta}^*(4), \mathbf{S}(4)$
5	$y(4), y(3), y(2), \mathbf{u}(3)$ $\boldsymbol{\theta}^*(3), \mathbf{S}(3)$	$y(5), \mathbf{u}(4)$	$\mathbf{z}(4)$ $\boldsymbol{\theta}^*(5), \mathbf{S}(5)$
⋮	⋮	⋮	⋮
t	$y(t-1), y(t-2), y(t-3)$ $\mathbf{u}(t-2)$ $\boldsymbol{\theta}^*(t-1), \mathbf{S}(t-1)$	$y(t), \mathbf{u}(t-1)$	$\mathbf{z}(t-1), \boldsymbol{\theta}^*(t), \mathbf{S}(t)$

Hence, the total amount of memory needed at instant t is $\{1 + 2n + n(n+1)/2\}$, which is independent of t. Similarly, the amount of computation needed for updating $\boldsymbol{\theta}^*(t-1)$, $\mathbf{S}(t-1)$ to $\boldsymbol{\theta}^*(t)$ and $\mathbf{S}(t)$ is fixed and independent of t.

Initial Conditions. Let $\boldsymbol{\theta}^*(0)$ be the available a priori estimate of $\boldsymbol{\theta}^0$ and let $\mathbf{S}(0)$ be any suitable positive-definite weighting matrix reflecting confidence in the estimate $\boldsymbol{\theta}^*(0)$. If an a priori estimate of $\boldsymbol{\theta}^*(0)$ is not available, one can use any arbitrary vector for $\boldsymbol{\theta}^*(0)$. Alternatively, for $t \geq (N_1 + 1)$, algorithm (C1) may be used with the following initial values of $\boldsymbol{\theta}^*(N_1)$ and $\mathbf{S}(N_1)$:

$$\begin{aligned} \mathbf{S}(N_1) &= \left[\sum_{j=n_1+1}^{N_1} \mathbf{z}(j-1)\mathbf{z}^{\mathrm{T}}(j-1) \right]^{-1} \\ \boldsymbol{\theta}^*(N_1) &= \mathbf{S}(N_1) \sum_{j=n_1+1}^{N_1} f_0(y(j))\mathbf{z}(j-1). \end{aligned} \tag{6d.1.1}$$

The integer N_1 should be sufficiently large so that the indicated inverse exists.

6d.2. Real Time Algorithms for the Multiple Output System

Case (i). Consider the general system in Eq. (6a.1.2) with $\boldsymbol{\rho}$ known. The algorithm for computing the Bayes estimate in (6b.3.2) and (6b.3.3) for multiple output systems is given below.

Algorithm

$$\begin{aligned} \boldsymbol{\theta}^*(t) &= \boldsymbol{\theta}^*(t-1) + \bar{\mathbf{S}}(t)\mathbf{Z}^{\mathrm{T}}(t-1)\boldsymbol{\rho}^{-1}[\mathbf{f}_0(\mathbf{y}(t)) - \mathbf{Z}(t-1)\boldsymbol{\theta}^*(t-1)] \\ \bar{\mathbf{S}}(t) &= \bar{\mathbf{S}}(t-1) - \bar{\mathbf{S}}(t-1)\mathbf{Z}^{\mathrm{T}}(t-1)[\mathbf{Z}(t-1)\bar{\mathbf{S}}(t-1)\mathbf{Z}^{\mathrm{T}}(t-1) + \boldsymbol{\rho}]^{-1} \\ &\quad \times \mathbf{Z}(t-1)\bar{\mathbf{S}}(t-1), \qquad t = 2, 3, 4, \ldots. \end{aligned} \tag{C2}$$

Initialization: $\boldsymbol{\theta}^*(1) = \boldsymbol{\theta}_0$, $\bar{\mathbf{S}}(1) = \mathbf{S}_0 > 0$.

The equivalence of (C2) to (6b.3.2) and (6b.3.3) is proved in Appendix 6.3. There are two important distinctions between algorithms (C1) and (C2). Algorithm (C2) involves the inversion of an $m \times m$ matrix in every updating, whereas in (C1) the inversion is not required. Second, (C2) involves the knowledge of the covariance matrix $\boldsymbol{\rho}^0$, whereas (C1) does not. Algorithm (C2) also belongs to the family of Kalman (1963) filter algorithms.

Case (ii). Consider the special system in (6a.4.4) and (6a.4.5) with $\boldsymbol{\rho}^0$ unknown. Here the estimate $\boldsymbol{\theta}^*$ can be expressed as $((\boldsymbol{\theta}_1^*)^{\mathrm{T}}, \ldots, (\boldsymbol{\theta}_m^*)^{\mathrm{T}})$. The expression for $\boldsymbol{\theta}_i^*$ is given in (6a.4.6). It does not involve the expressions for $\boldsymbol{\theta}_j^*$, $j \neq i$. This expression can be computed in real time using algorithm (C1).

Case (iii). Consider the general system in (6a.1.2) with $\boldsymbol{\rho}^0$ unknown. We cannot obtain the exact CML estimates without using the iterative method. But $\boldsymbol{\rho}^0$ can be estimated in real time by the usual technique and this estimate can then be used instead of $\boldsymbol{\rho}$ in (C2). The modified algorithm is given as (C3):

Algorithm

$$\bar{\boldsymbol{\theta}}(t) = \bar{\boldsymbol{\theta}}(t-1) + \bar{\mathbf{S}}(t)\mathbf{Z}^{\mathrm{T}}(t-1)\bar{\boldsymbol{\rho}}^{-1}(t-1)[\mathbf{f}_0(\mathbf{y}(t)) - \mathbf{Z}(t-1)\bar{\boldsymbol{\theta}}(t-1)]$$

$$\bar{\mathbf{S}}(t) = \bar{\mathbf{S}}(t-1) - \bar{\mathbf{S}}(t-1)\mathbf{Z}^{\mathrm{T}}(t-1) \times [\mathbf{Z}(t-1)\bar{\mathbf{S}}(t-1)\mathbf{Z}^{\mathrm{T}}(t-1) + \bar{\boldsymbol{\rho}}(t-1)]^{-1}\mathbf{Z}(t-1)\bar{\mathbf{S}}(t-1) \qquad \mathbf{C3}$$

$$\bar{\boldsymbol{\rho}}(t) = \bar{\boldsymbol{\rho}}(t-1) + \frac{1}{t}[\mathbf{f}_0(\mathbf{y}(t)) - \mathbf{Z}(t-1)\bar{\boldsymbol{\theta}}(t-1)] \times [\mathbf{f}_0(\mathbf{y}(t)) - \mathbf{Z}(t-1)\boldsymbol{\theta}(t-1)]^{\mathrm{T}}$$

One can show that $\bar{\boldsymbol{\theta}}(t)$ tends to $\boldsymbol{\theta}^*(t)$ as t tends to infinity. One can also argue that $\bar{\boldsymbol{\theta}}(t)$ is approximately equal to $E[\boldsymbol{\theta}|\boldsymbol{\xi}(t)]$, i.e., $\bar{\boldsymbol{\theta}}(t)$ is the Bayes estimator of $\boldsymbol{\theta}$ based on all the knowledge at time t.

6d.3. An Iterative Method for Computing CML Estimates in Multiple Output Systems

We mentioned earlier that the CML estimate of the parameter $\boldsymbol{\theta}^0$, in general, cannot be expressed in closed form, unlike the CML estimator in the single output system. We will develop an iterative method of evaluating the estimate. We assume that $\boldsymbol{\theta} \in \mathscr{R}^{n_0}$ and $\boldsymbol{\rho} \in \mathscr{R}^{m'}$ where $m' = m(m+1)/2$. The observation history $\boldsymbol{\xi}(N)$ is fixed. Using the observation set, we will construct a sequence of estimates $(\boldsymbol{\theta}^i, \boldsymbol{\rho}^i)$, $i = 1, 2, \ldots$, so that the sequence converges to $(\boldsymbol{\theta}^*, \boldsymbol{\rho}^*)$, the CML estimates of $\boldsymbol{\theta}^0$ and $\boldsymbol{\rho}^0$ based on $\boldsymbol{\xi}(N)$.

It is important to distinguish the iterative method from the recursive or real time computational methods. The latter methods involve updating, i.e., the estimate based on the history $\boldsymbol{\xi}(N+1)$ is to be computed by using an estimate based on the history $\boldsymbol{\xi}(N)$. In the former, the available observations are fixed in size, the different estimates $(\boldsymbol{\theta}^i, \boldsymbol{\rho}^i)$, $i = 1, 2, \ldots$, are all based on the same knowledge $\boldsymbol{\xi}(N)$, and the limit of the sequence $(\boldsymbol{\theta}^i, \boldsymbol{\rho}^i)$ tends to the required CML or any other estimate. Only a few estimators are real time computable.

We recall that the log likelihood function has the form

$$\ln g_2(\boldsymbol{\xi}(N), \boldsymbol{\theta}, \boldsymbol{\rho}) = -\frac{Nn_0}{2}\ln 2\pi - \frac{N}{2}|\det \boldsymbol{\rho}| - \tfrac{1}{2}\sum_{t=1}^{N} \|\mathbf{f}_0(\mathbf{y}(t)) - \mathbf{Z}(t-1)\boldsymbol{\theta}\|^2_{\boldsymbol{\rho}^{-1}}.$$

Iteration Procedure. Let $(\boldsymbol{\theta}^i, \boldsymbol{\rho}^i)$ represent the estimates obtained at the ith iteration. We will give a method of computing $(\boldsymbol{\theta}^{i+1}, \boldsymbol{\rho}^{i+1})$ from $(\boldsymbol{\theta}^i, \boldsymbol{\rho}^i)$.

(A) Obtain $\boldsymbol{\theta}^{i+1}$ by maximizing the likelihood function in $g_2(\boldsymbol{\xi}(N), \boldsymbol{\theta}, \boldsymbol{\rho}^i)$ with respect to $\boldsymbol{\theta}$:

$$\boldsymbol{\theta}^{i+1} = \left[\sum_{i=1}^{N} \mathbf{Z}^{\mathrm{T}}(t-1)(\boldsymbol{\rho}^i)^{-1}\mathbf{Z}(t-1)\right]^{-1} \sum_{t=1}^{N} \mathbf{Z}^{\mathrm{T}}(t-1)(\boldsymbol{\rho}^i)^{-1}\mathbf{f}_0(\mathbf{y}(t)).$$

(B) Obtain the following estimate of $\boldsymbol{\rho}^0$ by utilizing $\boldsymbol{\theta}^{i+1}$, the latest estimate of $\boldsymbol{\theta}^0$. It is obtained by maximizing $g_2(\boldsymbol{\xi}(N), \boldsymbol{\theta}^{i+1}, \boldsymbol{\rho})$ with respect to $\boldsymbol{\rho}$:

$$\boldsymbol{\rho}^{i+1} = \frac{1}{N}\sum_{t=1}^{N} (\mathbf{f}_0(\mathbf{y}(t)) - \mathbf{Z}(t-1)\boldsymbol{\theta}^{i+1})(\mathbf{f}_0(\mathbf{y}(t)) - \mathbf{Z}(t-1)\boldsymbol{\theta}^{i+1})^{\mathrm{T}}.$$

To start the iteration, we need an initial value of $\boldsymbol{\theta}^1$ and $\boldsymbol{\rho}^1$. We will set $\boldsymbol{\theta}^1$ to be the LIML estimate of $\boldsymbol{\theta}^0$. Then $\boldsymbol{\rho}^1$ can be determined from $\boldsymbol{\theta}^0$ by step B.

Steps A and B are alternated a number of times till there is no significant difference between the successive estimates $\boldsymbol{\theta}^i$ and $\boldsymbol{\theta}^{i+1}$. Typically, we stop the iteration at the kth stage if

$$\|\boldsymbol{\theta}^k - \boldsymbol{\theta}^{k-1}\|^2 < \varepsilon_1,$$

where ε_1 is a prespecified value. Alternatively, we can use a stopping rule that uses the log likelihood function itself, and stop at the kth iteration if

$$|\ln g_2(\boldsymbol{\xi}(N), \boldsymbol{\theta}^k, \boldsymbol{\rho}^k) - \ln g_2(\boldsymbol{\xi}(N), \boldsymbol{\theta}^{k-1}, \boldsymbol{\rho}^{k-1})| < \varepsilon_2.$$

Typically ε_1 is chosen to be 0.005.

Convergence of the Iteration Process. We reason that the sequence of estimators $(\boldsymbol{\theta}^i, \boldsymbol{\rho}^i)$ converges to the corresponding CML estimates as follows:

By step A,

$$\ln g_2(\boldsymbol{\xi}(N), \boldsymbol{\theta}^{i+1}, \boldsymbol{\rho}^i) \geq \ln g_2(\boldsymbol{\xi}(N), \boldsymbol{\theta}^i, \boldsymbol{\rho}^i).$$

By step B,

$$\ln g_2(\boldsymbol{\xi}(N), \boldsymbol{\theta}^{i+1}, \boldsymbol{\rho}^{i+1}) \geq \ln g_2(\boldsymbol{\xi}(N), \boldsymbol{\theta}^{i+1}, \boldsymbol{\rho}^i).$$

Thus,

$$\ln g_2(\boldsymbol{\xi}(N), \boldsymbol{\theta}^{i+1}, \boldsymbol{\rho}^{i+1}) \geq \ln g_2(\boldsymbol{\xi}(N), \boldsymbol{\theta}^i, \boldsymbol{\rho}^i).$$

Since g_2 is a concave function in $\boldsymbol{\theta}$ and $\boldsymbol{\rho}$, it is reasonable to conclude that the sequence $(\boldsymbol{\theta}^i, \boldsymbol{\rho}^i)$ will converge to the supremum of $\ln g_2$ with respect to $\boldsymbol{\theta}$ and $\boldsymbol{\rho}$.

6e. Combined Parameter Estimation and Prediction

6e.1. The Statement of the Problem

We will consider forecasting the value of a process one step ahead by using all the knowledge of the observations available at the current instant of time. But unlike the discussion in Chapter II, we will not assume that knowledge of the coefficients in the difference equation obeyed by the process is available. We will only assume that the process $\mathbf{y}$ belongs to a class $[\mathscr{M}, \mathscr{H}, \Omega]$ where $\mathscr{M}$ is the difference equation, $\mathscr{H}$ is the set of all possible values that can be assumed by $\boldsymbol{\theta}$, the vector of coefficients in $\mathscr{M}$, and Ω is the set of values that can be assumed by the covariance of the noise $\mathbf{w}$.

We can obtain an ad hoc solution to the problem as indicated in Fig. 6e.1.1 (Kashyap and Rao, 1973; Wittenmark, 1974). We can estimate the unknown parameters by any of the methods described earlier and then forecast the value by the formula of Chapter II with the system parameters replaced by their corresponding estimates.

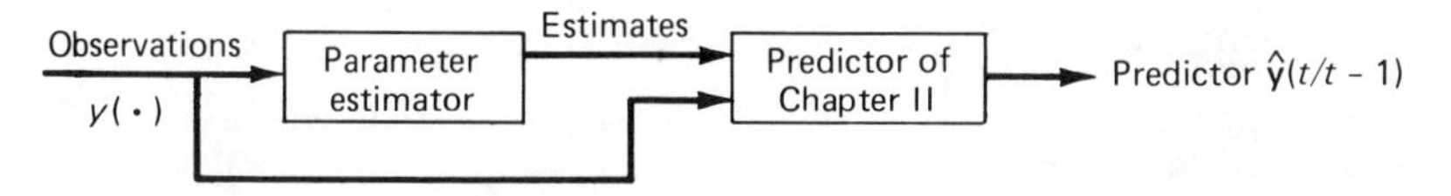

FIG. 6e.1.1. *Ad hoc* one-step-ahead predictor for an unknown process in the class $[\mathcal{M}, \mathcal{H}, \Omega]$.

Suppose we have a particular criterion such as the quadratic function for measuring the performance of the predictor. We can ask whether another predictor which is better than the ad hoc predictor given above can be found according to this criterion. The answer is affirmative in the sense that viewing parameter estimation and forecasting as interrelated parts of a single problem leads to a better predictor than the ad hoc predictor, which treats these two aspects completely independently. Toward this end we will employ the Bayesian approach. Any available prior knowledge about the process $\mathbf{y}$ can be incorporated in the Bayesian approach in the form of a prior distribution for $\boldsymbol{\theta}^0$. We will use the quadratic loss function for measuring the deviation between an estimate and the corresponding true value $\boldsymbol{\theta}^0$. We will find a predictor $\mathbf{y}^*(t|t-1)$ that minimizes the quadratic risk function $J(\hat{\mathbf{y}})$:

$$\begin{aligned} J(\hat{\mathbf{y}}) &= E[\|\mathbf{y}(t) - \hat{\mathbf{y}}(t|t-1)\|^2] \\ &= \int E[\|\mathbf{y}(t) - \hat{\mathbf{y}}(t|t-1)\|^2|\boldsymbol{\theta}]p(\boldsymbol{\theta})\,|d\boldsymbol{\theta}|. \end{aligned} \tag{6e.1.1}$$

The expectation in (6e.1.1) is evaluated by regarding $\boldsymbol{\theta}$ as a random variable with the assigned prior distribution $p(\boldsymbol{\theta})$.

Minimization of $J(\hat{\mathbf{y}})$ with respect to $\hat{\mathbf{y}}$ is relatively easy; the optimal predictor is

$$\mathbf{y}^*(t|t-1) = E[\mathbf{y}(t)|t-1] = \int E[\mathbf{y}(t)|t-1, \boldsymbol{\theta}]p(\boldsymbol{\theta}|\boldsymbol{\xi}(t-1))\,|d\boldsymbol{\theta}|, \tag{6e.1.2}$$

where $p(\boldsymbol{\theta}|\boldsymbol{\xi}(t-1))$ is the posterior density of $\boldsymbol{\theta}$ given the observation history $\boldsymbol{\xi}(t-1)$, and $E[\mathbf{y}(t)|t-1, \boldsymbol{\theta}]$ is the conditional expected value of $\mathbf{y}(t)$ based on all the knowledge until $t-1$ and the value of the coefficient vector $\boldsymbol{\theta}$ in $\mathcal{M}$. The latter predictor was discussed in Chapter II.

The predictor formula in (6e.1.2) involves the storage of all the observations available until time $t-1$ and their manipulation to obtain the predictor $\mathbf{y}^*(t|t-1)$. Computationally it may be a difficult problem. Consequently, we should consider the possibility of making the predictor real time computable, i.e., we should investigate whether the relevant information contained in the semi-infinite observation history $\boldsymbol{\xi}(t-1)$ can be expressed in the form of a finite-dimensional vector $\boldsymbol{\theta}^*(t-1)$ and whether this vector can be updated into $\boldsymbol{\theta}^*(t)$ on the arrival of additional observations at time t *without* having to go into the entire history $\boldsymbol{\xi}(t-1)$ again. Such a procedure will involve only a finite amount of computation at time t to determine $\mathbf{y}^*(t|t-1)$; the amount of memory needed is also finite. But such recursive procedures can be obtained only under appropriate assumptions, some of which are discussed below.

6e.2. Optimal Predictor in Single Output Systems

Let equation $\mathscr{I}$ be as in (6a.1.1) along with its associated assumptions. Let $p(\boldsymbol{\theta}) \sim N(\boldsymbol{\theta}_0, \mathbf{S}_0\rho)$, where $\boldsymbol{\theta}_0$, $\mathbf{S}_0$, and ρ are *known*, ρ being the noise covariance. By the theory in Section 6b,

$$p(\boldsymbol{\theta}|t-1) \sim N(\boldsymbol{\theta}^*(t-1), \mathbf{S}(t-1)\rho), \tag{6e.2.1}$$

$$E[y(t)|t-1, \boldsymbol{\theta}] = E[f_0^{-1}(\boldsymbol{\theta}^{\mathrm{T}}\mathbf{z}(t-1) + w(t))|t-1, \boldsymbol{\theta}]. \tag{6e.2.2}$$

We will consider some special cases.

Case (i). Let $f_0(y) = y$. Hence

$$f_0^{-1}(y) = y$$

$$E[y(t)|t-1, \boldsymbol{\theta}] = E[(\boldsymbol{\theta}^{\mathrm{T}}\mathbf{z}(t-1)) + w(t)|t-1, \boldsymbol{\theta}] = \boldsymbol{\theta}^{\mathrm{T}}\mathbf{z}(t-1),$$

and taking expectation over $\boldsymbol{\theta}$,

$$y^*(t|t-1) = E[\boldsymbol{\theta}^{\mathrm{T}}\mathbf{z}(t-1)|t-1] = (\boldsymbol{\theta}^*(t-1))^{\mathrm{T}}\mathbf{z}(t-1). \qquad (6e.2.3),$$

The predictor y^* is real time computable since $\boldsymbol{\theta}^*$ is real time computable by algorithm (C1). The evaluation of $\boldsymbol{\theta}^*$ does not involve knowledge of the variance ρ. Moreover, the optimal predictor is the same as the ad hoc predictor mentioned in Section 6e.1.

We can also obtain the mean square prediction error of the predictor in (6e.2.3):

$$\begin{aligned} E[(y(t) - y^*(t|t-1))^2] &= E[\{(\boldsymbol{\theta} - \boldsymbol{\theta}^*(t-1))^{\mathrm{T}}\mathbf{z}(t-1) + w(t)\}^2] \\ &= E[\{(\boldsymbol{\theta} - \boldsymbol{\theta}^*(t-1))^{\mathrm{T}}\mathbf{z}(t-1)\}^2] + E[w^2(t)], \\ &\quad \text{by (A1).} \end{aligned}$$

But

$$\begin{aligned} &E[\{(\boldsymbol{\theta} - \boldsymbol{\theta}^*(t-1))^{\mathrm{T}}\mathbf{z}(t-1)\}^2] \\ &\qquad = E[\mathbf{z}^{\mathrm{T}}(t-1)\{\boldsymbol{\theta} - \boldsymbol{\theta}^*(t-1)\}\{\boldsymbol{\theta} - \boldsymbol{\theta}^*(t-1)\}^{\mathrm{T}}\mathbf{z}(t-1)] \\ &\qquad = E[\mathbf{z}^{\mathrm{T}}(t-1)\mathbf{S}(t-1)\mathbf{z}(t-1)\rho], \qquad \text{by (6e.2.2).} \end{aligned}$$

Thus

$$E[(y(t) - y^*(t|t-1))^2] = \rho[1 + E\{\mathbf{z}^{\mathrm{T}}(t-1)\mathbf{S}(t-1)\mathbf{z}(t-1)\}]. \tag{6e.2.4}$$

If we assume that there are no growing trend terms in $\mathbf{z}(t)$ and $\mathbf{u}(t)$ is weak stationary, then the following relationships may be shown to be valid:

$$\left.\begin{aligned} E[\mathbf{z}^{\mathrm{T}}(t-1)\mathbf{S}(t-1)\mathbf{z}(t-1)] &\approx \frac{n}{t-1} \\ E[\{y(t) - y^*(t|t-1)\}^2] &\approx \rho\left(1 + \frac{n}{t-1}\right), \end{aligned}\right. \tag{6e.2.5}$$

where n is the dimension of $\boldsymbol{\theta}$. If $\boldsymbol{\theta}$ were known, the mean square prediction error would be ρ. When $\boldsymbol{\theta}$ is unknown and has to be estimated, the mean square prediction error is increased by a factor $n/(t-1)$.

Case (ii). Let

$$\begin{aligned} f_0(y) &= \ln y, \qquad f_0^{-1}(y) = \exp(y) \\ E[y(t)|t-1, \boldsymbol{\theta}] &= E[\exp(\boldsymbol{\theta}^T\mathbf{z}(t-1) + w(t))|t-1, \boldsymbol{\theta}] \\ &= \exp[\boldsymbol{\theta}^T\mathbf{z}(t-1)]E(e^{w(t)}|t-1, \boldsymbol{\theta}) \\ &= \exp(\boldsymbol{\theta}^T\mathbf{z}(t-1))\exp(\rho/2). \end{aligned}$$

Averaging over $\boldsymbol{\theta}$ using (6e.2.1) we obtain

$$\begin{aligned} y^*(t|t-1) &= \int [\exp(\boldsymbol{\theta}^T\mathbf{z}(t-1))\exp(\rho/2)]p(\boldsymbol{\theta}|t-1)\,|d\boldsymbol{\theta}|\,|d\rho| \\ &= \exp(\rho/2)\int \exp(\boldsymbol{\theta}^T\mathbf{z}(t-1))p(\boldsymbol{\theta}|t-1)\,|d\boldsymbol{\theta}| \\ &= \exp\left(\frac{\rho}{2}\right)\exp[(\boldsymbol{\theta}^*(t-1))^T\mathbf{z}(t-1) + \mathbf{z}^T(t-1)\mathbf{S}(t-1)\mathbf{z}(t-1)\rho]. \end{aligned} \tag{6e.2.6}$$

The ad hoc predictor in this case would be

$$\bar{y}(t|t-1) = E[y(t)|t-1, \boldsymbol{\theta} = \boldsymbol{\theta}^*] = \exp(\rho/2)\exp[(\boldsymbol{\theta}^*)^T\mathbf{z}(t-1)]. \tag{6e.2.7}$$

Thus the optimal predictor is quite different from the ad hoc predictor.

It is difficult to find an explicit expression for the mean square error of the predictor y^*. However, we can compare the optimal predictor y^* and the ad hoc predictor $\bar{y}$ in terms of their respective mean square errors:

$$\begin{aligned} E[(y(t) - \bar{y}(t|t-1))^2] &= E[(y(t) - y^*(t|t-1) + y^*(t|t-1) - \bar{y}(t|t+1))^2] \\ &= E[(y(t) - y^*(t|t-1))^2] \\ &\quad + E[(y^*(t|t-1) - \bar{y}(t|t-1))^2] \\ &\quad + 2E[(y(t) - y^*(t|t-1))(y^*(t|t-1) - \bar{y}(t|t-1))]. \end{aligned} \tag{6e.2.8}$$

The third term on the right-hand side of (6e.2.8) is zero since $E(y(t)|t-1) = y^*(t|t-1)$. Simplifying (6e.2.8) by using the expressions for y^* and $\bar{y}$ in (6e.2.6) and (6e.2.7), we obtain the following expression for the difference in the mean square errors of y^* and $\bar{y}$:

$$\begin{aligned} &E[(y(t) - \bar{y}(t|t-1))^2] - E[(y(t) - y^*(t|t-1))^2] \\ &\quad = E[(y^*(t|t-1) - \bar{y}(t|t-1))^2], \qquad \text{by (6e.2.8)} \\ &\quad = E[\exp\{\rho + 2\mathbf{z}^T(t-1)\boldsymbol{\theta}^*(t-1)\} \\ &\qquad \times \{\exp(\mathbf{z}^T(t-1)\mathbf{S}(t-1)\mathbf{z}(t-1)\rho) - 1\}^2], \\ &\qquad \text{by using (6e.2.6) and (6e.2.7).} \end{aligned} \tag{6e.2.9}$$

We saw earlier that the expression $\mathbf{S}(t-1)$ is of the order $O(1/t-1)$. Thus the squared term in (6e.2.9) can be expected to go to zero as t tends to infinity.

Numerical methods may be used to evaluate the mean square errors of the predictors y^* and $\bar{y}$.

The predictor in (6e.2.6) is recursively computable if ρ is known since $\boldsymbol{\theta}^*$ and $\mathbf{S}$ are recursively computable. If ρ is not known, then we can replace it by $\rho^*(t-1)$, the CML estimate of ρ:

$$\rho^*(t-1) = \frac{1}{t-1}\sum_{j=1}^{t-1} (f_0(y(j)) - \mathbf{z}^{\mathrm{T}}(j-1)\boldsymbol{\theta}^*(t-1))^2. \quad (6e.2.10)$$

But the estimate $\rho^*(t-1)$ is not real time computable. A recursive estimate $\bar{\rho}$ can be written as

$$\bar{\rho}(t) = \bar{\rho}(t-1) + \frac{1}{t}\,[f_0(y(t)) - (\boldsymbol{\theta}^*(t-1))^{\mathrm{T}}\mathbf{z}(t-1)]. \qquad (6e.2.11)$$

The predictor in (6e.2.6) with ρ replaced by $\bar{\rho}(t-1)$ in (6e.2.10) is recursively computable. The predictor in (6e.2.6) with ρ replaced by $\bar{\rho}$ or ρ^* is not optimal, i.e., it does not minimize any reasonable risk function.

6e.3. Optimal Predictor in Multiple Output Systems

Consider the multivariate system in (6a.1.3) along with its associated assumptions (E1′)–(E3′) and $\mathbf{f}_0[\mathbf{y}(t)] = \mathbf{y}(t)$. We assume that the covariance matrix $\boldsymbol{\rho}$ is known. Then $p(\boldsymbol{\theta}|t-1) \sim N(\boldsymbol{\theta}^*(t-1), \bar{\mathbf{S}}(t-1))$. From (6a.1.3), $E[\mathbf{y}(t)|t-1, \boldsymbol{\theta}] = \mathbf{Z}(t-1)\boldsymbol{\theta}$. Taking expectation over $\boldsymbol{\theta}$, the required predictor is $\mathbf{y}^*(t|t-1) = \mathbf{Z}(t-1)\boldsymbol{\theta}^*(t-1)$.

The mean square prediction error matrix

$$\begin{aligned}
&= E[(\mathbf{y}(t) - \mathbf{y}^*(t|t-1))(\mathbf{y}(t) - \mathbf{y}^*(t|t-1))^{\mathrm{T}}] \\
&= E[(\mathbf{Z}(t-1)(\boldsymbol{\theta} - \boldsymbol{\theta}^*(t-1)) + \mathbf{w}(t))(\mathbf{Z}(t-1)(\boldsymbol{\theta} - \boldsymbol{\theta}^*(t-1)) + \mathbf{w}(t))^{\mathrm{T}}] \\
&= E[\mathbf{Z}(t-1)\bar{\mathbf{S}}(t-1)\mathbf{Z}^{\mathrm{T}}(t-1)] + \boldsymbol{\rho}, \qquad \text{by (A1).} \qquad (6e.3.1)
\end{aligned}$$

Unlike the scalar output case, the optimal predictor explicitly involves $\boldsymbol{\rho}$. Through the calculation of $\boldsymbol{\theta}^*(t-1)$, the predictor $\mathbf{y}^*(t|t-1)$ can be recursively computed since $\boldsymbol{\theta}^*(t)$ can be recursively computed as in Section 6d.

If $\boldsymbol{\rho}$ is not known, one can replace it either by the estimate $\boldsymbol{\rho}^*$ or $\bar{\boldsymbol{\rho}}(t)$, both of which are based on the knowledge of observation until time $t-1$:

$$\bar{\boldsymbol{\rho}}(t) = \frac{1}{t-1}\sum_{j=1}^{N} (\mathbf{y}(j) - \mathbf{Z}(j-1)\boldsymbol{\theta}^*(j-1))(\mathbf{y}(j) - \mathbf{Z}(j-1)\boldsymbol{\theta}^*(j-1))^{\mathrm{T}} \qquad (6e.3.2)$$

$$\begin{aligned}
\boldsymbol{\rho}^*(t) &= \boldsymbol{\rho}^*(t-1) \\
&\quad + \frac{1}{t}\,(\mathbf{y}(t) - \mathbf{Z}(t-1)\boldsymbol{\theta}^*(t-1))(\mathbf{y}(t) - \mathbf{Z}(t-1)\boldsymbol{\theta}^*(t-1))^{\mathrm{T}}. \qquad (6e.3.3)
\end{aligned}$$

The estimate $\bar{\boldsymbol{\rho}}$ is more accurate, but $\boldsymbol{\rho}^*$ is recursively computable. When $\boldsymbol{\rho}$ is not known, one can replace $\boldsymbol{\rho}$ in (6e.3.1) by $\boldsymbol{\rho}^*$ in (6e.3.3) and the corresponding predictor $\bar{\mathbf{y}}$ is still recursively computable.

One can perform a similar operation if the left-hand side of Eq. (6a.1.3) was a nonlinear function of **y** instead of being **y** itself. Even though the exact method of developing predictors in such systems is clear, the general form of predictors is usually complex and one has to analyze them individually.

6f. Systems with Slowly Varying Coefficients

6f.1. Description of the Model

Difference equations in which the coefficients vary with time must be considered if we find that the constant coefficient models cannot fully explain the data. We shall begin with the model

$$y(t) = a_1(t) + \sum_{j=2}^{n_1} a_j(t)y(t-j+1) + \sum_{j=n_1+1}^{n_1+n_2} a_j y(t-j+1) + w(t), \quad (6f.1.1)$$

where a_j, $j = n_1+1, \ldots, n_1+n_2$, are constants and $a_j(t)$, $j = 1, \ldots, n_1$, are the time-varying coefficients.

Let

$$a_j(t) = b_j(t) + c_j, \qquad j = 1, \ldots, n_1, \quad (6f.1.2)$$

where c_j, $j = 1, \ldots, n_1$, are constants and $\{b_j(t)\}$, $j = 1, \ldots, n_1$, are sequences of zero mean random variables which are independent of $w(\cdot)$ and obey

$$b_j(t) = \gamma_j b_j(t-1) + \eta_j(t), \quad (6f.1.3)$$

where $\{\eta_j(t)\}$ are IID sequences with zero mean and variance q_j. It is convenient for estimation to express the system in a state variable form. Let

$$\begin{aligned}
\boldsymbol{\theta}^{\mathrm{T}}(t) &= [b_1(t), \ldots, b_{n_1}(t), c_1, \ldots, c_{n_1}, a_{n_1+1}, \ldots, a_{n_1+n_2}] \\
&= n\text{-vector}, \qquad n = 2n_1 + n_2 \\
\mathbf{z}^{\mathrm{T}}(t-1) &= [1, y(t-1), \ldots, y(t-n_1+1), 1, y(t-1), \ldots, y(t-n_1+1), \\
&\qquad y(t-n_1), \ldots, y(t-n_1-n_2)].
\end{aligned}$$

Then Eqs. (6f.1.3) and the fact that c_i, $i = 1, \ldots, n_1$, and a_i, $i \geq n_1+1$, are constants can be expressed as

$$\boldsymbol{\theta}(t) = \mathbf{A}\boldsymbol{\theta}(t-1) + \boldsymbol{\eta}(t), \quad (6f.1.4)$$

$\boldsymbol{\theta}(0)$ is given, and

$$\mathbf{A} = \mathrm{diag}[\underbrace{\gamma_1, \ldots, \gamma_{n_1}}_{n_1}, \underbrace{1, \ldots, 1}_{n_1+n_2}]$$

$$\boldsymbol{\eta}^{\mathrm{T}}(t) = [\eta_1(t), \ldots, \eta_n(t)].$$

$\{\boldsymbol{\eta}(t)\}$ is a sequence of IID random vectors with zero mean and covariance matrix $\mathbf{Q}$:

$$\mathbf{Q} = \operatorname{diag}[\underbrace{q_1, \ldots, q_{n_1}}_{n_1}, \underbrace{0, \ldots, 0}_{n_1+n_2}].$$

Equations (6f.1.1) and (6f.1.2) can be written as

$$y(t) = \boldsymbol{\theta}^{\mathrm{T}}(t-1)\mathbf{z}(t-1) + w(t). \tag{6f.1.5}$$

Equations (6f.1.4) and (6f.1.5) are in standard form for the application of Kalman–Bucy theory (Kalman, 1963). We may mention that there has been very little study on the properties of system (6f.1.4).

6f.2. Estimation of Parameters

It is easy to write down the expression for the likelihood function

$$\begin{aligned}
&p[y(N), \ldots, y(1), \boldsymbol{\theta}(N-1), \ldots, \boldsymbol{\theta}(1) \mid y(0), \ldots, y(-n_1-n_2+1), \boldsymbol{\theta}(0)] \\
&\quad = \prod_{t=n_1+n_2+1}^{N} p(w(t)) \prod_{i=1}^{n_1} \prod_{t=1}^{N} p(\eta_i(t)) \\
&\quad = \prod_{t=n_1+n_2+1}^{N} \frac{1}{(2\pi\rho)^{1/2}} \exp\left[-\frac{1}{2\rho}(y(t) - \boldsymbol{\theta}^{\mathrm{T}}(t-1)\mathbf{z}(t-1))^2\right] \\
&\qquad \times \left(\prod_{i=1}^{n_1} \prod_{t=1}^{N} \frac{1}{(2\pi q_i)^{1/2}} \exp\left[-\frac{1}{2q_i}(\theta_i(t) - \gamma_i\theta_i(t-1))^2\right]\right).
\end{aligned}$$

To obtain the maximum likelihood estimates of the unknowns using the given observations $y(-n_1-n_2+1), \ldots, y(N)$, we have to maximize the above expressions not only with respect to the unknown constants $\theta_i(0)$ $[=\theta_i(t)]$, $i = n_1+1, \ldots, n$, but also the random variables $\theta_i(t)$, $i = 1, \ldots, n$, $t = 1, \ldots, N$.

This is a formidable maximization problem in general. When γ_i, $i = 1, \ldots, n_1$, and $q_i\rho^{-1}$ are known, an elegant recursive solution can be given in terms of the Kalman theory. In the general case, we can give only approximate solutions since there does not exist a computer program that can carry out the maximization even when N is as small as 200. But the results do not seem to be very sensitive to the choice of γ_i and q_i, i.e., approximate knowledge of γ_i and q_i is sufficient to give us good estimates of the constants $\theta_i(0)$, $i = n_1 + 1, \ldots, n$.

Case (*i*). The constants γ_i and q_i, $i = 1, \ldots, n_1$, and ρ are known. Let $\bar{q}_i = q_i\rho^{-1}$, $\bar{\mathbf{Q}} = \mathbf{Q}\rho^{-1}$, $\hat{\boldsymbol{\theta}}(t)$ be the CML estimate of $\boldsymbol{\theta}(t)$ based on $y(j), j \leq t$, and $\mathbf{S}(t)\rho$ be the conditional covariance matrix of $\hat{\boldsymbol{\theta}}(t)$ equal to

$$E[(\boldsymbol{\theta}(t) - \hat{\boldsymbol{\theta}}(t))(\boldsymbol{\theta}(t) - \hat{\boldsymbol{\theta}}(t)) \mid y(j), j \leq t].$$

The recursive discrete Kalman filter algorithm for the computation of $\hat{\boldsymbol{\theta}}(t)$ and $\mathbf{S}(t)$ is as follows:

$$\begin{aligned}
\hat{\boldsymbol{\theta}}(t) &= \mathbf{A}\hat{\boldsymbol{\theta}}(t-1) + \mathbf{S}(t)\mathbf{z}(t-1)[y(t) - (\hat{\boldsymbol{\theta}}(t-1))^{\mathrm{T}}\mathbf{z}(t-1)], \\
\mathbf{S}(t) &= \{(\mathbf{A}\mathbf{S}(t-1)\mathbf{A}^{\mathrm{T}} + \bar{\mathbf{Q}})^{-1} + \mathbf{z}(t-1)\mathbf{z}^{\mathrm{T}}(t-1)\}^{-1}, \qquad t = 1, 2, \ldots, N.
\end{aligned} \tag{C4}$$

The algorithm needs initial conditions $\hat{\boldsymbol{\theta}}(0)$ and $\mathbf{S}(0)$. One choice is as follows:

$$\hat{\theta}_i(0) = \begin{cases} 0, & \text{if } i = 1, \ldots, n \\ \text{the corresponding estimate obtained from the constant} \\ \text{coefficient system using all observations} & \text{if } i = n_1 + 1, \ldots, n. \end{cases}$$

$$\mathbf{S}(0) = k\mathbf{I} \qquad \text{where } k \text{ is a large integer.}$$

Case (ii). γ_i, q_i, and ρ are not known. When ρ is not known, the easiest way to obtain an approximate estimate of ρ is by estimating the corresponding quantity in the corresponding constant coefficient system. If the system is slowly varying, the two estimates will not be far apart.

We will estimate the unknown elements of $\mathbf{A}$ in real time, treating the estimates $\hat{\theta}_i(t)$. $i = 1, \ldots, n_1$, as if they were the actual values $\theta_i(t)$, i.e.,

$$\hat{\gamma}_i(t) = \sum_{j=1}^{t} \hat{\theta}_i(j)\hat{\theta}_i(j-1) \Big/ \sum_{j=1}^{t} (\hat{\theta}_i(j-1))^2, \qquad i = 1, \ldots, n_1,$$

$$\hat{q}_i(t) = \sum_{j=2}^{t} (\hat{\theta}_i(j) - \gamma_i(j-1)\hat{\theta}_i(j-1))^2/t - 1, \qquad i = 1, \ldots, n_1.$$

Let

$$\hat{\mathbf{Q}}(t) = \operatorname{diag}[\hat{q}_1(t)(\hat{\rho})^{-1}, \ldots, \hat{q}_{n_1}(t)(\hat{\rho})^{-1}, 0, \ldots, 0]$$
$$\hat{\mathbf{A}}(t) = \operatorname{diag}[\hat{\gamma}_1(t), \ldots, \hat{\gamma}_{n_1}(t), 1, \ldots, 1].$$

The algorithm

$$\begin{aligned} \hat{\boldsymbol{\theta}}(t) &= \hat{\mathbf{A}}(t-1)\hat{\boldsymbol{\theta}}(t-1) + \mathbf{S}(t)\mathbf{z}(t-1)[y(t) - (\hat{\boldsymbol{\theta}}(t-1))\mathbf{z}(t-1)] \\ \mathbf{S}(t) &= [\{\hat{\mathbf{A}}(t-1)\mathbf{S}(t-1)(\hat{\mathbf{A}}(t-1))^{\mathrm{T}} + \hat{\mathbf{Q}}(t-1)\}^{-1} + \mathbf{z}(t-1)\mathbf{z}^{\mathrm{T}}(t-1)]^{-1} \end{aligned} \qquad \mathbf{C4'}$$

which can be initialized as before.

6f.3. Properties of the Estimate

In Case (i), the conditional covariance of the estimate $\hat{\boldsymbol{\theta}}(t)$ given $\boldsymbol{\xi}(t-1)$ is $\mathbf{S}(t)\rho$. In Case (ii), this statement is only approximately true. The evaluation of the Cramer–Rao bounds is difficult and hence is not considered here.

In Case (i), the sequence $[(y(t) - \hat{\boldsymbol{\theta}}(t-1)\mathbf{z}(t-1)), t = 1, 2, 3, \ldots]$ is white if the model is valid. This sequence plays the same role as the residual sequence in constant coefficient systems. Testing this sequence for whiteness is a check on the validity of the model.

6g. Robust Estimation in AR Models

For some empirical time series, the autoregressive model whose coefficients are obtained from the algorithm (C1) has certain inadequacies. For instance, consider the AR model obtained from N observations of a speech waveform (where $N = 400$, for example). Its spectral density does not display some of the dominant frequencies that would be evident in other methods of analysis such as spectrogram [Markel, 1972]. One possible reason for such behavior may be that the frequency $(1/N)$ implicitly introduced by the length of the data is not explicitly used in the model. Such a deficiency can be corrected by the use of a "window" function. Typically, if $\{y(1), \ldots, y(N)\}$ is the given observation set, we construct a new set $\{x(1), \ldots, x(N)\}$ and fit an AR model for the sequence $\{x\}$, where $x(t) = W(t)y(t)$. The window function $W(t)$ is chosen arbitrarily as follows: $W(t) = a - b\cos(2\pi t/(N-1))$. The new AR model does not suffer from the earlier-mentioned deficiency [Markel, 1972].

If the time series $\{y(1), \ldots, y(N)\}$ obtained from the system (6a.1.1) has several members with unusually large values, i.e., values that deviate from the mean by many standard deviations, then the estimates of the coefficients given by the algorithm (C1) are not appropriate. The quadratic loss function is inappropriate in such cases, since it attaches undue weight to the residuals from such "untypical" observations. In such cases, it is convenient to use the following loss function [Nasburg and Kashyap (1975)]

$$J(\boldsymbol{\theta}) = \sum_{t=1}^{N} F(w_1(t, \boldsymbol{\theta})), \qquad w_1(t, \boldsymbol{\theta}) = (f_0(y(t)) - \boldsymbol{\theta}^{\mathrm{T}}\mathbf{z}(t-1))/\rho^{1/2}$$

$$\begin{aligned} F(x) &= x^2 && \text{if } |x| < 1 \\ &= x && \text{if } x > 1 \\ &= -x && \text{if } x < -1 \end{aligned}$$

One can develop an iterative method for choosing $\boldsymbol{\theta}$ to minimize $J(\boldsymbol{\theta})$. One can show that the sequence $\{\boldsymbol{\theta}^k\}$ given by following second-order gradient methods tends to the value of $\boldsymbol{\theta}$ that minimizes $J(\boldsymbol{\theta})$

Algorithm.

$$\boldsymbol{\theta}^{k+1} = \boldsymbol{\theta}^k + a\rho_1^{1/2}\mathbf{H}^{-1}(\boldsymbol{\theta}^k)\mathbf{h}(\boldsymbol{\theta}^k)$$

where

$$\begin{aligned} \mathbf{h}(\boldsymbol{\theta}) &= \sum_{t=1}^{N} \operatorname{sat}(w_1(t, \boldsymbol{\theta}))\mathbf{z}(t-1) \\ \mathbf{H}(\boldsymbol{\theta}) &= \sum_{t=1}^{N} \mathbf{z}(t-1)\mathbf{z}^{\mathrm{T}}(t-1)F_0(w_1(t, \boldsymbol{\theta})) \\ w_1(t, \boldsymbol{\theta}) &= (f_0(y(t)) - \boldsymbol{\theta}^{\mathrm{T}}\mathbf{z}(t-1))/\rho_1^{1/2} \\ F_0(x) &= \begin{cases} 1 & \text{if } |x| < c \\ 0 & \text{otherwise} \end{cases} \end{aligned}$$

$$\operatorname{sat} x = \begin{cases} +1 & \text{if } x \geq c \\ x & \text{if } |x| < c \\ -1 & \text{if } x < c \end{cases}$$

a is a constant, usually 0.5; c is any constant between 1 and 2. The algorithm is initialized by the estimate $\boldsymbol{\theta}'$ obtained from the algorithm C1. The preliminary residual variance ρ_1 is obtained from this estimate, i.e.,

$$\rho_1 = 1/N \sum_{t=1}^{N} (f_0(y(t)) - (\boldsymbol{\theta}')^{\mathrm{T}}\mathbf{z}(t-1))^2$$

6h. Conclusions

This chapter is devoted to the estimation of parameters in autoregresssive and generalized autoregressive processes in both single output systems (6a.1.1) and multiple output systems (6a.1.2). We first begin with various types of maximum likelihood estimators such as the full information ML, conditional ML, and QML, and discuss their properties. Next we consider the Bayesian estimators so that any available prior information on the parameters can be utilized in the estimation process.

We consider three different types of computational procedures. First, we consider the iterative estimation of parameters in a single equation system. In some instances, the algorithm can be rewritten so that the computation can be performed and the estimates can be updated in real time. Second, we discuss a compact method of fitting a number of AR models of different orders for the same data. Third, we discuss the parameter estimation in time-varying systems.

Finally we consider the problem of simultaneous prediction and parameter estimation, i.e., forecasting of processes whose difference equations involve unknown parameters.

Appendix 6.1. Proofs of Theorems in Section 6a

A. Proof of Theorem 6a.1

Theorem 6a.1. Consider the multivariate system in (6a.4.4) and (6a.4.5). Let

$$\begin{aligned} J_1(\boldsymbol{\theta}, \boldsymbol{\rho}) &\triangleq \ln g_2(\boldsymbol{\xi}(N), \boldsymbol{\theta}, \boldsymbol{\rho}) \\ &= -\frac{mN}{2}\ln 2\pi - \frac{N}{2}\ln\det\boldsymbol{\rho} - \tfrac{1}{2}\|\mathbf{f}_0(\mathbf{y}(t)) - \mathbf{Z}(t-1)\boldsymbol{\theta}\|^2_{\boldsymbol{\rho}^{-1}}. \end{aligned} \tag{1}$$

Let $\boldsymbol{\theta}^*$ and $\boldsymbol{\rho}^*$ be defined as in (6a.4.6) and (6a.4.7). Then

(i) $J_1(\boldsymbol{\theta}^*, \boldsymbol{\rho}) \geq J_1(\boldsymbol{\theta}, \boldsymbol{\rho}), \quad \forall \boldsymbol{\rho} \in \Omega$ and $\boldsymbol{\theta} \in \mathscr{H}$.

(ii) $J_1(\boldsymbol{\theta}^*, \boldsymbol{\rho}^*) \geq J_1(\boldsymbol{\theta}^*, \boldsymbol{\rho}), \quad \forall \boldsymbol{\rho} \in \Omega$.

Proof of Part (*i*). Let

$$J_2(\boldsymbol{\theta}, \boldsymbol{\rho}) = \sum_{t=1}^{N} \|\mathbf{f}_0(\mathbf{y}(t)) - \mathbf{Z}(t-1)\boldsymbol{\theta}\|^2_{\boldsymbol{\rho}^{-1}}$$

and

$$\mathbf{f}_0(t) \triangleq \mathbf{f}_0(\mathbf{y}(t)).$$

We can rewrite $\mathbf{Z}(t-1)\boldsymbol{\theta}$ as $\boldsymbol{\Phi}\mathbf{z}(t-1)$, where $\mathbf{z}(\cdot)$ is an $\bar{n}$-vector defined as in (6a.4.4) and $\boldsymbol{\Phi}$ is an $m \times \bar{n}$ matrix, $\boldsymbol{\theta}$ being a vector of dimension n_0, $n_0 = m\bar{n}$,

$$\boldsymbol{\Phi} = \begin{bmatrix} \boldsymbol{\theta}_1^{\mathrm{T}} \\ \vdots \\ \boldsymbol{\theta}_m^{\mathrm{T}} \end{bmatrix}.$$

Similarly, $\boldsymbol{\Phi}^*$ is defined from $\boldsymbol{\theta}^*$. Let

$$\begin{aligned}
J_3(\boldsymbol{\Phi}, \boldsymbol{\rho}) \triangleq J_2(\boldsymbol{\theta}, \boldsymbol{\rho}) &= \operatorname{trace} \sum_{t=1}^{N} \boldsymbol{\rho}^{-1}\{\mathbf{f}_0(t) - \boldsymbol{\Phi}\mathbf{z}(t-1)\}\{\mathbf{f}_0(t) - \boldsymbol{\Phi}\mathbf{z}(t-1)\} \\
&= \operatorname{trace} \sum_{t=1}^{N} \boldsymbol{\rho}^{-1}\{\mathbf{f}_0(t) - \boldsymbol{\Phi}^*\mathbf{z}(t-1) + (\boldsymbol{\Phi}^* - \boldsymbol{\Phi})\mathbf{z}(t-1)\} \\
&\quad \times \{\mathbf{f}_0(t) - \boldsymbol{\Phi}^*\mathbf{z}(t-1) + (\boldsymbol{\Phi}^* - \boldsymbol{\Phi})\mathbf{z}(t-1)\}^{\mathrm{T}} \\
&= \operatorname{trace}\left[\sum_{t=1}^{N} \boldsymbol{\rho}^{-1}\{\mathbf{f}_0(t) - \boldsymbol{\Phi}^*\mathbf{z}(t-1)\}\{\mathbf{f}_0(t) - \boldsymbol{\Phi}^*\mathbf{z}(t-1)\}^{\mathrm{T}}\right] \\
&\quad + \operatorname{trace}\left[\sum_{t=1}^{N} \boldsymbol{\rho}^{-1}\{\mathbf{f}_0(t) - \boldsymbol{\Phi}^*\mathbf{z}(t-1)\}\mathbf{z}^{\mathrm{T}}(t-1)(\boldsymbol{\Phi}^* - \boldsymbol{\Phi})^{\mathrm{T}}\right] \\
&\quad + \operatorname{trace}\left[\sum_{t=1}^{N} \boldsymbol{\rho}^{-1}(\boldsymbol{\Phi}^* - \boldsymbol{\Phi})\mathbf{z}(t-1)\mathbf{z}^{\mathrm{T}}(t-1)(\boldsymbol{\Phi}^* - \boldsymbol{\Phi})^{\mathrm{T}}\right].
\end{aligned} \tag{2}$$

The second term in (2)

$$\begin{aligned}
&= \operatorname{trace}\left[\boldsymbol{\rho}^{-1}\left\{\sum_{t=1}^{N} \mathbf{f}_0(t)\mathbf{z}^{\mathrm{T}}(t-1) - \boldsymbol{\Phi}^* \sum_{t=1}^{N} \mathbf{z}(t-1)\mathbf{z}^{\mathrm{T}}(t-1)\right\}(\boldsymbol{\Phi}^* - \boldsymbol{\Phi})^{\mathrm{T}}\right] \\
&= 0,
\end{aligned}$$

by the definition of $\boldsymbol{\Phi}^*$. By inspection, the third term in (2) is nonnegative. The first term in (1) is $J_3(\boldsymbol{\Phi}^*, \boldsymbol{\rho})$. Hence $J_3(\boldsymbol{\Phi}^*, \boldsymbol{\rho}) \leq J_3(\boldsymbol{\Phi}, \boldsymbol{\rho})$, i.e., $J_2(\boldsymbol{\theta}^*, \boldsymbol{\rho}) \leq J_2(\boldsymbol{\theta}, \boldsymbol{\rho})$, $\forall \boldsymbol{\theta} \in \mathscr{H}$, $\boldsymbol{\rho} \in \Omega$. Hence $J_1(\boldsymbol{\theta}^*, \boldsymbol{\rho}) \geq J_1(\boldsymbol{\theta}, \boldsymbol{\rho})$.

Proof of Part (ii). We want to maximize $J_1(\boldsymbol{\theta}^*, \boldsymbol{\rho})$ with respect to $\boldsymbol{\rho}$. Equivalently, we can maximize $J_1(\cdot)$ with respect to $\boldsymbol{\rho}^{-1}$ since it is easier to do so:

$$\begin{aligned}
&\frac{\partial J_1(\boldsymbol{\theta}, \boldsymbol{\rho})}{\partial \boldsymbol{\rho}^{-1}} \\
&= -\frac{N}{2}\left\{\boldsymbol{\rho}^{-1} - \frac{1}{N}\sum_{t=1}^{N} [\mathbf{f}_0(\mathbf{y}(t)) - \mathbf{Z}(t-1)\boldsymbol{\theta}^*][\mathbf{f}_0(\mathbf{y}(t)) - \mathbf{Z}(t-1)\boldsymbol{\theta}^*]^{\mathrm{T}}\right\},
\end{aligned} \tag{3}$$

$$\left.\frac{\partial J_1(\boldsymbol{\theta}^*, \boldsymbol{\rho})}{\partial \boldsymbol{\rho}^{-1}}\right|_{\boldsymbol{\rho}=\boldsymbol{\rho}^*} = 0, \tag{4}$$

by definition of $\boldsymbol{\rho}^*$. To verify the second derivative condition we need to change the notation slightly. Let

$$\boldsymbol{\alpha} = [(\rho^{-1})_{11}, (\rho^{-1})_{12}, \ldots, (\rho^{-1})_{1n}, \ldots, (\rho^{-1})_{nn}]^{\mathrm{T}}.$$

Equation (3) can be rewritten as

$$\frac{\partial J_1(\boldsymbol{\theta}^*, \boldsymbol{\rho})}{\partial \boldsymbol{\alpha}} = -\frac{N}{2}\boldsymbol{\alpha} + \text{a function independent of } \boldsymbol{\alpha}. \tag{5}$$

Differentiating (5) with respect to $\boldsymbol{\alpha}$, we obtain

$$\frac{\partial^2 J_1(\boldsymbol{\theta}^*, \boldsymbol{\rho})}{\partial \boldsymbol{\alpha}^2} = -\frac{N}{2}\mathbf{I} \tag{6}$$

$$< \mathbf{0}. \tag{7}$$

Equations (4) and (7) imply part (ii) of Theorem 6a.1.

B. Properties of the CML Estimators in Multiple Output Systems

(i) *To prove:*

$$E[\boldsymbol{\theta}^*] = \boldsymbol{\theta}^0 + O(1/N^{1/2}), \tag{8}$$

where

$$\boldsymbol{\theta}^* = \left[\sum_{t=1}^{N} \mathbf{Z}^{\mathrm{T}}(t-1)\boldsymbol{\rho}^{-1}\mathbf{Z}(t-1)\right]^{-1}\left[\sum_{t=1}^{N} \mathbf{Z}^{\mathrm{T}}(t-1)\boldsymbol{\rho}^{-1}\mathbf{f}_0(\mathbf{y}(t))\right]. \tag{9}$$

Proof. Substituting for $\mathbf{f}_0(\mathbf{y}(t))$ in (9) and simplifying, we obtain

$$E[\boldsymbol{\theta}^*] = \boldsymbol{\theta}^0 + E\left[\bar{\mathbf{S}}(N)\sum_{t=1}^{N} \mathbf{Z}^{\mathrm{T}}(t-1)\boldsymbol{\rho}^{-1}\mathbf{w}(t)\right], \tag{10}$$

where $\bar{\mathbf{S}}(N)$ is the inverse matrix in (9). Let

$$\mathbf{G}(N) = \sum_{t=1}^{N} \mathbf{Z}^{\mathrm{T}}(t-1)\boldsymbol{\rho}^{-1}\mathbf{w}(t).$$

Recall that $\mathbf{w}(t)$ is independent of $\mathbf{Z}(t-1)$. Hence, the sum $\mathbf{G}(N)$ is made up of pairwise independent random vectors. Consequently $E[\|\mathbf{G}(N)\|^2] = O(N)$. We argued in Chapter IV that $\bar{\mathbf{S}}(N)$ is of the order $O(1/N)$, under the assumptions in Section 6a. Hence,

$$\|E(\boldsymbol{\theta}^*) - \boldsymbol{\theta}^0\|^2 \le E\|\bar{\mathbf{S}}(N)\|^2 \cdot E\|\mathbf{G}(N)\|^2 = O(1/N).$$

(ii) *To prove:*

$$E[(\boldsymbol{\theta}^* - \boldsymbol{\theta}^0)(\boldsymbol{\theta}^* - \boldsymbol{\theta}^0)^{\mathrm{T}}] \approx \left[E\left\{\sum_{t=1}^{N} \mathbf{Z}^{\mathrm{T}}(t-1)(\boldsymbol{\rho}^0)^{-1}\mathbf{Z}(t-1)\right\}\right]^{-1}. \tag{11}$$

Proof. From (9),

$$\boldsymbol{\theta}^* = \boldsymbol{\theta}^0 + \bar{\mathbf{S}}(N)\sum_{t=1}^{N} \mathbf{Z}^{\mathrm{T}}(t-1)(\boldsymbol{\rho}^0)^{-1}\mathbf{w}(t)$$

$$\begin{aligned} E[(\boldsymbol{\theta}^* - \boldsymbol{\theta}^0)(\boldsymbol{\theta}^* - \boldsymbol{\theta}^0)^{\mathrm{T}}] &= E\left[\bar{\mathbf{S}}(N)\left(\sum_{t=1}^{N} \mathbf{Z}^{\mathrm{T}}(t-1)((\boldsymbol{\rho}^0)^{-1}\mathbf{w}(t))\right)\right. \\ &\quad \times \left.\left(\sum_{t=1}^{N} \mathbf{Z}^{\mathrm{T}}(t-1)((\boldsymbol{\rho}^0)^{-1}\mathbf{w}(t))\right)\bar{\mathbf{S}}(N)\right] \\ &\triangleq E[\bar{\mathbf{S}}(N)\mathbf{Q}(N)\bar{\mathbf{S}}(N)], \end{aligned} \tag{12}$$

where $\mathbf{Q}(N)$ is the middle term in (12).

To simplify the manipulations we will make the following two assumptions,

(a) $\bar{\mathbf{S}}(N)$ is asymptotically independent of $\mathbf{Q}(N)$.
(b) $[E(\bar{\mathbf{S}}(N))]^{-1}$ is asymptotically approximately equal to $E[(\bar{\mathbf{S}}(N))^{-1}]$.

It should be noted that the result in (11) can be proved without explicitly making these assumptions, but it would be lengthy.

Using assumption (a),

$$E[(\boldsymbol{\theta}^* - \boldsymbol{\theta}^0)(\boldsymbol{\theta}^* - \boldsymbol{\theta}^0)^{\mathrm{T}}] = E[\bar{\mathbf{S}}(N)]E[\mathbf{Q}(N)]E[\bar{\mathbf{S}}(N)]$$

$$\begin{aligned} E[\mathbf{Q}(N)] &= E\left[\sum_{t_1=1}^{N}\sum_{t_2=1}^{N} \mathbf{Z}^{\mathrm{T}}(t_1 - 1)(\boldsymbol{\rho}^0)^{-1}\mathbf{w}(t_1)\mathbf{w}^{\mathrm{T}}(t_2)(\boldsymbol{\rho}^0)^{-1}\mathbf{Z}(t_2 - 1)\right] \\ &= \Bigg[\underbrace{\sum_{t_1=1}^{N}\sum_{t_2=1}^{N}}_{t_1>t_2} + \underbrace{\sum_{t_1=1}^{N}\sum_{t_2=1}^{N}}_{t_2>t_1} + \underbrace{\sum_{t_1=1}^{N}\sum_{t_2=1}^{N}}_{t_1=t_2}\Bigg] \\ &\quad \times [\mathbf{Z}^{\mathrm{T}}(t_1 - 1)(\boldsymbol{\rho}^0)^{-1}\mathbf{w}(t_1)\mathbf{w}^{\mathrm{T}}(t_2)(\boldsymbol{\rho}^0)^{-1}\mathbf{Z}(t_2 - 1)] \\ &= \mathrm{I} + \mathrm{II} + \mathrm{III} \end{aligned} \tag{13}$$

$$\mathrm{I} = \underbrace{\sum_{t_1=1}^{N}\sum_{t_2=1}^{N}}_{t_1>t_2} E[\mathbf{Z}^{\mathrm{T}}(t_1 - 1)(\boldsymbol{\rho}^0)^{-1}E(\mathbf{w}(t_1))\mathbf{w}^{\mathrm{T}}(t_2)(\boldsymbol{\rho}^0)^{-1}\mathbf{Z}(t_2 - 1)].$$

Since $\mathbf{w}(t_1)$ is independent of $\mathbf{w}(t_2)$, $\mathbf{Z}^{\mathrm{T}}(t_2 - 1)$, and $\mathbf{Z}^{\mathrm{T}}(t_1 - 1)$ when $t_1 > t_2$, and $E[\mathbf{w}(t_1)] = 0$, term I in (13) is zero. Similarly, term II in (13) is zero. Term III in (13) is $\sum_{t=1}^{N} E[\mathbf{Z}^{\mathrm{T}}(t - 1)(\boldsymbol{\rho}^0)^{-1}E[\mathbf{w}(t)\mathbf{w}^{\mathrm{T}}](\boldsymbol{\rho}^0)^{-1}\mathbf{Z}(t - 1)]$. Since $\mathbf{w}(t)$ is independent of $\mathbf{Z}(t - 1)$,

$$\mathrm{III} = \sum_{t=1}^{N} E[\mathbf{Z}^{\mathrm{T}}(t - 1)(\boldsymbol{\rho}^0)^{-1}\mathbf{Z}(t - 1)].$$

Since $\boldsymbol{\rho}^0 = E[\mathbf{w}(t)\mathbf{w}^{\mathrm{T}}(t)]$, asymptotically we have

$$\mathrm{III} = E[(\bar{\mathbf{S}}(N))^{-1}] \approx [E(\bar{\mathbf{S}}(N))]^{-1},$$

Hence,

$$E[(\boldsymbol{\theta}^* - \boldsymbol{\theta}^0)(\boldsymbol{\theta}^* - \boldsymbol{\theta}^0)^{\mathrm{T}}] = E[\bar{\mathbf{S}}(N)] + o(1/N).$$

The need for the correction factor $o(1/N)$ can be demonstrated as in the proof of part (i), above.

C. To Show That the Expression for the MSE Given in Part (ii) Is Asymptotically Valid Even for the Estimate of $\boldsymbol{\theta}$ Determined without the Knowledge of $\boldsymbol{\rho}$

$$\begin{aligned} \boldsymbol{\theta}^* &= \left[\sum_{t=1}^{N} \mathbf{Z}^{\mathrm{T}}(t - 1)(\boldsymbol{\rho}^*)^{-1}\mathbf{Z}(t - 1)\right]^{-1}\left[\sum_{t=1}^{N} \mathbf{Z}^{\mathrm{T}}(t - 1)(\boldsymbol{\rho}^*)^{-1}\mathbf{f}_0(\mathbf{y}(t))\right] \\ &= \boldsymbol{\theta}^0 + \left[\sum_{t=1}^{N} \mathbf{Z}^{\mathrm{T}}(t - 1)(\boldsymbol{\rho}^*)^{-1}\mathbf{Z}(t - 1)\right]^{-1}\left[\sum_{t=1}^{N} \mathbf{Z}^{\mathrm{T}}(t - 1)(\boldsymbol{\rho}^*)^{-1}\mathbf{w}(t)\right], \end{aligned} \tag{14}$$

where $\boldsymbol{\rho}^*$ is the CML estimate of $\boldsymbol{\rho}^0$. Let $\boldsymbol{\rho}^* = \boldsymbol{\rho}^0 + \Delta\boldsymbol{\rho}$, or

$$(\boldsymbol{\rho}^*)^{-1} \approx (\boldsymbol{\rho}^0)^{-1} - (\boldsymbol{\rho}^0)^{-1}\Delta\boldsymbol{\rho}(\boldsymbol{\rho}^0)^{-1}. \tag{15}$$

Substituting (15) in (14) and simplifying, we obtain

$$\begin{aligned}\boldsymbol{\theta}^* - \boldsymbol{\theta}^0 &= \left[\sum_{t=1}^{N} \mathbf{Z}^{\mathrm{T}}(t-1)(\boldsymbol{\rho}^0)^{-1}\mathbf{Z}(t-1)\right]^{-1} \sum_{t=1}^{N} \mathbf{Z}^{\mathrm{T}}(t-1)(\boldsymbol{\rho}^0)^{-1}\mathbf{w}(t) \\ &\quad + \text{a function involving } \mathbf{Z}(t-1),\ \boldsymbol{\rho}^0,\ \mathbf{w}(t), \text{ and } \Delta\boldsymbol{\rho}. \\ &= \mathrm{I} + \mathrm{II} \end{aligned} \tag{16}$$

The contribution of the II term in (16) to the mean square matrix of $(\boldsymbol{\theta}^* - \boldsymbol{\theta}^0)$ is $o(1/N)$ since the mean square matrix of $\Delta\boldsymbol{\rho}$ is of the order $O(1/N)$. Consequently, term I is the dominant contributor to the mean square matrix of $(\boldsymbol{\theta}^* - \boldsymbol{\theta}^0)$ leading to the expression in the earlier section.

Appendix 6.2. The Expressions for the Posterior Densities

A. Theorem 6b.3

For the multivariate system in (6a.1.3) with additional assumptions (E1′)–(E3′), the posterior density of $\boldsymbol{\theta}$ is

$$p(\boldsymbol{\theta}|\boldsymbol{\xi}) = N(\boldsymbol{\theta}^*(N), \bar{\mathbf{S}}(N)),$$

where

$$\boldsymbol{\xi} = [\underbrace{\mathbf{y}(N), \ldots, \mathbf{y}(1)}_{\boldsymbol{\xi}_2}, \underbrace{\mathbf{y}(0), \ldots, \mathbf{y}(-m_1), \mathbf{u}(N-1), \ldots, \mathbf{u}(-m_3)}_{\boldsymbol{\xi}_1}]$$

$$\bar{\mathbf{S}}(N) = \left[\sum_{t=1}^{N} \mathbf{Z}^{\mathrm{T}}(t-1)\boldsymbol{\rho}^{-1}\mathbf{Z}(t-1) + \bar{\mathbf{S}}_0^{-1}\right]^{-1}$$

$$\boldsymbol{\theta}^*(N) = \bar{\mathbf{S}}(N)\left[\sum_{t=1}^{N} \mathbf{Z}^{\mathrm{T}}(t-1)\boldsymbol{\rho}^{-1}\mathbf{f}_0(\mathbf{y}(t)) + \bar{\mathbf{S}}_0^{-1}\boldsymbol{\theta}^0\right].$$

Proof.

$$\begin{aligned} p(\boldsymbol{\theta}|\boldsymbol{\xi}) &= p(\boldsymbol{\theta}, \boldsymbol{\xi})/p(\boldsymbol{\xi}) = p(\boldsymbol{\theta}, \mathbf{y}(1), \ldots, \mathbf{y}(N), \boldsymbol{\xi}_1)/p(\boldsymbol{\xi}) \\ &= \left[\prod_{t=1}^{N} p(\mathbf{y}(t)|\mathbf{y}(t-1), \ldots, \mathbf{y}(1), \boldsymbol{\theta}, \boldsymbol{\xi}_1) p(\boldsymbol{\theta}|\boldsymbol{\xi}_1) p(\boldsymbol{\xi}_1)/p(\boldsymbol{\xi})\right] \\ &= \left[\prod_{t=1}^{N} p(\mathbf{y}(t)|\mathbf{y}(t-1, \ldots, \mathbf{y}(1), \boldsymbol{\theta}, \boldsymbol{\xi}_1) p(\boldsymbol{\theta})/p(\boldsymbol{\xi}|\boldsymbol{\xi}_1)\right], \quad \text{by (E1′)}. \end{aligned} \tag{17}$$

$$\begin{aligned} &p(\mathbf{y}(t)|\mathbf{y}(t-1), \ldots, \mathbf{y}(1), \boldsymbol{\xi}_1, \boldsymbol{\theta}) \\ &\quad = p_w[(\mathbf{f}_0(\mathbf{y}(t)) - \mathbf{Z}(t-1)\boldsymbol{\theta})|\boldsymbol{\theta}] \\ &\quad = (2\pi) \det \boldsymbol{\rho}^{-1/2} \exp[-\tfrac{1}{2}\|\mathbf{f}_0(\mathbf{y}(t)) - \mathbf{Z}(t-1)\boldsymbol{\theta}\|^2_{\boldsymbol{\rho}^{-1}}], \end{aligned} \tag{18}$$

where $p_w(\cdot)$ denotes the probability density of the random variable $\mathbf{w}$ which is $N(0, \boldsymbol{\rho})$, $\boldsymbol{\rho}$ being known. Substituting (18) and the expression in (E3′) for $p(\boldsymbol{\theta})$ into (17), we obtain

$$\begin{aligned} p(\boldsymbol{\theta}|\boldsymbol{\xi})p(\boldsymbol{\xi}|\boldsymbol{\xi}_1) &= \prod_{t=1}^{N} (2\pi)^{-m/2}|\det \boldsymbol{\rho}|^{-1/2} \exp[-\tfrac{1}{2}\|\mathbf{f}_0(\mathbf{y}(t)) - \mathbf{Z}(t-1)\boldsymbol{\theta}\|^2_{\boldsymbol{\rho}^{-1}}] \\ &\quad \times (2\pi)^{-n_0/2}|\det \bar{\mathbf{S}}_0|^{-1/2} \exp[-\tfrac{1}{2}\|\boldsymbol{\theta} - \boldsymbol{\theta}^0\|^2_{\bar{\mathbf{S}}_0^{-1}}] \\ &= (2\pi)^{-(Nm+n_0)/2}|\det \boldsymbol{\rho}|^{-N/2}|\det \bar{\mathbf{S}}_0|^{-1/2} \\ &\quad \times \exp[-\tfrac{1}{2}\|\mathbf{f}_0(\mathbf{y}(t)) - \mathbf{Z}(t-1)\boldsymbol{\theta}\|^2_{\boldsymbol{\rho}^{-1}} - \tfrac{1}{2}\|\boldsymbol{\theta} - \boldsymbol{\theta}^0\|^2_{\bar{\mathbf{S}}_0^{-1}}]. \end{aligned} \tag{19}$$

We notice that the argument of the exponent in (19) is a quadratic function of $\boldsymbol{\theta}$. We can conclude that the posterior density of $\boldsymbol{\theta}$ is normal and determine it by completing the squares.

The argument of the exponent on the right-hand side of (19)

$$\begin{aligned} &= -\tfrac{1}{2}\|\mathbf{f}_0(\mathbf{y}(t)) - \mathbf{Z}(t-1)\boldsymbol{\theta}\|^2_{\boldsymbol{\rho}^{-1}} - \tfrac{1}{2}\|\boldsymbol{\theta} - \boldsymbol{\theta}^0\|^2_{\bar{\mathbf{S}}_0^{-1}} \\ &= -\tfrac{1}{2}\Bigg[\boldsymbol{\theta}^T\left(\sum_{t=1}^{N} \mathbf{Z}^T(t-1)\boldsymbol{\rho}^{-1}\mathbf{Z}(t-1) + \bar{\mathbf{S}}_0^{-1}\right)\boldsymbol{\theta} \\ &\quad - 2\boldsymbol{\theta}^T\left(\sum_{t=1}^{N} \mathbf{Z}^T(t-1)\boldsymbol{\rho}^{-1}\mathbf{f}_0(\mathbf{y}(t)) + \bar{\mathbf{S}}_0^{-1}\boldsymbol{\theta}^0\right) + \text{terms not involving } \boldsymbol{\theta}\Bigg] \\ &= -\tfrac{1}{2}[\boldsymbol{\theta}^T[\bar{\mathbf{S}}^{-1}(N)]^{-1}\boldsymbol{\theta} - 2\boldsymbol{\theta}^T[\bar{\mathbf{S}}^{-1}(N)]^{-1}\boldsymbol{\theta}^*(N) + \text{terms not involving } \boldsymbol{\theta}] \end{aligned}$$

by the definition of $\bar{\mathbf{S}}(N)$ and $\boldsymbol{\theta}^*(N)$,

$$= -\tfrac{1}{2}[\|\boldsymbol{\theta} - \boldsymbol{\theta}^*(N)\|^2_{(\bar{\mathbf{S}}(N))^{-1}} + \text{terms not involving } \boldsymbol{\theta}]. \tag{20}$$

Equation (20) yields the required probability density.

B. Theorem 6b.2

For the single output system, $p(\boldsymbol{\theta}|\boldsymbol{\xi}(N)) = N(\boldsymbol{\theta}^*(N), \mathbf{S}(N)\rho)$.

Proof. It follows from Theorem 6b.3 by setting $m = 1$, $\bar{\mathbf{S}}(N) = \mathbf{S}(N)\rho$, and $\bar{\mathbf{S}}_0 = \mathbf{S}_0\rho$.

Appendix 6.3. The Derivation of Computational Algorithms

A. Derivation of Algorithm (C2) from (21) and (22)

$$\boldsymbol{\theta}^*(N) = \bar{\mathbf{S}}(N)\left\{\sum_{t=1}^{N} \mathbf{Z}^T(t-1)\boldsymbol{\rho}^{-1}\mathbf{f}_0(\mathbf{y}(t)) + \mathbf{S}_0^{-1}\boldsymbol{\theta}_0\right\} \tag{21}$$

$$\bar{\mathbf{S}}(N) = \left[\sum_{t=1}^{N} \mathbf{Z}^T(t-1)\boldsymbol{\rho}^{-1}\mathbf{Z}(t-1) + \mathbf{S}_0^{-1}\right]^{-1}. \tag{22}$$

We need the following lemma:

Lemma 1.

$$\bar{\mathbf{S}}(N-1)\mathbf{Z}^T(N-1)[\mathbf{Z}(N-1)\bar{\mathbf{S}}(N-1)\mathbf{Z}^T(N-1) + \boldsymbol{\rho}]^{-1} = \bar{\mathbf{S}}(N)\mathbf{Z}^T(N-1)\boldsymbol{\rho}^{-1}.$$

Proof of (C2). By (22), we have

$$(\bar{\mathbf{S}}(N))^{-1} = \bar{\mathbf{S}}^{-1}(N-1) + \mathbf{Z}^{\mathrm{T}}(N-1)\boldsymbol{\rho}^{-1}\mathbf{Z}(N-1). \tag{23}$$

We will apply the matrix inversion lemma (Rao, 1965) to (23) to obtain

$$\begin{aligned}\bar{\mathbf{S}}(N) = \bar{\mathbf{S}}(N-1) - \bar{\mathbf{S}}(N-1)\mathbf{Z}^{\mathrm{T}}(N-1) \\ \times [\mathbf{Z}(t-1)\bar{\mathbf{S}}(N-1)\mathbf{Z}^{\mathrm{T}}(N-1) + \boldsymbol{\rho}]^{-1}\mathbf{Z}(N-1)\bar{\mathbf{S}}(N-1).\end{aligned} \tag{24}$$

To obtain a recursive algorithm for $\boldsymbol{\theta}^*(N)$, we start with (21):

$$\begin{aligned}\boldsymbol{\theta}^*(N) &= \bar{\mathbf{S}}(N)\mathbf{Z}^{\mathrm{T}}(N-1)\boldsymbol{\rho}^{-1}\mathbf{f}_0(\mathbf{y}(N)) \\ &\quad + \bar{\mathbf{S}}(N)\left\{\sum_{t=1}^{N-1}\mathbf{Z}^{\mathrm{T}}(t-1)\boldsymbol{\rho}^{-1}\mathbf{Z}(t-1)\mathbf{f}_0(\mathbf{y}(t)) + \mathbf{S}_0^{-1}\boldsymbol{\theta}_0\right\}.\end{aligned}$$

By substituting for the expression in braces, using (21) with N replaced by $N-1$, we obtain

$$\boldsymbol{\theta}^*(N) = \bar{\mathbf{S}}(N)\mathbf{Z}^{\mathrm{T}}(N-1)\boldsymbol{\rho}^{-1}\mathbf{f}(\mathbf{y}(N)) + \bar{\mathbf{S}}(N)(\bar{\mathbf{S}}(N-1))^{-1}\boldsymbol{\theta}^*(N-1). \tag{25}$$

Substituting for $\bar{\mathbf{S}}(N)$ in the second term above from (24), we obtain the following expression for $\boldsymbol{\theta}^*(N)$:

$$\begin{aligned}\boldsymbol{\theta}^*(N) &= \bar{\mathbf{S}}(N)\mathbf{Z}^{\mathrm{T}}(N-1)\boldsymbol{\rho}^{-1}\mathbf{f}_0(\mathbf{y}(N)) + \boldsymbol{\theta}^*(N-1) \\ &\quad - \bar{\mathbf{S}}(N-1)\mathbf{Z}^{\mathrm{T}}(N-1)[\mathbf{Z}(N-1)\bar{\mathbf{S}}(N-1)\mathbf{Z}^{\mathrm{T}}(N-1) + \boldsymbol{\rho}]^{-1} \\ &\quad \times \mathbf{Z}(N-1)\boldsymbol{\theta}^*(N-1).\end{aligned} \tag{26}$$

Using Lemma 1 to simplify the third term above, we have

$$\begin{aligned}\boldsymbol{\theta}^*(N) &= \bar{\mathbf{S}}(N)\mathbf{Z}^{\mathrm{T}}(N-1)\boldsymbol{\rho}^{-1}\mathbf{f}_0(\mathbf{y}(N)) + \boldsymbol{\theta}^*(N-1) \\ &\quad - \bar{\mathbf{S}}(N)\mathbf{Z}^{\mathrm{T}}(N-1)\boldsymbol{\rho}^{-1}\mathbf{Z}(N-1)\boldsymbol{\theta}^*(N-1) \\ &= \boldsymbol{\theta}^*(N-1) + \bar{\mathbf{S}}(N)\mathbf{Z}^{\mathrm{T}}(N-1)\boldsymbol{\rho}^{-1}[\mathbf{f}_0(\mathbf{y}(N)) - \mathbf{Z}(N-1)\boldsymbol{\theta}^*(N-1)],\end{aligned}$$

by rearranging the terms.

Proof of Lemma 1.

$$\begin{aligned}\bar{\mathbf{S}}(N)\mathbf{Z}^{\mathrm{T}}(N-1)\boldsymbol{\rho}^{-1} &= \bar{\mathbf{S}}(N-1)\mathbf{Z}^{\mathrm{T}}(N-1)\boldsymbol{\rho}^{-1} - \bar{\mathbf{S}}(N-1)\mathbf{Z}^{\mathrm{T}}(N-1) \\ &\quad \times [\mathbf{Z}(N-1)\bar{\mathbf{S}}(N-1)\mathbf{Z}^{\mathrm{T}}(N-1) + \boldsymbol{\rho}]^{-1} \\ &\quad \times \{\mathbf{Z}(N-1)\bar{\mathbf{S}}(N-1)\mathbf{Z}^{\mathrm{T}}(N-1)\}\boldsymbol{\rho}^{-1}\end{aligned}$$

by (24). Adding and subtracting a term $\boldsymbol{\rho}$ to the expression in braces and simplifying, we get the identity in Lemma 1:

$$\begin{aligned}\bar{\mathbf{S}}(N)\mathbf{Z}^{\mathrm{T}}(N-1)\boldsymbol{\rho}^{-1} &= \bar{\mathbf{S}}(N-1)\mathbf{Z}^{\mathrm{T}}(N-1)\boldsymbol{\rho}^{-1} - \bar{\mathbf{S}}(N-1)\mathbf{Z}^{\mathrm{T}}(N-1)\boldsymbol{\rho}^{-1} \\ &\quad - \bar{\mathbf{S}}(N-1)\mathbf{Z}^{\mathrm{T}}(N-1)[\mathbf{Z}(N-1)\bar{\mathbf{S}}(N-1)\mathbf{Z}^{\mathrm{T}}(N-1)] \\ &= \bar{\mathbf{S}}(N-1)\mathbf{Z}^{\mathrm{T}}(N-1)[\mathbf{Z}(N-1)\bar{\mathbf{S}}(N-1)\mathbf{Z}^{\mathrm{T}}(N-1) + \boldsymbol{\rho}]^{-1}.\end{aligned}$$

B. Derivation of Algorithm (C1)

Algorithm (C1) is a special case of algorithm (C2), as indicated below:

$$m = 1, \qquad \bar{\mathbf{S}}(N) = \mathbf{S}(N)\rho, \qquad \bar{\mathbf{S}}_0 = \mathbf{S}_0\rho.$$

Appendix 6.4. Evaluation of the Cramér–Rao Lower Bound in Multivariate AR Systems

We will show that the CML estimate of $\boldsymbol{\theta}^0$ has asymptotically minimum variance, even when $\boldsymbol{\rho}^0$ is not known, the variance being given by the Cramér–Rao lower bound or inequality. We considered the case when $\boldsymbol{\rho}^0$ is known in Chapter IV:

$$\begin{aligned}\ln p(\mathbf{f}_0(\mathbf{y}(N))|\boldsymbol{\xi}(N)) &= \frac{-Nm}{2}\ln 2\pi - \frac{N}{2}\ln\det(\boldsymbol{\rho}^0)\\ &\quad - \tfrac{1}{2}\sum_{t=1}^{N}\|\mathbf{f}_0(\mathbf{y}(t)) - \mathbf{Z}(t-1)\boldsymbol{\theta}^0\|^2_{(\boldsymbol{\rho}^0)^{-1}}\\ &\triangleq g(\boldsymbol{\xi}(N-1), \boldsymbol{\theta}^0, \boldsymbol{\rho}^0).\end{aligned}$$

We will regard $(\boldsymbol{\rho}^0)^{-1}$ as the unknown quantity instead of $\boldsymbol{\rho}^0$. Let

$$\phi^0_{ij} = [(\rho^0)^{-1}]_{ij}, \qquad \phi_{ij} = [(\rho)^{-1}]_{ij}.$$

Let

$$\boldsymbol{\phi}^0 = (\phi^0_{11}, \phi^0_{12}, \ldots, \phi^0_{mm}), \qquad \boldsymbol{\phi} = (\phi_{11}, \phi_{12}, \ldots, \phi_{mm})$$

and

$$g_1(\boldsymbol{\theta}, \boldsymbol{\rho}) = g(\boldsymbol{\xi}(N), \boldsymbol{\theta}, \boldsymbol{\rho}).$$

Let

$$\mathscr{I}_N(\boldsymbol{\theta}, \boldsymbol{\phi}) = -E\begin{Bmatrix}\dfrac{\partial^2 \ln g_1}{\partial\boldsymbol{\theta}^2} & \dfrac{\partial^2 \ln g_1}{\partial\boldsymbol{\theta}\,\partial\boldsymbol{\phi}}\\[2ex] \dfrac{\partial^2 \ln g_1}{\partial\boldsymbol{\phi}\,\partial\boldsymbol{\theta}} & \dfrac{\partial^2 \ln g_1}{\partial\boldsymbol{\phi}^2}\end{Bmatrix}$$

$$\frac{\partial \ln g_1}{\partial\boldsymbol{\theta}} = \sum_{t=1}^{N}\mathbf{Z}^{\mathrm{T}}(t-1)(\boldsymbol{\rho}^0)^{-1}(\mathbf{f}_0(\mathbf{y}(t)) - \mathbf{Z}(t-1)\boldsymbol{\theta})$$

$$\frac{\partial^2 \ln g_1}{\partial\boldsymbol{\theta}^2} = -\sum_{t=1}^{N}\mathbf{Z}^{\mathrm{T}}(t-1)(\boldsymbol{\rho}^0)^{-1}\mathbf{Z}(t-1)$$

$$\frac{\partial^2 \ln g_1}{\partial\boldsymbol{\theta}\,\partial\phi_{ij}} = \sum_{t=1}^{N}\sum_{p=1}^{m}\sum_{q=1}^{m}(\mathbf{Z}^{\mathrm{T}}(t-1))_{pi}(\mathbf{f}_0(\mathbf{y}(t)) - \mathbf{Z}(t-1)\boldsymbol{\theta})_{pq}.$$

Clearly,

$$E\left[\frac{\partial^2 \ln g_1}{\partial\boldsymbol{\theta}\,\partial\phi_{ij}}\middle|\boldsymbol{\theta} = \boldsymbol{\theta}^0\right] = 0, \qquad \forall i, j$$

since $(\mathbf{f}_0(\mathbf{y}(t)) - \mathbf{Z}(t-1)\boldsymbol{\theta}^0) = \mathbf{w}^0(t)$ and $\mathbf{w}^0(t)$ is independent of $\mathbf{Z}(t-1)$. Consequently,

$$\mathscr{I}_N(\boldsymbol{\theta}, \boldsymbol{\phi}) = -\begin{bmatrix} E\left[\dfrac{\partial^2 \ln g_1}{\partial\boldsymbol{\theta}^2}\right] & 0\\[2ex] 0 & E\left[\dfrac{\partial^2 \ln g_1}{\partial\boldsymbol{\phi}^2}\right]\end{bmatrix}.$$

By the Cramér–Rao inequality

$$E\left[\begin{array}{c:c} (\boldsymbol{\theta}^* - \boldsymbol{\theta}^0)(\boldsymbol{\theta}^* - \boldsymbol{\theta}^0)^{\mathrm{T}} & (\boldsymbol{\theta}^* - \boldsymbol{\theta}^0)(\boldsymbol{\phi}^* - \boldsymbol{\phi}^0)^{\mathrm{T}} \\ \hdashline (\boldsymbol{\theta}^* - \boldsymbol{\theta}^0)(\boldsymbol{\phi}^* - \boldsymbol{\phi}^0)^{\mathrm{T}} & (\boldsymbol{\phi}^* - \boldsymbol{\phi}^0)(\boldsymbol{\phi}^* - \boldsymbol{\phi}^0)^{\mathrm{T}} \end{array}\right] \geq \mathscr{I}_N^{-1}(\boldsymbol{\theta}, \boldsymbol{\phi}) + o(1/N),$$

or

$$E[(\boldsymbol{\theta}^* - \boldsymbol{\theta}^0)(\boldsymbol{\theta}^* - \boldsymbol{\theta}^0)^{\mathrm{T}}] \geq \left[E\left\{\sum_{t=1}^{N} \mathbf{Z}^{\mathrm{T}}(t-1)(\boldsymbol{\rho}^0)^{-1}\mathbf{Z}(t-1)\right\}\right]^{-1} + o(1/N).$$

Problems

1. Consider the process $\mathbf{y}$ obeying $\mathbf{y}(t) = \mathbf{A}\mathbf{y}(t-1) + \mathbf{w}(t)$, where $\mathbf{A} = \left[\begin{smallmatrix} 0.8 & 0.4 \\ 0.2 & 0.5 \end{smallmatrix}\right]$ and $\boldsymbol{\rho} = \operatorname{cov}[\mathbf{w}] = \mathbf{I}$, $\mathbf{w}$ normal. Simulate the process on a computer; obtain N observations and find the FIML, CML, and LIML estimates of the elements of $\mathbf{A}$ and $\boldsymbol{\rho}$ with $N = 50, 100, 200, 500$.

2. Let $y(t) = y(t-1)(\theta_1 + w(t))$, where $w(\cdot)$ is normal and obeys (A1) with variance ρ. Obtain the CML estimates of θ_1 and ρ from the observations $y(1), \ldots, y(N)$, where $y(t) \neq 0$, $\forall t = 1, \ldots, N$. Given ρ, what is the appropriate prior density for θ_1, yielding an expression for the posterior density of θ_1 without involving any integrals?

3. A common method of modeling a process $y(\cdot)$ that is suspected to have undergone a rather drastic change at time T, is given below, where T is an integer.

$$y(t) = \theta_0 + \theta_1 u(t-T) + (\theta_2 + \theta_3 u(t-T))y(t-1) + w(t),$$

where $u(t) = 0$ if $t \leq 0$ and 1 if $t > 0$. Given $y(1), \ldots, y(N)$ and T, obtain the CML estimates of θ_i, $i = 0, 1, 2, 3$, and ρ. How do you estimate T if T is unknown? Is a Bayesian estimate easier to compute than a CML estimate?

4. Consider the following system with time-varying coefficients:

$$\begin{aligned} y(t) &= \alpha_1(t)y(t-1) + w_1(t) \\ \alpha_1(t) &= \theta_1\alpha_1(t-1) + \theta_2 + w_2(t). \end{aligned}$$

Let $\mathbf{w}^{\mathrm{T}}(t) = (w_1(t), w_2(t))$ be normal and obey (A1) with covariance matrix $\boldsymbol{\rho}$. Obtain CML estimates of θ_1, θ_2, and $\boldsymbol{\rho}$ using only $y(1), \ldots, y(N)$.

5. Consider a process $y(\cdot)$ obeying an AR(2) process where $w(\cdot)$ obeys (A1) and has variance 1:

$$y(t) = 1.35y(t-1) - 0.67y(t-2) + w(t).$$

Define a new smoothed process $y_1(t) = \frac{1}{4}\sum_{j=1}^{4} y((t-1)4 + j)$. Construct a linear predictor for $y_1(t)$ in terms of $y_1(t-1)$ and $y_1(t-2)$. Does $y_1(\cdot)$ obey an AR(n) process for some n? If so, give the appropriate value of n. Compare the spectral densities of y and y_1.

Chapter VII | Parameter Estimation in Systems with Both Moving Average and Autoregressive Terms

Introduction

In the preceding chapter, we considered the estimation of parameters in systems governed by difference equations without moving average terms. This restriction is removed in this chapter, i.e., we will consider the estimation of unknown coefficients in difference equations that have moving average terms in addition to the autoregressive terms and terms corresponding to the observable inputs discussed in Chapter VI. The difference equation should be linear in the unknown coefficients, but it could be nonlinear in the other variables. We consider parameter estimation in systems governed by a single difference equation as well as in systems consisting of a set of simultaneous difference equations.

We remarked that if we are interested in fitting a model to the given data only for the purposes of prediction, the need for the inclusion of moving average (MA) terms in the model arises infrequently since a model with a sufficiently large number of AR terms is often (but not always) sufficient. However, there are other reasons for dealing with systems with MA terms. In engineering systems, such as aircraft and hydroelectric regulation, we have a good idea of the structure of the systems, which often involve moving average terms. It is clearly not prudent to ignore these terms in the model. The parameters in such systems can be estimated with the aid of the techniques described in this chapter.

We will consider only two approaches to parameter estimation: namely, the maximum likelihood and the limited information techniques. The maximum likelihood methods are treated first in some detail because they yield asymptotically unbiased and minimum variance estimates when the noise distribution is normal. The evaluation of these estimators, especially in multivariate systems in canonical form I or pseudocanonical form II, involves considerable computation. Even though other comparatively simple methods of estimation such as the least squares or the limited information methods are available, one still has to study the ML estimates and evaluate them in a number of problems to understand the tradeoff between computational complexity and accuracy of these estimators.

Next we consider the limited information (LI) methods. The attractive feature about the limited information methods is that the unknowns in every one of the m equations constituting the multivariate system can be estimated individually. Thus, the complex problem of estimation in a multivariate system is reduced to m relatively simple problems of estimation of parameters in single equation systems. The accuracy of these estimates appears to be only slightly less than that of the conditional maximum likelihood (CML) estimates, and the computational effort in evaluating them is only a small fraction of that in CML estimation.

Besides these two approaches, there are many other methods of estimation, such as generalized least squares (Clarke, 1967; Soderstrom, 1972; Hsia, 1975), two-stage least squares (Pandya, 1974), recursive least squares (Panuska, 1969), instrumental variable methods (Wong and Polak, 1967; Finnegan and Rowe, 1974; Young, 1970), correlation methods (Box and Jenkins, 1970; Whittle, 1951), real time computable methods (Gertler and Banyasz, 1974; Soderstrom, 1973), and others. All these methods are adequately documented in the literature and have been reviewed recently (Kashyap and Nasburg, 1974; Kashyap and Rao, 1975). Hence we will not discuss these approaches.

7a. Maximum Likelihood Estimators

7a.1. Statement of the Problem

7a.1.1. Single Output Process

Consider the following difference equation for the scalar variable y:

$$\mathscr{S}: f_0(y(t)) = \sum_{j=1}^{m_1} A_j f_j(y(t-j)) + \sum_{j=1}^{m_3} G_j u(t-j) + \sum_{j=1}^{l_2} F_j \psi_j(t-1) + w(t) + \sum_{j=1}^{m_2} B_j w(t-j). \tag{7a.1.1}$$

where f_i, $i = 0, 1, \ldots, m_1$, are known functions and the remaining variables were defined earlier. Let

$$\boldsymbol{\theta}^0 = (A_1, \ldots, A_{m_1}, G_1, \ldots, G_{m_3}, F_1, \ldots, F_{l_2}, B_1, \ldots, B_{m_2})^{\mathrm{T}}$$

dimension of $\boldsymbol{\theta}^0 \triangleq n = m_1 + m_2 + m_3 + l_2$.

Let $\mathscr{H}$ be the set of all values that can be assumed by $\boldsymbol{\theta}^0$. Our first concern is to impose a number of assumptions on the set $\mathscr{H}$, and (7a.1.1) so that given a process y, there is at most one vector in $\mathscr{H}$ which together with $\mathscr{S}$ represents the process.

Let the random input $w(\cdot)$ have normal distribution and obey (A1), the observable input $u(\cdot)$ obey (A2) and (A3), and the trend vector $\boldsymbol{\psi}(\cdot)$ obey (A4). A number of assumptions are necessary on the parameter $\boldsymbol{\theta}^0$, i.e., the assumptions on the set $\mathscr{H}$. First, assumption (A5′) is imposed on $\boldsymbol{\theta}^0$ to assure the asymptotic stability of the system. If the functions f_i, $i = 0, 1, \ldots, m_1$, are all identical, then (A5) implies (A5′). Next, to ensure the invertibility of the system we need assumption (A6). Finally, we have to assume that Eq. (7a.1.1) cannot be further simplified; i.e., its order cannot be decreased. If all f_i, $i = 0, \ldots, m_1$, are identical, then this condition is equivalent to (A7).

Let the available observation set at instant N be denoted by $\xi(N)$,

$$\xi(N) = \{y(N), \ldots, y(0), y(-1), \ldots, y(-m_1), u(N-1), \ldots, u(-m_3)\}.$$

It is desired to obtain maximum likelihood estimates of the unknowns $\boldsymbol{\theta}^0$ and ρ characterizing the model which is obeyed by the given empirical process y, based on the observation set $\xi(N)$. The computation of CML estimates of the coefficients in single equation systems has been treated comprehensively by Astrom *et al.* (1965).

7a.1.2. The Vector Process

As before, let the vector process $\mathbf{y}$ obey

$$\mathscr{M}: \mathbf{f}_0(\mathbf{y}(t)) = \sum_{j=1}^{m_1} \mathbf{A}_j \mathbf{f}_j(\mathbf{y}(t-j)) + \sum_{j=1}^{m_3} \mathbf{G}_j \mathbf{u}(t-j) + \mathbf{F}\boldsymbol{\psi}(t-1) + \mathbf{w}(t) + \sum_{j=1}^{m_2} \mathbf{B}_j \mathbf{w}(t-j), \qquad (7a.1.2)$$

where

$$\mathbf{f}_j(\mathbf{y}(t)) = [f_{j1}(\mathbf{y}(t)), f_{j2}(\mathbf{y}(t)), \ldots, f_{jm}(\mathbf{y}(t))]^{\mathrm{T}}, \qquad j = 0, 1, \ldots, m_1,$$
$$\mathbf{w}(t) = (w_1(t), \ldots, w_m(t))^{\mathrm{T}}.$$

The functions $\mathbf{f}_j(\cdot)$ are known. The input $\mathbf{u}$ and trend $\boldsymbol{\psi}(\cdot)$ are the same as in Chapter IV. The vector parameter $\boldsymbol{\theta}^0$ is made up of all components of the matrices $\mathbf{A}_j$, $j = 1, \ldots, m_1$, $\mathbf{G}_j$, $j = 1, \ldots, m_3$, $\mathbf{F}$, and $\mathbf{B}_j$, $j = 1, \ldots, m_2$. The dimension of $\boldsymbol{\theta}^0 = n_0 = m^2(m_1 + m_2) + m(m_3 l_1 + l_2)$.

Let $\mathscr{H}$ be the set of values that can be assumed by $\boldsymbol{\theta}^0$. As before, we need to impose assumptions so that the value of $\boldsymbol{\theta}^0$ in (7a.1.2) uniquely characterizes the given empirical process $\mathbf{y}$. The inputs $\mathbf{w}$ and $\boldsymbol{\psi}$ obey assumptions (A1)–(A4) and, in addition, $\mathbf{w}$ is Gaussian. We also need the multivariate version of (A5′) to ensure asymptotic stability. If $f_i(y_i) = y_i$, then (A5′) is implied by (A5). Similarly, assumption (A6) is needed for intertibility. Further, we need to restrict ourselves to difference equations characterized by the 4-tuples ($\mathbf{A}$, $\mathbf{B}$, $\mathbf{G}$, $\boldsymbol{\rho}$) which are either in one of the canonical forms (I, II, or III) or in the pseudocanonical form II and to make the corresponding additional assumption, discussed in Chapter V. For instance, if we are working in canonical form I, we need assumption, (A9) and (A7) in addition to assumptions (A1)–(A6) in order to ensure the uniqueness of the vector $\boldsymbol{\theta}^0$ characterizing the parameters of the system. Similarly, if we are working with canonical form II or III or the pseudocanonical form II, we need the additional assumptions (A10), (A11),

(A7′) and (A8). While dealing with multivariate difference equations, with the 4-tuple $(\mathbf{A}, \mathbf{B}, \mathbf{G}, \boldsymbol{\rho})$ obeying canonical form II or III, the question of estimability can be handled by using only the single equation theory of Chapter IV.

Our intention is to estimate the unknown coefficient vector $\boldsymbol{\theta}^0$ and the covariance matrix $\boldsymbol{\rho}^0$ from the observation history $\boldsymbol{\xi}(N)$ defined earlier. Kashyap (1970b) and Mehra (1971) have discussed the estimation of parameters in such systems.

7a.2. The Full Information Maximum Likelihood (FIML) and Conditional Maximum Likelihood (CML) Estimators

Let the available observation set at time N be $\boldsymbol{\xi}(N) = (y(N), u(N))$, where $y(N) = \{\mathbf{y}(N), \ldots, \mathbf{y}(1 - m_1)\}$ and $u(N) = \{\mathbf{u}(N-1), \ldots, \mathbf{u}(1 - m_3)\}$. Let the vector of all the unknown parameters in (7a.1.2) be $\boldsymbol{\phi}^0 = \{\boldsymbol{\theta}^0, \boldsymbol{\rho}^0\}$. Similarly, let $\mathbf{w}^0(t)$ denote the unknown disturbances in (7a.1.2) which gave rise to the given observations y, u. Let $p[y(N), u(N); \boldsymbol{\phi}^0]$ denote the probability density of the available observation set at time N. The FIML estimate of $\boldsymbol{\phi}^0$ is the value of $\boldsymbol{\phi}$ that maximizes the function $\ln p[y(N), u(N); \boldsymbol{\phi}]$ with respect to $\boldsymbol{\phi}$:

$$\ln p[y(N), u(N); \boldsymbol{\phi}] = \ln p[y(N) | u(N), \boldsymbol{\phi}] + \ln p[u(N)].$$

By assumption (A3), the second term in this expression is independent of $\boldsymbol{\phi}$. Hence, the FIML estimate is obtained by maximizing $J_1(\boldsymbol{\phi})$

$$J_1(\boldsymbol{\phi}) = \ln p[\mathbf{y}(N), \ldots, \mathbf{y}(1 - m_1) | u(N), \boldsymbol{\phi}].$$

The evaluation of $J_1(\boldsymbol{\phi})$ is difficult for reasons similar to those discussed in Chapter VI. It is easier to evaluate $J_2'(\boldsymbol{\phi})$:

$$J_2'(\boldsymbol{\phi}) = \ln p[\mathbf{y}(N), \ldots, \mathbf{y}(1) | \mathbf{y}(0), \ldots, \mathbf{y}(1 - m_1), u(N), \boldsymbol{\phi}].$$

The estimate obtained by maximizing $J_2'(\boldsymbol{\phi})$ with respect to $\boldsymbol{\phi}$ is called the conditional maximum likelihood (CML) estimate. Asymptotically, CML and FIML estimates are identical for the same reasons discussed in Chapter VI. Utilizing the normality assumption, the log likelihood function $J_2'(\boldsymbol{\phi})$ divided by N can be labeled $J_2(\boldsymbol{\theta}, \boldsymbol{\rho})$ and written

$$J_2(\boldsymbol{\theta}, \boldsymbol{\rho}) = -\frac{m}{2} \ln(2\pi) - \tfrac{1}{2} \ln |\det \boldsymbol{\rho}| - \frac{1}{2N} \sum_{t=1}^{N} \mathbf{w}^{\mathrm{T}}(t, \boldsymbol{\theta}) \boldsymbol{\rho}^{-1} \mathbf{w}(t, \boldsymbol{\theta}), \tag{7a.2.1}$$

where $\mathbf{w}(t, \boldsymbol{\theta})$ are the recursively generated estimates of $\mathbf{w}^0(t)$, the unknown disturbance in (7a.1.2) computed by using the observations y, u and the dummy parameter vector $\boldsymbol{\theta}$:

$$\begin{aligned} \mathbf{w}(t, \boldsymbol{\theta}) = \mathbf{f}_0(\mathbf{y}(t)) &- \sum_{j=1}^{m_1} \mathbf{A}_j \mathbf{f}_j(\mathbf{y}(t-j)) - \sum_{j=1}^{m_3} \mathbf{G}_j \mathbf{u}(t-j) - \mathbf{F}\boldsymbol{\psi}(t-1) \\ &- \sum_{j=1}^{m_2} \mathbf{B}_j \mathbf{w}(t-j, \boldsymbol{\theta}). \end{aligned} \tag{7a.2.2}$$

The CML estimates of $\boldsymbol{\theta}^0$ and $\boldsymbol{\rho}^0$ are denoted by $\boldsymbol{\theta}^*$ and $\boldsymbol{\rho}^*$, respectively:

$$(\boldsymbol{\theta}^*, \boldsymbol{\rho}^*) = \arg\left\{\sup_{\boldsymbol{\theta},\boldsymbol{\rho}} J_2(\boldsymbol{\theta}, \boldsymbol{\rho})\right\}. \tag{7a.2.3}$$

In order to find $\boldsymbol{\rho}^*$, we need to differentiate $J_2(\boldsymbol{\theta}, \boldsymbol{\rho})$ with respect to $\boldsymbol{\rho}$. This operation is rather difficult. But it is easier to differentiate $J_2(\boldsymbol{\theta}, \boldsymbol{\rho})$ with respect to $\boldsymbol{\rho}^{-1}$ rather than $\boldsymbol{\rho}$, as indicated:

$$\frac{\partial J_2(\boldsymbol{\theta}, \boldsymbol{\rho})}{\partial \boldsymbol{\rho}^{-1}} = \tfrac{1}{2}\left[\boldsymbol{\rho} - \frac{1}{N}\sum_{t=1}^{N} \mathbf{w}(t, \boldsymbol{\theta})\mathbf{w}^{\mathrm{T}}(t, \boldsymbol{\theta})\right].$$

Setting this matrix of derivatives to zero, we can solve for $\hat{\boldsymbol{\rho}}$:

$$\hat{\boldsymbol{\rho}} = \frac{1}{N}\sum_{t=1}^{N} \mathbf{w}(t, \boldsymbol{\theta})\mathbf{w}^{\mathrm{T}}(t, \boldsymbol{\theta}). \tag{7a.2.4}$$

By inspection, $\hat{\boldsymbol{\rho}}$ must also be the value of $\boldsymbol{\rho}$ which renders the matrix of derivatives $\partial J_2(\boldsymbol{\theta}, \boldsymbol{\rho})/\partial \boldsymbol{\rho}$ zero. We will obtain the concentrated likelihood function given below from (7a.2.4) by replacing the matrix $\boldsymbol{\rho}$ in (7a.2.1) by $\hat{\boldsymbol{\rho}}$ in (7a.2.4):

$$\begin{aligned} J_3(\boldsymbol{\theta}) &\triangleq J_2(\boldsymbol{\theta}, \boldsymbol{\rho} = \hat{\boldsymbol{\rho}}) \\ &= -\frac{m}{2}\ln 2\pi - \tfrac{1}{2}\ln\left|\det \frac{1}{N}\sum_{t=1}^{N} \mathbf{w}(t, \boldsymbol{\theta})\mathbf{w}^{\mathrm{T}}(t, \boldsymbol{\theta})\right| - \frac{m}{2}. \end{aligned} \tag{7a.2.5}$$

We can show the concavity of the function $J_3(\boldsymbol{\theta})$ in the region around $\boldsymbol{\theta}^0$, under appropriate assumptions (Kashyap and Rao, 1975). Hence, the function $J_3(\boldsymbol{\theta})$ possesses a unique supremum, say at $\boldsymbol{\theta} = \boldsymbol{\theta}^*$, which is the CML estimate of $\boldsymbol{\theta}^0$ under $\boldsymbol{\xi}(N)$. The estimate $\boldsymbol{\theta}^*$ can be evaluated only by iterative methods, which are discussed later. The corresponding CML estimate of $\boldsymbol{\rho}^0$ is obtained from (7a.2.4) by replacing the variable $\boldsymbol{\theta}$ by $\boldsymbol{\theta}^*$. Let

$$J_5(\boldsymbol{\theta}) = \frac{1}{N}\sum_{t-1}^{N} \mathbf{w}^{\mathrm{T}}(t, \boldsymbol{\theta})\mathbf{w}(t, \boldsymbol{\theta}). \tag{7a.2.6}$$

An elementary consequence of expression (7a.2.5) is that the estimate of $\boldsymbol{\theta}$ obtained by minimizing $J_5(\boldsymbol{\theta})$, the sum of the residual squares, is not a maximum likelihood estimate!

7a.3. Properties of the CML Estimate

7a.3.1. Probability Distribution of $\boldsymbol{\theta}^*(N)$

It is widely known that the CML estimate $\boldsymbol{\theta}^*(N)$ is consistent and asymptotically normal with the following mean and variance:

$$E[\boldsymbol{\theta}^*(N)|\bar{\mathbf{S}}(N); \boldsymbol{\theta}^0, \boldsymbol{\rho}^0] = \boldsymbol{\theta}^0 + O(1/N^{1/2}) \tag{7a.3.1}$$

$$\operatorname{cov}[\boldsymbol{\theta}^*(N)|\bar{\mathbf{S}}(N); \boldsymbol{\theta}^0, \boldsymbol{\rho}^0] = \bar{\mathbf{S}}(N) + o(1/N) \tag{7a.3.2}$$

where

$$\bar{\mathbf{S}}(N) \triangleq \left[\tfrac{1}{2}\,\nabla^2_{\boldsymbol{\theta}\boldsymbol{\theta}} \sum_{t=1}^{N} \mathbf{w}^{\mathrm{T}}(t, \boldsymbol{\theta})(\boldsymbol{\rho}^0)^{-1}\mathbf{w}(t, \boldsymbol{\theta})\big|_{\boldsymbol{\theta}=\boldsymbol{\theta}^*}\right]^{-1}. \tag{7a.3.3}$$

We can easily find the marginal covariance matrix of $\boldsymbol{\theta}^*(N)$ by averaging over the matrix $\bar{\mathbf{S}}(N)$.

$$\operatorname{cov}[\boldsymbol{\theta}^*(N)|\boldsymbol{\theta}^0, \boldsymbol{\rho}^0] = E[\bar{\mathbf{S}}(N)|\boldsymbol{\theta}^0, \boldsymbol{\rho}^0] \triangleq \mathscr{I}_N^{-1}(\boldsymbol{\theta}^0, \boldsymbol{\rho}^0), \tag{7a.3.4}$$

where

$$\mathscr{I}_N(\boldsymbol{\theta}^0, \boldsymbol{\rho}^0) \triangleq E[(\bar{\mathbf{S}}(N))^{-1}|\boldsymbol{\theta}^0, \boldsymbol{\rho}^0]. \tag{7a.3.5}$$

The estimate $\boldsymbol{\theta}^*(N)$ has asymptotically minimum variance; i.e., the asymptotic variance given above is identical to the asymptotic lower bound given by the Cramér–Rao inequality. The proof of this statement is similar to that of a similar property for QML estimates in Chapters IV and V.

The CML estimate is found to be consistent if the distribution of $\mathbf{w}(\cdot)$ belongs to a wider class than just the class of normal distributions. Further, the CML estimate is known to be consistent even if the assumption (A3) (independence of $\mathbf{u}$ and $\mathbf{w}$) is not satisfied, provided all the other assumptions hold, and we know a small neighborhood that contains the unknown $\boldsymbol{\theta}^0$. It is needless to add that the estimate $\boldsymbol{\theta}^*$ is not necessarily asymptotically efficient under these relaxed assumptions.

7a.3.2. The Probability Distribution of $\hat{\rho}(N)$

We will consider only the single output case, i.e., $\hat{\rho}(N)$ is a scalar. Let $\hat{\rho}_1(N) = N\hat{\rho}(N)/\rho^0$. Then $\hat{\rho}_1(N)$ has asymptotically chi-squared distribution with $(N - n)$ degrees of freedom. Hence,

$$E[\hat{\rho}(N)] = \rho^0[(N - n)/N]. \tag{7a.3.8}$$

Thus the estimate $\hat{\rho}$ has a bias of the order $O(1/N)$, which tends to zero as N tend to infinity.

$$\begin{aligned} \operatorname{var}(\hat{\rho}(N)|\rho^0) &= E\left[(\rho(N) - \rho^0)\left(\frac{N-n}{N}\right)^2 \middle| \rho^0\right] \\ &= \frac{(\rho^0)^2}{N^2} \operatorname{var}(\rho_1(N)|\rho^0) \\ &= \frac{2(\rho^0)^2}{N^2}(N - n) \end{aligned} \tag{7a.3.9}$$

or

$$E[(\rho(N) - \rho^0)^2|\rho^0] = (\rho^0)^2[(2N - 2n + n^2)/N^2]. \tag{7a.3.10}$$

7b. Numerical Methods for CML Estimation

There have been some attempts at using the standard optimization routines for performing the numerical optimization of functions $J_3(\boldsymbol{\theta})$ or $J_2(\boldsymbol{\theta}, \boldsymbol{\rho})$. However, such attempts have not been very successful in the multivariate case, primarily because of the dimensionality of the problem and its corresponding complexity. These standard routines do not utilize all the available information about the function being optimized. We need gradient or Newton–Raphson methods of iterative optimizations which exploit the structure of the particular loss function. A number of algorithms are available depending on whether the function to be maximized is $J_2(\boldsymbol{\theta}, \boldsymbol{\rho})$ or $J_3(\boldsymbol{\theta})$ and whether the gradients are evaluated directly using the given observations or the fourier transformed observations (Hannan, 1969; Akaike, 1973). We will give here one particular method which has been found to be very successful in practice. Other variants such as the algorithms of Durbin (1959) and Walker (1962) are discussed in the review papers by Kashyap and Nasburg (1974) and Kashyap and Rao (1975).

7b.1. A Method for Evaluating $\boldsymbol{\theta}^*(N)$

We will iteratively maximize the function $J_2(\boldsymbol{\theta}, \boldsymbol{\rho})$. Let $(\boldsymbol{\theta}^k, \boldsymbol{\rho}^k)$ denote the estimate of $\boldsymbol{\theta}^0$ and $\boldsymbol{\rho}^0$ at the kth iteration. We construct the sequence $(\boldsymbol{\theta}^k, \boldsymbol{\rho}^k)$, $k = 1, 2, \ldots$, so that the sequence tends to the required estimate $(\boldsymbol{\theta}^*, \boldsymbol{\rho}^*)$. By using the estimate $(\boldsymbol{\theta}^k, \boldsymbol{\rho}^k)$, available at the end of the kth iteration the estimate $(\boldsymbol{\theta}^{k+1}, \boldsymbol{\rho}^{k+1})$ will be derived with the aid of the following two steps:

Step (i). Using $(\boldsymbol{\theta}^k, \boldsymbol{\rho}^k)$ available from the kth iteration, compute the next iterate $\boldsymbol{\theta}^{k+1}$ so as to satisfy

$$J_2(\boldsymbol{\theta}^{k+1}, \boldsymbol{\rho}^k) > J_2(\boldsymbol{\theta}^k, \boldsymbol{\rho}^k). \tag{7b.1.1}$$

In order to do so, recall the expression for $J_2(\boldsymbol{\theta}, \boldsymbol{\rho})$:

$$J_2(\boldsymbol{\theta}, \boldsymbol{\rho}) = -\frac{m}{2} \ln 2\pi - \tfrac{1}{2} \ln|\det \boldsymbol{\rho}| - \tfrac{1}{2} J_4(\boldsymbol{\theta}, \boldsymbol{\rho}), \tag{7b.1.2}$$

where,

$$J_4(\boldsymbol{\theta}, \boldsymbol{\rho}) = \frac{1}{2N} \sum_{t=1}^{N} \mathbf{w}^{\mathrm{T}}(t, \boldsymbol{\theta}) \boldsymbol{\rho}^{-1} \mathbf{w}(t, \boldsymbol{\theta}). \tag{7b.1.3}$$

Therefore, locally maximizing $J_2(\boldsymbol{\theta}, \boldsymbol{\rho})$ with respect to $\boldsymbol{\theta}$ is equivalent to locally minimizing $J_4(\boldsymbol{\theta}, \boldsymbol{\rho})$ with respect to $\boldsymbol{\theta}$. The latter operation is carried out iteratively by using the Newton–Raphson method to yield the following expression for $\boldsymbol{\theta}^{k+1}$:

$$\boldsymbol{\theta}^{k+1} = \boldsymbol{\theta}^k - c[\nabla^2_{\boldsymbol{\theta\theta}} J_4(\boldsymbol{\theta}, \boldsymbol{\rho})]^{-1} \nabla_{\boldsymbol{\theta}} J_4(\boldsymbol{\theta}, \boldsymbol{\rho})|_{\boldsymbol{\theta}=\boldsymbol{\theta}^k, \boldsymbol{\rho}=\boldsymbol{\rho}^k}, \tag{7b.1.4}$$

where c is a suitable constant, such as 1 or 0.5, and

$$\nabla_{\boldsymbol{\theta}} J_4(\boldsymbol{\theta}, \boldsymbol{\rho}) = \frac{1}{N} \sum_{t=1}^{N} \sum_{i,j=1}^{m} (\nabla_{\boldsymbol{\theta\theta}} w_i(t, \boldsymbol{\theta}))(\boldsymbol{\rho}^{-1})_{ij} w_j(t, \boldsymbol{\theta})$$

$$\begin{aligned} \nabla^2_{\boldsymbol{\theta\theta}} J_4(\boldsymbol{\theta}, \boldsymbol{\rho}) = \frac{1}{N} \sum_{t=1}^{N} \sum_{i,j=1}^{m} [&(\nabla^2_{\boldsymbol{\theta\theta}} w_i(t, \boldsymbol{\theta}))(\boldsymbol{\rho}^{-1})_{ij} w_j(t, \boldsymbol{\theta}) \\ &+ (\nabla_{\boldsymbol{\theta}} w_i(t, \boldsymbol{\theta}))(\nabla_{\boldsymbol{\theta}} w_j(t, \boldsymbol{\theta}))^{\mathrm{T}} (\boldsymbol{\rho}^{-1})_{ij}], \end{aligned} \tag{7b.1.5}$$

$$\nabla_{\boldsymbol{\theta}} w_i(t, \boldsymbol{\theta}) = \left(\frac{\partial w_i(t, \boldsymbol{\theta})}{\partial \theta_1}, \ldots, \frac{\partial w_i(t, \boldsymbol{\theta})}{\partial \theta_m} \right)^{\mathrm{T}}, \tag{7b.1.6}$$

$$\nabla_{\boldsymbol{\theta\theta}} w_i(t, \boldsymbol{\theta}) = \{\nabla^2_{\boldsymbol{\theta}_k \boldsymbol{\theta}_l} w_i(t, \boldsymbol{\theta})\}, \qquad k, l = 1, 2, \ldots, n.$$

Step (*ii*). Using $\boldsymbol{\theta}^{k+1}$, compute $\boldsymbol{\rho}^{k+1}$ by maximizing $J_2(\boldsymbol{\theta}^{k+1}, \boldsymbol{\rho})$ with respect to $\boldsymbol{\rho}$. The result is

$$\boldsymbol{\rho}^{k+1} = \frac{1}{N}\sum_{t=1}^{N} \mathbf{w}(t, \boldsymbol{\theta}^{k+1})\mathbf{w}^{\mathrm{T}}(t, \boldsymbol{\theta}^{k+1}). \tag{7b.1.7}$$

Steps (i) and (ii) complete the computations needed for the $(k + 1)$ iteration.

To evaluate the derivatives $\nabla_{\boldsymbol{\theta}} w_i(t, \boldsymbol{\theta})$, $\nabla^2_{\boldsymbol{\theta\theta}} w_i(t, \boldsymbol{\theta}), \ldots$, we first obtain difference equations for them by considering the difference equation for w_i and differentiating it throughout once or twice, as the case may be, with respect to the appropriate element or elements of $\boldsymbol{\theta}$.

For instance, let the equation for w_1 be as follows:

$$w_1(t, \boldsymbol{\theta}) = y_1(t) - \theta_1 y_1(t-1) - \theta_2 y_2(t-1) - \theta_3 w_1(t-1, \boldsymbol{\theta}) - \theta_4 w_2(t-1, \boldsymbol{\theta}).$$

Then the difference equations for the various gradients are

$$\begin{aligned}
\nabla_{\theta_1} w_1(t, \boldsymbol{\theta}) &= -y_1(t-1) - \theta_3 \nabla_{\theta_1} w_1(t-1, \boldsymbol{\theta}) - \theta_4 \nabla_{\theta_1} w_2(t-1, \boldsymbol{\theta}), \\
\nabla_{\theta_2} w_1(t, \boldsymbol{\theta}) &= -y_2(t-1) - \theta_3 \nabla_{\theta_2} w_1(t-1, \boldsymbol{\theta}) - \theta_4 \nabla_{\theta_2} w_2(t-1, \boldsymbol{\theta}), \\
\nabla_{\theta_3} w_1(t, \boldsymbol{\theta}) &= -w_1(t-1, \boldsymbol{\theta}) - \theta_3 \nabla_{\theta_3} w_1(t-1, \boldsymbol{\theta}) - \theta_4 \nabla_{\theta_3} w_2(t-1, \boldsymbol{\theta}), \\
\nabla_{\theta_4} w_1(t, \boldsymbol{\theta}) &= -w_2(t-1, \boldsymbol{\theta}) - \theta_3 \nabla_{\theta_4} w_1(t-1, \boldsymbol{\theta}) - \theta_4 \nabla_{\theta_4} w_2(t-1, \boldsymbol{\theta}).
\end{aligned} \tag{7b.1.8}$$

Similarly, we can write the difference equations for the second derivatives also. It is appropriate to remember that n_0, the dimension of $\boldsymbol{\theta}$, is relatively high even for simple problems. For instance, for a system with $m = 3$, $m_1 = m_2 = m_3 = 1$, $n_0 = 27$. The evaluation of the first derivative vector $\nabla_{\boldsymbol{\theta}} J_4(\boldsymbol{\theta})$ involves the evaluation of the $mn_0 = 81$ gradients $\nabla_{\theta_j} w_i(t, \boldsymbol{\theta})$, for each t. Similarly, if we do not make simplifications or approximations, the second derivative matrix $\nabla^2_{\boldsymbol{\theta\theta}} J_4(\boldsymbol{\theta})$ will involve the evaluation of the $\frac{1}{2}m(n_0{}^2 + n_0) = 1134$ second derivatives on the matrices $\nabla^2_{\boldsymbol{\theta\theta}} w_i(t, \boldsymbol{\theta})$, $i = 1, \ldots, m$. It is easy to see that the computation problem becomes difficult without further simplification of the second derivative matrix.

We will now show that the first term in (7b.1.5) tends to zero if N tends to infinity and $\boldsymbol{\theta} = \boldsymbol{\theta}^0$. From Eq. (7b.1.8), we see that $\nabla_{\boldsymbol{\theta}} w_j(t, \boldsymbol{\theta}^0)$ and $\nabla^2_{\boldsymbol{\theta\theta}} w_j(t, \boldsymbol{\theta}^0)$ depend only on their own past values and $\mathbf{w}(i, \boldsymbol{\theta}^0)$, $i < t - 1$, and consequently they are independent of $w_k(t, \boldsymbol{\theta}^0)$ for all k, j, and t. The first term on the right-hand side of (7b.1.5) is

$$\begin{aligned}
&\sum_{i,j=1}^{m} \frac{1}{N} \sum_{t=1}^{N} (\nabla^2_{\boldsymbol{\theta\theta}} w_i(t, \boldsymbol{\theta}))(\boldsymbol{\rho}^{-1})_{ij} w_j(t, \boldsymbol{\theta}) \\
&\qquad \approx \sum_{i,j=1}^{m} E(\nabla^2_{\boldsymbol{\theta\theta}} w_i(t, \boldsymbol{\theta}) w_j(t, \boldsymbol{\theta}))(\boldsymbol{\rho}^{-1})_{ij}, \qquad \text{for large } N.
\end{aligned}$$

Since $E(w_j) = 0$ and the $\nabla^2_{\boldsymbol{\theta\theta}} w_i(t, \boldsymbol{\theta}^0)$ are independent of $w_j(t, \boldsymbol{\theta}^0)$, the expectation above is zero if $\boldsymbol{\theta}$ is not far from $\boldsymbol{\theta}^0$ and hence one can ignore the first term on the right-hand side of (7b.1.5). Such an approximation will not cause any difficulties in inverting the corresponding matrix, since the second term on the right-hand side of (7b.1.5) is a positive-definite matrix whether or not $\boldsymbol{\theta} = \boldsymbol{\theta}^0$. Further, this approximation should not cause any numerical difficulties in implementing the Newton–Raphson scheme since it is well known that the Newton–Raphson algorithm works satisfactorily even when only an approximate value of the second derivative is available. Thus, the Newton–Raphson algorithm in (7b.1.4) can be rewritten in the following form, *without* involving the *second derivatives* of $w(t, \boldsymbol{\theta})$:

$$\boldsymbol{\theta}^{k+1} = \boldsymbol{\theta}^k - [H(\boldsymbol{\theta}, \boldsymbol{\rho})]^{-1} \nabla_{\boldsymbol{\theta}} J_4(\boldsymbol{\theta}, \boldsymbol{\rho})|_{\boldsymbol{\theta}=\boldsymbol{\theta}^k, \boldsymbol{\rho}=\boldsymbol{\rho}^k}, \tag{7b.1.9}$$

where

$$\mathbf{H}(\boldsymbol{\theta}, \boldsymbol{\rho}) = \frac{1}{N} \sum_{t=1}^{N} \sum_{i,j=1}^{m} (\nabla_{\boldsymbol{\theta}} w_i(t, \boldsymbol{\theta})(\nabla_{\boldsymbol{\theta}} w_j(t, \boldsymbol{\theta})^{\mathrm{T}})(\rho^{-1})_{ij}, \tag{7b.1.10}$$

$$\nabla_{\boldsymbol{\theta}} J_4(\boldsymbol{\theta}, \boldsymbol{\rho}) = \frac{1}{N} \sum_{t=1}^{N} \sum_{i,j=1}^{m} (\nabla_{\boldsymbol{\theta}} w_i(t, \boldsymbol{\theta}))(\rho^{-1})_{ij} w_j(t, \boldsymbol{\theta}), \tag{7b.1.11}$$

correspondingly

$$\boldsymbol{\rho}^{k+1} = \frac{1}{N} \sum_{t=1}^{N} \mathbf{w}(t, \boldsymbol{\theta}^{k+1}) \mathbf{w}^{\mathrm{T}}(t, \boldsymbol{\theta}^{k+1}). \tag{7b.1.12}$$

Thus, we avoid the computation of the $\frac{1}{2}m(n_0^2 + n_0)$ second derivatives of $\mathbf{w}(t, \boldsymbol{\theta})$ for all $t = 1, \ldots, N$. The required algorithm is contained in Eqs. (7b.1.9)–(7b.1.12), and is summarized below:

The Algorithm

(i) We begin the iteration with a pair $(\boldsymbol{\theta}^1, \boldsymbol{\rho}^1)$ which can be chosen arbitrarily or obtained from a relatively simple technique like least squares or recursive least squares (Kashyap and Nasburg, 1974).

(ii) By using the quantity $\boldsymbol{\theta}^k$ available at the end of the kth iteration, compute the quantities $\mathbf{H}(\boldsymbol{\theta}, \boldsymbol{\rho})$ and $\nabla_{\boldsymbol{\theta}} J_4(\boldsymbol{\theta}, \boldsymbol{\rho})$ by the Fourier techniques or time domain techniques (described in Section 7b.2).

(iii) Obtain $\boldsymbol{\theta}^{k+1}$ and $\boldsymbol{\rho}^{k+1}$ by (7b.1.9) and (7b.1.12).

(iv) The iteration is stopped if the absolute values of the differences between the estimates of $\boldsymbol{\theta}^*$ in successive iterations are less than a small prespecified value. Alternatively, the iteration is stopped if the absolute value of the sum of the components of the derivative vector $\nabla_{\boldsymbol{\theta}} J_4(\boldsymbol{\theta}, \boldsymbol{\rho})$ is less than a small prespecified value.

Comment. All gradient or gradient-related methods are very sensitive to the choice of the starting value of the estimates for the iteration process; i.e., the gradient methods converge to the correct minimizing value only when the starting values of the estimates are not "too far" from the correct value. The computational procedure for ML estimates is no exception to this general rule. We illustrate the phenomenon by a simulation example.

Example 7b.1. Let the process $y(\cdot)$ obey the moving average process

$$y(t) = w(t) + \theta_1 w(t-1) + \theta_2 w(t-2) + \theta_3 w(t-3)$$
$$\theta_1 = 0.8, \qquad \theta_2 = 0.6, \quad \theta_3 = 0.4,$$

TABLE 7b.1.1. Results of Algorithm for the Maximization of $J_2(\boldsymbol{\theta}, \boldsymbol{\rho})$

	Initial estimates			Final estimates				Likelihood function
Case	$\hat{\theta}_1(0)$	$\hat{\theta}_2(0)$	$\hat{\theta}_3(0)$	$\hat{\theta}_1(N)$	$\hat{\theta}_2(N)$	$\hat{\theta}_3(N)$	Iterations	$J_2(\hat{\boldsymbol{\theta}}(N), \hat{\boldsymbol{\rho}}(N))$
1	0.7289	0.5265	0.4497	0.6898	0.4969	0.3631	4	61.15
2	0	0	0	0.6896	0.4961	0.3629	8	61.16
3	−0.27	−0.27	−0.27	0.6896	0.4964	0.3630	9	61.16
4	−0.3	−0.3	−0.3	0.6283	0.5331	−0.1943	3	210.90
5	−0.45	0.45	0.45	Did not converge			—	—

where $w(\cdot)$ is a sequence of independent $N(0, 1)$ variables. This equation was simulated on a digital computer and 150 observations $y(t)$, $t = 1, \ldots, 150$, were generated. These observations were used to determine the ML estimates by the computation procedure discussed in this section. Different starting values were chosen for the algorithm, and the algorithm was terminated when the absolute change in the value of the estimate from one iteration to the next iteration was less than 0.001 (i.e., $|\hat{\theta}_i(k + 1) - \hat{\theta}_i(k)| < 0.001$ for every $i = 1, 2, 3$). In the majority of cases, the algorithm converged to the correct result. However, in Table 7b.1.1, we give a few cases in which the algorithm did not converge or converged to the wrong value. We have listed the final value of the estimate in the last iteration before termination, the starting values and the number of iterations before termination. In case 5, the algorithm was diverging; i.e., the absolute (numerical) value of one of the estimates was oscillating. In case 4, the algorithm converged to the wrong value. The initial value for case 1 is the estimate given by the recursive algorithm of Panuska (1969). Note that the convergence in this case needs the smallest number of iterations.

In conclusion, we may state that it may be advisable to compute the ML estimates with more than one initial value to remove the possibility of obtaining spurious extrema.

7b.2. Numerical Evaluation of the Variance of the Estimates

7b.2.1. The Conditional Covariance of $\boldsymbol{\theta}^*(N)$

We recall that the conditional covariance of $\hat{\boldsymbol{\theta}}(N)$ given $\boldsymbol{\xi}(N - 1)$ is $\mathbf{S}(N)$ where

$$\begin{aligned}
[\bar{\mathbf{S}}(N)]^{-1} &= \tfrac{1}{2}\nabla^2_{\boldsymbol{\theta\theta}} \sum_{t=1}^{N} \mathbf{w}^{\mathrm{T}}(t, \boldsymbol{\theta})(\boldsymbol{\rho}^0)^{-1}\mathbf{w}(t, \boldsymbol{\theta})\big|_{\boldsymbol{\theta}=\boldsymbol{\theta}^*(N)} \\
&= \tfrac{1}{2}\sum_{t=1}^{N}\sum_{k,j=1}^{m} [\nabla_{\boldsymbol{\theta}} w_k(t, \boldsymbol{\theta})\ \nabla_{\boldsymbol{\theta}} w_j(t, \boldsymbol{\theta})(\boldsymbol{\rho}^0)^{-1}_{kj} \\
&\quad + \nabla^2_{\boldsymbol{\theta\theta}} w_k(t, \boldsymbol{\theta})((\boldsymbol{\rho}^0)^{-1}_{ij}) w_j(t, \boldsymbol{\theta})]\big|_{\boldsymbol{\theta}=\boldsymbol{\theta}^*(N)}.
\end{aligned} \tag{7b.2.1}$$

In view of the comments of the preceding section, the second term in (7b.2.1) can be neglected. Then we can approximately evaluate the matrix $\bar{\mathbf{S}}(N)$ from

(7b.2.1) by replacing $\boldsymbol{\rho}^0$ by $\boldsymbol{\rho}^*(N)$. Note that the first term in (7b.2.1) is evaluated at every iteration of the algorithm in Section 7b.1. As such, the evaluation of variance of the estimate does not involve any additional computational effort.

7b.2.2. The Marginal Covariance of $\boldsymbol{\theta}^*(N)$

Recall that the covariance of $\boldsymbol{\theta}^*(N)$ is $\mathscr{I}_N^{-1}(\boldsymbol{\theta}^0, \boldsymbol{\rho}^0)$ defined in (7a.3.5). There are three possible methods of computing the matrix $\mathscr{I}(\boldsymbol{\theta}^0, \boldsymbol{\rho}^0)$. The first method consists of evaluating it by direct simulation; i.e., the difference equations for $\nabla_{\boldsymbol{\theta}}\mathbf{w}(t, \boldsymbol{\theta})$ are programmed on a digital computer with $\boldsymbol{\theta} = \boldsymbol{\theta}^0$ and the sample mean of $[\nabla_{\boldsymbol{\theta}} w_i(t, \boldsymbol{\theta}) \nabla_{\boldsymbol{\theta}} w_j(t, \boldsymbol{\theta})]$ is determined by using a large number of samples. Moreover, we can simulate different "runs" of the same model by using different sequences of $\{\mathbf{w}(\cdot)\}$ and $\{\mathbf{y}(\cdot)\}$ with the same statistical properties, find the average of the sample means of the corresponding quantity over different runs, and hence evaluate $\mathscr{I}(\boldsymbol{\theta}^0, \boldsymbol{\rho}^0)$ to any desired degree of accuracy.

In the second method, we can obtain a linear difference equation for the quantity $E[\nabla_{\boldsymbol{\theta}} w_i(t, \boldsymbol{\theta})\ \nabla_{\boldsymbol{\theta}} w_j(t, \boldsymbol{\theta})]$. This difference equation can be solved numerically, the steady state value of which gives the required function $\mathscr{I}(\boldsymbol{\theta}^0, \boldsymbol{\rho}^0)$. We will illustrate the method by an example. A variant of this method can be found in the paper by Astrom (1967).

Example 7b.2. Let the system equation be

$$y(t) = ((1 + \theta_2{}^0 D)/(1 - \theta_1{}^0 D))w(t), \qquad \boldsymbol{\theta} = (\theta_1, \theta_2)^{\mathrm{T}}, \tag{7b.2.2}$$

where $\{w(\cdot)\}$ is the usual Gaussian $N(0, \rho^0)$ sequence. The equation for $w(t, \boldsymbol{\theta})$ is

$$w(t, \boldsymbol{\theta}) = y(t) - \theta_1 y(t-1) - \theta_2 w(t-1, \boldsymbol{\theta}). \tag{7b.2.3}$$

Let

$$\nabla_{\theta_1} w(t, \boldsymbol{\theta}) \triangleq z_1(t, \boldsymbol{\theta}); \qquad \nabla_{\theta_2} w(t, \boldsymbol{\theta}) \triangleq z_2(t, \boldsymbol{\theta}).$$

From (7b.2.3) one can derive difference equations for $z_1(\cdot)$ and $z_2(\cdot)$:

$$\begin{aligned} z_1(t, \boldsymbol{\theta}) &= -\theta_2 z_1(t-1, \boldsymbol{\theta}) - y(t-1), \\ z_2(t, \boldsymbol{\theta}) &= -\theta_2 z_2(t-1, \boldsymbol{\theta}) - w(t-1, \boldsymbol{\theta}). \end{aligned} \tag{7b.2.4}$$

which have an input $y(\cdot)$ that is not white. Hence, we will rewrite (7b.2.4) in state variable form using (7b.2.2) so that the new set of equations is driven only by a white noise sequence. Let

$$\mathbf{x}(t) = [x_1(t), x_2(t), z_1(t), z_2(t)]^{\mathrm{T}}$$

where $x_1(t) \triangleq \theta_2 w(t)$ and $x_2(t) \triangleq y(t)$. Then $x(\cdot)$ obeys

$$\mathbf{x}(t) = \mathscr{A}\mathbf{x}(t-1) + \mathscr{B}\mathbf{w}(t) \tag{7b.2.5}$$

with

$$\mathscr{A} = \begin{bmatrix} 0 & 0 & 0 & 0 \\ 1 & \theta_1 & 0 & 0 \\ 0 & -1 & -\theta_2 & 0 \\ -\dfrac{1}{\theta_2} & 0 & 0 & -\theta_2 \end{bmatrix}, \qquad \mathscr{B} = \begin{bmatrix} \theta_2 \\ 1 \\ 0 \\ 0 \end{bmatrix}.$$

$$\mathbf{P}(t) \triangleq \operatorname{cov} \mathbf{x}(t) = \left[\begin{array}{c:c} \mathbf{P}_{11}(t) & \mathbf{P}_{12}(t) \\ \hdashline \mathbf{P}_{12}^{\mathrm{T}}(t) & \mathbf{P}_{22}(t) \end{array}\right].$$

Equation (7b.2.5) then yields

$$\mathbf{P}(t) = \mathscr{A}\mathbf{P}(t-1)\mathscr{A}^{\mathrm{T}} + \mathscr{B}\mathscr{B}^{\mathrm{T}}\rho^0. \tag{7b.2.6}$$

We can solve Eq. (7b.2.6) numerically or analytically and obtain $\mathbf{P}(\infty)$:

$$\mathscr{I}_N(\boldsymbol{\theta}^0, \rho^0) = N \lim_{t\to\infty} E[\mathbf{z}(t)\mathbf{z}^{\mathrm{T}}(t)]/\rho^0 = N\mathbf{P}_{22}(\infty)/\rho^0,$$

where $\mathbf{z}^{\mathrm{T}}(t) = (z_1(t), z_2(t))$.

$$\operatorname{cov}[\boldsymbol{\theta}^*(N)] = \mathscr{I}_N^{-1}(\boldsymbol{\theta}^0, \rho^0) + o(1/N) = \frac{1}{N}\frac{1}{(\theta_1{}^0 + \theta_2{}^0)^2}$$

$$\times \begin{bmatrix} (1-(\theta_1{}^0)^2)(1+\theta_1{}^0\theta_2{}^0)^2, & -(1-(\theta_1{}^0)^2)(1-(\theta_2{}^0)^2)(1+\theta_1{}^0\theta_2{}^0) \\ -(1-(\theta_1{}^0)^2)(1-(\theta_2{}^0)^2)(1+\theta_1{}^0\theta_2{}^0), & (1-(\theta_2{}^0)^2)(1+\theta_1{}^0\theta_2{}^0)^2 \end{bmatrix} + o\left(\frac{1}{N}\right),$$

$$\operatorname{cov}[\hat{\rho}(N)] = 2(\rho^0)^2/N + o(1/N).$$

Comment: Effect of Pole Zero Cancellation. Note that if $\theta_1{}^0$ and $-\theta_2{}^0$ in Example 7b.1 are very close to each other, i.e., if system (7b.2.2) has a nearly common pole and zero, then the variance of the estimates of $\theta_1{}^0$ and $\theta_2{}^0$ will be very large. For instance, let $\theta_1{}^0 = 0.5$, $\theta_2{}^0 = -0.501$, $\rho^0 = 1$.

$$\operatorname{cov}[(\hat{\theta}_1(N), \hat{\theta}_2(N))] = (1/N)\begin{bmatrix} 1.17 \times 10^6 & -0.702 \times 10^6 \\ -0.702 \times 10^6 & 1.17 \times 10^6 \end{bmatrix}.$$

Hence, even if we have 10^6 observations, the variance of the estimate of $(\theta_1{}^0 + \theta_2{}^0)$ will be very high compared with the number $(|\theta_1{}^0 + \theta_2{}^0|)$. Thus, from the estimates alone we will not be able to say whether the true values of $\theta_1{}^0$ and $-\theta_2{}^0$ are different. This feature is common to all systems for which even a single pole and a zero are close to each other. The best way to handle such a system is to begin with a lower order system. Usually, such a lower order fit is all that is required for control and prediction purposes.

7c. Limited Information Estimates

One of the principal reasons for the high degree of computational complexity of the FIML and CML estimators in multivariate systems with moving average terms is that *all* the unknowns in the system are estimated simultaneously. A method of reducing the computational complexity is to consider the possibility of separately estimating the unknowns in every one of the m individual difference equations; i.e., we consider the possibility of replacing a large and complex estimation problem with m relatively simple estimation problems.

The decomposition can be performed provided we are working with difference equations in canonical form II or III. The limited information estimates have some similarity to the limited information methods found in economic literature. We will discuss the limited information estimates with canonical forms II and III separately.

7c.1. Estimation of Parameters in Canonical Form II

Consider Eq. (7a.1.2) in canonical form II rewritten here as

$$\mathbf{f}_0(\mathbf{y}(t)) = \sum_{j=1}^{m_1} \mathbf{A}_j \mathbf{f}_j(\mathbf{y}(t-j)) + \sum_{j=1}^{m_3} \mathbf{G}_j \mathbf{u}(t-j) + \mathbf{F}\boldsymbol{\psi}(t-1) + \mathbf{w}(t) + \sum_{j=1}^{m_2} \mathbf{B}_j \mathbf{w}(t-j), \tag{7c.1.1}$$

where $\mathbf{f}_j(\mathbf{y}(t)) = [f_{j1}(\mathbf{y}(t)), \ldots, f_{jm}(\mathbf{y}(t))]^{\mathrm{T}}$. Since the matrices $\mathbf{B}_j$, $j = 1, \ldots, m_2$, are diagonal by definition, the vector equation (7c.1.1) can be rewritten as m separate scalar equations such that the ith equation, i.e., the equation for $f_{0i}(\mathbf{y}(t))$, will involve only the noise variables $w_i(t-j)$ for various j but not the noise variable w_k, $k \neq i$. The ith equation corresponding to (7c.1.1) is written as

$$f_{0i}(\mathbf{y}(t)) = \mathbf{x}_i^{\mathrm{T}}(t-1)\boldsymbol{\theta}_i^0 + w_i(t), \qquad i = 1, \ldots, m, \tag{7c.1.2}$$

where

$$\mathbf{x}_i(t-1) = [\mathbf{f}_1^{\mathrm{T}}(\mathbf{y}(t-1)), \ldots, \mathbf{f}_{m_1}^{\mathrm{T}}(\mathbf{y}(t-m_1)), \mathbf{u}^{\mathrm{T}}(t-1), \ldots, \mathbf{u}^{\mathrm{T}}(t-m_3), \boldsymbol{\psi}^{\mathrm{T}}(t-1), w_i(t-1), \ldots, w_i(t-m_2)]^{\mathrm{T}}. \tag{7c.1.3}$$

In (7c.1.2), the vector $\mathbf{x}_i$ is of dimension n_i and does not have any noise variables w_k, $k \neq i$. The vector $\boldsymbol{\theta}_i^0$ in (7c.1.2) is the corresponding vector of unknown coefficients in the ith equation only, and it is also of dimension n_i. The vector of all the unknowns, $\boldsymbol{\theta}^0$, can be expressed as $\boldsymbol{\theta}^0 = [(\boldsymbol{\theta}_1^0)^{\mathrm{T}}, \ldots, (\boldsymbol{\theta}_m^0)^{\mathrm{T}}]^{\mathrm{T}}$ where the dimension of $\boldsymbol{\theta}^0 \triangleq n_0 = \sum_{i=1}^m n_i$. We can regard (7c.1.2) as a single output generalized ARMA equation, the output being $f_{0i}(\mathbf{y}(t))$. The terms $\mathbf{f}_1(\mathbf{y}(t-1)), \ldots, \mathbf{f}_{m_1}(\mathbf{y}(t-m_1))$ in it can be treated as exogenous variables. We can generate the estimators $w_i(t, \boldsymbol{\theta}_i)$ of $w_i(t)$ as follows by utilizing the dummy $\boldsymbol{\theta}_i$ and Eq. (7c.1.2):

$$w_i(t, \boldsymbol{\theta}_i) = f_{0i}(\mathbf{y}(t)) - \mathbf{x}_i^{\mathrm{T}}(t-1, \boldsymbol{\theta}_i)\boldsymbol{\theta}_i, \tag{7c.1.4}$$

where

$$\mathbf{x}(t-1, \boldsymbol{\theta}_i) = [\mathbf{f}_1^{\mathrm{T}}(\mathbf{y}(t-1)), \ldots, \mathbf{f}_{m_1}^{\mathrm{T}}(\mathbf{y}(t-m_1)), \mathbf{u}^{\mathrm{T}}(t-1), \ldots, \mathbf{u}^{\mathrm{T}}(t-m_1), \boldsymbol{\psi}^{\mathrm{T}}(t-1), w_i(t-1, \boldsymbol{\theta}_i), w_i(t-2, \boldsymbol{\theta}_i), \ldots, w_i(t-m_2, \boldsymbol{\theta}_i)]^{\mathrm{T}}.$$

We can estimate $\boldsymbol{\theta}_i^0$ separately by minimizing the loss function $J_i(\boldsymbol{\theta}_i)$ with respect to $\boldsymbol{\theta}_i$ where $J_i(\boldsymbol{\theta}_i) = \sum_{t=1}^{N} w_i^2(t, \boldsymbol{\theta}_i)$. For numerically minimizing the function $J_i(\boldsymbol{\theta}_i)$ we need only the estimates $w_i(t, \boldsymbol{\theta}_i)$, $t = 1, \ldots, n$, computed from (7c.1.4), which are derived from the ith equation in (7c.1.2). We do not need the information contained in any other equation, nor in the unknowns $\boldsymbol{\theta}_j^0$, $j \neq i$. It is easy to show that the estimate $\hat{\boldsymbol{\theta}}_i$ obtained by minimizing $J_i(\boldsymbol{\theta}_i)$ with respect to $\boldsymbol{\theta}_i$ is consistent.

In computing the estimate $\hat{\boldsymbol{\theta}}_i$, we have entirely ignored the correlations between the noise inputs w_i and w_j; this could result in the estimate $\hat{\boldsymbol{\theta}}_i$ being inefficient. The term limited information comes from the fact that we have ignored the knowledge of correlations between noise inputs. However, there is another reason for $\hat{\boldsymbol{\theta}}_i$ being inefficient. In discussing canonical form II in Chapter V we mentioned that, sometimes all the components of the unknown vector $\boldsymbol{\theta}^0$ are not independent; i.e., if we expressed the same system in canonical form I instead of II, the corresponding vector of unknowns in canonical form I will (usually) be of a smaller dimension than $\boldsymbol{\theta}^0$ in form II. The dependency among the components of $\boldsymbol{\theta}^0$ can be expressed as

$$g_i(\boldsymbol{\theta}^0) = 0, \qquad i = 1, \ldots, n_5. \tag{7c.1.5}$$

Clearly, the information contained in (7c.1.5) is ignored while computing the estimates $\hat{\boldsymbol{\theta}}_1, \ldots, \hat{\boldsymbol{\theta}}_m$ individually, and this feature makes the estimate $\hat{\boldsymbol{\theta}}$ inefficient. We emphasize that dependency like (7c.1.5) does not occur in every problem.

But we cannot obtain explicit analytical expressions for the loss of efficiency of the LI estimates and hence we have to resort to numerical experiments. Our numerical experiments indicate that the loss of accuracy in the LI estimate due to the first cause is not substantial. This feature is demonstrated in the example discussed in Section 7d.1. The loss of accuracy due to the second cause mentioned above is difficult to estimate. Preliminary numerical results presented in Section 7d.1 indicate that the loss of accuracy due to the second cause is also not substantial.

7c.2. Estimation of Parameters in Canonical Form III

Recall that in canonical form III, $\mathbf{A}_0$ is lower triangular with unit diagonal elements $\mathbf{B}(D)$, and $\boldsymbol{\rho}$ is diagonal. For notational simplicity we will restrict the functions $\mathbf{f}_j$ in (7a.1.2) as follows:

$$f_{kj}(\mathbf{y}(t)) = y_j(t), \qquad k = 0, 1, \ldots, m_1; \quad j = 1, \ldots, m.$$

As in Section 7c.1, the equation for $y_j(t)$ can be written as follows, involving only the noise variables w_i:

$$y_i(t) = \mathbf{x}_i^{\mathrm{T}}(t-1)\boldsymbol{\theta}_i^0 + w_i(t), \tag{7c.2.1}$$

where n_i is the dimension of $\mathbf{x}_i$ and

$$\mathbf{x}_i(t-1) = [y_1(t), \ldots, y_{i-1}(t), \mathbf{y}^{\mathrm{T}}(t-1), \ldots, \mathbf{y}^{\mathrm{T}}(t-m_1), \mathbf{u}^{\mathrm{T}}(t-1), \ldots, \\ \mathbf{u}^{\mathrm{T}}(t-m_3), \boldsymbol{\psi}^{\mathrm{T}}(t-1), w_i(t-1), \ldots, w_i(t-m_2)]^{\mathrm{T}}.$$

Note that in $\mathbf{x}_i(t-1)$, the $y_i(t)$ component is absent. As before, we can generate the estimates $w_i(t, \boldsymbol{\theta}_i)$ of $w_i(t)$ by using the dummy variable $\boldsymbol{\theta}_i$ and Eq. (7c.2.1). For generating $w_i(t, \boldsymbol{\theta}_i)$ we do not need any information regarding $w_j(t, \boldsymbol{\theta}_j)$, $j \neq i$. As before, the limited information estimate $\hat{\boldsymbol{\theta}}_i$ of $\boldsymbol{\theta}_i^0$ is constructed by minimizing the function $J_i(\boldsymbol{\theta}_i) = \sum_{t=1}^{N} w_i^2(t, \boldsymbol{\theta}_i)$. The estimates $\hat{\boldsymbol{\theta}}_i$, $i = 1, 2, \ldots, m$, can be evaluated independently of one another.

To get another interpretation of these LI estimates, let the part of the conditional likelihood function that depends explicitly on $\boldsymbol{\theta}$ be rewritten as

$$J(\boldsymbol{\theta}_1^0, \ldots, \boldsymbol{\theta}_m^0) = \frac{1}{2\rho_{ii}} \sum_{i=1}^{m} J_i(\boldsymbol{\theta}_i^0). \tag{7c.2.2}$$

The simplification given in (7c.2.2) is possible because $\boldsymbol{\rho}^0$ is diagonal by the definition of canonical form III. Let $\hat{\boldsymbol{\theta}} = ((\hat{\boldsymbol{\theta}}_1)^{\mathrm{T}}, \ldots, (\hat{\boldsymbol{\theta}}_m)^{\mathrm{T}})^{\mathrm{T}}$. From (7c.2.2), it is clear that the LI estimate $\hat{\boldsymbol{\theta}}$ is made up of the individually evaluated component vectors $\hat{\boldsymbol{\theta}}_i$, $i = 1, \ldots, m$, which minimize $J_i(\boldsymbol{\theta})$ with respect to $\boldsymbol{\theta}_i$, respectively, where $J_i(\cdot)$ is the effective part of the log likelihood function of the disturbances $\{w_i\}$. Hence, the LI estimates of this section can be called limited information maximum likelihood (LIML) estimates.

But the LIML estimate is not, in general, equal to the corresponding CML estimate even asymptotically, because of the possible redundancy in canonical form III as indicated in Chapter V.

Unlike the estimation with canonical form II, there is no loss of efficiency due to the correlation among the components of $\mathbf{w}(t)$ since $\boldsymbol{\rho}$ is diagonal. We will illustrate the method by a numerical example in Section 7d.

7d. Numerical Experiments with Estimation Methods

Numerical comparison of the various methods of estimation in univariate and multivariate systems are considered in the papers by Gustauson (1972), Bohlin (1971), and Kashyap and Nasburg (1974). Here we will numerically compare only the CML estimates with the LI estimates and compare the different canonical forms for their ability to represent empirical data.

7d.1. The Comparison of Estimation Methods in Multiple Output Systems

Consider the bivariate process ($m = 2$) given below:

$$\mathbf{y}(t) = \mathbf{A}_1\mathbf{y}(t-1) + \mathbf{A}_2\mathbf{y}(t-2) + \mathbf{B}_1\mathbf{w}(t-1) + \mathbf{w}(t). \tag{7d.1.1}$$

The true numerical values of the matrices $\mathbf{A}_1$, $\mathbf{A}_2$, $\mathbf{B}_1$, and $\boldsymbol{\rho}$, the covariance matrix of the noise, are given in Table 7d.1.1. To demonstrate the power of the limited information estimates, we use a $\boldsymbol{\rho}$ having a correlation coefficient of 0.994 between noise variables w_1 and w_2. If the LI estimates in this example are close to the true values, then the limited information estimates in systems

TABLE 7d.1.1. CML Estimates and LI Estimates with Canonical Form II

	$\mathbf{A}_1$		$\mathbf{A}_2$		$\mathbf{B}_1$		$\boldsymbol{\rho}$		Comp. time (sec)	$J_2'(\boldsymbol{\theta}^*, \boldsymbol{\rho}^*)$
True values	1.2	−0.5	−0.6	0.5	0.6	0	20	26		
	0.6	0.3	0.5	0.6	0	0.8	26	34		
CML	1.23	−0.49	−0.59	0.49	0.58	—	18.17	23.62	66.6	670.4
	0.65	0.27	0.54	0.61	—	0.81	23.62	30.86		
LI form II	1.23	−0.59	−0.52	0.58	0.71	—	18.51	24.08	14.3	757.6
	0.66	0.41	0.43	0.49	—	0.70	24.08	31.73		

TABLE 7d.1.2. CML and LIML Estimates with Form III

	$\mathbf{A}_0'$		$\mathbf{A}_1'$		$\mathbf{A}_2'$	
True values	1	−0.765	−0.058	−0.270	−0.297	−0.496
	0	1	0.6	0.3	0.5	0.6
Converted CML	1	−0.765	−0.073	−0.260	−0.208	−0.505
	0	1	0.656	0.413	0.434	0.490
LIML form III	1	−0.764	−0.053	−0.277	−0.306	−0.486
	0	1	0.656	0.413	0.434	0.490

with other values of $\boldsymbol{\rho}$ will also be reasonable. The corresponding CML estimates of $\mathbf{A}_1$, $\mathbf{A}_2$, $\mathbf{B}_1$, $\boldsymbol{\rho}$ based on $N = 200$ observations of the process $\mathbf{y}(\cdot)$ are also given in Table 7d.1.1. In the table, we have also listed the log likelihood value $J_2'(\boldsymbol{\theta}^*, \boldsymbol{\rho}^*)$ which equals $NJ_2(\boldsymbol{\theta}^*, \boldsymbol{\rho}^*)$, where J_2 is defined in (7a.2.1). It should be noted that the CML estimates are very close to the true values even with such a relatively small observation set. Since Eq. (7d.1.1) is in canonical form II, we can compute the unknowns in it by the limited information method; i.e., we estimate each set of five coefficients in the two equations in (7d.1.1) separately instead of estimating all of them at once; the results are given in Table 7d.1.1 along with the time required for computation (on a CDC 6500 computer) which is one-sixth of the corresponding time for computing CML estimates. The CML and LI estimates are close to each other.

To illustrate the LIML estimates using canonical form III, we first convert the dynamical system in (7d.1.1) into canonical form III as given below:

$$\begin{aligned}\mathbf{A}_0\mathbf{y}(t) = \mathbf{A}_1'\mathbf{y}(t-1) + \mathbf{A}_2'\mathbf{y}(t-2) + \mathbf{A}_3'\mathbf{y}(t-3) + \mathbf{w}(t) \\ + \mathbf{B}_1'\mathbf{w}(t-1) + \mathbf{B}_2'\mathbf{w}(t-2),\end{aligned}$$

where $\mathbf{B}_1'$, $\mathbf{B}_2'$ are diagonal and $\mathbf{w}$ has diagonal covariance matrix. The numerical values of the matrices $\mathbf{A}_0'$, $\mathbf{A}_1'$, $\mathbf{A}_2'$, $\mathbf{A}_3'$, $\mathbf{B}_1'$, $\mathbf{B}_2'$ and the corresponding LIML estimates based on only 200 observations are given in Table 7d.1.2 along with the CML estimates. The latter estimates are derived from the corresponding CML estimates of Table 7d.1.1 since a CML estimate of a function of a parameter is also a CML estimate.

The three sets of estimates can be compared on the basis of the relative values of the index $J_2'(\boldsymbol{\theta}^*, \boldsymbol{\rho}^*)$, i.e., the negative log likelihood function of the observations. The values of the index for the CML estimates, LI estimates in form II, and LI estimates in form III are 670.4, 757.6, and 679.10, respectively. The relative closeness of the values of J_2' given by the CML and LIML estimates computed using canonical form III indicates the higher accuracy of the LI estimate obtained by using form III over the LI estimate computed by using form II, as mentioned earlier. The relatively large difference between the J_2' values of the CML and LI estimates of form II can be ascribed to the strong

$\mathbf{A}_3'$	$\mathbf{B}_1'$	$\mathbf{B}_2'$	$\boldsymbol{\rho}$	Comp. time (sec)
−0.709 0.125	1.4 —	0.480 —	0.118 —	—
0 0	— 0.8	— 0	— 34.0	
−0.752 0.123	1.389 —	0.469 —	0.092 —	66.6
0 0	— 0.695	— 0	— 30.86	
−0.703 0.120	1.311 —	0.399 —	0.102 —	14.9
0 0	— 0.695	— 0	— 31.72	

correlation between w_1 and w_2, which makes the matrix $\boldsymbol{\rho}$ almost singular. If the correlation has a more reasonable value, such as 0.5 or less, then the discrepancy between J_2' values of the CML and LI estimates using form II would not be so large.

7d.2. Fitting of Dynamic Models to Data

Autoregressive models or generalized AR models are popularly used for fitting data because of their computational simplicity. Even though there have been conjectures regarding the relative roles of ARMA and pure AR models in data fitting, the computational difficulties associated with the parameter estimation in the ARMA models have precluded the analysis of these conjectures.

We suggest that models in (7a.1.2) in canonical form II or III be considered for fitting data instead of using AR models only. The computational complexity with the models in canonical form II or III is not excessive if limited information methods are used. Moreover, the problem of parameter redundancy with canonical forms II and III mentioned in Chapter V may not be important in data fitting problems. If we detect any redundancy in the fitted model, we can obtain a better fit by reducing the order of the model. We will give an illustration.

Consider the ARMA($m_1 = 2, m_2 = 1$) process in canonical form II in (7d.1.1). If we express it in canonical form III, it will be an ARMA($m_1 = 3$, $m_2 = 2$) process. We will try to fit various models to the observations obtained from the process in (7d.1.1). Some of these models are ARMA(3, 2), ARMA(2, 1), ARMA(3, 0), ARMA(4, 0), and ARMA(5, 0), all of form III. The estimated values of the coefficients and J_2' are given in Table 7d.2.1 along with the computation time required for fitting. If we evaluate the different models on the basis of the value of J_2', ARMA(3, 0) is a better fit than ARMA(2, 1), although neither is a very good fit to the observed process considering the difference between the J_2' value for the ARMA(3, 2), ARMA(3, 0), and ARMA(2, 1) models. The ARMA(5, 0) fit may be satisfactory on the basis of the value of J_2', but it is not known how well the fitted ARMA(5, 0) model reflects the various statistical characteristics of the original process such as the spectrum and correlogram.

TABLE 7d.2.1. The Various Fitted Models

	$\mathbf{A}_0$		$\mathbf{A}_1$		$\mathbf{A}_2$		$\mathbf{A}_3$		$\mathbf{A}_4$	
True values	1	−0.765	−0.058	−0.270	−0.297	−0.496	−0.709	0.125	—	—
	0	1	0.6	0.3	0.5	0.6	0	0	—	—
Fitted to true model [ARMA(3, 2)]	1	−0.764	−0.053	−0.277	−0.306	−0.486	−0.703	0.120	—	—
	0	1	0.656	0.413	0.434	0.490	0	0	—	—
ARMA(2, 1)	1	−0.757	0.736	−0.868	−0.879	0.179	—	—	—	—
	0	1	0.656	0.413	0.434	0.490	—	—	—	—
AR(3)	1	−0.759	0.969	−1.043	−1.058	0.394	0.196	−0.058	—	—
	0	1	−0.044	1.567	0.537	−0.795	−0.587	0.217	—	—
AR(4)	1	−0.761	1.039	−1.1	−1.154	0.493	0.293	−0.131	−0.026	0.028
	0	1	−0.16	1.661	−1.304	0.551	0.710	−1.273	−1.795	0.191
AR(5)	1	−0.763	1.126	−1.173	−1.286	0.626	0.332	−0.142	−0.022	−0.016
	0	1	−0.691	2.086	−0.245	−0.461	−0.244	−0.569	−1.191	−0.072

	A_5		B_1		B_2		ρ		Comp. time (sec)	J_2
True values	—	—	1.4	—	0.48	—	0.118	—	—	—
	—	—	—	0.8	—	0	—	34.0		
Fitted to true model [ARMA(3, 2)]	—	—	1.311	—	0.399	—	0.102	—	14.9	679.1
	—	—	—	0.695	—	—	—	31.72		
ARMA(2, 1)	—	—	0.245	—	—	—	0.143	—	14.5	711.6
	—	—	—	0.695	—	—	—	31.72		
AR(3)	—	—	—	—	—	—	0.34	—	1.5	703.6
	—	—	—	—	—	—	—	31.89		
AR(4)	—	—	—	—	—	—	0.122	—	1.7	689.5
	—	—	—	—	—	—	—	30.80		
AR(5)	−0.092	−0.003	—	—	—	—	0.111	—	2.0	677.9
	0.098	0.133	—	—	—	—	—	30.38		

7g. Conclusions

We have discussed the two principal methods of estimation in multivariate systems involving both moving average and autoregressive parameters, namely the maximum likelihood, and limited information methods. From the point of view of computation, ML methods are more difficult than the least squares or limited information methods. But the ML methods also yield estimates with the highest possible accuracy. The limited information methods are a good compromise from the points of view of accuracy and complexity of computation.

Problems

1. Consider the ARMA(2, 1) process

$$\mathbf{y}(t) = \mathbf{A}_1\mathbf{y}(t-1) + \mathbf{A}_2\mathbf{y}(t-2) + \mathbf{w}(t) + \mathbf{B}_1\mathbf{w}(t-1), \tag{*}$$

where

$$\mathbf{A}_1 = \begin{bmatrix} 0.4 & 0.1 \\ 0.2 & 0.5 \end{bmatrix}, \quad \mathbf{A}_2 = \begin{bmatrix} 0.6 & 0.2 \\ 0.3 & 0.4 \end{bmatrix}, \quad \mathbf{B}_1 = \begin{bmatrix} 0.8 & 0 \\ 0 & 0.7 \end{bmatrix}$$

and

$$\boldsymbol{\rho} = \mathrm{cov}[\mathbf{w}] = \begin{bmatrix} 2 & 2 \\ 2 & 4 \end{bmatrix}.$$

Generate $\mathbf{y}(1), \ldots, \mathbf{y}(N)$ by simulating (*) on a computer. Using only these observations, construct autoregressive approximations to (*) of orders 3, 4, and 5. Compare the one-step forecasts of $\mathbf{y}$ given by these approximations with that given by (*) itself. In addition, compare the spectral densities of the processes obeying the various AR models with that of (*).

2. Using $\mathbf{y}(1), \ldots, \mathbf{y}(N)$ generated in Problem 1, obtain (a) The CML and LIML estimates of $\mathbf{A}_1$, $\mathbf{A}_2$, $\mathbf{B}_1$, and $\boldsymbol{\rho}$, (b) The least squares estimate of $\mathbf{A}_1$ and $\mathbf{A}_2$ using the method of Clarke (1967).

3. Consider any ARMA(3, 2) for the two-variable process $\mathbf{y}$ in canonical form I. As in Problem 1 obtain approximations to it in the form of ARMA(4, 1) and ARMA(4, 2) models in canonical form II. Compare the approximations to the correct model from the point of view of one-step-ahead forecasts and spectral densities.

4. Consider the sequence $\{y(\cdot)\}$ obeying $y(t) = w(t) + \theta w(t-1)$ where $w(\cdot)$ is the usual IID $N(0, \rho)$ sequence. Obtain an expression for the asymptotic variance of the CML estimate of θ based on N observations $y(\cdot), \ldots, y(N)$ in terms of ρ, θ, and N.

Chapter VIII | Class Selection and Validation of Univariate Models

Introduction

In the preceding chapters, we considered the estimation problem, i.e., the choice of a best fitting model for the given empirical time series in a given class of models. The only restriction on the class of models was that the difference equation characterizing all the members of the class must be linear in the coefficients. In this chapter, we consider the choice of an appropriate class of models for a given time series and obtain a best fitting model in the selected class for the given time series by the estimation techniques discussed in the earlier chapters. We will next develop a set of criteria for testing whether the given model satisfactorily represents the given empirical series to the desired level of accuracy. This procedure is referred to as validation of the model.

All the possible candidate classes of models are obtained by the inspection of data as outlined in Chapter III. For instance, if the data have systematic oscillations in them, we should consider the weak stationary AR or ARMA models, the seasonal ARIMA models, and the covariance stationary models having sinusoidal terms. The number of terms in the difference equations characterizing the various classes should be chosen consistent with the size of the available data. As an illustration let us consider the problem of developing models for annual river flow data with only 60 observations. In this problem, there is no need to consider more than three or four autoregressive terms. On the other hand, if we want to model the daily river flow data involving 15,000 observations, we may need an AR model of order 20 or more if the model is to represent the data satisfactorily. In other words, when we have more data, we can attempt to obtain a relatively refined model that is necessarily complex. On the other hand, if we have limited data, we cannot think of constructing a very refined model and hence the level of complexity in it will not be high. We may also mention that we do not separately discuss the class of purely deterministic models. The selection and validation methods used here will also indicate whether a purely deterministic model is appropriate for the given empirical series.

In choosing an appropriate class of models among a number of possible candidates we need a suitable criterion or goal. This goal may be specified by the user of the model. For instance, the goal may be that the best fitting model in the chosen class must have good forecasting ability. This aim is quite common in time series models. Another common use of the model is to generate synthetic data that possess similar statistical characteristics as the observed data.

Many common criteria such as least squares may not lead to good models. For instance, consider the problem of fitting a polynomial (in time) for an empirical process where we want to choose the order of the polynomial. If we choose the least squares criterion, then the larger the degree of the polynomial, the better is the fit. But often such a high-order polynomial fit (or often any polynomial fit) rarely passes the validation tests and has relatively poor prediction capability. Hence, we will work with a more sensitive criterion such as the likelihood or prediction error.

An ideal criterion for the selection of the appropriate class of models is that the chosen class be such that the best fitting model in this class for the given observation set should pass all validation tests at the required significance level. The model should yield satisfactory predictions also. Then one can be sure that the fitted model satisfactorily represents the data, thereby confirming the appropriateness of the choice of the particular class of models. However, such a procedure is impractical. In practical problems, the number of candidate classes can be relatively large. Without any direct method for the selection of the appropriate class of models, we have to find the best fitting model in each class for the given data set and apply the set of validation tests in each case until we find the appropriate class whose best fitting model obeys the validation tests. This is computationally impractical. Hence we have to consider separately the problems of class selection and validation.

Still, the interplay between the two problems is strong. For instance, suppose we find that the best fitting model in the class of models given by the selection method does not pass some of the validation tests. Then we have to choose a new class of models in light of this information and repeat the process till a validated model is obtained. The problems of class selection and validation are discussed in Sections 8a and 8c.

In this chapter, we have restricted ourselves to the comparison of finite difference equation models. In Chapter X, we will handle the annual river flow modeling problem where we need to compare a finite difference equation model with an *infinite moving average model*. The methods developed in this chapter are not directly applicable for such comparisons. However, we can modify the tests given here so that they can be used in such comparisons, as discussed in Chapter X.

8a. The Nature of the Selection Problem

We recall that a class of models is a triple $[\mathscr{s}, \mathscr{H}, \Omega]$ where $\mathscr{s}$ is a stochastic difference equation involving a vector of coefficients $\boldsymbol{\theta} \in \mathscr{H}$ and the corresponding covariance $\rho \in \Omega$:

$$\mathscr{s}: f_0(y(t)) = \boldsymbol{\theta}^{\mathrm{T}}\mathbf{x}(t-1) + w(t).$$

In this equation the components of $\mathbf{x}(t-1)$ are known functions of a finite number of components among $y(t-j), j \geq 1, w(t-j), j \geq 1, \mathbf{u}(t-j), j \geq 1$, and the deterministic trend functions $\psi_j(\cdot), j = 1, \ldots, l_2$. Whenever explicitly stated, the vector $\mathbf{x}(t-1)$ can also have terms such as $w(t-j), j = 1, 2, \ldots$, i.e., the system equation $\mathscr{s}$ can have moving average terms. $f_0(\cdot)$ is a known function of $y(t)$ [or $y(t)$ and $y(t-j), j > 1$] such as $f_0(y(t)) = y(t)$, or $\ln y(t)$, or $(y(t) - y(t-1))$, etc. The set $\mathscr{H}$ describes the set of values that can be assumed by $\boldsymbol{\theta}$.

A particular member in the class $[\mathscr{s}, \mathscr{H}, \Omega]$ is denoted by $(\mathscr{s}, \boldsymbol{\theta}, \rho)$. We define the set $\mathscr{H}$ such that every component of any of its members must be nonzero. This condition ensures that any two classes of models $[\mathscr{s}_1, \mathscr{H}_1, \Omega_1]$ and $[\mathscr{s}_2, \mathscr{H}_2, \Omega_2]$ are nonoverlapping as long as the corresponding equations $\mathscr{s}_1$ and $\mathscr{s}_2$ differ in at least one member. By this definition, the class of all AR(2) models is nonoverlapping with the class of all AR(1) models. It should be emphasized that we cannot define a class such as the class of AR models of order less than or equal to m_1 since we insist on every coefficient in the equation corresponding to the model in a class to be nonzero.

We are given r nonoverlapping classes $C_0, \ldots, C_{r-1}, C_i = [\mathscr{s}_i, \mathscr{H}_i, \Omega_i]$, and a set of empirical observations $\{y(1), \ldots, y(N)\} \triangleq \boldsymbol{\xi}$. Given that the set $\boldsymbol{\xi}$ could have come from a member of any one of the classes $C_i, i = 0, \ldots, r-1$, it is desired to determine the "most likely" or "most probable" class that could have produced this set of observations.

The problem of comparison of different classes has not received sufficient attention in the literature. Usually, a very narrow aspect of the problem, namely the choice of the orders of the autoregressive and moving average parts in an ARMA class of models, is considered. It is needless to say that the ARMA class is not the most appropriate class for every time series. For instance, consider the problem of developing a model for a biological population of whooping cranes (Fig. 11a.4.1). The prediction error of the best autoregressive model for this population is much greater than that of the best fitting multiplicative (or logarithmic) model. This feature cannot be inferred directly by inspection of the data. Such important information can be obtained only by comparing the two classes of models in detail, namely, the additive and the multiplicative classes of models. Another illustration is the modeling of the series of annual catch of Canadian lynx (Fig. 3b.1.2). The lynx series indicates a strong oscillatory component of about eight- to nine-year periods. Since the days of Yule (1927), it has been customary to model such a time series by using only autoregressive terms. In Chapters X and XI we show that, in many instances, a model involving autoregressive terms and sinusoidal terms reproduces the statistical characteristics of the observed data much better than a pure AR process.

The current practice has been to discuss the class selection problem mainly in terms of the correlation coefficients of the data (Box and Jenkins, 1970). Such a procedure may not always be appropriate since it implies that we are effectively restricting ourselves to the class of ARMA models and excluding the class of covariance stationary or the multiplicative models.

Another aspect of the class selection problem is whether the class of AR$(n + 1)$ models always yields a better fit to a given time series than the class of AR(n) models. (Recall that both classes are nonoverlapping.) This question can be generalized as follows. If we have two classes of models, $[\mathscr{I}_1, \mathscr{H}_1, \Omega_1]$ and $[\mathscr{I}_2, \mathscr{H}_2, \Omega_2]$, where equation $\mathscr{I}_2$ has all the terms in the equation $\mathscr{I}_1$, does $[\mathscr{I}_2, \mathscr{H}_2, \Omega_2]$ *always* yield a better fit, in some sense, than $[\mathscr{I}_1, \mathscr{H}_1, \Omega_1]$? The answer is no. If the best fitting model from the smaller class $[\mathscr{I}_1, \mathscr{H}_1, \Omega_1]$ passes all the validation tests, then the best fitting model in class $[\mathscr{I}_2, \mathscr{H}_2, \Omega_2]$ has poorer prediction ability than the best fitting model from class $[\mathscr{I}_1, \mathscr{H}_1, \Omega_1]$. The accuracy of the estimates in the model from $[\mathscr{I}_1, \mathscr{H}_1, \Omega_1]$ is lower than that in the model from $[\mathscr{I}_2, \mathscr{H}_2, \Omega_2]$. This result is demonstrated in Appendix 8.1.

In developing the methods of class selection, we should note that we can never establish with finite data that a model completely represents the process under consideration. At best, we can come up with a model such that the discrepancy between the characteristics of the model and those of the data is within the range of sampling error (i.e., the error caused by the finiteness of data). As such, purely deterministic methods of selection which are based on the fact that a process exactly obeys a deterministic model are of limited use. Typically these methods involve the determination of the rank of a certain matrix, and in double precision arithmetic almost all matrices can be shown to be full rank.

One of the popular methods for the comparison of classes is the method of hypothesis testing. We will discuss it in some detail. Even though the theory is elegant, it involves arbitrary quantities such as significance levels. Further, it has limited applicability in the sense that it can handle essentially two classes of models at a time and the classes can have only generalized AR models in them. Hence, we will develop two other methods of comparison, namely the likelihood approach and the prediction approach. Both of these approaches can handle more than two classes at a time, and these classes are not restricted to those having only generalized AR models. In the case of AR models, the decision rule given by the likelihood approach is identical to that given by the hypothesis testing approach for a particular significance level.

In addition, we can compare the various classes by directly comparing the best fitting models from each of the classes for the given data. We give two such methods, namely the prediction and the Bayesian methods. We will discuss the relative merits of the various methods toward the end of Section 8b.

8b. The Different Methods of Class Selection

8b.1. The Likelihood Approach

This approach can be used to compare a number of classes simultaneously. The system equations can have moving average terms. The basic idea is to compute the likelihood of the given observations coming from each of the r classes C_i, $i = 0, \ldots, r-1$, and to choose the class that yields the maximum value among them. But, evaluation of the likelihood is not obvious since every class has a number of models in it with different probability density functions and we do not know the identity of the member in each class which may have produced the observations. In the next section, we will obtain an explicit expression for the likelihood value associated with a family of probability density functions instead of a single density function.

8b.1.1. The Log Likelihood Value Associated with a Class of Models

Consider a class $C = [\mathscr{s}, \mathscr{H}, \Omega]$. Let the given set of observations, $\boldsymbol{\xi}$, come from a model $(\mathscr{s}, \boldsymbol{\theta}^0, \rho^0) \in C$, where $\boldsymbol{\phi}^0 \triangleq (\boldsymbol{\theta}^0, \rho^0)$ is *unknown*. Let the probability density of the available observation set $\boldsymbol{\xi}$ be $p(\boldsymbol{\xi}; \boldsymbol{\phi}^0)$, which is a known function of $\boldsymbol{\xi}$ and $\boldsymbol{\phi}^0$. Our problem is to find an approximation for the function $\ln p(\boldsymbol{\xi}; \boldsymbol{\phi}^0)$ in terms of $\boldsymbol{\xi}$ only.

Theorem 8b.1. Let $\boldsymbol{\phi}^*$ be the CML estimate of $\boldsymbol{\phi}^0$ based on the observation set $\boldsymbol{\xi}$. Then

$$E[\ln p(\boldsymbol{\xi}; \boldsymbol{\phi}^0) | \boldsymbol{\phi}^*] = L + E[O(\|\boldsymbol{\phi}^0 - \boldsymbol{\phi}^*\|^3) | \boldsymbol{\phi}^*]$$

where $L = \ln p(\boldsymbol{\xi}; \boldsymbol{\phi}^*) - n_{\boldsymbol{\phi}}$ and $n_{\boldsymbol{\phi}}$ is the dimension of $\boldsymbol{\phi}^0$.

Proof. We will expand $\ln p(\boldsymbol{\xi}; \boldsymbol{\phi}^0)$ in a Taylor series about $\boldsymbol{\phi}^*$:

$$\begin{aligned} \ln p(\boldsymbol{\xi}; \boldsymbol{\phi}^0) = {} & \ln p(\boldsymbol{\xi}; \boldsymbol{\phi}^*) + (\nabla_{\boldsymbol{\phi}} \ln p(\boldsymbol{\xi}; \boldsymbol{\phi}^*))^{\mathrm{T}} (\boldsymbol{\phi}^0 - \boldsymbol{\phi}^*) \\ & + \tfrac{1}{2} (\boldsymbol{\phi}^0 - \boldsymbol{\phi}^*)^{\mathrm{T}} \nabla^2_{\boldsymbol{\phi}\boldsymbol{\phi}} \ln p(\boldsymbol{\xi}; \boldsymbol{\phi}^*) (\boldsymbol{\phi}^0 - \boldsymbol{\phi}^*) \\ & + O(\|\boldsymbol{\phi}^0 - \boldsymbol{\phi}^*\|^3). \end{aligned} \tag{8b.1.1}$$

Now let us take conditional expectation of (8b.1.1) given $\boldsymbol{\xi}$. The expectation of the second term on the right-hand side of (8b.1.1) is zero in view of the definition of the CML estimate, i.e.,

$$\nabla_{\boldsymbol{\phi}} \ln p(\boldsymbol{\xi}; \boldsymbol{\phi}) |_{\boldsymbol{\phi} = \boldsymbol{\phi}^*} = 0. \tag{8b.1.2}$$

We recall the following expression from Section 7a.3.1:

$$E[(\boldsymbol{\phi}^* - \boldsymbol{\phi}^0)(\boldsymbol{\phi}^* - \boldsymbol{\phi}^0)^{\mathrm{T}} | \mathbf{S}(N)] = \mathbf{S}(N), \tag{8b.1.3}$$

where

$$\mathbf{S}(N) = -[\tfrac{1}{2} \nabla^2_{\boldsymbol{\phi}\boldsymbol{\phi}} \ln p(\boldsymbol{\xi}, \boldsymbol{\phi}) |_{\boldsymbol{\phi} = \boldsymbol{\phi}^*}]^{-1}. \tag{8b.1.4}$$

$$\begin{aligned} & E[\text{III term in the right-hand side of (8b.1.1)} \ |\boldsymbol{\phi}^*] \\ & \quad = \tfrac{1}{2} E[\operatorname{trace}\{\nabla^2_{\boldsymbol{\phi}\boldsymbol{\phi}} \ln p(\boldsymbol{\xi}; \boldsymbol{\phi}^*)(\boldsymbol{\phi}^0 - \boldsymbol{\phi}^*)(\boldsymbol{\phi}^0 - \boldsymbol{\phi}^*)^{\mathrm{T}}\} | \boldsymbol{\phi}^*] \\ & \quad = \operatorname{trace}[E[(\tfrac{1}{2} \nabla^2_{\boldsymbol{\phi}\boldsymbol{\phi}} \ln p(\boldsymbol{\xi}; \boldsymbol{\phi}^*)) E\{(\boldsymbol{\phi}^0 - \boldsymbol{\phi}^*)(\boldsymbol{\phi}^0 - \boldsymbol{\phi}^*)^{\mathrm{T}} | \mathbf{S}(N)\} | \boldsymbol{\phi}^*]] \\ & \quad = -\operatorname{trace}\{E[\operatorname{trace}\{\mathbf{S}^{-1}(N)\mathbf{S}(N)\} | \boldsymbol{\phi}^*]\}, \qquad \text{by (8b.1.3) and (8b.1.4)} \\ & \quad = -n_{\boldsymbol{\phi}}. \end{aligned} \tag{8b.1.5}$$

Q.E.D.

We will regard L as the required approximation of $\ln p(\boldsymbol{\xi}, \boldsymbol{\phi}^0)$ and label it the *log likelihood of class C* with the observation set $\boldsymbol{\xi}$.

We note that $\ln p(\boldsymbol{\xi}; \boldsymbol{\phi}^*)$ would have been the correct log likelihood value if $\boldsymbol{\xi}$ had come from a model characterized by $\boldsymbol{\phi}^*$. But $\boldsymbol{\phi}^*$ is an estimate of the unknown $\boldsymbol{\phi}^0$ computed using the observation set $\boldsymbol{\xi}$ itself. Hence, this additional ignorance about $\boldsymbol{\phi}^0$ manifests itself in a reduction in the likelihood by a quantity $n_{\boldsymbol{\phi}}$.

Now we are in a position to give the refined decision rule for the selection of an appropriate class for the given observation set $\boldsymbol{\xi}$. This decision rule was proposed by Akaike (1972, 1974) based on prediction error and information theoretic considerations. The derivation given here is relatively new.

Decision Rule 1

(i) For every class C_i, $i = 0, \ldots, r-1$, find the conditional maximum likelihood estimate $\hat{\boldsymbol{\phi}}_i$ of $\boldsymbol{\phi}_i^0$ given that $\boldsymbol{\phi}_i^0 \in \mathscr{H}_i$ using the given observation set $\boldsymbol{\xi}$. Compute the corresponding class likelihood function L_i:

$$L_i = \ln p(\boldsymbol{\xi}; \hat{\boldsymbol{\phi}}_i) - n_i, \quad \hat{\boldsymbol{\phi}}_i = (\hat{\boldsymbol{\theta}}_i, \hat{\rho}_i), \tag{8b.1.6}$$

where n_i is the dimension of the vector $\boldsymbol{\phi}_i^0 \triangleq (\boldsymbol{\theta}_i^0, \rho_i^0)$.

(ii) Choose that class which yields the *maximum* value of L among $\{L_i, i = 0, \ldots, r-1\}$. If there is more than one class yielding the maximum value, choose one of them according to some secondary criterion.

The decision rule is very versatile and is applicable for a variety of systems including those having moving average terms. It is also relatively easy to compute in many classes of systems as discussed in the next section. The decision rule can be justified by Bayesian reasoning, as will be demonstrated later. This decision rule implicitly incorporates the notion of parsimony in it. A class whose difference equation involves a large number of coefficients will have a correspondingly large reduction in the log likelihood function caused by the term $(-n_i)$ in (8b.1.6).

Our next task is the development of explicit forms of L and the decision functions for particular sets of classes.

8b.1.2. The Simplified Form of L for a Class of Nontransformed Models

Let $C = [\mathscr{s}, \mathscr{H}, \Omega]$, where

$$\mathscr{s}: y(t) = \boldsymbol{\theta}^{\mathrm{T}}\mathbf{x}(t-1) + w(t). \tag{8b.1.7}$$

In (8b.1.7), the components of $\mathbf{x}(t-1)$ are functions of a finite number of $y(t-j)$, $j > 0$, $w(t-j)$, $j > 0$, or deterministic trend functions. $w(\cdot)$ is the usual zero mean IID Gaussian noise with variance ρ. Note that (8b.1.7) subsumes all the usual nontransformed models such as the ARMA and the generalized ARMA models, the ARIMA and the seasonal ARIMA models, and covariance stationary models:

$$\ln p(y(1), \ldots, y(N); \boldsymbol{\phi}) = \ln p(y(N), \ldots, y(m_1 + 1) | y(1), \ldots, y(m_1); \boldsymbol{\phi}) + \ln p(y(1), \ldots, y(m_1); \boldsymbol{\phi}), \tag{8b.1.8}$$

where m_1 is the maximum lag of AR terms in the equation $\mathscr{o}$ of class C, i.e., there are no terms like $y(t - j)$ where $j > m_1$ on the right-hand side of equation $\mathscr{o}$.

$$\begin{aligned} \ln p(y(N), \ldots, y(m_1 + 1) | y(m_1), \ldots, y(1); \hat{\boldsymbol{\phi}}) &= \prod_{i=m_1+1}^{N} \ln p(y(i) | y(i - 1), \ldots, y(i - m); \hat{\boldsymbol{\phi}}) \\ &= -\frac{N - m_1}{2} \ln (2\pi\hat{\rho}) - \frac{N - m_1}{2}, \end{aligned} \tag{8b.1.9}$$

where $\hat{\rho}$ is the CML estimate of ρ,

$$\hat{\rho} = \frac{1}{N - m_1} \sum_{t=m_1+1}^{N} (y(t) - (\hat{\boldsymbol{\theta}})^{\mathrm{T}} \mathbf{x}(t - 1))^2.$$

Next, we can approximate the expression $p(y(1), \ldots, y(m_1); \boldsymbol{\phi}^*)$ by regarding $y(1), \ldots, y(m_1)$ to be independent Gaussian variables with zero mean and variance ρ_y:

$$\ln p(y(1), \ldots, y(m_1)) = -(m_1/2) \ln 2\pi\rho_y - \tfrac{1}{2} \sum_{i=1}^{m_1} y^2(i)/\rho_y. \tag{8b.1.10}$$

Substituting (8b.1.9) and (8b.1.10) into (8b.1.8) and substituting the resulting expression for (8b.1.8) into (8b.1.6), we obtain the following expression for L:

$$\begin{aligned} L = \left\{ -\frac{N}{2} \ln \hat{\rho} - n_{\boldsymbol{\phi}} \right\} + \left\{ -\frac{N}{2} \ln 2\pi + \frac{m_1}{2} \ln \left(\frac{\hat{\rho}}{\rho_y} \right) - \frac{N}{2} \right. \\ \left. + \frac{1}{2} \left(m_1 - \frac{(\sum_{t=1}^{m_1} y^2(t))}{\rho_y} \right) \right\} = \bar{L} + \bar{\bar{L}}. \end{aligned} \tag{8b.1.11}$$

When we are comparing different classes of models with $m_1 \ll N$ the variation in $\bar{\bar{L}}$ in (8b.1.11) from class to class is negligible as compared with the variation in $\bar{L}$. Hence, for the purposes of comparison, it is enough if we consider $(\bar{L})$ in (8b.1.11) in such cases:

$$\begin{aligned} \bar{L} &= -\frac{N}{2} \ln \hat{\rho} - n_{\boldsymbol{\phi}} \\ &= -\left[\frac{\text{(number of observations)}}{2} \right] \ln \text{(residual variance)} \\ &\quad - \text{total number of parameters to be estimated in the equation } \mathscr{o}. \end{aligned} \tag{8b.1.12}$$

Thus, decision rule 1 reduces to the choice of the class with the maximum value of $\bar{L}$ among all possible classes. The estimate $\hat{\rho}$ is the CML estimate mentioned in Eq. (7a.2.4).

8b.1.3. Comparison of Two Classes with Nontransformed Models

We will specialize the result given above for the selection between two classes labeled 1 and 2. In this case decision rule 1 reduces to comparing the $\bar{L}$ values for the two classes. Let the residual variances and the number of parameters in the two classes be $\hat{\rho}_i$, $i = 1, 2$, and n_i, $i = 1, 2$. Let $n_2 \geq n_1$. Then

$$\bar{L}_i = -\frac{N}{2} \ln \hat{\rho}_i - n_i, \qquad i = 1, 2.$$

We prefer class 1 to class 2 if $\bar{L}_1 > \bar{L}_2$, i.e.,

$$\ln(\hat{\rho}_1/\hat{\rho}_2) < 2(n_2 - n_1)/N. \tag{8b.1.13}$$

If the difference $|\hat{\rho}_1 - \hat{\rho}_2|$ is small as compared to $\hat{\rho}_1$, we can further approximate (8b.1.13) as follows:

$$\ln \frac{\hat{\rho}_1}{\hat{\rho}_2} = \ln\left(\frac{\hat{\rho}_2 + (\hat{\rho}_1 - \hat{\rho}_2)}{\hat{\rho}_2}\right) \approx \frac{\hat{\rho}_1 - \hat{\rho}_2}{\hat{\rho}_2}.$$

The decision rule simplifies as follows:

$$N(\hat{\rho}_1 - \hat{\rho}_2)/\hat{\rho}_2 \begin{cases} \leq 2(n_2 - n_1) \Rightarrow \text{prefer class 1} \\ > 2(n_2 - n_1) \Rightarrow \text{prefer class 2.} \end{cases} \tag{8b.1.14}$$

Let us discuss the decision rule in (8b.1.14). Usually, the larger parameter class (i.e., Class 2) yields smaller residual variance than the smaller parameter class. For instance, if Classes 1 and 2 are classes of AR models of orders 1 and 2, then

$$\hat{\rho}_1 = \min_{\theta_1} \frac{1}{N-1} \sum_{t=2}^{N} (y(t) - \theta_1 y(t-1))^2,$$

$$\hat{\rho}_2 = \min_{\theta_1, \theta_2} \frac{1}{N-2} \sum_{t=3}^{N} (y(t) - \theta_1 y(t-1) - \theta_2 y(t-2))^2.$$

Barring the cases when $y(2)$ is typically very large or small, $\hat{\rho}_2$ should be smaller than $\hat{\rho}_1$ since we are minimizing over a larger number of parameters in the former case. Hence, we can state that when additional terms are added to a model, the residual variance usually decreases. The question is whether the decrease is statistically significant, i.e., whether the decrease can be attributed solely to sampling variations caused by the finiteness of N or whether it can be attributed to the newly introduced terms in the model. When N is small, we expect a large sampling variation in the difference $(\hat{\rho}_1 - \hat{\rho}_2)$ as compared to the case of large N. Hence, we need an effective measure of the contribution of the newly added terms to the decrease $(\hat{\rho}_1 - \hat{\rho}_2)$. According to the rule in (8b.1.14), this effective contribution is measured by the product $N(\hat{\rho}_1 - \hat{\rho}_2)/\hat{\rho}_2$. If this quantity is greater than a threshold value equal to $2(n_2 - n_1)$, then we declare that the newly added terms are important.

The decision rule in (8b.1.14) is similar in form to the one given by the theory of hypothesis testing in Section 8b.2. As such, in some special cases, we can also get an asymptotic expression for the probability of error caused by the decision rule in (8b.1.14), i.e., the probability of rejecting the smaller model when it is the correct model. We will discuss the details later.

8b.1.4. The Simplified Likelihood Expression in Some Transformed Models

We will consider the evaluation of the expression L in a class of multiplicative models in y, i.e., models that are additive in $\ln y$. In such cases, equation $\mathscr{s}$ can be written as

$$\ln y(t) = \boldsymbol{\theta}^{\mathrm{T}}\mathbf{x}(t-1) + w(t), \tag{8b.1.15}$$

where the components of $\mathbf{x}(t-1)$ are known functions of a finite number of $y(t-j), j \geq 1$, $w(t-j), j \geq 1$, and deterministic trends or other inputs. We will obtain an explicit expression for the likelihood. As before, we can split the joint probability density $p(y(1), \ldots, y(N); \boldsymbol{\phi})$ as in (8b.1.8) and evaluate the two terms in it separately. Consider the first term in (8b.1.8):

$$\begin{aligned} &p(y(N), \ldots, y(m_1+1) | y(m_1), \ldots, y(1); \boldsymbol{\phi}) \\ &\quad = \prod_{i=m_1+1}^{N} p(y(i) | y(i-1), \ldots, y(1); \boldsymbol{\phi}) \\ &\quad = \prod_{i=m_1+1}^{N} \frac{1}{y(i)}\, p(w(i); \boldsymbol{\phi}), \end{aligned} \tag{8b.1.16}$$

where $p(w(i))$ is the normal probability density of $w(\cdot)$ and the factor $1/y(i)$ is caused by the transformation of y into $\ln y$:

left-hand side of (8b.1.16)

$$= \prod_{i=m_1+1}^{N} \frac{1}{y(i)} \frac{1}{(2\pi\rho)^{1/2}} \exp\left[-\frac{1}{2}\frac{(\ln y(i) - (\boldsymbol{\theta})^{\mathrm{T}}\mathbf{x}(i-1))^2}{\rho}\right].$$

Taking logarithms

$$\begin{aligned} &\ln p(y(N), \ldots, y(m_1+1) | y(m_1), \ldots, y(1); \hat{\boldsymbol{\phi}}) \\ &\quad = -\sum_{i=m_1+1}^{N} \ln y(i) - \left(\frac{N-m_1}{2}\right) \ln 2\pi\hat{\rho} - \left(\frac{N-m_1}{2}\right), \end{aligned} \tag{8b.1.17}$$

where $\hat{\rho}$ is the residual variance. As a first approximation we can assume $y(1), \ldots, y(m_1)$ to be independent and log-normally distributed with parameters 0 and ρ_y to evaluate $\ln p(y(1), \ldots, y(m_1); \boldsymbol{\phi})$. Hence

$$\ln p(y(1), \ldots, y(m_1); \boldsymbol{\phi}) = -\sum_{i=1}^{m_1} \ln y(i) - (m_1/2) \ln 2\pi\rho - \tfrac{1}{2} \sum_{i=1}^{m} y^2(i)/\rho_y. \tag{8b.1.18}$$

Hence, substituting (8b.1.17) and (8b.1.18) into (8b.1.8), we obtain the following expression for L:

$$L = \ln p(y(N), \ldots, y(1); \hat{\boldsymbol{\phi}}) - n_{\phi}$$
$$= \left\{ -\sum_{i=1}^{N} \ln y(i) - \frac{N}{2} \ln \hat{\rho} - n_{\phi} \right\}$$
$$+ \left\{ -\frac{N}{2} \ln 2\pi + \frac{m_1}{2} (\ln \hat{\rho} - \ln \rho_y) - \frac{N}{2} + \frac{1}{2} \left(m_1 - \sum_{t=1}^{m_1} \frac{y^2(t)}{\rho_y} \right) \right\}$$
$$= \bar{L} + \bar{\bar{L}}. \tag{8b.1.19}$$

As before, the variation in the second term $\bar{\bar{L}}$ in (8b.1.19) from class to class is small in comparison with that in the first term if $m_1 \ll N$. Hence the effective part of the statistic L is $\bar{L}$:

$$\bar{L} = -\sum_{t=1}^{N} \ln y(t) - \frac{N}{2} \ln \hat{\rho} - n_{\phi}$$
$$= -N\hat{E}(\ln y(\cdot)) - N/2 \ln \hat{\rho} - n_{\phi}, \tag{8b.1.20}$$

where

$$\hat{E}(\ln y(\cdot)) \triangleq (1/N) \sum_{t=1}^{N} \ln y(t). \tag{8b.1.21}$$

Thus decision rule 1 reduces to the choice of the class having the largest value of $\bar{L}$ among the various classes. The expression in (8b.1.20) is also valid if we are dealing with a logarithmic differenced or ARIMA model.

8b.1.5. Comparison of Transformed and Nontransformed Models

Suppose we have to compare class 1, a class of nontransformed models obeying equation $\mathscr{s}$ as in (8b.1.7) with class 2, a class having transformed models obeying equation $\mathscr{s}$ as in (8b.1.15), with $f_0(y) = \ln y$. We can compare them by the statistic $\bar{L}$:

$$\bar{L}_1 = -\frac{N}{2} \ln \hat{\rho}_1 - n_1. \tag{8b.1.22}$$

$$\bar{L}_2 = -\frac{N}{2} \ln \hat{\rho}_2 - n_2 - N\hat{E}(\ln y(\cdot)). \tag{8b.1.23}$$

By decision rule 1, we prefer class 1 to class 2 if $\bar{L}_1$ is greater than $\bar{L}_2$. The important term in expression (8b.1.23) is $N\hat{E}(\ln y(\cdot))$. If this additional term in (8b.1.23) were absent, we would always have to prefer class 2 to class 1 because log y always has a smaller variance than y. The decision rule says that even if $n_1 = n_2$, we have to compare $\ln \hat{\rho}_1$ with $(\ln \hat{\rho}_2) + \hat{E}(\ln y(\cdot))/2$ for determining the correct class. The comparison of the two classes of models given above is beyond the scope of the classical hypothesis testing methods.

Some examples of simultaneous comparisons of different classes of models are given in Chapter XI.

8b.2. Hypothesis Testing

A general discussion of hypothesis testing can be found in the books by Rao (1965) and Lehmann (1959) and others, and its application to time series can be found in the texts by Anderson (1971), Box and Jenkins (1970), Whittle (1951), and in other books and papers. Interesting critiques of the classical hypothesis testing are given by Pratt *et al.* (1965) and Birnbaum (1969), among others.

8b.2.1. Development of the Decision Rule

The classical theory of hypothesis testing is applicable for comparing only two classes of models, say C_0 and C_1. Let the two classes be

$$C_0 = [\mathscr{I}_0, \mathscr{H}_0, \Omega], \qquad C_1 = [\mathscr{I}_1, \mathscr{H}_1, \Omega].$$

The difference equation $\mathscr{I}_0$ in C_0 has fewer terms than that in C_1. Hence we can call the class C_0 the simpler of the two classes.

We want to find a decision rule d that maps the given observation set $\boldsymbol{\xi}$ into one of the two classes C_0 and C_1. More specifically, by using the observation set $\boldsymbol{\xi}$ we want to find a statistic, say $\eta_d(\boldsymbol{\xi})$, a function of $\boldsymbol{\xi}$, and a threshold η_0, such that

$$\text{decision rule } d\colon \eta_d(\boldsymbol{\xi})\begin{cases} \leq \eta_0 \to \text{accept } C_0 \\ > \eta_0 \to \text{accept } C_1 \end{cases} \tag{8b.2.1}$$

has reasonable properties. In a stochastic situation, there is always the possibility of a wrong decision, i.e., assigning the given empirical process to the wrong class. There are two types of errors, I and II, namely rejection of class C_0 when it is the correct class and rejection of class C_1 when it is the correct class:

$$\begin{aligned} &\text{probability of error I} \\ &\qquad = \text{prob}\{\eta_d(\boldsymbol{\xi}) > \eta_0 | \boldsymbol{\xi} \text{ obeys } (\mathscr{I}_0, \boldsymbol{\theta}_0, \rho) \in C_0\} = g_0(\boldsymbol{\theta}_0, d) \\ &\text{probability of error II} \\ &\qquad = \text{prob}\{\eta_d(\boldsymbol{\xi}) \leq \eta_0 | \boldsymbol{\xi} \text{ obeys } (\mathscr{I}_1, \boldsymbol{\theta}_1, \rho) \in C_1\} = g_1(\boldsymbol{\theta}_1, d). \end{aligned} \tag{8b.2.2}$$

Note that the two probabilities of error are functions not only of the rule d but also of the parameter vector $\boldsymbol{\theta}_0$ or $\boldsymbol{\theta}_1$ characterizing the process.

A decision rule that leads to as few errors as possible, i.e., a rule d that would make $g_0(\boldsymbol{\theta}_0, d)$ small for every $\boldsymbol{\theta}_0 \in \mathscr{H}$ and $g_1(\boldsymbol{\theta}_1, d)$ small for every $\boldsymbol{\theta}_1 \in \mathscr{H}_1$, would be ideal. However, this problem is not well posed, because making $g_0(\boldsymbol{\theta}_0, d)$ small does not also render $g_1(\boldsymbol{\theta}_1, d)$ small for every $\boldsymbol{\theta}_1 \in \mathscr{H}_1$. For instance, if we consider the case of $\eta_0 = \infty$, then $g_0(\boldsymbol{\theta}_0, d)$ is zero for all $\boldsymbol{\theta}_0 \in \mathscr{H}_0$ but $g_1(\boldsymbol{\theta}_1, d)$ will be unity for all $\boldsymbol{\theta}_1 \in \mathscr{H}_1$. So we must have a compromise.

In the Neyman–Pearson theory, we first restrict ourselves to only those decision rules which yield a probability of error I not greater than a certain prespecified value ε_0. Therefore, we consider all decision rules d that satisfy

$$g_0(\boldsymbol{\theta}, d) \leq \varepsilon_0, \qquad \forall \boldsymbol{\theta} \in \mathscr{H}_0. \tag{8b.2.3}$$

Among this class of decision rules, we select a particular rule d^*, which satisfies

$$g_1(\boldsymbol{\theta}, d^*) \leq g_1(\boldsymbol{\theta}, d), \qquad \forall \boldsymbol{\theta} \in \mathscr{H}_1 \quad \text{and} \quad d \quad \text{obeying (8b.2.3)}, \qquad (8b.2.4)$$

provided it exists. If such a rule d^* exists, we call it the uniformly most powerful (UMP) decision rule. However, such a UMP rule d^* rarely exists since it is difficult to conceive of a decision function d^* that satisfies (8b.2.4) for every $\boldsymbol{\theta} \in \mathscr{H}_1$.

Therefore, we consider a still smaller class of decision rules, the class of rules that obey

$$g_0(\boldsymbol{\theta}, d) = \varepsilon_0, \qquad \forall \boldsymbol{\theta} \in \mathscr{H}_0, \qquad (8b.2.5)$$

where $(1 - \varepsilon_0)$ is called the significance level, usually expressed as a percentage.

Strange as it may seem, usually there does exist a decision rule d that satisfies (8b.2.5) asymptotically. We do not have to worry about the choice of decision function d when there exists more than one d satisfying (8b.2.5) since the chances of such an occurrence are very small.

Next we have to turn our attention to obtaining a decision rule d obeying (8b.2.5). An intuitively appealing candidate for a rule obeying (8b.2.5) is the likelihood ratio test. Let $p_i(\boldsymbol{\xi}, \boldsymbol{\theta}, \rho | \boldsymbol{\theta} \in \mathscr{H}_i)$ be the probability density of the observation set $\boldsymbol{\xi}$ obeying the model $(\mathscr{I}, \boldsymbol{\theta}, \rho) \in C_i$. Consider $\eta_1(\boldsymbol{\xi})$, the likelihood ratio:

$$\eta_1(\boldsymbol{\xi}) = \sup_{\boldsymbol{\theta} \in \mathscr{H}_1, \rho} p_1(\boldsymbol{\xi}; \boldsymbol{\theta}, \rho) / \sup_{\boldsymbol{\theta} \in \mathscr{H}_0, \rho} p_0(\boldsymbol{\xi}; \boldsymbol{\theta}, \rho). \qquad (8b.2.6)$$

The difference between the comparison of likelihoods in (8b.2.6) and that done in Section 8b.1 is that the former approach does not make allowance for the relative sizes of the sets $\mathscr{H}_0$ and $\mathscr{H}_1$. The likelihood ratio test is given as

$$d_0: \begin{cases} \eta_1(\boldsymbol{\xi}) \leq \eta_2 \to \text{classify } \boldsymbol{\xi} \text{ in } C_0 \\ \eta_1(\boldsymbol{\xi}) > \eta_2 \to \text{classify } \boldsymbol{\xi} \text{ in } C_1, \end{cases} \qquad (8b.2.7)$$

where η_2 is a suitable threshold. We will first simplify the function $\eta_1(\boldsymbol{\xi})$. Assuming the normality of the observations $\boldsymbol{\xi}$, let $\hat{\rho}_0$ be the residual variance of the best fitting model in the class C_0 for the given data $\boldsymbol{\xi}$, and let $\hat{\rho}_1$ be the corresponding residual variance of the best fitting model in the class C_1 also based on $\boldsymbol{\xi}$. Since the difference equation for a model in C_1 has more coefficients than that in C_0 we usually have $\hat{\rho}_0 \geq \hat{\rho}_1$. We have shown the validity of the following result in Chapter VI:

$$\sup_{\boldsymbol{\theta} \in \mathscr{H}_i, \rho} p_i(\boldsymbol{\xi}; \boldsymbol{\theta}, \rho) \approx \left\{ \frac{\exp(-N/2)}{(2\pi)^{N/2}} \right\} \left\{ \frac{1}{(\hat{\rho}_i)^{N/2}} \right\}.$$

Hence,

$$\eta_1(\boldsymbol{\xi}) = (\hat{\rho}_1 / \hat{\rho}_0)^{N/2}.$$

Thus the decision rule in (8b.2.7) involves only the statistic $(\hat{\rho}_1/\hat{\rho}_0)$. It can be rewritten in a slightly different manner as in (8b.2.8) so that its probability distribution under C_0 has a standard form. The decision rule d_0 is rewritten as decision rule 2.

Decision Rule 2 (Residual Variance Ratio Test)

$$d_0 : \eta_d(\boldsymbol{\xi}) = \frac{\hat{\rho}_0 - \hat{\rho}_1}{\hat{\rho}_1} \cdot \frac{(N - m_1)}{m'} \begin{cases} \leq \eta_0 \rightarrow \text{choose } C_0 \\ < \eta_0 \rightarrow \text{choose } C_1, \end{cases} \tag{8b.2.8}$$

where N is the number of observations in $\boldsymbol{\xi}$, m_1 is the number of nonzero coefficients in the equation $\mathscr{s}_1$ in C_1, and m' is the number of additional coefficients in equation $\mathscr{s}_1$ in C_1 which are not present in equation $\mathscr{s}_0$ in C_0.

We *conjecture* that the probability distribution of the statistic η_d when $\boldsymbol{\xi}$ obeys $(\mathscr{s}, \boldsymbol{\theta}, \rho) \in C_0$ is an F-distribution with m' and $N - m_1$ degrees of freedom. The truth of this conjecture stated above *can be verified* in certain special cases, such as the classes of nontransformed AR models (Whittle, 1952) or nontransformed AR models with sinusoidal terms in them. Further, there is some empirical evidence for the validity of the conjecture when $m' = 1$. The important point about the conjecture is that the probability distribution is independent of both $\boldsymbol{\theta}$ and ρ as long as $\boldsymbol{\theta} \in \mathscr{H}_0$ and $\rho \in \Omega_0$. Since the F-distribution is tabulated, we can find the threshold η_0 to yield the particular value of the error I probability ε_0.

But we have no way of determining the probability of type II error. This is one of the principal drawbacks of this theory. We have to use an arbitrary value for the probability of error ε_0, such as 0.05, and hope that the corresponding probability of error II is not large. But we do not have any idea of the tradeoff involved between choosing $\varepsilon_0 = 0.05$ and choosing, say, $\varepsilon_0 = 0.02$. Thus, even though the decision rule in (8b.2.8) is similar to the one in Section 8a, based on the likelihood approach, the rule in Section 8a is to be preferred because it does not involve any arbitrary parameters such as ε_0.

One of the incidental advantages of the theory of this section is that we get an approximate expression for the probability of error of type I in using the decision rule given by the likelihood approach of Section 8a for the two-class problem, the two classes having the special structure mentioned in this section.

The ratio $\{(\hat{\rho}_0 - \hat{\rho}_1)/\hat{\rho}_1\}$ is the fractional decrease in the residual variance caused by the additional terms in the difference equations in class C_1 which are absent in class C_0. To illustrate the importance of this ratio, we consider the following special case.

Example 8b.1. Let the two classes be C_0 and C_1:

$$C_i = \{\mathscr{s}_i, \mathscr{H}_i, \Omega\}, \qquad i = 0, 1$$

$$\mathscr{s}_0 : y(t) = \sum_{i=1}^{m_1} \theta_i x_i(t) + w(t)$$

$$\mathscr{s}_1 : y(t) = \sum_{i=1}^{m_1+1} \theta_i x_i(t) + w(t).$$

Given an observation set $\boldsymbol{\xi} = \{y(1), \ldots, y(N)\}$ we want to find the appropriate class for it between C_0 and C_1. Specifically, we want to find whether the term $x_{m_1+1}(t)$ is necessary for modeling y.

In the present problem,

$$\hat{\rho}_0 = \min_{\theta_1, \ldots, \theta_{m_1}} \frac{1}{N} \sum_{t=1}^{N} \{y(t) - \theta_1 x_1(t) - \cdots - \theta_{m_1} x_{m_1}(t)\}^2,$$

$$\hat{\rho}_1 = \min_{\theta_1, \ldots, \theta_{m_1+1}} \frac{1}{N} \sum_{t=1}^{N} \{y(t) - \theta_1 x_1(t) - \cdots - \theta_{m_1+1} x_{m_1+1}(t)\}^2.$$

We can interpret the ratio $\{(\hat{\rho}_0 - \hat{\rho}_1)/\hat{\rho}_1\}$ as the partial autocorrelation coefficient between $y(t)$ and $x_{m_1+1}(t)$ after eliminating the effect of the intermediate variables $x_1(t), \ldots, x_{m_1}(t)$. Suppose we fit two separate models for $y(t)$ and $x_{m_1+1}(t)$ in terms of $x_1(t), \ldots, x_{m_1}(t)$:

$$y(t) = \beta_1 x_1(t) + \cdots + \beta_{m_1} x_{m_1}(t) + w_1(t),$$
$$x_{m_1+1}(t) = \gamma_1 x_1(t) + \cdots + \gamma_{m_1} x_{m_1}(t) + w_2(t).$$

Then the ordinary correlation coefficient between the residual variables $w_1(t)$ and $w_2(t)$ is the ratio $(\hat{\rho}_0 - \hat{\rho}_1)/\hat{\rho}_1$. Another way of expressing the significance of $(\hat{\rho}_0 - \hat{\rho}_1)$ is in the language of the analysis of variance. We can split up the variance of y $(=\hat{R}_0)$ as follows:

$$\hat{R}_0 = (\hat{R}_0 - \hat{\rho}_0) + (\hat{\rho}_0 - \hat{\rho}_1) + (\hat{\rho}_1),$$

or

var y = variance contribution of $x_1, \ldots, x_{m_1}$ + variance contribution of x_{m_1+1} + unexplained variance.

If we assume that $y(\cdot)$ is also IID and $x_i(t)$, $i = 1, \ldots, m_1 + 1$, and all t are mutually independent random variables or linearly independent functions of time or both, then the conjecture mentioned earlier regarding the probability distribution of $(\hat{\rho}_0 - \hat{\rho}_1)/\hat{\rho}_1$ is true. If the ratio $(\hat{\rho}_0 - \hat{\rho}_1)/\hat{\rho}_1$ is significant at a reasonable level, such as $\varepsilon = 0.05$, the term $x_{m_1+1}(t)$ can be considered significant for modeling $y(t)$.

8b.2.2. The Choice of ε, Probability of Type I Error

In order to understand the decision rule in (8b.2.8), we have to understand the nature of the F-distribution. Let η be distributed as $F(m, N)$. Then

$$E[\eta | H_0] = \frac{N}{N-2}, \quad \text{if} \quad N > 4$$
$$\approx 1 \quad \text{for large } N.$$
$$\mathrm{Var}[\eta | H_0] = \frac{2N^2(N + m - 2)}{(N-4)(N-2)^2 m}, \quad \text{if} \quad N > 4$$
$$\approx \frac{2}{m}, \quad \text{for large } N.$$

The mode of probability density of η is located at $(1 - 2/m)$, if $m \geq 2$ and N is large. The choice of the significance level ε is arbitrary. If ε is assumed to be too small, the error II probability may be large. Experience suggests that the value of $\varepsilon = 0.05$ is usually satisfactory. The corresponding value of η_0 is given by

$$\eta_0 \approx E[\eta|H_0] + 2[\mathrm{Var}(\eta|H_0)]^{1/2} \approx 1 + 2(2/m)^{1/2}.$$

Typically, when a value of $\eta_d(\xi)$ is observed to be in the tail of the distribution, C_0 is declared invalid. If $\eta_d(\xi)$ is near the mode, then C_0 is accepted. If $\eta_d(\xi)$ falls in the intermediate region, the probability in the tail of the distribution to the right of this value is computed using tables of the F-distribution. The test then declares the acceptance of C_0 at the level ε_0 if the computed probability is less than ε_0, i.e., if $\eta_d \leq \eta'$. The threshold η' is listed below for $\varepsilon_0 = 0.05$.

N	100	100	100	∞	∞	∞
m	1	2	3	1	2	3
η'	3.94	3.00	2.61	3.84	2.99	2.60

We see from the table that the decision rule (8b.1.14) in the two-class problem with the likelihood approach corresponds to a probability of type I error of 0.05.

In order to see the interrelationship between the acceptance of the hypothesis and the sample size, let us continue with Example 8b.1.

Example 8b.1 (continued). Suppose

$$\hat{\rho}_0 = 0.2, \qquad \hat{\rho}_1 = 0.18,$$

Without specifying the sample size N, the significance of $\eta_d(\xi)$ cannot be specified.

(i) If the values of $\hat{\rho}_0$, $\hat{\rho}_1$ above were obtained with $N = 100$ and $m_1 = 3$, then from (8b.2.8)

$$\eta_d(\xi) = \left(\frac{\hat{\rho}_0 - \hat{\rho}_1}{\hat{\rho}_1}\right)\left(\frac{N - m_1 - 1}{1}\right) = 10.67.$$

The value of $\eta_d(\xi)$ in this case is very much greater than the threshold 3.94 corresponding to 5% of error I probability. Hence C_1 is accepted for ξ.

(ii) If the same values of $\hat{\rho}_0$, $\hat{\rho}_1$ were obtained with $N = 20$, then

$$\eta_d(\xi) = \left(\frac{\hat{\rho}_0 - \hat{\rho}_1}{\hat{\rho}_1}\right)\left(\frac{N - 4}{1}\right) = 1.778.$$

This value of $\eta_d(\xi)$ is much smaller than the threshold and hence at the 5% level the class C_0 may be accepted for the given process.

On the other hand, if we want to work with an error I probability equal to 0.005, we would accept C_0 in both cases.

The example clearly illustrates the strong interrelationship between the chosen *class* of models, the *available data* set, and the *desired accuracy*. If we fix ε, the error I probability, then we need to go to complex models (i.e., models having large numbers of parameters) only if N is relatively large. The complexity of the model given by the decision rule is a function of the number of observations involved.

The rationale for the choice of ε is not at all clear. One often finds in statistical literature statements such as "our test was performed at 98% level ($\varepsilon = 0.02$) and hence it is superior to other tests which were done at 95% significance level ($\varepsilon = 0.05$)." Such statements are justifiable only if we feel that the error of type I is very much more dangerous than error of type II and hence we are better off working with as small a value of ε as possible, even though we know that decreasing ε implies an increase in the probability of type II error (Pratt *et al.* 1965). For example, consider a hypothesis C_0, "whiskey cures snake bite," and let C_1 be its negation. A Puritan may feel that acceptance of C_0 when it is not true is much more dangerous than rejection of C_0 when it is true. In the former case, there may not only be deaths from snake bites but also more drunks in the street who consume alcohol in view of its supposed medicinal properties. In the latter case, one would look for alternative cures for snake bite. In this example rejection of hypothesis C_0 at level $\varepsilon = 0.02$ can be regarded as superior to rejection at level $\varepsilon = 0.05$.

But such logic is quite unwarranted in model building problems. We have to regard errors of both types I and II to be important. Hence we cannot say that testing with $\varepsilon = 0.02$ is better than testing with $\varepsilon = 0.05$ because in the former case we have greater probability of error II. This feature points to the importance of the likelihood approach which does not involve the use of subjective quantities such as ε.

8b.3. Class Selection Based on Recursive (or Real Time) Prediction

This method allows us to compare a number of different classes of models C_i, $i = 0, 1, \ldots, r-1$, simultaneously, where $C_i = [\mathscr{I}_i, \mathscr{H}_i, \Omega_i]$, provided they *do not* have any moving average terms (Kashyap, 1971). Some of the models can have equations with time-varying coefficients as well, provided they do not have any moving average terms. The basic tool used in the method is the adaptive predictor of $y(t)$ based on the available history until that time, namely $\{y(1), \ldots, y(t-1)\}$. Suppose for the moment that the given process $y(\cdot)$ obeys class C_i. Even though we do not know the particular model in class C_i obeyed by y, knowledge of class C_i itself is enough to design an optimal one-step predictor of $y(t)$ based on $y(1), \ldots, y(t-1)$, by using a suitable criterion. Let this predictor be labeled $\hat{y}_i(t|t-1)$. If the given data are $\{y(1), \ldots, y(N)\}$, then by using only the first measurement $y(1)$ the value of $y(2)$ is predicted, and the predictor is labeled $\hat{y}_i(2|1)$. The prediction error $(y(2) - \hat{y}_i(2|1)) = \tilde{y}(2)$ can also be computed. Next, the two observations $[y(1), y(2)]$ are used to predict $y(3)$, the predicted value of which is labeled $\hat{y}_i(3|2)$, and the prediction error $\tilde{y}(3) = y(3) - \hat{y}_i(3|2)$. The procedure is similarly repeated so that in general, by using $y(1), \ldots, y(N-1)$, we predict values of $y(t)$, $t = 2, \ldots, N$. By using all the prediction errors $\tilde{y}(2), \ldots, \tilde{y}(N)$, we can compute the mean square prediction error J_i:

$$J_i = \frac{1}{N-1} \sum_{t=2}^{N} \{y(t) - \hat{y}_i(t|t-1)\}^2.$$

In computing J_i, the prediction capability of the class C_i has been tested $(N-1)$ times. We compute the indices J_i, $i = 0, 1, \ldots, r-1$, for the classes C_i, $i = 0, 1, \ldots, r-1$, respectively. Then the following decision rule may be considered:

Decision Rule 3

Consider the indices J_i, $i = 0, 1, \ldots, r-1$, for the r classes obtained by using the given data. If there is only one class C_{i_0} such that the index J_{i_0} is the smallest among the set $\{J_i, i = 0, \ldots, r-1\}$, then we assign the data to class C_{i_0}. If the minimum value in the set is achieved by more than one member, then the given data are assigned to one of these classes according to a subsidiary criterion such as minimal complexity.

The intuitive reason for the rule is as follows. The consequence of y belonging to the class C_i has been tested by performing $(N-1)$ prediction experiments. If the index J_{i_0} is the smallest in the set $(J_i, i = 0, \ldots, r-1)$, it means that the predictor $y_{i_0}(\cdot)$ is the best among $\hat{y}_0(\cdot), \ldots, \hat{y}_{r-1}(\cdot)$. Since all the predictors are designed by using the same quadratic criterion function, the differences in their performance should be ascribed to the fact that the assumptions made in finding the predictor $\hat{y}_{i_0}(t|t-1)$ have greater validity than the assumptions behind the other predictors. In other words, class C_{i_0} is more plausible than the others for the given data.

We will now develop the details of the predictors and give other reasons for the adoption of the decision rule.

Suppose the given empirical process belongs to the class C_i where equation $\mathscr{o}$ has the form

$$\mathscr{o}: f_0\{y(t-1)\} = \sum_{j=1}^{n} \theta_j x_j(t-1) + w(t), \qquad \boldsymbol{\theta} = (\theta_1, \ldots, \theta_n)^{\mathrm{T}} \in \mathscr{H}_i, \quad (8b.3.1)$$

where $f_0(\cdot)$ is a known function, $x_j(t-1), j = 1, \ldots, n$, are functions of lagged values of $y(t)$, of trend terms, or of observable input terms, and $w(\cdot)$ is a normally distributed $N(0, \rho)$ IID sequence. We would like to compute the one-step-ahead predictor of $y(t)$ based on the past history up to the instant $(t-1)$. Such a predictor should be some function of the observations up to time $(t-1)$ and should not involve, either explicitly or implicitly, the observation $y(t)$ or later observations. Since the numerical values of the parameter $\boldsymbol{\theta}$ that characterizes the process y in C_i are unknown, the Bayesian approach is taken and the vector $\boldsymbol{\theta}$ is regarded as a random variable with preassigned prior distribution. Let $\hat{y}_i(t|t-1)$ denote the Bayesian one-step-ahead predictor of

$y(t)$ according to the mean square prediction error criterion based on all observations until time $t - 1$, assuming that it belongs to the ith class:

$$\begin{aligned} \hat{y}_i(t|t-1) &= E[y(t)|t-1; y \text{ obeys } C_i] \\ &= E\left[f_0^{-1}\left(\sum_{j=1}^{n} \theta_j x_j(t-1) + w(t)\right)\Big| t-1\right]. \end{aligned} \tag{8b.3.2}$$

The evaluation of this expectation involves the aforementioned prior distributions of $\boldsymbol{\theta}$. The predictor above has the property

$$E[\{y(t) - \hat{y}_i(t|t-1)\}^2] \leq E[\{y(t) - \bar{y}_i(t|t-1)\}^2] \tag{8b.3.3}$$

as established in Section 6e where $\bar{y}_i(t|t-1)$ is any other predictor of $y(t)$ based on the history until time $(t-1)$. By using the given observation set $y(1), \ldots, y(N)$ we can evaluate the indices J_i, $i = 0, 1, \ldots$.

Equation (8b.3.3) allows us to state the following lemma.

Lemma. If the given empirical process $y(\cdot)$ belongs to the class C_i, then the following inequality is true:

$$E[J_i] \leq E[J_j], \qquad \forall j = 0, 1, \ldots, r-1,$$

where the expectations are evaluated in the Bayesian manner using the prior distributions of the parameter $\boldsymbol{\theta}$ belonging to the class C_i.

This lemma can be proved using the method indicated in Appendix 8.1. This lemma and typical inductive reasoning (namely, $A \to B$; if B is true, then A is very likely true) lead to the decision rule given earlier. The lemma suggests that the probabilities of the two types of errors caused by the decision rule may not be excessive. The following comments are in order.

Comment 1. The predictors $\hat{y}_i(t|t-1)$ can be computed recursively by using the techniques of Chapter VI. In computing the predictor $\hat{y}_i$, we use the Bayesian estimate $\hat{\boldsymbol{\theta}}(t)$ of the corresponding parameter vector $\boldsymbol{\theta}$ based on the observations up to time t. When the moving average terms are absent, the estimate $\hat{\boldsymbol{\theta}}(t+1)$ can be computed from $\hat{\boldsymbol{\theta}}(t)$ in real time using either algorithm (C1) in constant coefficient system or (C4) in the system with time-varying coefficients.

Comment 2. In the decision rule we can use the ad hoc predictor $\bar{y}_i(t|t-1)$ of Section 6e which involves the likelihood estimates instead of the Bayesian predictor $\hat{y}_i(t|t-1)$. In certain circumstances, for example, if ρ is unknown, such a predictor may be relatively easier to compute than the Bayesian predictor. One can prove the asymptotic validity of a lemma similar to the one mentioned earlier in this section when the Bayesian predictor $\hat{y}_i(t|t-1)$ is replaced by the predictor $\bar{y}_i(t|t-1)$ in the decision rule.

Comment 3. The approach given here can be used to simultaneously compare a number of classes whose difference equations may involve transformed variables such as $\ln y$ as well. Such a comparison is not possible in the hypothesis testing approach.

8b.4. Methods Based on Comparison of the Best Fitting Models

In this section, we will obtain the best fitting models in each of the r classes for the given data and compare the r best fitting models directly. This is in contrast to the methods used earlier where we directly compared the classes and not the models. Even though we used quantities such as the CML estimate of parameters in each class, the reasoning was not explicitly based on comparing individual models.

8b.4.1. The Prediction Approach

This is a relatively simple method for comparison of a number of classes and is commonly used in pattern recognition problems (Duda and Hart, 1973). It is based on the comparison of the relative prediction abilities of the r best fitting models. In contrast to the method in Section 8b.3 we do not require that the estimates of the parameters be updated in real time. Hence, the decision rule is considerably easier to evaluate than that in Section 8b.3.

Let N be even. The given data set is divided into, for instance, two equal parts, the first part of which, $\{y(1), \ldots, y(N/2)\}$, is used to obtain a best fitting model for the process in each class. Suppose the best fitting model in class C_i by the CML method is $(\mathscr{s}_i, \boldsymbol{\theta}_i^*, \rho_i^*)$. For this *particular model* a least squares one-step-ahead predictor, $y_i^*(t|t-1)$, is then obtained. The coefficients of this predictor are constants and are obtained from the vector $\boldsymbol{\theta}_i^*$. This predictor is used to obtain the best one-step-ahead predictors of $y(t)$ given all observations until time $t-1$, $t = N/2 + 1, \ldots, N$. Thus, when t is greater than $N/2$, $y_i^*(t+1|t)$ is a function of all the observations $y(1), \ldots, y(t)$ only, and hence is a legitimate predictor for $y(t+1)$, whereas $y_i^*(N/2|N/2-1)$ is not a legitimate predictor of $y(N/2)$ since it depends explicitly on $y(N/2)$ used to obtain the estimate $\boldsymbol{\theta}_i^*$. An index of predictive ability P_i, similar to that discussed earlier, may now be computed:

$$P_i = \frac{1}{N - [N/2]} \sum_{t=[N/2]+1}^{N} \{y(t) - y_i^*(t|t-1)\}^2. \tag{8b.4.1}$$

Decision Rule 4

We prefer the class C_{i_0} if P_{i_0} is the only member having the smallest value in the set $\{P_0, \ldots, P_{r-1}\}$. If there is more than one member leading to the same minimum value, choose one among them according to a secondary criterion.

The decision rule can be justified by showing that the following is asymptotically valid if C_{i_0} is the correct class:

$$E[P_{i_0}] \leq E[P_j], \qquad \forall j = 0, 1, \ldots, r-1. \tag{8b.4.2}$$

We will give an outline of the proof. Let $C_i = [\mathscr{s}_i, \mathscr{H}_i, \Omega]$. There are two separate cases. Let us compare the class C_{i_0} with a class C_j in which equation $\mathscr{s}_j$ does not possess at least one term in $\mathscr{s}_{i_0}$. In this case, the result in (8b.4.2) can be obtained easily, at least asymptotically. Next, compare C_{i_0} with the class C_j in which equation $\mathscr{s}_j$ contains all the terms in $\mathscr{s}_{i_0}$. Then the result (8b.4.2) can be proved as shown in Appendix 8.1.

The procedure given here is computationally easier than the recursive prediction method of Section 8b.3, but it is also less powerful than the method of Section 8b.3 when N is small. This assertion can be demonstrated as follows. First, in computing the mean square prediction error of Section 8b.3 we perform the prediction operation $(N - 1)$ times, whereas in the method given in this section we perform the prediction operation only $[N/2]$ times. Correspondingly, the estimates of the mean square prediction error in this section are less accurate than those of Section 8b.3. Second, in the prediction formula of Section 8b.3 we use more accurate estimates of parameters than the corresponding quantities in this section. For example, in Section 8b.3, the forecast $y_i(t|t - 1)$ is based on the parameter estimates computed on the basis of $(t - 1)$ observations. But the forecast $y_i(t|t - 1)$ in this section is based only on the parameter estimates computed on the basis of $[N/2]$ observations, regardless of the value of t.

8b.4.2. The Bayesian Approach

Let the best fitting models from the r classes be denoted by M_i, $i = 0, 1, \ldots, r - 1$. We compute the posterior probability of every one of these models being the correct model given the observation history and prefer that model having the highest posterior probability. The preferred class is the one associated with the preferred model.

To compute the posterior probabilities, we need the prior probabilities of the r models, which are chosen as follows: If the ith model has n_i coefficients in it, then

$$p[M_i] = k \exp[-n_i], \qquad i = 0, 1, \ldots, r - 1,$$

where k is a normalizing constant. The intuitive justification for this choice is the principle of parsimony; i.e., we should prefer to use a model with as few terms as possible. An additional advantage of this choice is that the choice of the final class by this method becomes identical to that given by the likelihood approach for large N with normal noise variables.

The prior distribution mentioned here belongs to the family of objective prior probability functions. But subjective Bayesians (Savage, 1962) believe that the prior distribution should contain all the available descriptive information about the class. The prior distribution given here has both descriptive and prescriptive roles.

Next we will compute the posterior probabilities. Let $\xi = \{y(1), \ldots, y(N)\}$ and

$$p(M_i|\xi) = \frac{p(\xi|M_i)p(M_i)}{\sum_{i=0}^{r-1} p(\xi|M_i)p(M_i)}.$$

The probability density $p(\boldsymbol{\xi}|M_i)$ is known by definition, i.e., $p(\boldsymbol{\xi}|M_i) \triangleq p(\boldsymbol{\xi}; \boldsymbol{\phi}_i^*)$, where $\boldsymbol{\phi}_i^*$ is the CML (or Bayesian) estimate of the unknown parameter vector $\boldsymbol{\phi}_i^0$ in the ith class characterizing the given observation set $\boldsymbol{\xi}$:

$$\ln p(M_i|\boldsymbol{\xi}) = \ln p(\xi, \boldsymbol{\phi}_i^*) - n_i + \text{a term independent of } i.$$

Thus, the class having the higher posterior probability is the same as the class given by the likelihood approach of Section 8b.1.1.

We have presented only one aspect of the Bayesian theory. For other viewpoints on the use of Bayesian theory in time series analysis, we refer to Box and Tiao (1973) and Zellner (1971).

8b.5. Discussion of the Various Class Selection Methods

Among all the methods, the likelihood approach is very versatile, theoretically sound, and appears to give reasonable results in practice; i.e., often the best fitting models from the classes chosen by this decision rule pass all validation tests and give reasonable forecasts. It can simultaneously handle a number of classes, including those having moving average terms or log transformed terms. Since the decision rule in the two-class problem has the same form as that in the hypothesis testing approach, we can also know the probability of type I error given by the decision rule of the likelihood approach whenever such information is available in the hypothesis testing methods. For instance, while comparing two AR models or generalized AR models we can obtain the probability of type I error caused by the decision rule given by the likelihood approach. One of the important advantages of the likelihood approach is that it does not involve the use of arbitrary quantities such as the significance levels and it can be justified using Bayesian reasoning as well. One shortcoming of the likelihood approach for the determination of the order of the AR fit to the data is that the order of the AR fit obtained from this approach is usually larger than is necessary for passing validation tests.

The hypothesis testing approach is more ambitious since there is an attempt to obtain a decision rule with certain prespecified probability of error. But in practice, it can handle only two classes at a time and even these two classes must be made up of generalized AR models. For instance, it cannot be used for comparing a nontransformed and a transformed model. Moreover, the F-distribution of the statistic is only asymptotically true. When we are working with sample sizes of $N = 50$ or 100, it is not clear whether the F-distribution is valid. The problem of modeling the lynx population in Chapter XI illustrates some of the difficulties. The difficulty in specifying the significance level has already been mentioned.

The prediction approach of Section 8b.4 is important because the predictor designed on the basis of one batch of data is evaluated by testing it on another batch of data. This approach is valid for systems having moving average terms also, and is the one taken frequently in practice. It is instructive to analyze the difference between the estimates of the mean square prediction error obtained during the design of the predictor by using the first batch of data and that obtained during testing by using the second batch of data. The differences between the two mean square errors are examined to determine whether they are due to sampling variations only or to the poor quality of the model.

The real time (or recursive) prediction approach of Section 8b.3 is especially useful with systems in which some of the parameters may vary with time. It is particularly useful in ascertaining the causal structure in systems with time-varying coefficients.

One disadvantage of the prediction methods is that they are apt to give models that may not pass the validation tests. The reason is that the emphasis is mainly on forecasting. The approach tends to deemphasize those models having terms such as the sinusoidal terms, which may not contribute much to forecasting, but which may be necessary in "reproducing" the statistical characteristics of the given data such as the correlogram or spectral density. In other words, the prediction approaches often yield classes that are appropriate in obtaining forecasts. The likelihood approach leads to a class whose best fitting model often passes the validation tests. Hence, we should use both the prediction and the likelihood approaches to understand the difference in the best fitting models in the two methods.

8c. Validation of Fitted Models

8c.1. The Nature of Model Validation

By using the given data and the class of models given by the class selection methods, we can obtain a best fitting model in that class. Next we will investigate how well the model represents the given data; this is referred to as validation of the model. Specifically, we want to know the limitations of the model. For instance, consider the best fitting model for the daily flow of a river based on the history of the daily flows. Can we use this model for getting one-month-ahead or one-year-ahead predictions? The answer is usually no. In other words, a model based on monthly data usually gives better one-month-ahead predictions than a daily model. The next natural question concerns the extent to which the model reproduces the characteristics of the data. For instance, if we take the residuals $\bar{w}(j)$ from the model, what is the maximum lag j for which $\bar{w}(t)$ and $\bar{w}(t-j)$ can be said to be independent of each other? Such questions must be answered during the validation of the model.

It is important to note the fact that a best fitting model has been obtained from the optimal class given by the class selection methods does not imply that the corresponding best fitting model always adequately represents the data. We recall that the class selection methods yield the best class among the previously chosen candidate classes. It is often true that we may overlook the relevant class of models on account of the current fashions. For instance, in the early 1920s it was common to look only for models with sinusoidal terms to explain oscillatory phenomena. Now the fashion appears to be to explain all processes in terms of AR or ARMA models. If we carry out the validation tests, we will get a good idea of the strengths and weaknesses of the fitted model obtained from the various classes.

The first approach for validation is to check the validity of the assumptions made in the model by using the given observation set ξ. The principal assumptions made in the model are that the noise input $w(\cdot)$ has zero mean, a common variance, and $w(t)$ is independent of $w(j)$ for $t \neq j$ and with $y(t-j), j > 0$. To verify the assumptions, we begin with the residual series obtained by the best fitting model in the given class to the given set of observations and check whether the resulting series of residuals obeys these assumptions. If the model has been chosen from the class having time-varying coefficients, then we have to consider the corresponding whitened observation sequence instead of the residual series. Since we are in a probabilistic domain, we can only verify whether the assumptions are checked at the required level of significance. The corresponding tests can be obtained with the aid of the theory of hypothesis testing. The criticism made earlier on the methods of hypothesis testing such as the arbitrariness of the significance level does not matter here because we are interested in checking whether a model is satisfactory and *not* in choosing *a* particular model from a number of models.

The second approach in validation is to test whether the various theoretical statistical characteristics of the *output* of the model are close enough to the corresponding statistical characteristics of the empirical data such as the correlogram and spectral density. However, a quantitative measure of the degree of fit between the statistical characteristics, commensurate with the available observation size, is to be obtained. The validity of the model should be decided not only on the numerical divergence between the two sets of characteristics but also on the sample size. This can be achieved by means of tests designed with the aid of the theory of hypothesis testing. The validation must be carried out by means of a variety of tests so that any deficiency of the model may be isolated and appropriately corrected.

The two approaches complement each other. The tests based on the residuals are relatively easy to carry out but they can point out only the major errors in the models. Comparison of the various statistical characteristics of the output of the model and data is definitive in the validation of the model, but it is computationally quite time-consuming. These topics are considered in detail later.

Model validation is a relative concept. The existence of a validated model for the given data does not preclude the possibility of another model that gives a better fit to the given data than the current model. This new model may involve additional input variables that may not have been included in the first model.

8c.2. Validation Tests Using Residuals

Using the given model and the observation history ξ, we can generate the residuals $x(j)$, $j = 1, \ldots, N$, which are the estimates of the noise variables

$w(j), j = 1, \ldots, N$, respectively. We will not repeat the method of generation of residuals since it has been discussed in Chapters VI and VII. Note that residual $x(j)$ was denoted by the symbol $w(t, \boldsymbol{\theta}^*)$ earlier. We want to test whether this sequence could be considered as a zero mean independent sequence with normal distribution $N(0, \rho)$, where ρ is unknown.

We have seen earlier that specific tests can be developed only for testing whether a particularly finite parameter vector belongs to one of a finite number of classes. Specifically, testing whether or not a sequence is independent is not meaningful unless the alternative types of dependency are specified. Consequently, we will follow the sequential procedure listed below:

1. Assume the given sequence to be independent with normal distribution, $N(\theta, \rho)$. Test whether $\theta = 0$ without knowing ρ, by using test 1 developed below.

2. Assume the given sequence to be normal and independent obeying (8c.2.1) where $w(\cdot)$ is a zero mean IID $N(0, \rho)$ sequence. This sequence $w(\cdot)$ is not the same sequence $w(\cdot)$ appearing in the Section 8a.

$$x(t) = w(t) + \sum_{j=1}^{m_2} (\alpha_j \cos \omega_j t + \beta_j \sin \omega_j t). \tag{8c.2.1}$$

Test whether $\alpha_j = 0$, $\beta_j = 0$, $\forall j = 1, 2, \ldots, m_2$, by using tests 2, 3, or 4 discussed below.

3. Assume the given sequence to be normally distributed with zero mean and to obey

$$x(t) = \sum_{j=1}^{m_1} \alpha_j x(t-j) + w(t), \tag{8c.2.2}$$

where $\{w(t)\}$ is a sequence of zero mean normal random variables $N(0, \rho)$, with $\rho > 0$, and otherwise unknown. Test whether $\alpha_j = 0$, $\forall j = 1, \ldots, m_1$, by using tests 5 and 6.

4. Assume that the given sequence $\{x(\cdot)\}$ is independent. Test to see if its distribution is normal.

We will develop separate tests for each of these items and use the same level of significance in all of them. If the tests of items 1–3 are successful, then the corresponding model M from which the $\{x(1), \ldots, x(N)\}$ were derived will be considered to be valid. If not, the deficiency in the model M is located from the failure of a test and the model is accordingly modified. On many occasions, the test in item 4 does not pass. This only implies that the given sequence is not normally distributed. Since normality is used in tests of items 1–3, one may wonder whether the results of the tests may be in jeopardy by this result. This is not so since the tests developed for items 1–3 are approximately valid even when the distribution is not normal.

The tests to be presented can be derived from the hypothesis testing approach or likelihood approach mentioned in Section 8b and are treated by Whittle

(1951, 1952), Anderson (1971), Box and Jenkins (1970), and Quenouille (1957), among others.

The use of these tests can hardly be underestimated, especially in view of their simplicity. A model building procedure involves considerable trial and error and in each iteration we can use the tests to verify that the model obtained at that iteration is satisfactory.

8c.2.1. Test for the Zero Mean (Test 1)

Let the two classes be

$$C_0 = [\mathscr{I}_0, \mathscr{H}_0, \Omega], \qquad C_1 = [\mathscr{I}_1, \mathscr{H}_1, \Omega]$$
$$\mathscr{I}_0: x(t) = w(t); \qquad \mathscr{I}_1: x(t) = \theta + w(t),$$

where $w(\cdot)$ is a sequence of zero mean IID random variables, with distribution $N(0, \rho)$, $\rho \in \Omega = (0, \infty)$, $\mathscr{H}_0$ is the null set, and $\mathscr{H}_1 = \{\theta: -\infty < \theta < \infty, \theta \neq 0\}$. On the basis of residuals $\{x(1), \ldots, x(N)\}$ we have to choose one of the two classes C_0 and C_1:

$$\bar{x} = \frac{1}{N}\sum_{t=1}^{N} x(t), \qquad \hat{\rho} = \frac{1}{N-1}\sum_{t=1}^{N} \{x(t) - \bar{x}\}^2.$$

The test statistic is $\eta(x) = (N)^{1/2}\bar{x}/(\hat{\rho})^{1/2}$.

Under C_0, η is t-distributed with $(N-1)$ degrees of freedom independent of ρ. Consequently, we can choose the following *decision rule*:

$$|\eta(x)| \begin{cases} < \eta_0 \to \text{accept } C_0 \\ \geq \eta_0 \to \text{reject } C_0. \end{cases}$$

The threshold η_0 is chosen from a table of t-distributions corresponding to a given level of significance ε_0 such as 0.05. This test is known to be locally most powerful.

We will now show the relation of this test to the test in Section 8b.2. Let $\hat{\rho}$ and $\hat{\rho}_1$ be the residual variances of the best fitting models in C_0 and C_1 respectively

$$\hat{\rho}_0 = \frac{1}{N}\sum_{t=1}^{N} x^2(t), \qquad \hat{\rho}_1 = \frac{1}{N}\sum_{t=1}^{N} \{x(t) - \bar{x}\}^2$$

Then the statistic in the decision rule (8b.2.8) has the form

$$\eta_1(x) = \frac{(\hat{\rho}_0 - \hat{\rho}_1)}{\hat{\rho}_1} N = \frac{1}{N}(\bar{x})^2/\hat{\rho}_1.$$

It should be noted that $\eta_1(x)$ is proportional to $\eta^2(x)$. If $\eta(x)$ has a t-distribution, then it is easy to show that η_1 is approximately F-distributed.

A few values of the threshold η_0 are listed below for various values of N and ε_0:

N	50	50	100	100
ε_0	0.05	0.02	0.05	0.02
η_0	2.01	2.4	1.98	2.36

8c.2.2. Test for the Absence of Sinusoidal Terms

We will present three different methods for testing whether the given residual sequence is free of deterministic sinusoidal components.

Test 2: Frequency of the Given Sequence is Known (Anderson, 1971). Let the two classes be C_0 and C_1:

$$C_i = [\mathscr{S}_i, \mathscr{H}_i, \Omega]$$
$$\mathscr{S}_1: x(t) = \alpha \cos \omega_1 t + \beta \sin \omega_1 t + w(t)$$
$$\mathscr{S}_0: x(t) = w(t).$$

In $\mathscr{S}_0$ and $\mathscr{S}_1$, $w(t)$ is an independent $N(0, \rho)$ sequence, $\rho \in \Omega$, the frequency ω_1 is known, $\mathscr{H}_0$ is the null set, and $\mathscr{H}_1 = \{(\alpha, \beta): -\infty < \alpha, \beta < \infty, \alpha \neq 0, \beta \neq 0\}$. We can use the likelihood ratio test to derive the test statistic. Moreover, the probability distribution of the test statistic can also be determined explicitly:

$$\hat{\alpha} = \frac{2}{N}\sum_{t=1}^{N} x(t) \cos \omega_1 t, \qquad \hat{\beta} = \frac{2}{N}\sum_{t=1}^{N} x(t) \sin \omega_1 t.$$

Let $\gamma^2 = (\hat{\alpha})^2 + (\hat{\beta})^2$. Let $\hat{\rho}_1$ and $\hat{\rho}_0$ be the residual variances of the best fitted models in classes C_1 and C_0 respectively

$$\hat{\rho}_1 = \min_{\alpha,\beta} \frac{1}{N}\sum_{t=1}^{N} \{x(t) - \alpha \cos \omega_1 t - \beta \sin \omega_1 t\}^2,$$

$$\hat{\rho}_0 = \frac{1}{N}\sum_{t=1}^{N} x^2(t) \approx \hat{\rho}_1 + \gamma^2/2.$$

The expression above is exact if $\sum_{t=1}^{N} \cos 2\omega_1 t = \sum_{t=1}^{N} \sin 2\omega_1 t = 0$. Otherwise it is an approximation.

The test statistic $\eta(x)$ can be written as

$$\eta(x) = \frac{\hat{\rho}_0 - \hat{\rho}_1}{\hat{\rho}_1} \cdot \frac{N - (\text{number of coefficients estimated in a model in } \mathscr{S}_1)}{\text{additional terms in model } \mathscr{S}_1 \text{ not in } \mathscr{S}_0}$$

$$= \frac{\hat{\rho}_0 - \hat{\rho}_1}{\hat{\rho}_1}\left\{\frac{N-2}{2}\right\}$$

$$= \frac{\gamma^2(N-2)}{4\hat{\rho}_1} \sim F(2, N-2) \qquad \text{if} \quad x \in C_0.$$

The *decision rule* is

$$\eta(x)\begin{cases} \leq \eta_0 \rightarrow \text{accept } C_0 \\ > \eta_0 \rightarrow \text{accept } C_1, \end{cases}$$

where η_0 is selected in accordance to the corresponding significance level.

Test 2 cannot be used if the frequency ω_1 is not known. In such a case we have to use tests 3 and 4. In these tests, we check for the presence of sinusoidal components of *all* possible frequencies. If only N residuals are available, then only the frequencies that are multiples of $(1/N)$ are to be considered in view of Nyquist's theorem. Consequently, only the frequencies $\omega_j = 2\pi j/N$, $j = 1, \ldots, N - 1$, are of interest. Since only the frequency is of interest and not the phase, the frequency $\omega_{N-j} = 2\pi(1 - j/N)$ is redundant if the frequency ω_j has been considered. Thus, the list of the possible frequencies reduces to $\omega_j = 2\pi j/N, j = 1, \ldots, \lfloor N/2 \rfloor$, where

$$\begin{aligned} \lfloor N/2 \rfloor &= \frac{N}{2} \quad \text{if} \quad N \text{ is even} \\ &= \frac{(N-1)}{2} \quad \text{if} \quad N \text{ is odd.} \end{aligned}$$

For example, if $N = 9$ monthly observations of a process are available, the relevant frequencies are $2\pi/9$, $4\pi/9$, $6\pi/9$, and $8\pi/9$ radians per month and the corresponding periods are 9, 9/2, 9/3, 9/4 months. We will now state the Fisher test.

Test 3: Fisher Test. Here we assume that the sequence $\{x(\cdot)\}$ is either white or a mixture of white noise and sinusoidal components of only *one* frequency among ω_j, $j = 1, \ldots, \lfloor N/2 \rfloor$. Define the classes $C_i = (\mathscr{I}_i, \mathscr{H}_i, \Omega)$, $i = 0, 1$, as before except that $\mathscr{I}_1$, $\mathscr{I}_0$ have the following form:

$$\begin{aligned} \mathscr{I}_1\colon x(t) &= \sum_{j=1}^{\lfloor N/2 \rfloor} (\alpha_j \cos \omega_j t + \beta_j \sin \omega_j t) + w(t), \\ \mathscr{I}_0\colon x(t) &= w(t), \end{aligned}$$

where $\{w(t)\}$ is an IID $N(0, \rho)$ sequence and $\omega_j = 2\pi j/N$, $j = 1, \ldots, \lfloor N/2 \rfloor$. Let $\boldsymbol{\theta} = \{\alpha_j, \beta_j, j = 1, 2, \ldots, \lfloor N/2 \rfloor\}$. The equation $\mathscr{I}_1$ can have sinusoidal terms of only *one* frequency. This is assured by defining the set $\mathscr{H}_1$ as shown below. The set $\mathscr{H}_1$ has $\lfloor N/2 \rfloor$ distinct points in it:

$$\begin{aligned} \mathscr{H}_1 = \{\boldsymbol{\theta}\colon \alpha_j = \beta_j = 0, \forall j = 1, 2, \ldots, \lfloor N/2 \rfloor, \text{ except } j = j_1, \alpha_{j_1} \neq 0, \\ \beta_{j_1} \neq 0; j_1 \text{ can be any integer in } (1, 2, \ldots, \lfloor N/2 \rfloor)\}. \end{aligned}$$

We will obtain the test statistic from the likelihood ratio considerations. The test statistic $\eta(x)$ has the form

$$\eta(x) = \max[\gamma_1^2, \gamma_2^2, \ldots, \gamma_{\lfloor N/2 \rfloor}^2] \Big/ \sum_{j=1}^{\lfloor N/2 \rfloor} \gamma_j^2,$$

where

$$\gamma_k^2 = \left(\frac{2}{N} \sum_{t=1}^{N} x(t) \cos \omega_k t\right)^2 + \left(\frac{2}{N} \sum_{t=1}^{N} x(t) \sin \omega_k t\right)^2.$$

When $x(\cdot)$ obeys class C_0, the statistic $\eta(x)$ does not have an F-distribution, unlike the previous cases, because of the maximum operation in the definition of $\eta(x)$. The correct probability distribution of $\eta(x)$, determined by Fisher, is given as

$$\text{prob}[\eta(x) \geq \eta_1 | x \text{ obeys } C_0] \sim \lfloor N/2 \rfloor (1 - \eta_1)^{\lfloor N/2 \rfloor - 1}.$$

The probability density of η has a one-sided tail. The *decision rule* is

$$\eta(x)\begin{cases} \leq \eta_1 \to \text{accept } C_0 \\ > \eta_1 \to \text{accept } C_1. \end{cases}$$

The threshold η_1 is again found by the corresponding significance level.

We stress that the decision rule given above cannot be interpreted as a likelihood ratio test when equation $\mathscr{s}_1$ has sinusoidal terms of more than one frequency. In such a case, the distribution given above is also not valid. The following example (Andel and Balek, 1971) will illustrate this feature.

Example 8.2. Consider an empirical series with $N = 74$ having a single dominant frequency $2\pi k_1/N$. Let the significance level be 98%. The threshold η_1 is calculated by solving

$$37(1 - \eta_1)^{37-1} = 0.02 = \text{prob}[\eta(x) \geq \eta_1 | C_0].$$

The solution of the equation yields the value $\eta_1 = 0.189$. Let

$$K = \max_k \gamma_k^2 \triangleq \gamma_{k_1}^2, \qquad G = \sum_k^{\lfloor N/2 \rfloor} \gamma_k^2 - K.$$

By definition, $\eta(x) = K/(K + G)$. Hence, the test detects the presence of a sinusoidal trend term if $K/(K + G) > 0.189$, or $G < 4.81K$.

Now let us add another sinusoidal component of frequency $2\pi k_2/N$ to the series. Let us suppose that $\gamma_{k_1}^2$ has the maximum value among γ_k^2, $k = 1, 2, \ldots,$ as before. Let $\gamma_{k_1}^2$ and $\gamma_{k_2}^2$ have the values

$$\gamma_{k_1}^2 = 1.1K; \qquad \gamma_{k_2}^2 = K; \qquad \sum_k \gamma_k^2 = G' + 1.1K + K.$$

Let G', the contribution of the nondominant frequency terms, be the same as in the earlier case, i.e., G' is chosen such that $K/(K + G') = 0.189$ or $G' = 4.81K$. The test statistic η has the form

$$\eta = \frac{1.1K}{1.1K + K + G'} = \frac{1}{6.71} = 0.159.$$

The value of η is clearly less than the threshold value of 0.189. Moreover,

$$P[\eta(x) \geq 0.159 | x \text{ obeys } C_0] = 37(1 - 0.149)^{36} = 0.111,$$

which is very much greater than 0.05, let alone 0.02. Thus the test does not indicate the presence of even a single sinusoidal trend term even though the sequence has sinusoidal terms of two frequencies in it. If only one of the frequencies had been present, the test would have detected it.

The following test does not suffer from the deficiency of test 3.

Test 4: Cumulative Periodogram Test (Bartlett, 1966). Define equations $\mathscr{s}_0$, $\mathscr{s}_1$, the set $\mathscr{H}_0$, and $\boldsymbol{\theta}$ as in test 3. The set $\mathscr{H}_1$ is defined as

$$\mathscr{H}_1 = \{\boldsymbol{\theta}\text{: at least one component of } \boldsymbol{\theta} \text{ is not zero}\}.$$

This test is administered in a different manner than the preceding ones. We compute the following statistics g_k, $k = 1, \ldots, \lfloor N/2 \rfloor$:

$$g_k = \sum_{j=1}^{k} \gamma_j^2 \Big/ \sum_{j=1}^{\lfloor N/2 \rfloor} \gamma_j^2,$$

where γ_j^2 is defined in test 3.

The graph of g_k vs. k is known as the normalized cumulative periodogram. If C_0 is the correct class, the graph of $E(g_k)$ vs. k is a straight line joining (0, 0) to (0.5, 1) and hence the normalized cumulative periodogram must be closely intertwined around the line joining (0, 0) to (0.5, 1). The probability that the entire periodogram lies within a strip bounded by lines parallel to the line joining (0, 0) to (0.5, 1) at distances $\pm\lambda/\sqrt{\lfloor N/2 \rfloor}$ is

$$\sum_{j=-\infty}^{\infty} (-1)^j \exp(-2\lambda^2 j^2),$$

For instance, the probability is 0.95 if $\lambda = 1.35$, and if $\lambda = 1.63$ it is 0.99. Suppose we are working at the 5% level of significance. Then if the entire cumulative periodogram falls within the 95% strip, we accept C_0. Otherwise, we reject C_0.

This test is quite different from the other tests discussed earlier in that the test involves the computation of the $\lfloor N/2 \rfloor$ test statistics g_k, $k = 1, \ldots, \lfloor N/2 \rfloor$. Also, it does not have the limitations of the Fisher test mentioned earlier. Furthermore, when the test fails, it also gives an indication of the dominant frequencies in the given observations. If the cumulative periodogram overshoots the boundary at frequencies $\omega_{k_1}, \omega_{k_2}, \ldots$, then some of these frequencies are the dominant frequencies in the observations.

8c.2.3. Test for Serial Independence

We will determine whether the given sequence $\{x(\cdot)\}$ is serially correlated (Whittle, 1951, 1952).

Test 5. Let the two classes be

$$C_i = \{\mathscr{I}_i, \mathscr{H}_i, \Omega_i\}, \qquad i = 0, 1$$

$$\mathscr{I}_0: x(t) = w(t), \qquad \mathscr{I}_1: x(t) = \sum_{j=1}^{n_1} \theta_j x(t-j) + w(t),$$

where $\{w(\cdot)\}$ is an IID $N(0, \rho)$ sequence, $\rho \in \Omega$, $\boldsymbol{\theta} = (\theta_1, \theta_2, \ldots, \theta_{n_1})^T$ and $\mathscr{H}_1 = \{\boldsymbol{\theta}: \boldsymbol{\theta} \neq 0\}$, $\mathscr{H}_0 = \{0\}$, n_1 is a prespecified integer. We will comment later on its choice.

We can use the likelihood ratio test as in test 2. The validity of the F-distribution of the corresponding test statistic can be established for large N.

Let $\hat{\rho}_0$, $\hat{\rho}_1$ be the residual variances of the best fitting models for the given data in the two classes C_0 and C_1, respectively. Let $\hat{R}_k$ be the empirical covariance at lag k,

$$\hat{R}_k = \frac{1}{N-k} \sum_{j=k+1}^{N} x(j)x(j-k).$$

Then

$$\hat{\rho}_0 = \hat{R}_0; \qquad \hat{\rho}_1 = \det \mathbf{\Gamma}_{n_1} / \det \mathbf{\Gamma}_{n_1 - 1},$$

where $\mathbf{\Gamma}_{n_1} = n_1 \times n_1$ matrix, $(\mathbf{\Gamma}_{n_1})_{ij} = \hat{R}_{|i-j|}, \quad i, j = 1, 2, \ldots, n_1$.
The test statistic is

$$\begin{aligned}\eta(x) &= \left(\frac{N}{n_1} - 1\right)\left(\frac{\hat{\rho}_0}{\hat{\rho}_1} - 1\right) \\ &\sim F(n_1, N - n_1) \qquad \text{for large } N, \quad \text{if } x \text{ obeys } C_0.\end{aligned}$$

As before, the *decision rule* is

$$\eta(x)\begin{cases} \leq \eta_1 \rightarrow \text{choose } C_0 \\ > \eta_1 \rightarrow \text{choose } C_1, \end{cases}$$

where η_1 is chosen by the corresponding significance level.

The test is unbiased, i.e., if x obeys any model in C_1, then its probability of error II is less than $(1 - \varepsilon)$, where ε is the significance level, for any $\boldsymbol{\theta}$ and ρ.

Consider the integer n_1. Typically when we are working with data sets where N is of the order of several hundred, n_1 is chosen to be either $0.1N$ or $0.05N$. If the model does not pass the test at such high values of n_1, we must find the maximum value of n_1 at which the test does pass. This integer n_1 gives us a good idea of the limitations of the model.

Test 6: Portmanteau Test (Box and Jenkins, 1970). This is a goodness of fit test. As such only the class C_0 is well defined. As before, C_0 has only the zero mean white noise models with variance ρ, $0 < \rho < \infty$. The class C_1 is the class of all models excluding those in C_0.

Let R_j, $j = 0, 1, 2, \ldots$, be the correlation coefficients of the process x. The corresponding estimated spectral density is $S(\omega)$:

$$S(\omega) = \sum_{j=-\infty}^{\infty} Re^{-ij\omega}.$$

If $x(\cdot)$ is white, then $S(\omega) = S(0)$, $\forall \omega$, and the mean square deviation of $S(\omega)$ from the value $S(0)$ is

$$\frac{1}{\pi} \int_{-\pi}^{\pi} (S(\omega) - S(0))^2 \, d\omega = \sum_{j=1}^{\infty} R_j^2.$$

We can test the whiteness of a sequence $\{x(\cdot)\}$ by evaluating $\sum_{j=1}^{\infty} R_j^2$. In practice, we have to truncate the series and replace the correlation coefficient R_j by its estimate $\hat{R}_j$ defined earlier.

The required *test statistic* $\eta(x)$ is $\eta(x) = (N - n_1) \sum_{k=1}^{n_1} (\hat{R}_k)^2/(\hat{R}_0)^2$. The integer n_1 is usually chosen to be about $0.1N$ or $0.01N$, depending on the size of N. If x obeys C_0, then $\eta(x)$ has the χ^2-distribution with n_1 degrees of freedom. The *decision rule* is

$$\eta(x)\begin{cases} <\eta_1 \to \text{accept } C_0 \\ \geq \eta_1 \to \text{reject } C_0, \end{cases}$$

where η_1 is a threshold computed from the corresponding level of significance $(1 - \varepsilon)$, i.e., η_1 is found from the equation

$$\text{prob}[\eta(x) \geq \eta_1 | x \text{ obeys } C_0] = \varepsilon.$$

The test is realtively easy to implement as compared with test 5. When we are dealing with observation size N of the order of several thousands, test 5 is impractical but test 6 is feasible. The price we have to pay in using the easier portmanteau test is that it may lead to a higher probability of error as compared with test 5. We will illustrate such a possibility when $n_1 = 1$.

Example. Consider the two classes C_i, $i = 0, 1$, $C_i = \{\mathscr{I}_i, \mathscr{H}_i, \Omega\}$,

$$\mathscr{I}_0: x(t) = w(t), \qquad \mathscr{I}_1: x(t) = \theta x(t-1) + w(t),$$

where $w(\cdot)$ is the usual $N(0, \rho)$ sequence, $\mathscr{H}_1 = \{\theta: -1 < \theta < 1, \theta \neq 0\}$, and $\mathscr{H}_0$ is the null set. The *decision rule* in test 6 is

$$\eta_1 - \eta_{01}\begin{cases} \geq 0 \to \text{reject } C_0 \\ <0 \to \text{accept } C_0, \end{cases}$$

where $\eta_1(x) = (N-1)R_1{}^2/R_0{}^2 = (N-1)(\hat{\theta})^2 \sim \chi^2(1)$, if C_0 is true. $\hat{\theta}$ is the maximum likelihood estimate of θ when x obeys C_1 and η_{01} is the threshold.

Decision rule in test 5: Let us use $n_1 = 1$:

$$\eta_2 - \eta_{02}\begin{cases} \geq 0 \to \text{reject } C_0 \\ <0 \to \text{accept } C_0, \end{cases}$$

where η_{02} is a threshold and

$$\eta_2(x) = (N-1)[\{1/(1-(\hat{\theta})^2)\} - 1] \sim F(1, N-1), \qquad \text{when } x \text{ obeys } C_0.$$

It is well known that the $F(1, N-1)$-distribution is asymptotically identical to the chi-squared distribution with one degree of freedom. Consequently, η_{01} and η_{02} are equal to each other, say η_0, provided the same level of significance is used in both tests: $P_{2,i}(\theta)$ is the probability of error II with test i, given that θ is the true value of the AR parameter, $i = 5, 6$ [i.e., x obeys $(\mathscr{I}_1, \theta, \rho) \in C_1$]:

$$\begin{aligned} P_{2,5}(\theta) &= \text{prob}[\eta_2 \leq \eta_0 | C_1, \theta] \\ &= \text{prob}[|\hat{\theta}| < [(\eta_0/(N-1))/(1 + \eta_0/(N-1))]^{1/2} | C_1, \theta]. \end{aligned}$$

Similarly,

$$\begin{aligned} P_{2,6}(\theta) &= \text{prob}[\eta_1 < \eta_0 | C_1, \theta] \\ &= \text{prob}[|\hat{\theta}| \leq [\eta_0/(N-1)]^{1/2} | C_1, \theta]. \end{aligned}$$

But

$$\left(\frac{\eta_0}{N-1}\right)^{1/2} > \left[\left(\frac{\eta_0}{N-1}\right)\left(\frac{1}{1+\eta_0/(N+1)}\right)\right]^{1/2}.$$

Hence $P_{2,5}(\theta) \leq P_{2,6}(\theta)$ for every $\theta \in \mathscr{H}_1$, i.e., test 6 has uniformly smaller power than test 5.

The Test for Normality. We will omit the details of the normality tests since they can be found in any standard text in statistics.

8c.3. Validation Tests Based on Comparison of the Various Characteristics of Model and Data

In these tests, we will directly compare the theoretical characteristics of the output of the model with the corresponding empirical characteristics of the given data. Of course, we can compare only a few characteristics. The typical characteristics chosen here are the correlogram, spectral density, and rescaled range–lag characteristic. On specific occasions, we may need to compare some other characteristics as in Chapter X.

Before getting the statistical characteristics, we may simply want to simulate the model on a computer, obtaining the $\{w(\cdot)\}$ sequence from a pseudorandom number generator and checking whether the synthetic (or simulated or generated) data display some of the major characteristics in the data such as growth or systematic oscillations. We recall that one of the purposes of a model may be the generation of synthetic data that approximately reproduce the major statistical characteristics of the observed data.

We may have difficulties in simulating seasonal ARIMA models, because they are marginally stable; i.e., if the equation is written as $A(D)y(t) = B(D)w(t)$ where $A(D) = A_1(D)(1 - D^L)$, then the polynomial $A(D)$ has L zeros on the unit circle. Hence there is a possibility of the output of the model going to infinity on account of the buildup of the roundoff errors.

8c.3.1. Comparison of Correlograms

Let

$$\bar{R}_k(N) = \frac{1}{N}\sum_{j=1}^{N} (y(j) - \bar{y}(N))(y(j-k) - \bar{y}(N))$$

$$\bar{y}(N) = \frac{1}{N}\sum_{j=1}^{N} y(j)$$

$$R_k = \lim_{N\to\infty} \bar{R}_k(N), \qquad \sigma_k(N) = \{E[(\bar{R}_k(N) - R_k)^2]\}^{1/2}.$$

The graph of $\bar{R}_k(N)$ vs. k, for fixed N, is called the empirical correlogram of the sequence $\{y(1), \ldots, y(N)\}$, whereas the graph of R_k vs. k is called the theoretical correlogram of the process $y(\cdot)$. The theoretical correlogram of the output of the model must be compared with the corresponding correlogram of the data. The degree of fit between the two can be quantitatively expressed consistent with the available observation size.

From the techniques given in Chapter II it is possible to find the analytic expressions for R_k and σ_k for the output of the given model. However, the derivation of these expressions is cumbersome, especially when sinusoidal terms are present in the model. Alternatively, the theoretical correlogram R_k and the corresponding standard deviation $\sigma_k(N)$ of the model can be estimated by simulating the stochastic difference equation. Suppose 100 independent sequences of the output of the model, each having N observations, are generated. Let $R_k^{(i)}$ denote the empirical correlogram for the ith sequence. Let

$$R_k^M = \frac{1}{100} \sum_{j=1}^{100} R_k^{(i)}$$

$$(\sigma_k^M) = \left[\frac{1}{100} \sum_{j=1}^{100} (R_k^{(i)} - R_k^M)^2\right]^{1/2}.$$

The graph of the sample mean R_k^M vs. k should give us a good estimate of the true correlogram of the model, and the quantity $\sigma_k{}^M$ should give us a good idea of the standard deviation of the empirical correlogram $\bar{R}_k(N)$ of the output of the model (with N observations) caused by the finiteness of the sample.

The empirical correlogram $\bar{R}_k(N)$ of the given data knowing N observations can be computed as usual. If the following relation is satisfied ($N = 100$ in our example),

$$R_k^M - 2\sigma_k^M \leq \bar{R}_k(N) \leq R_k^M + 2\sigma_k^M, \qquad \forall k = 1, 2, \ldots, N-1,$$

then the empirical correlogram of the data can be regarded as being a good fit to the theoretical correlogram of the model, and hence the model can be considered as adequate in representing the correlogram of the data.

There have been suggestions, especially in the hydrologic literature, that the model should be chosen so that $\bar{R}_1(N)$, the lag 1 correlation coefficient of the data, must be approximately equal to the theoretical lag 1 correlation coefficient of the model. There seems to be little justification for such a requirement, considering the fact that the $\bar{R}_1(N)$ is only one of the many statistical characteristics of the process.

8c.3.2. Comparison of Spectral Estimates

We will discuss two important spectral characteristics, namely the periodogram and the spectral density. In contrast with the correlogram, the estimates of the spectral functions are less accurate and, consequently, when we match the spectral characteristics of the given data and those of the model, we look only for a *qualitative* and not a quantitative fit.

If the given a posteriori model has only sinusoidal terms of frequency $\omega^{(1)}$,

$\omega^{(2)}, \ldots$, then the theoretical periodogram, i.e., $E[\gamma_N^2(k)]$, will be zero for all frequencies except $\omega = \omega^{(1)}, \omega^{(2)}, \ldots$,

$$\gamma_N^2(k) = \left(\frac{2}{N}\sum_{j=1}^{N} y(j) \cos 2\pi f_k j\right)^2 + \left(\frac{2}{N}\sum_{j=1}^{N} y(j) \sin 2\pi f_k j\right)^2.$$

By using the given N observations of the data, we can obtain the observed periodogram [i.e., the graph of $\gamma_N^2(j)$ vs. $f_j, f_j = j/N$]. We can regard the fit between the observed and the theoretical periodograms to be good if the observed periodogram has relatively sharp peaks *only* at the frequencies $\omega^{(1)}$, $\omega^{(2)}, \ldots$.

The spectral density of the model and of the given data may be compared as follows. Let the given model for $y(\cdot)$ be as given below in operator notation:

$$A(D)y(t) = B(D)w(t) + \sum_{j=1}^{n_2} \alpha_j \cos \omega^{(j)}t + \beta_j \sin \omega^{(j)}t.$$

$S_{yy}(e^{i\omega})$, the theoretical spectral density of the output of the model, is given as

$$S_{yy}(e^{i\omega}) = \rho^0 B(e^{-i\omega})B(e^{i\omega})/\{A(e^{i\omega})A(e^{-i\omega})\}, \qquad \omega = 2\pi f.$$

In order to obtain an estimate of the empirical spectral density of a process based only on the observed data $\{y(1), \ldots, y(N)\}$, the sinusoidal trend terms corresponding to the frequencies $\omega^{(j)}, j = 1, 2, \ldots$, must be removed by Fourier techniques. Let the detrended data be denoted $\bar{y}(i)$, $i = 1, \ldots, N$. By using $\bar{y}(\cdot)$, we can compute the following two estimates of the spectral density using Bartlett's ($\bar{\mathbf{S}}_1(\omega)$) and Daniell's ($\bar{\mathbf{S}}_2(\omega)$) windows (Wold, 1965):

$$\bar{\mathbf{S}}_1(\omega) = \omega + 2\sum_{k=1}^{m} \bar{R}_k\left(1 - \frac{k}{m}\right)\left(\frac{\sin(\omega k)}{k}\right), \qquad m < N,$$

$$\bar{\mathbf{S}}_2(\omega) = \omega + 2\sum_{k=1}^{m} \bar{R}_k\left(1 - \frac{k}{m}\right)\left(\frac{\sin(\omega k)}{k}\right)\left(\frac{\sin hk}{hk}\right); \qquad m < N, \quad h = \frac{m}{N}.$$

The truncation parameter m can be chosen at our discretion. The two estimates above can be compared with the theoretical spectral density mentioned earlier. Alternatively, we could use the spectral estimates obtained from AR models (Markel, 1972).

8c.3.3. Comparison of Rescaled Range (R/σ) Characteristics

We can also compare the output of the model and the observed data from the point of view of the probabilistic characteristics of the extreme values; i.e., for instance, how close is the average of the maximum value of the output of the model in a given time interval t to the corresponding quantity for the given data? It is better to consider the rescaled range–lag characteristic involving the cumulative range of values in a certain period of time.

The estimation of the R/σ versus the time lag s characteristic from the given data along with the corresponding standard deviations has been mentioned in

Chapter II. If the R/σ vs. s values of the given empirical data fall within the one standard deviation limits of the R/σ vs. s characteristic obtained from the simulated data, we can claim that the model "preserves" the R/σ characteristic.

8c.4. Validation of a Model with Time-Varying Coefficients

Validation of a model with time-varying coefficients by comparing the characteristics of the data and output of the model is not different from the methods discussed in Section 8c.3. Similarly, testing the whiteness of residuals in the presence of sinusoidal terms in a model with time-varying coefficients is identical to the methods in Section 8c.2. Hence, the only difference between the models with time-varying coefficients and other models is in the generation of the residuals.

Consider the system in (6f.1.4) and (6f.1.5). If $\hat{\boldsymbol{\theta}}(t-1)$ is the estimate of $\boldsymbol{\theta}(t-1)$ based on all observations until time $(t-1)$, then $(y(t) - (\hat{\boldsymbol{\theta}}(t-1))^{\mathrm{T}}\mathbf{z} \times (t-1))$, $t = 1, 2, 3, \ldots,$ is theoretically a zero mean white sequence. This sequence must be tested for whiteness using the tests of Section 8c.2.

8d. Discussion of Selection and Validation

8d.1. Guidelines for Model Building

The class selection methods give only the best class among the list of chosen classes. There is no guarantee that the best fitting model from the best class given by the class selection methods is the appropriate model, i.e., it may not pass the validation tests. If the fitted model does not pass the validation tests, the implication is that the proper class of models needed for the given time series has been left out in the list of classes considered for the selection process. We should make every attempt to consider all the possible classes relevant for the given series as mentioned in Chapter III. In Chapters X and XI we have given several illustrations to demonstrate how different time series which may appear similar to one another require quite different models for satisfactorily representing them.

In validating a model the tests involving the residuals and those involving the comparison of the output characteristics of the model and data are both important. The residual tests are relatively easy to perform, but they may not detect the presence of sinusoidal terms in the series, as will be shown in modeling the population of Canadian lynx in Chapter XI. The reason is not hard to find. Residual tests 2–4 are designed to distinguish between white noise models and white noise models plus sinusoidal trends. They may not yield good results in discriminating between models in which the corresponding noise is not white but correlated. In such cases, tests based on comparing the correlograms or spectral density of the model and data are very useful in showing the limitations of the model.

It must be emphasized that the validation tests do not guarantee that the models which pass the tests are parsimonious. There may be other models from other classes which pass all the tests and have fewer parameters than the one under consideration. If we have doubts about the necessity of certain terms in the difference equation, we should omit those terms in them and check whether the simpler model will pass the validation tests.

8d.2. Meaning of Validation, Complexity of Model, and Size of the Observation History

We will first clarify the meaning of a validated model by an example. Consider a relatively simple stationary time series with, say, 100 observations in it and suppose an AR(2) model, fitted to it, satisfies all the validation tests at a suitable significance level. Can we say that the given physical process obeys an AR(2) model? An affirmative answer is *not* warranted. If the underlying physical process does obey an AR(2) model, then the statistical characteristics of the model, such as correlogram and spectral density, should be close to the corresponding empirical characteristics obtained with an additional, say, 10,000 observations of the process at a reasonable significance level. Such a situation does not occur often. For instance, consider the modeling of a daily river flow process. Usually an IAR(2) model is sufficient if we are dealing with about 100 observations. But the characteristics of the fitted IAR(2) model will be drastically different from the corresponding empirical characteristics of an additional 10,000 observations of the flow. Thus, strictly speaking, we are not entitled to say that the daily river flow process can be represented by an IAR(2) model. What we can say is that an IAR(2) model can adequately explain the 100 observations used in its construction. To put it another way, the empirical correlogram, spectral density and other characteristics of the 100 observations lie within the two standard deviation limits of the corresponding characteristics of the IAR(2) model, but we cannot say on the basis of the given observation set whether the observed discrepancy between the corresponding characteristics of the model and the data is caused by the inappropriateness of the *model* or by the sampling variation in the estimates of the characteristics caused by the finiteness of the observation.

Thus the validation concept considered in this chapter has a tentative character. The validation tests only imply that the given model adequately represents the given observation set. Of course, the larger the observation set used in the construction of the model, the greater is the possibility of the model adequately explaining the data to be gathered in the future.

Let us denote the minimal complexity of a validated model constructed on the basis of N observations by $C(N)$. A rough estimate of $C(N)$ is the number of terms in a parsimonious model which can pass all the validation tests with the given observation set of size N. The foregoing discussion implies that, as the available number of observations increases, we may need more and more refined models. Recall that in correlogram tests the deviation caused by sampling is of the order $O(1/\sqrt{N})$. Hence, the larger the N, the smaller should be the discrepancy between the characteristics of the model and the data, at the same significance level.

There are two possibilities for the behavior of $C(N)$ with N: (i) It can tend to a constant value as N tends to infinity. (ii) It may grow with N.

Case (i) requires no comment. If case (ii) is valid, then it is an indication that none of the classes of the models considered is appropriate. Perhaps the process is nonstationary and/or does not obey a finite stochastic difference model. One such example is the fractional noise process.

All the foregoing comments are valid only if we are looking for a model that passes all the validation tests. In many cases, we may be interested only in forecasting and not in the generation of synthetic data so that validation of the model by using synthetic data is superfluous. In such cases, difference equations of relatively small orders are often quite sufficient in providing adequate forecasting formulas and one need consider complex models only occasionally.

8e. Conclusions

Given a number of possible classes of models, a number of methods have been developed for selecting a suitable class for the given time series. Next we developed a number of tests, the so-called validation tests, to determine whether the best fitting model from this class adequately represents the given series. Guidelines were suggested for the construction of models using the selection and validation methods. The relation of the complexity of the validated model to the available observation size was also discussed.

Appendix 8.1. Mean Square Prediction Error of Redundant Models

We will give an example to show that the mean square prediction error of the overspecified model may be larger than that of the parsimonious model.

Example. Let the process $y(\cdot)$ obey

$$\begin{aligned} y(k) &= (\boldsymbol{\theta}^0)^{\mathrm{T}}\mathbf{z}(k-1) + w(k) \\ (\boldsymbol{\theta}^0)^{\mathrm{T}} &= (\theta_1{}^0, \ldots, \theta_n{}^0), \qquad \theta_n{}^0 \neq 0 \\ \mathbf{z}(k-1) &\triangleq (y(k-1), \ldots, y(k-n))^{\mathrm{T}}. \end{aligned} \tag{1}$$

where $\{w(\cdot)\}$ is IID, $N(0, \rho)$ and $w(i)$ is independent of $y(i-j)$ for all $j > 0$. Let the redundant model be

$$\begin{aligned} y(k) &= (\boldsymbol{\beta}^0)^{\mathrm{T}}\bar{\mathbf{z}}(k-1) + w(k), \\ (\boldsymbol{\beta}^0)^{\mathrm{T}} &= (\beta_1{}^0, \ldots, \beta_n{}^0, \beta_{n+1}^0, \ldots, \beta_{m+n}^0), \\ \bar{\mathbf{z}}(k-1) &= [y(k-1), \ldots, y(k-(n+m))]. \end{aligned} \tag{2}$$

Let $\hat{\boldsymbol{\theta}}$ and $\hat{\boldsymbol{\beta}}$ be the least squares estimates of $\boldsymbol{\theta}^0$ and $\boldsymbol{\beta}^0$ based on N observations $y(1), \ldots, y(N)$. The estimates $\hat{\boldsymbol{\theta}}$ and $\hat{\boldsymbol{\beta}}$ will be used to design predictors for another run of the process $\{y(\cdot)\}$ which is statistically independent of the observations that were used in the computation of $\hat{\boldsymbol{\theta}}$ and $\hat{\boldsymbol{\beta}}$. Let $\hat{y}_1(k|k-1)$ and $\hat{y}_2(k|k-1)$ be the one-step-ahead predictors of $y(\cdot)$ based on Eqs. (1) and (2), respectively:

$$\hat{y}_1(k|k-1) = (\hat{\boldsymbol{\theta}})^{\mathrm{T}}\mathbf{z}(k-1), \qquad \hat{y}_2(k|k-1) = (\hat{\boldsymbol{\beta}})^{\mathrm{T}}\bar{\mathbf{z}}(k-1).$$

Let R_1 and R_2 be the mean square prediction errors associated with the predictors y_1 and y_2:

$$\begin{aligned} R_1 &= E[(y(k) - \hat{y}_1(k|k-1))^2] = E[(w(k) + (\boldsymbol{\theta}^0 - \hat{\boldsymbol{\theta}})^{\mathrm{T}}\mathbf{z}(k-1))^2] \\ &= E[w^2(k)] + \operatorname{trace} E[(\boldsymbol{\theta}^0 - \hat{\boldsymbol{\theta}})(\boldsymbol{\theta}^0 - \hat{\boldsymbol{\theta}})^{\mathrm{T}}\mathbf{z}(k-1)\mathbf{z}^{\mathrm{T}}(k-1)] \\ &\quad + 2E[(\boldsymbol{\theta}^0 - \hat{\boldsymbol{\theta}})^{\mathrm{T}} w(k)\mathbf{z}(k-1)]. \end{aligned} \tag{3}$$

The estimate $\hat{\boldsymbol{\theta}}$ is independent of $y(\cdot)$ as affirmed earlier. Hence,

$$E[(\boldsymbol{\theta} - \hat{\boldsymbol{\theta}})^{\mathrm{T}} w(k)\mathbf{z}(k-1)] = (E[(\boldsymbol{\theta} - \hat{\boldsymbol{\theta}})])^{\mathrm{T}} E(w(k)\mathbf{z}(k-1)) = 0, \tag{4}$$

since the second term is zero. Further, the estimate $\hat{\boldsymbol{\theta}}$ has the following property since the estimate has been computed by the least squares technique:

$$E[(\boldsymbol{\theta}^0 - \hat{\boldsymbol{\theta}})(\boldsymbol{\theta}^0 - \hat{\boldsymbol{\theta}})^{\mathrm{T}}] \approx \frac{1}{N}\{E[\mathbf{z}(k-1)\mathbf{z}(k-1)]^{\mathrm{T}}\}^{-1}\rho. \tag{5}$$

Substituting (4) into (3), we obtain

$$\begin{aligned} R_1 &= E[w^2(k)] + \operatorname{trace}\{E[(\boldsymbol{\theta} - \hat{\boldsymbol{\theta}})(\boldsymbol{\theta} - \hat{\boldsymbol{\theta}})^{\mathrm{T}}\mathbf{z}(k-1)\mathbf{z}(k-1)^{\mathrm{T}}]\} \\ &= E[w^2(k)] + \operatorname{trace}\{(E(\boldsymbol{\theta} - \hat{\boldsymbol{\theta}})(\boldsymbol{\theta} - \hat{\boldsymbol{\theta}})^{\mathrm{T}})(E(\mathbf{z}(k-1)\mathbf{z}^{\mathrm{T}}(k-1)))\} \\ &\qquad \text{since } \mathbf{z}(k-1) \text{ is independent of } \hat{\boldsymbol{\theta}}, \\ &= \rho + \rho \operatorname{trace}\{\mathbf{I}/N\}, \qquad \text{by (5)} \\ &= \rho(1 + n/N). \end{aligned}$$

Similarly,

$$R_2 = E[(y(k) - \hat{y}_2(k|k-1))^2] = \rho(1 + (n+m)/N).$$

Hence, $R_2 > R_1$.

Problems

1. Generate 20 independent sets of data $y(1), \ldots, y(N)$ from the AR(3) model (*). With each set with $N = 100$, determine the appropriate model among the following AR(i), $i = 1, 2, \ldots, 6$, classes by (a) the class likelihood method, (b) the recursive prediction method of Section 8b.3, (c) the prediction method of Section 8b.4.1, and (d) the Bayesian method with the prior probability for the class AR(i) chosen as $c \exp[-i^2]$, where c is the normalizing constant. Repeat the experiment with $N = 50$. Do you expect the number of cases among the 20 sets in which the best fitted model is AR(3) to be the same when $N = 50$ and $N = 100$?

$$y(t) = 0.8y(t-1) + 0.5y(t-2) + 0.2y(t-3) + w(t). \tag{*}$$

2. Study the sensitivity of the various methods of class selection to deviations from the normality assumption made on w by repeating the experiment in Problem 1 where $w(\cdot)$ is drawn from a mixture distribution $0.95N(0, 0.5) + 0.05N(0, 10)$ which has a relatively long tail. While using the class likelihood method, derive the decision rule assuming the distribution of w to be normal.

3. Consider a set of 120 observations $y(1), \ldots, y(120)$ from any real life process, such as monthly sunspots (Waldemeir, 1961). Generate the following aggregated sequence $y_1(1), \ldots, y_1(30)$:

$$y_1(t) = \tfrac{1}{4} \sum_{j=1}^{4} y((t-1)4 + j).$$

(a) Construct the best fitting AR models for the sequences $\{y(\cdot)\}$ and $\{y_1(\cdot)\}$. Validate the best fitting models.
(b) Verify that if $y(\cdot)$ exactly obeys an AR model of order n, then $y_1(\cdot)$ cannot obey exactly an AR(m) model for any m.
(c) Let the orders of the best fitting AR models in part (a) above be named n and n_1, respectively. Suppose these best fitting models have passed all the validation tests. Does this result contradict the assertion in part (b) above?
(d) Suppose $y(\cdot)$ obeys exactly the best fitted model AR(n) obtained above. Using the AR(n) model, derive the linear least squares predictor $y_1(t)$ based only on the finite past $y_1(t-j), j = 1, \ldots, n_1$. Is this predictor different from the one given by the best fitting AR(n_1) model obtained directly from the observed sequence $\{y_1(\cdot)\}$?

Chapter IX | Class Selection and Validation of Multivariate Models

Introduction

We will discuss the selection and validation of multivariate models. Needless to say, all the problems in univariate models, such as the selection of trend terms, appear in multivariate models also, and these must be similarly handled. In addition, there are other problems that are related to multivariate models only.

While discussing the problem of class selection in Chapter VIII, the classes to be compared were assumed given. They could have been chosen by means of the knowledge of the underlying physical process, intuition, or trial and error procedure. In the multivariate case, the variety of possible models is so great that there is a clear need for some guidelines in selecting the classes of models that have to be compared. For instance, let us consider a multivariate model in canonical form II and let the equation for the variable y_i be represented as

$$y_i(t) = \sum_{j=1}^{m} a_{ij}(D)y_j(t-1) + b_i(D)w_i(t), \qquad i = 1, \ldots, m,$$

where $a_{ij}(D)$, $b_i(D)$ are polynomials in D.

The inclusion of all the variables $y_j, j \neq i$, in the y_i equation often results in a highly redundant model. Therefore, a method is sought for choosing the appropriate variables among y_j, $j \neq i$, needed in the equation for $y_i(t)$. The choice of appropriate variables y_i is intimately related to the question of causality and is discussed later. The question of causality frequently occurs in engineering and economic problems. By considering such questions, we can obtain a class of models that can be considered further by the selection techniques which are analogous to the corresponding methods in the univariate models.

The question of validation in multivariable systems is clearly more difficult than that in univariate systems. One can always design individual validation tests for a given problem based on prior knowledge of the system. However, there are only a few detailed empirical studies of the validation of multivariate models (Quenouille, 1957; Klein and Evans, 1968). As such, it is difficult to judge the relative merits of the various suggested methods. We will present a few methods of validation which are generalizations of the corresponding methods for univariate systems.

The question of causality is of importance and is discussed in some detail in many papers dealing with the modeling of economic and other time series. One of the problems in such studies is that constant coefficient models may not always be valid. It is needless to say that any discussion of causality relations which is not based on a valid model of the underlying process is questionable. In addition to the discussion on causality presented below, we present in Chapter XI two case studies dealing with different aspects of causal relations.

9a. Nature of the Selection Problem

There are several reasons for investigating multivariate models. First of all, there is a possibility of increasing the accuracy of forecasts of a variable by utilizing the observational history of other variables. The second reason is that if we are interested in constructing models for control or for generating synthetic data, it is important that properties such as cross spectrum between different variables be reproduced by the model.

However, in many of the complex multivariate models reported in the literature there is an implicit assumption that the complex multivariate models are *always* superior to the simple univariate models from the point of view of forecasting. We have seen that even in a univariate model, addition of too many terms results in a reduction of the forecasting ability of the model. A similar result must be true for the multivariate models as well. For instance, consider the FRB–MIT–PENN econometric model for forecasting variables such as the gross national product. The model has 171 equations. Several investigators (Nelson, 1973) have found that in some cases, the simpler univariate model for a variable gives better forecasts than the complex multivariate model. In view of these findings it has been suggested that a weighted combination of the forecasts given by the complex multivariate model and that of the simpler univariate model be used to obtain a good forecast for that variable. We may consider another example from an area in which we have more experience. Let us consider models of monthly river flows. Undoubtedly, rainfall is the principal contributor to the river flow. But rainfall measured at a station gives relatively little information about the monthly river flow in a large watershed of thousands of square miles. As such, the forecast of the river flow variable obtained by regressing the flow variable on the rainfall variables at a number of stations is often considerably worse than that obtained by the univariate model of the flow. The forecasts of river flow obtained by using the past histories of both flow and rainfall data will not be much better than those obtained using only the past values of the flows. However, when we deal with the problem of forecasting the runoff from small watersheds, the rainfall information is very useful for predicting flows, especially the mean daily flows. (Tao *et al.*, 1975).

As before, we define a class of models to be a triple $[\mathscr{M}, \mathscr{H}, \Omega]$ where $\mathscr{M}$ is a stochastic difference equation involving a vector of coefficients $\boldsymbol{\theta} \in \mathscr{H}$ and the covariance matrix of the noise $\mathbf{w}$ is $\boldsymbol{\rho} \in \Omega$:

$$\mathscr{M}: \mathbf{f}_0(\mathbf{y}(t)) = \boldsymbol{\theta}^{\mathrm{T}}\mathbf{x}(t-1) + \mathbf{w}(t), \tag{9a.1.1}$$

where $\mathbf{x}(t-1)$ is an $m \times n$ matrix made up of terms such as $\mathbf{y}(t-j)$, $j = 1, \ldots, m$, the trend terms, and moving average terms such as $\mathbf{w}(t-j)$, $j = 1, \ldots, m_2$,

$$\mathbf{f}_0(\mathbf{y}(t)) = (f_{01}(y_1(t)), \ldots, f_{0m}(y_m(t)))^{\mathrm{T}},$$

where the functions $f_{0j}(\cdot)$ are known.

The problem of selection can be posed as follows: Given r classes $C_i = \{\mathscr{M}_i, \mathscr{H}_i, \Omega_i\}$, $i = 0, \ldots, r-1$, it is desired to find that class which is most appropriate for characterizing the given set of observations $\mathbf{y}(1), \ldots, \mathbf{y}(N)$.

A natural question is the choice of the candidate classes C_i, $i = 0, 1, \ldots, r-1$. In the univariate cases discussed in Chapters VIII, X, and XI, the candidate classes are chosen by a combination of intuition, knowledge of the physical laws of the process, and a trial and error procedure as mentioned earlier. Since the number of possible classes in the multivariate case is several orders larger than the corresponding number of univariate classes, it is of utmost importance that we develop a systematic method of determining the possible classes. This is best done by considering the equations for the individual variables $y_1, \ldots, y_m$ separately. Ideally we begin with the "best" univariate model for a variable y_i and investigate the variables y_j, $j \neq i$, needed in the equation for y_i without considering any other equation. However, we have to make sure that the equations for the various variables in combination will be in some canonical form. Otherwise it will not be possible to establish the uniqueness of the system of equations to represent the given multivariate process. Hence we find that it is convenient to work in canonical form II. If we have m variables, say $y_1, y_2, \ldots, y_m$, the multivariate equations would be

$$f_{0k}(y_k(t)) = \sum_{j=1}^{m} a_{kj}(D) f_{kj}(y_j(t-1)) + b_{kk}(D) w_k(t), \tag{9a.1.2}$$

where a_{kj}, b_{kk} are polynomials in D, $b_{kk}(0) = 1$, and

$$\mathbf{w}(t) = (w_1(t), w_2(t), \ldots, w_m(t))^{\mathrm{T}}$$

is a vector zero mean white noise sequence with covariance matrix $\boldsymbol{\rho}$, which may not necessarily be diagonal. We can work with the equations for y_k, $k = 1, 2, \ldots, m$, separately and inquire whether or not the terms of the variables y_j, $j \neq k$, are needed in the equation for y_k. This will be the procedure followed here. We can have constant or trend terms in (9a.1.2) if necessary. The determination of the appropriate terms needed for the equation $y_k(t)$ can be determined by the methods of Section 8b. We describe some of the relevant details in the next section. It should be noted that the results obtained by separately treating the individual equations for the various variables $y_1, \ldots, y_m$ cannot be definitive because the corresponding parameter estimates are not efficient. The results of Section 9b suggest the various multivariate classes that have to be considered further. These classes have to be compared with each other directly. (However, all the classes need not be in the same canonical form.) This aspect of direct comparison of classes is treated in Section 9c.

It should be noted that in Eq. (9a.1.2) for $y_k(t)$, the terms $y_j(t), j \neq k$, are not considered. In economic literature, it is usual to allow for the presence of terms $y_j(t), j = k$, in the equation for $y_k(t)$. These models are therefore called simultaneous equation models. We have already seen that by using Eq. (9a.1.1) we can obtain the corresponding canonical form III which allows for the presence of the simultaneous terms, albeit in a particular manner. If we insert all the $y_j(t), j \neq k$, terms in the equation for $y_k(t)$, for every $k = 1, \ldots, m$, we cannot ensure, in general, the uniqueness of the overall system. Hence we will work only with the form in (9a.1.1), repeating that there is no loss of generality in omitting the simultaneous terms. In general, by allowing for simultaneous terms we can render the covariance matrix $\boldsymbol{\rho}$ diagonal.

9b. Causality and the Construction of Preliminary Models

9b.1. The Definition of Causality

The concept of causality is commonly used in experimental sciences where we can directly verify whether one variable is the cause of another. In this book we are dealing with the techniques of model construction based on data that are the end products of complex processes which are not easily amenable for experimentation. In particular, we cannot set one of the variables at some fixed value and observe the changes in the other variables. Hence we cannot be completely certain about the pattern of influence among the different variables. Thus we have to introduce a definition of causality which is weaker than that employed in the experimental sciences. One such definition is given below. Other equivalent definitions are given by Sims (1972) and Caines and Chan (1974).

"y_2 will be said to be causal for y_1 (and denoted by $y_2 \to y_1$) if the accuracy of the forecast of $y_1(t)$ obtained by using the history $y_1(j)$, $j < t$, and $y_2(j)$, $j < t$, is greater than that obtained using only the past history of y_1." This definition can be formalized as follows:

Definition (Granger, 1963). A variable y_2 is called causal for a variable y_1 and denoted by the relation $y_2 \to y_1$ if the following inequality is satisfied:

$$\operatorname{cov}[y_1(t) | y_1(t-j), y_2(t-j), j > 1] < \operatorname{cov}[y_1(t) | y_1(t-j), j > 1]. \quad (9b.1.1)$$

The implication of the definition for modeling is given by the following theorem:

Theorem 9b.1. $y_2 \to y_1$ *iff* the coefficient $a_{12}(D)$ in the minimum forecast error equation for $y_1(t)$ in the family of models (9a.1.2) involving at most two variables y_1 and y_2 is nonzero.

The proof of Theorem 9b.1 is trivial.

Thus, checking for causality while dealing with only two variables is easy. We first model the variable $y_1(t)$ using only its past values. Next we model it using both its past values and the past values of $y_2(t)$. If the coefficients of the $y_2(\cdot)$ terms are statistically significant, we deduce $y_2 \to y_1$ in the sense of Granger (1963).

However, the situation becomes more complex if there are more than two variables. For illustration consider the case of three variables y_1, y_2, and y_3 such that $y_2 \to y_1$ and $y_3 \to y_1$. Let us also assume that

$$\begin{aligned}&\operatorname{cov}[y_1(t)|y_1(t-j), y_2(t-j), y_3(t-j), j > 1]\\ &\quad = \operatorname{cov}[y_1(t)|y_1(t-j), y_3(t-j), j > 1].\end{aligned}$$

In such a case, the right-hand side of the equation for y_1, which yields the minimum mean square forecast error, has only the terms $y_1(t-j)$, $y_3(t-j)$, $j > 1$, but does not have any term $y_2(t-j)$, $j > 1$. Hence, if there are more than two variables, say $y_1, \ldots, y_m$, the fact that $y_i \to y_1$ by the definition given above does not imply that the term $y_i(t-j)$, for some $j > 1$, should be present in the equation for y_1 which gives minimum mean square forecast error.

If we have two variables y_1 and y_2 such that $y_2 \to y_1$ but $y_1 \not\to y_2$, then the input y_2 in the validated model for y_1 is said to be exogenous for the variable y_1. Another synonym for exogenous input is feedback-free input. The determination of whether or not an input is exogenous is important in parameter estimation problems.

The definition of causality given above is satisfactory as long as we limit ourselves to forecasting. Some other aspects need further exploration. For example, suppose we find $y_2 \to y_1$, and construct a validated model for y_1 involving y_2 terms. From the resulting equation, can we say that a certain change in the y_2 variable in the equation will cause a prespecified change in $y_1(t)$? The answer is clearly no, unless we have performed additional tests on the system to validate such a claim. The difficulty is the possible multiplicity of the models. For instance, there may be two equations for $y_1(t)$ which yield the same mean square forecast error, one of which has the lagged values $y_1(t-j)$ and $y_2(t-j)$ whereas the other has lagged terms $y_1(t-j)$ and $y_3(t-j)$. By the first model, a change in $y_3(t-1)$ has no effect on $y_1(t)$, but this statement is contradicted by the second model. Of course, we can try to compare the two models by means of the validation tests and draw inferences from only the validated model. While dealing with relatively simple or isolated physical or natural processes, the validated model seems quite useful in predicting changes in the variables caused by changes in the other variables. However, in the more complex processes such as weather modification experiments or those in econometrics, it is very difficult to precisely predict changes in a variable caused by the changes in the other variables.

9b.2. Tests for Causality

9b.2.1. The Case with Two Variables

We are interested in forecasting y_1 and hence would like to test whether $y_2 \to y_1$. A possible approach is to fit two AR models for $y_1(t)$, one of the models having the past values of y_1 only and the other model having the past values of both y_1 and y_2. The residual variances of the two models could be compared using regression theory, and then tested to see if the addition of the $y_2(\cdot)$ terms in the second model brings down the residual variance substantially. This method is not general enough for two reasons. First of all, the autoregressive model is not always the most appropriate model. We may need log transformed models or IAR models, etc. When we add the variable y_2, we do not know whether we should consider a term such as $y_2(t-1)$ or $\ln y_2(t-1)$. Inferences based on a model with incorrect structure can be only of dubious value. Second, the regression theory and the associated statistics are valid only if independent and identically distributed observations are used for both regressed and regressor variables. Hence, it is convenient to work with whitened processes derived from y_1 and y_2 instead of dealing with y_1 and y_2 directly.

Let us consider the validated univariate models for both y_1 and y_2

$$\begin{aligned} f_{01}(y_1(t)) &= (\boldsymbol{\theta}_1)^{\mathrm{T}}\mathbf{x}_1(t-1) + w_1(t) \\ f_{02}(y_2(t)) &= (\boldsymbol{\theta}_2)^{\mathrm{T}}\mathbf{x}_2(t-1) + w_2(t), \end{aligned} \tag{9b.2.1}$$

where we use the past history of only the corresponding variable. The vector $\mathbf{x}_i(t-1)$ is made up of $y_i(t-j)$ for various j and trends, if any. $w_1(\cdot)$ and $w_2(\cdot)$ are the two "whitened processes" to be used presently. If $y_2 \to y_1$, then there must exist a j and an $\alpha_j \neq 0$ so as to satisfy

$$w_1(t) = \alpha_j w_2(t-j) + \eta(t), \tag{9b.2.2}$$

where $\eta(\cdot)$ is a zero mean IID sequence. If we want to show that $y_2 \to y_1$, we have to show that α_j is not significantly different from zero for any j. This can be done by using any of the tests of Section 8b. To avoid unnecessary repetition, we give only two tests, based on the hypotheses testing and the likelihood approach, respectively.

Test 1: Based on Hypothesis Testing (Regression Theory). Suppose we have N observation pairs $\{w_1(t), w_2(t), t = 1, \ldots, N\}$. Let

$$r = \left(\frac{1}{N-j}\sum_{t=j+1}^{N} w_1(t)w_2(t-j)\right)\bigg/\left(\frac{1}{N-j}\sum_{j=t+1}^{N} w_1{}^2(t)\right)\left(\frac{1}{N-j}\sum_{j=t+1}^{N} w_2{}^2(t)\right).$$

If $\alpha_j = 0$, then

$$(N - j - 1)\left(\frac{r^2}{1 - r^2}\right) \sim F(1, N - j). \tag{9b.2.3}$$

If we fix a suitable probability of type I error such as $(1 - \varepsilon)$, where $\varepsilon = 0.05$ or 0.01, then we get the corresponding threshold d from the table of F-distributions. The decision rule can be written in terms of r^2 as given below:

$$r^2 \begin{cases} \leq d/(N - j - 1 + d), & \text{accept hypothesis} \quad \alpha_j = 0 \\ > d/(N - j - 1 + d), & \text{reject hypothesis} \quad \alpha_j = 0 \end{cases} \tag{9b.2.4}$$

Test 2: Based on the Likelihood Approach. We compute the modified likelihood values J_i, $i = 1, 2$, for class 1 $(\alpha_j = 0)$ and class 2 $(\alpha_j \neq 0)$:

$$J_1 = -\frac{N}{2} \ln\left(\frac{1}{N} \sum_{t=1}^{N} w_1^2(t)\right)$$

$$J_2 = -\frac{N}{2} \ln\left(\frac{1}{N} \sum_{t=1}^{N} w_1^2(t)(1 - r^2)\right) - 1.$$

We choose the class that gives the larger value of J.

9b.2.2. The Case with More Than Two Variables

Suppose there are other variables, say $y_2, y_3, \ldots$, for possible use in forecasting y_1. A natural step would be to consider each variable separately for testing causality by the tests described above. But we may come across a curious occurrence called multicollinearity (Johnston, 1963; Goldberger, 1968). Consider the three residual variables w_1, w_2, and w_3 obtained from y_1, y_2, and y_3. Suppose we find by the earlier test that w_2 and w_3 are individually significant in modeling w_1. But if we use a multiple regression for w_1, using the lagged values of both w_2 and w_3, then only one of them may be significant. Similarly, we may find that w_2 may not be significant by itself, but it may be significant in conjunction with another significant variable, say w_3. Hence we must try all the various combinations of the auxiliary variables to find the effective combination that will determine whether or not a group of auxiliary variables is significant. We proceed as follows (note that in the earlier case we tested for the significance of only one term). Let the given model be

$$w_1(t) = \sum_{j=1}^{n_2} \alpha_j w_2(t - j) + \sum_{j=1}^{n_3} \beta_j w_3(t - j) + \eta(t), \qquad n_2 > n_3, \tag{9b.2.5}$$

$$\text{Null hypothesis:} \quad \begin{cases} \alpha_j = 0, j = 1, \ldots, n_2 \\ \beta_j = 0, j = 1, \ldots, n_3. \end{cases}$$

We want to test for the validity of the null hypothesis. We can use any of the tests of Section 8b to do this. We will give only two tests, namely the tests based on the likelihood approach and hypothesis testing.

Test 3: Based on Hypothesis Testing. Let ρ_1 be the estimated mean square value of $w_1(\cdot)$ and ρ_2 be the estimated mean square value of the residual $\eta(t)$ when the null hypothesis is not true:

$$\rho_1 = \frac{1}{N} \sum_{t=n_2+1}^{N} w_1^2(t),$$

$$\rho_2 = \frac{1}{(N-n_2)} \sum_{t=n_2+1}^{N} \left[w_1(t) - \sum_{j=1}^{n_2} a_j w_2(t-j) - \sum_{j=1}^{n_3} b_j w_3(t-j) \right]^2,$$

where a_j, b_k are the estimates of α_j, β_k, $j = 1, \ldots, n_2$, $k = 1, \ldots, n_3$, in (9b.2.5), obtained by using the observations $w_i(t)$, $i = 1, 2, 3$; $t = 1, \ldots, N$. If the null hypothesis is true, then the statistic χ has the following F-distribution:

$$\chi = \left(\frac{\rho_1 - \rho_2}{\rho_2} \right) \left(\frac{N - n_2 - 1}{n_2 + n_3} \right) \sim F(n_2 + n_3, N - n_2 - 1).$$

We can devise a suitable decision rule as before after choosing a suitable significance level.

Test 4: Modified Likelihood Approach. We compute the likelihood indices J_1 and J_2 under the null hypothesis and alternate, respectively:

$$J_1 = -\frac{N}{2} \ln \rho_1, \qquad J_2 = -\left(\frac{N - n_2}{2} \right) \ln \rho_2 - (n_2 + n_3).$$

We choose the class that yields the larger value of J.

9b.3. Choice of a Preliminary Model

We will give a method for the selection of a preliminary model for a variable, say y_1, when the history of two other variables, say y_2 and y_3, is available.

Step (i). Construct the best possible univariate model for every variable y_i, $i = 1, 2, 3$, involving the past history of that variable only. Let the corresponding whitened processes be labeled w_i, $i = 1, 2, 3$.

Step (ii). Consider a model

$$w_1(t) = \sum_{j=1}^{n_1} \alpha_j w_1(t-j) + \sum_{j=1}^{n_2} \beta_j w_2(t-j) + \sum_{j=1}^{n_3} \gamma_j w_3(t-j) + \eta(t).$$

Choose the appropriate combination of the lagged terms of $w_1(\cdot)$, $w_2(\cdot)$, and $w_2(\cdot)$ by the tests described earlier. Note that if all β_j and γ_j are not significant, α_j will not be significant since $\{w_1(t)\}$ is a white sequence. But some of the α_j may be significant when some of the β_j or γ_j are significant.

Step (iii). We can express the whitened variables w_i in terms of the variables y_i, $i = 1, 2, 3$.

We may repeat that for carrying out step (ii) we can use tests such as the one based on the recursive prediction described in Section 8b.3 in addition to those given in the preceding section.

9c. Direct Comparison of Multivariate Classes of Models

In this section we will develop methods for comparing r classes of multivariable models, namely $C_i = \{\mathscr{M}_i, \mathscr{H}_i, \Omega_i\}$, $i = 0, 1, \ldots, r - 1$. This section is similar to Section 8b where similar topics for single equation models were considered. The classes C_i can be chosen either by physical knowledge or intuition or on the basis of a preliminary model constructed earlier and the associated causality relations. As in Section 8b, we can give a number of methods, but we will discuss only a few of them.

The preliminary model constructed earlier in Section 9b-3 is not satisfactory in two ways. Since it has been constructed by considering the equations for the variables y_i, $i = 1, 2, 3, \ldots$, separately, the parameter estimates may not be efficient. Second, only direct comparison with other types of models can bring out the strengths and deficiencies of the various models.

9c.1. The Likelihood Approach

We can define the likelihood value associated with a class of multivariate models just as in the case of univariate models. Since Theorem 8b.1 is valid for multivariate observations as well, we can write down the following expression for the likelihood value associated with the class $C = \{\mathscr{M}, \mathscr{H}, \Omega\}$:

$$L = \ln p(\boldsymbol{\xi}; \hat{\boldsymbol{\phi}}) - n_{\boldsymbol{\phi}},$$

where $\boldsymbol{\xi}$ is the available history of observations, $p(\boldsymbol{\xi}, \boldsymbol{\phi})$ is the probability density of $\boldsymbol{\xi}$ characterized by $\boldsymbol{\phi} = (\boldsymbol{\theta}, \boldsymbol{\rho})$, $\boldsymbol{\theta} \in \mathscr{H}$ and $\boldsymbol{\rho} \in \Omega$, $\hat{\boldsymbol{\phi}}$ is the CML estimate of $\boldsymbol{\phi}^0$ based on $\boldsymbol{\xi}$, and $n_{\boldsymbol{\phi}}$ is the dimension of $\boldsymbol{\phi}$.

Decision Rule. Evaluate L for the various classes C_i, $i = 0, 1, \ldots, r - 1$, and choose that class which yields the maximum value of L. (If more than one class yields the same value of L, we need a secondary tie-breaking rule.)

Our next task is to obtain simplified expressions for L for certain classes of models. We will give only the final results since the derivations are similar to the corresponding ones in Chapter VIII.

Case (i): Nontransformed Equations. The equation $\mathscr{M}$ for the multivariate system has the following form which must be in *some* canonical form:

$$\mathscr{M}: \mathbf{y}(t) = \boldsymbol{\theta}^{\mathrm{T}}\mathbf{x}(t-1) + \mathbf{w}(t),$$

where $\boldsymbol{\theta} \in \mathscr{H}$, $\boldsymbol{\rho} \in \Omega$, $\boldsymbol{\rho}$ is the covariance matrix of $\mathbf{w}(\cdot)$, and $\mathbf{x}(t-1)$ is made up of $\mathbf{y}(t-j), j = 1, \ldots, m_1$, $\mathbf{w}(t-j), j = 1, \ldots, m_2$, and so on. $\mathbf{x}(t-1)$ can also have some components of $y_i(t)$ for appropriate i in the case of canonical forms. Let $\hat{\boldsymbol{\rho}}$ be the CML estimate of $\boldsymbol{\rho}$. Then

$$L = -\frac{N}{2} \ln \det \hat{\boldsymbol{\rho}} - n_{\boldsymbol{\phi}}, \qquad n_{\boldsymbol{\phi}} = \text{dimension of } (\boldsymbol{\theta}, \boldsymbol{\rho}).$$

Case (ii): Partially Transformed Equations. Suppose the equation $\mathcal{M}$ is as follows in some canonical form:

$$\mathcal{M}: \mathbf{f}_0(\mathbf{y}(t)) = \boldsymbol{\theta}^{\mathrm{T}}\mathbf{x}(t-1) + \mathbf{w}(t)$$

$$\mathbf{f}_0(\mathbf{y}(t)) = (\ln y_1(t), \ln y_2(t), \ldots, \ln y_{n_1}(t), y_{n_1+1}(t), \ldots, y_m(t))^{\mathrm{T}},$$

where $\mathbf{x}(\cdot)$ is the same as in Case (i). The vector $\boldsymbol{\theta} \in \mathcal{H}$ and $\boldsymbol{\rho} \in \Omega$, $\boldsymbol{\rho}$ being the covariance matrix of $\mathbf{w}(t)$, and $1 \leq n_1 \leq m$. Let $\boldsymbol{\rho}$ be the CML estimate of $\boldsymbol{\rho}$ based on the given observations $\mathbf{y}(j)$, $1 \leq j \leq N$. Then the class likelihood has the form

$$L = -\frac{N}{2} \ln \det \hat{\boldsymbol{\rho}} - N \sum_{i=1}^{n_1} \hat{E}\{\ln y_i(t)\},$$

where

$$\hat{E} \ln y_i(t) = \frac{1}{N} \sum_{t=1}^{N} \ln y_i(t).$$

Summing up, the likelihood approach enables us to compare entirely different classes which may be in different canonical forms and which may have entirely different structures. Moving average terms can also be present in the equations.

9c.2. The Recursive Prediction Approach

This method can be used to compare any two classes of models which are in the same canonical form and *do not* have any moving average terms. In addition, this method is very helpful in tracking down any time-varying parameters in the system and in exposing the deficiencies in the model.

Suppose we want to compare r classes $C_i = \{\mathcal{M}_i, \mathcal{H}_i, \Omega_i\}$, $i = 0, 1, \ldots, r-1$. Let the equation $\mathcal{M}_i$ of any one of them be

$$\mathcal{M}_i: \mathbf{f}_0^{(i)}(\mathbf{y}(t)) = \mathbf{x}_i^{\mathrm{T}}(t-1)\boldsymbol{\theta}_i^0 + \mathbf{w}(t).$$

The function $\mathbf{f}_0^{(i)}(\cdot)$ need not be the same in all equations of the classes.

Let $\hat{\boldsymbol{\theta}}_i(t)$ be the least squares estimate of the parameter $\boldsymbol{\theta}_i^0$ in the equation $\mathcal{M}_i$ based on the observations $\mathbf{y}(j), j \leq t$, and let $\hat{\mathbf{y}}_i(t|t-1)$ be the linear least squares of predictor of $\mathbf{y}(t)$ given $\mathbf{y}(j), j \leq (t-1)$, and that the process $\mathbf{y}$ obeys equation $\mathcal{M}_i$ with $\boldsymbol{\theta}_i^0$ replaced by the estimate $\hat{\boldsymbol{\theta}}_i(t-1)$. The statistic used for comparing the classes is J_i:

$$J_i = \sum_{t=n_1}^{N} \|\mathbf{y}(t) - \hat{\mathbf{y}}_i(t|t-1)\|^2, \qquad i = 0, 1, \ldots, r-1.$$

We choose the class that yields the smallest value of J. Note that the parameter estimate $\hat{\boldsymbol{\theta}}_i(t-1)$ can be computed in real time by using algorithm (C3) of Chapter VI.

9c.3. Hypothesis Testing

As in Chapter VIII, the hypothesis testing approach is limited to comparing pairs of AR models as in

$$\mathbf{y}(t) = \mathbf{A}_0 + \sum_{j=1}^{n} \mathbf{A}_j \mathbf{y}(t-j) + \mathbf{w}(t) \tag{9c.3.1}$$

$$\mathbf{y}(t) = \mathbf{A}_0 + \sum_{j=1}^{n+1} \mathbf{A}_j \mathbf{y}(t-j) + \mathbf{w}(t), \qquad \mathbf{A}_{n+1} \neq 0, \tag{9c.3.2}$$

which differ from one another by an AR term. There could be, at most, some additional deterministic trend terms common to both (9c.3.1) and (9c.3.2). We cannot compare a log transformed model with a nontransformed model as in Section 9c.1. The advantage of this approach is that we obtain a quantitative expression for the error probability caused by this decision rule.

9c.3.1. The Multivariate Version of the Variance Ratio Test

We will compare two autoregressive models of order n and $n + 1$ in (9c.3.1) and (9c.3.2). Depending on the causal connections between the variables, some of the elements of the matrices $\mathbf{A}_j$ may be identically zero so that all the elements of all the matrices $\mathbf{A}_j$ need not always be estimated.

Let $\overline{\mathbf{A}}_j$, $j = 0, \ldots, n$, be the CML estimates of $\mathbf{A}_j$, $j = 0, 1, \ldots, n$, in (9c.3.1), and $\mathbf{A}_j^*$, $j = 0, \ldots, n - 1$, be the CML estimates of $\mathbf{A}_j$, $j = 0, 1, \ldots, n + 1$, in (9c.3.2) based on the same set of observations $\{\mathbf{y}(1), \ldots, \mathbf{y}(N)\}$. Let

$$\mathbf{w}'(t) = \mathbf{y}(t) - \overline{\mathbf{A}}_0 - \sum_{j=1}^{n} \overline{\mathbf{A}}_j \mathbf{y}(t-j)$$

$$\mathbf{w}^*(t) = \mathbf{y}(t) - \mathbf{A}_0^* - \sum_{j=1}^{n+1} \mathbf{A}_j^* \mathbf{y}(t-j)$$

$$\boldsymbol{\rho}_n = \frac{1}{N-n} \sum_{j=n+1}^{N} \mathbf{w}'(j)(\mathbf{w}'(j))^{\mathrm{T}}$$

$$\boldsymbol{\rho}_{n+1} = \frac{1}{N-(n+1)} \sum_{j=n+2}^{N} \mathbf{w}^*(j)(\mathbf{w}^*(j))^{\mathrm{T}}.$$

Let p_n and p_{n+1} be the number of coefficients estimated in the nth and $(n + 1)$th order, respectively. Consider the following statistic χ:

$$\chi = \frac{\det \boldsymbol{\rho}_n - \det \boldsymbol{\rho}_{n+1}}{\det \boldsymbol{\rho}_{n+1}} \, \frac{N-(n+1)-p_{n+1}}{p_{n+1}-p_n}.$$

If the nth-order model is correct, then χ is claimed to obey the following distribution:

$$\chi \sim F(p_{n+1} - p_n, N - (n+1) - p_{n+1}). \tag{9c.3.3}$$

We can choose a suitable threshold based on the F-distribution. We can accept the validity of the nth-order model if the numerical value of χ is below the threshold, and reject it otherwise.

The claim in Eq. (9c.3.3) needs additional consideration. A complete proof of (9c.3.3) does not seem to exist. However, it is true under certain conditions. For instance, Eq. (9c.3.3) follows from the multivariate regression theory (Appendix 9.1) if we use the whitened observations $w_1, w_2, \ldots$, in place of the actual observations $y_1, y_2, \ldots$ (i.e., $\{w_1(t)\}$ is the sequence obtained by whitening the sequence $\{y_1(t)\}$, usually by a scalar AR process). Once we get the model in terms of the whitened variables, we can obtain the model in terms of the variables $y_1, y_3, \ldots$, by simple manipulation.

This method can be used to determine the appropriate order of the AR model. Beginning with $n = 1$, n is increased until it is found that further increase results in the inability of the model to pass the above-mentioned test. However, it should be noted that the model obtained by working with the whitened observations and then converting the whitened variables into the original variables y_i may be sometimes considerably different from the AR model obtained by using the observation y_i directly. Often, the model obtained from the whitened observations is of higher order than the one obtained directly. The reason for this is simple. Suppose n_1 is the order of the AR model for the whitened vector variable $\mathbf{w}(\cdot)$. Usually n_1 is not large. But when we replace the whitened variables $\mathbf{w}$ in terms of the direct variable $\mathbf{y}$, the order of the model in terms of the y_i variables is $(n_1 + n_2)$, where n_2 is the maximum order of the univariate AR models among the univariate AR models of the variables $y_1, y_2, \ldots$.

9c.3.2. Partial Autocorrelation and Related Tests

Suppose $\mathbf{y}(\cdot)$ obeys the nth-order AR process in (9c.3.1). Then it can be shown (Quenouille, 1957) that $\mathbf{y}(\cdot)$ also obeys another nth-order AR model:

$$\mathbf{y}(t) = \mathbf{B}_0 + \sum_{j=1}^{n} \mathbf{B}_j \mathbf{y}(t+j) + \boldsymbol{\eta}(t), \tag{9c.3.4}$$

where the $\{\boldsymbol{\eta}(\cdot)\}$ sequence is also a zero mean sequence and is independent of $\mathbf{y}(t+j)$, $j > 0$. It is customary to call (9c.3.4) the forward AR process to distinguish it from the usual or backward AR process in (9c.3.1). By definition $\mathbf{w}(t)$ is independent of $\mathbf{y}(t-j)$, $j > 0$. Hence $\mathbf{w}(t+n+s)$ is independent of $\boldsymbol{\eta}(t)$ for all $s > 1$ since $\boldsymbol{\eta}(t)$ is a linear function of the observations $\mathbf{y}(t)$, $\mathbf{y}(t+1)$, $\ldots$, $\mathbf{y}(t+n)$. Thus, a method of testing the validity of the order of AR processes is to check whether the two sequences $\{\mathbf{w}(t+n+s), t = 1, 2, \ldots\}$ and $\{\boldsymbol{\eta}(t), t = 1, 2, \ldots\}$ are uncorrelated for every $s \geq 1$. We can start with $n = 1$ and increase n until we find that these two sequences are uncorrelated. The numerical procedure can be summarized as follows (Quenouille, 1957):

1. Start with $n = 1$.
2. Fit an nth-order backward AR process as in (9c.3.1) to the observations $\mathbf{y}(1), \ldots, \mathbf{y}(N)$ and generate the residuals $\mathbf{w}(t)$, $t = 1, 2, \ldots$.

3. Fit an nth-order forward AR process as in (9c.3.4) and generate the residuals $\{\boldsymbol{\eta}(t)\}$.

4. Normalize the residual sequences $\{\mathbf{w}(\cdot)\}$ and $\{\boldsymbol{\eta}(\cdot)\}$ so that their empirical means are zero and empirical covariances form a unit matrix. In order to do this, let

$$\mathbf{P}_1 = \frac{1}{N-n}\sum_{t=n+1}^{N} \mathbf{w}(t)\mathbf{w}^{\mathrm{T}}(t), \qquad \mathbf{P}_2 = \frac{1}{N-n}\sum_{t=1}^{N=n} \boldsymbol{\eta}(t)\boldsymbol{\eta}^{\mathrm{T}}(t).$$

Factor the matrices $\mathbf{P}_1$ and $\mathbf{P}_2$ in terms of nonsingular triangular matrices $\mathbf{C}_1$ and $\mathbf{C}_2$:

$$\mathbf{P}_1 = \mathbf{C}_1\mathbf{C}_1^{\mathrm{T}}, \qquad \mathbf{P}_2 = \mathbf{C}_2\mathbf{C}_2^{\mathrm{T}}.$$

Let

$$\mathbf{w}'(t) = \mathbf{C}_1^{-1}\mathbf{w}(t), \qquad \boldsymbol{\eta}'(t) = \mathbf{C}_2^{-1}\boldsymbol{\eta}(t).$$

Then the empirical means and covariance matrices of $\{\mathbf{w}'(\cdot)\}$ and $\{\boldsymbol{\eta}'(\cdot)\}$ are $\mathbf{0}$ and $\mathbf{I}$.

5. Find the empirical cross covariance matrix $\mathbf{R}_s$:

$$\mathbf{R}_s = \frac{1}{N-n-s}\sum_{t=1}^{N-n-s} \mathbf{w}(t+n+s)\boldsymbol{\eta}^{\mathrm{T}}(t).$$

Then for large N, each element of $\mathbf{R}_s$ is an independent normal $N(0, 1)$ variable.

Let

$$\chi = \sum_{i,j=1}^{m} (\mathbf{R}_s)_{ij}^2.$$

Then if the order of the AR process is not greater than n, χ is distributed as a $\chi^2(m^2)$ distribution for every $s = 1, 2, \ldots$. Beginning with $n = 1$, test whether the observed matrices $\mathbf{R}_s$ can be considered to have come from a $\chi^2(m^2)$ distribution for $s = 1$. If so, $n = 1$ is the correct order. If not, increase n until we find that the corresponding statistic has the appropriate value. We can avoid the factorization of $\mathbf{P}_1$, $\mathbf{P}_2$ into triangular matrices by working in canonical form III. Then the residuals will automatically have zero mean and *diagonal* covariance matrix if the model is correct. Normalization of the residuals to have unit covariance matrix is trivial.

The case of $s = 1$ is usually called a partial autocorrelation test because the matrix $\mathbf{R}_1$ can be considered the partial autocovariance matrix between $\mathbf{y}(t)$ and $\mathbf{y}(t+n+1)$ after removing the effect of the terms $\mathbf{y}(t+1), \ldots, \mathbf{y}(t+n)$. In other words, we fit a regression for $\mathbf{y}(t+n+1)$ in terms of the $\mathbf{y}(t+n), \ldots, \mathbf{y}(t+1)$ and obtain the residuals $\mathbf{w}(\cdot)$. Similarly, we fit a regression of $\mathbf{y}(t)$ in terms of $\mathbf{y}(t+1), \ldots, \mathbf{y}(t+n)$ to obtain the residuals $\boldsymbol{\eta}(t)$. Then the covariance matrix $\mathbf{R}_1$ measures only the correlation between $\mathbf{y}(t)$ and $\mathbf{y}(t+n+1)$ after removing the effect of the intermediate terms $\mathbf{y}(t+1), \ldots, \mathbf{y}(t+n)$.

The case $s = 1$ seems to give good results. There are some problems with the tests with $s = 2, 3, \ldots$, which are usually referred to as Bartlett–Rajalaksman (1953) tests.

In the test described in this section, the probability distribution of the stitistic χ is not based on any heuristic arguments as in Section 9c.3.1.

9c.4. Discussion of the Various Comparison Methods

The advantages and disadvantages of the three methods given here follow the corresponding discussion in Chapter VIII. The likelihood approach is versatile and can be defended on Bayesian grounds as well. By using the likelihood approach different classes of models in different canonical forms can be compared. It involves the determinant of $\boldsymbol{\rho}$, which theoretically is the same for all canonical forms, and $n_{\boldsymbol{\phi}}$, the number of parameters. Hence the likelihood approach clearly indicates the compromises involved between the accuracy lost by the addition of new parameters and the increased quality of fit because of the addition of new parameters.

The real time prediction scheme is important not only for selection purposes but also for validation purposes. The only additional comment that need be made on the hypothesis testing method is that the probability distribution of the statistic mentioned is valid for large N, but it is not known how large N should be.

9d. Validation of Models

As in univariate models, we must validate the model both by testing the residuals for whiteness and by direct comparison of the output characteristics of the model with those of the observed data. We will discuss residual testing at some length.

Suppose $\mathbf{x}'(1), \ldots, \mathbf{x}'(N)$ are the N residual vectors. We can normalize them so that they have zero empirical mean and unit empirical covariance matrix. The normalized sequence is labeled $\mathbf{x}(1), \ldots, \mathbf{x}(N)$. We can consider the scalar sequences $\{x_k(t), t = 1, \ldots, N\}$, $k = 1, \ldots, m$, and apply the whiteness tests of Chapter VIII. In addition, we need to check the cross correlations between the different residual variables.

9d.1. Residual Tests

9d.1.1. Residual Tests Based on Time Domain Methods

Test 1. We can compute the empirical cross correlations

$$\frac{1}{N-k} \sum_{t=k+1}^{N} x_i(t) x_j(t-k)$$

for various i, j, and k, $i, j = 1, \ldots, m$, $i \neq j$, and $k = 0, 1, 2, \ldots$, and check whether they are significantly different from zero by using the test in Eq. (9b.2.3). Alternatively, we can simply check whether they are within the two standard deviation limits, i.e., within $\pm 2/\sqrt{N}$ limits of the zero value.

Test 2. This is the multivariate version of the test in Section 8c.2.3. Here we deal directly with the vector residuals $\mathbf{x}(1), \ldots, \mathbf{x}(N)$.

$$\begin{aligned} &\text{Null hypothesis:} && \mathbf{x}(t) = \boldsymbol{\eta}(t) \quad \text{(white noise)} \\ &\text{Alternate hypothesis:} && \mathbf{x}(t) \text{ obeys an AR process of order } n_1. \end{aligned}$$

We can use the test in Section 9c.3.1 to compare the two hypotheses. Let $\boldsymbol{\rho}_0$ be the covariance matrix of $\mathbf{x}(t) = \mathbf{I}$, and $\boldsymbol{\rho}_1$ the covariance of the matrix of residuals $\boldsymbol{\eta}(\cdot)$ after fitting an n_1-order AR model to the sequence $\{\mathbf{x}(i)\}$. If the null hypothesis is true, then

$$\frac{\det \boldsymbol{\rho}_0 - \det \boldsymbol{\rho}_1}{\det \boldsymbol{\rho}_1} \cdot \frac{N - 2n_1}{n_1 m^2} \sim F(n_1 m^2, N - 2n_1).$$

We can easily design a threshold after choosing a suitable significance level.

9d.1.2. Residual Tests Based on Spectral Methods

We will give spectral tests for determining whether two sequences $\{x_1(t)\}$ and $\{x_2(t)\}$ are uncorrelated. We need the following definitions in these tests. Let

$$\left.\begin{aligned} A_i(f) &= \sum_{t=1}^{N} x_i(t) \cos 2\pi f t, \\ B_i(f) &= \sum_{t=1}^{N} x_i(t) \sin 2\pi f t, \end{aligned}\right\} i = 1, 2, \qquad f = 1/N, 2/N, \ldots .$$

The empirical cross spectral density is given by

$$S_{12}(f) = (A_1 + jB_1)(A_2 - jB_2) = A_1A_2 + B_1B_2 + j(B_1A_2 - B_2A_1).$$

The phase of the spectral density is

$$F_{12}(f) = \tan^{-1} \frac{B_1A_2 - A_1B_2}{A_1A_2 + B_1B_2},$$

and the cospectrum is

$$L_{12}(f) = A_1A_2 + B_1B_2.$$

Test 3: Integrated Phase Test. This test is similar to the cumulative periodogram test in Section 8c.2.2. If x_1 and x_2 are uncorrelated, it can be shown (Jenkins and Watts, 1968) that $F_{12}(f)$ is approximately uniformly distributed in the range $[-\pi/2, \pi/2]$. Let

$$J_k = \sum_{j=1}^{k} F_{12}(f_j), \qquad f_j = \frac{j}{N}.$$

The graph of cumulative phase spectrum J_k vs. f_k should be a straight line if the two sequences are not correlated. Since $F_{12}(f)$ can assume both positive and negative values, the slope of the plot of the cumulative phase spectrum could be either $+1$ or -1. To judge the deviation from the straight line, the 95 and 75% confidence limits can be inserted at distances $\pm 1.36/\sqrt{N}$ and $\pm 1.02/\sqrt{N}$ from the theoretical cumulative spectrum.

Test 4: The Integrated Cospectrum Test

$$J_{12}(f_k) = \frac{2}{N} \sum_{j=1}^{k} L_{12}(f_j), \qquad k = 1, 2, \ldots.$$

If the process x_1 and x_2 are uncorrelated, the expected value of J_{12} should be zero for all f. Hence the graph of $J_{12}(f_k)$ vs. f_k must be a line close to the horizontal (or f) axis, and intertwined around it (Jenkins and Watts, 1968).

9d.2. Other Validation Tests

As in univariate models, the residual tests presented in Section 9d.1 are weak. The fact that a model does not pass the validation tests indicates that there is some gross error in the model. But the fact that a model passes the validation tests does not imply that it is automatically satisfactory. In particular, a constant coefficient model may pass all the validation tests mentioned above. A close analysis may still reveal that the coefficients vary considerably in time. Hence, we have to compare all the models that pass the validation tests before arriving at the final model. We cannot exclude the possibility of more than one model among several models, which are considerably different from one another, passing all the validation tests. To make sure that the constant coefficient model is appropriate, one should use the recursive (real time) parameter estimation algorithm (C3) to get the estimate $\hat{\boldsymbol{\theta}}(t)$, the estimate of $\boldsymbol{\theta}^0$ based on the observation history until time t. We can check whether the variation in the successive estimates $\hat{\boldsymbol{\theta}}(n_1), \ldots, \hat{\boldsymbol{\theta}}(N)$ is consistent with the corresponding theoretical standard deviations of the estimates.

We can directly compare the various statistical characteristics of the output of the model such as the covariance matrices and spectral density matrices, with the corresponding (matrix) characteristics of the empirical data. Such procedures are similar to those discussed in Chapter VIII.

9e. Conclusions

We have given a brief account of some of the validation procedures for constructing models of the multivariate time series, limiting ourselves to the exposition of those techniques that have been known to give adequate results in practice. It is possible to obtain a preliminary multivariate model by constructing the corresponding univariate models. This preliminary model suggests the various classes of models which have to be compared by using the class selection methods. We have also discussed at some length the problem of causality and the problem of the choice of appropriate variables in each equation. Direct methods are given for the comparison of the multivariate models and for the validation of the multivariate model.

There are other validation methods that are useful in individual problems. We have not discussed them because of lack of adequate information concerning their use.

Appendix 9.1. Geometry of Correlation and Regression

We will use a slightly different notation. The scalar random variables will be denoted by the upper case letters $Y, X_1, X_2, \ldots, X_m$. The corresponding lower case letters are the observations of the corresponding random variables. There are three types of correlation coefficients used in data analysis, namely the simple correlation coefficient, the partial autocorrelation coefficient, and the multiple correlation coefficient. The simple or ordinary correlation coefficient is considered only between pairs of variables and denoted by $r_{YX_{m+1}}$ for the pair (Y, X_{m+1}). The partial autocorrelation coefficient, say $r_{YX_{m+1}|X_1,\ldots,X_m}$, is also defined over a pair of variables (Y, X_{m+1}) given m other variables $X_1, \ldots, X_m$. The multiple correlation coefficient $R_{YX_1,\ldots,X_{m+1}}$ is defined for any variable, say Y, and a set of other variables $X_1, \ldots, X_{m+1}$. All these quantities can be defined in geometric terms (Wonnacot and Wonnacot, 1970; Rao, 1965). $Y, X_1, \ldots, X_m$ are zero mean.

Suppose we have N IID observations $(y_i, x_{1i}, \ldots, x_{mi})$, $i = 1, \ldots, N$. Let

$$\mathbf{y} = (y_1, \ldots, y_N)^{\mathrm{T}}, \qquad \mathbf{x}_i = (x_{i1}, \ldots, x_{iN})^{\mathrm{T}}, \qquad i = 1, \ldots, m.$$

Let all observations be normalized so that the sum of the squares of the N components is unity. Let the N-vectors $\mathbf{y}, \mathbf{x}_1, \mathbf{x}_2, \ldots, \mathbf{x}_{m+1}$ in the N-dimensional Euclidean space be the random variables $Y, X_1, \ldots, X_{m+1}$, and let $r_{YX_{m+1}}$ be the inner product of $\mathbf{y}$ and $\mathbf{x}_{m+1}$,

$$r_{YX_{m+1}} = \sum_{i=1}^{N} y_i x_{m+1,i}.$$

Let $\bar{\mathbf{y}}$ be the orthogonal projection of $\mathbf{y}$ on the space spanned by $\mathbf{x}_1, \ldots, \mathbf{x}_m$, $\hat{\mathbf{y}}$ the orthogonal projection of $\mathbf{y}$ on the space spanned by $\mathbf{x}_1, \ldots, \mathbf{x}_{m+1}$, and $\bar{\mathbf{x}}_{m+1}$ the orthogonal projection of $\mathbf{x}_{m+1}$ on the space spanned by $\mathbf{x}_1, \ldots, \mathbf{x}_m$. Let $R_{YX_1,\ldots,X_{m+1}}$ be the ordinary correlation coefficient between $\mathbf{y}$ and $\hat{\mathbf{y}}$ and $r_{YX_{m+1}|X_1,\ldots,X_m}$ be the ordinary correlation coefficient between $(\mathbf{y} - \bar{\mathbf{y}})$ and $(\mathbf{x}_{m+1} - \bar{\mathbf{x}}_{m+1})$. It is easy to show that

$$|r_{YX_{m+1}}| \le |R_{YX_1,\ldots,X_{m+1}}|,$$
$$|r_{YX_{m+1}|X_1,\ldots,X_m}| \le |R_{YX_1,\ldots,X_{m+1}}|$$

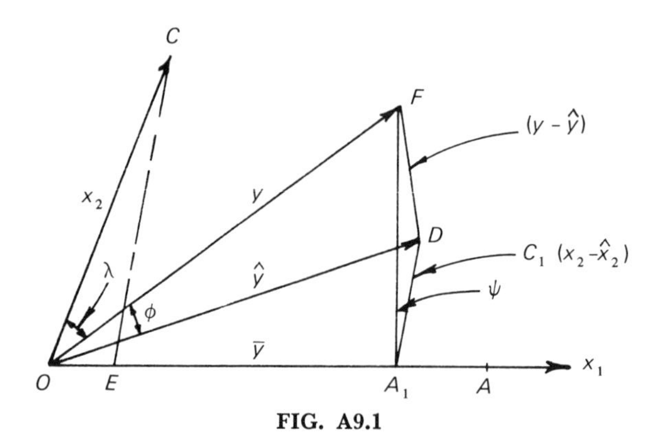

FIG. A9.1

Figure A9.1 gives a geometric interpretation of the various correlation coefficients for the case $n = 1$. Let $\mathbf{x}_1$ and $\mathbf{x}_2$ be represented along OA and OC. Let $OF = \mathbf{y}$. OA_1 is the orthogonal projection of $\mathbf{y}(OF)$ on $OA = \bar{\mathbf{y}}$, OD is the orthogonal projection of $\mathbf{y}$ on the plane $OAC = \hat{\mathbf{y}}$, OE is the orthogonal projection of $\mathbf{x}_2(OC)$ on $OA = \bar{\mathbf{x}}_2$, and $EC = \mathbf{x}_2 - \hat{\mathbf{x}}_2$. Let A_1D be parallel to EC. Then $r_{YX_2} = \cos\lambda$, $R_{YX_1X_2} = \cos\phi$, $r_{YX_2|X_1} = \cos\psi$.

We will also give a geometric interpretation of the test for the appropriateness of a regression term. Suppose we want to test the appropriateness of the regressor $\mathbf{x}_{n+1}$ for $\mathbf{y}$, having already chosen regressors $\mathbf{x}_1, \ldots, \mathbf{x}_n$. For the case $n = 1$, we have, from triangle A_1DF in Fig. A9.1,

$$\|\mathbf{y} - \bar{\mathbf{y}}\|^2 = \|\mathbf{y} - \hat{\mathbf{y}}\|^2 + (A_1D)^2 = \|\mathbf{y} - \hat{\mathbf{y}}\|^2 + C_1{}^2\|\mathbf{x}_2 - \hat{\mathbf{x}}_2\|^2,$$

where $C_1{}^2$ is a suitable constant. In general, an unexplained variation after Y is regressed on $X_1, \ldots, X_m$ equals the unexplained variation after Y is regressed on $X_1, \ldots, X_n$ plus the variation explained by X_{m+1}.

$$\text{The statistic } F = \frac{(\text{additional variation explained by introducing } X_{m+1})/1}{(\text{residual variation after fitting } X_1, \ldots, X_{m+1})/(N - m - 2)}.$$

We can explain the test in terms of the partial autocorrelation coefficient $r_{YX_{m+1}|X_1,\ldots,X_m}$. Let $r_{YX_{m+1}|X_1,\ldots,X_m} = \cos\psi$. Then by construction, the test statistic

$$F \triangleq (\cot\psi)^2(N - m - 2) = \frac{r^2_{YX_{m+1}|X_1,\ldots,X_m}(N - m - 2)}{1 - r^2_{YX_{m+1}|X_1,\ldots,X_m}}.$$

Notes

The literature on the validation of multivariate models is not as extensive as that of univariate models, and detailed empirical studies are scarce. Quenouille (1957) gives a detailed study of modeling agricultural data. An idea of the methods used in econometrics for model verification can be obtained from

Klein and Evans (1968). The problem of causality has been discussed by many authors, including Wold (1954), Granger (1963), Sims (1972), Caines and Chan (1974), and others. However, all these papers assume the validity of a constant coefficient model which is not often warranted.

Problems

1. Consider a pair of processes (y, u) obeying (*) which is in canonical form II:

$$\mathbf{A}(D)\begin{bmatrix} y \\ u \end{bmatrix} = \mathbf{B}(D)\mathbf{w} \qquad (*)$$

where $\mathbf{w}(t) = (w_1(t), w_2(t))^{\mathrm{T}}$, $\operatorname{cov}[\mathbf{w}(t)] = \boldsymbol{\rho}$ is nondiagonal,

$$\mathbf{A}(D) = \begin{bmatrix} A_{11}(D) & A_{12}(D) \\ A_{21}(D) & A_{22}(D) \end{bmatrix}, \qquad \mathbf{B}(D) = \operatorname{diag}(b_{11}(D), b_{22}(D))$$

$$\mathbf{A}(0) = \mathbf{I}, \qquad \mathbf{B}(0) = \mathbf{I}.$$

If $A_{21}(D) = 0$, then there is said to be no feedback from y to u. Show that $A_{21}(D) = 0$ if the linear least squares filter of $y(t)$ based on all $u(t-j), j \geq 0$, is identical to the linear least squares filter of $y(t)$ given all $u(t-j), j$ arbitrary.

2. Use the results of Problem 1 to design a test to determine whether $A_{21}(D) = 0$ using only the observation $[y(t), u(t), t = 1, \ldots, N]$.

3. Simulate a bivariate AR(3) model excited by a normal noise. Using only 100 observations from this simulated output obtain the best fitting AR model by the methods of Sections 9c.3.1 and 9c.3.2 and the class likelihood approach. Repeat the experiment with different sets of data from the same model and determine the approach which gives the AR model of correct lag more often than others.

4. Consider a weak stationary AR process $\mathbf{y}$ obeying (**) with covariance matrices $\mathbf{R}_s$, $s = 0, 1, 2$:

$$\mathbf{y}(t) = \sum_{j=1}^{n} \mathbf{A}_j \mathbf{y}(t-j) + \mathbf{w}(t). \qquad (**)$$

Show that y also obeys the "forward" AR process (***)

$$\mathbf{y}(t) = \sum_{j=1}^{n} \mathbf{B}_j \mathbf{y}(t+j) + \boldsymbol{\eta}(t), \qquad (***)$$

where $\{\boldsymbol{\eta}(\cdot)\}$ is also a white noise sequence with $E[\boldsymbol{\eta}(t)\boldsymbol{\eta}^{\mathrm{T}}(t)] = \boldsymbol{\rho}'$. Show that $\mathbf{A}_j$ and $\mathbf{B}_k$ are related to $\mathbf{R}_s$ as follows:

$$\mathbf{R}_s = \sum_{i=s}^{n} \mathbf{A}_i \mathbf{R}_{s-i} = \sum_{i=1}^{n} \mathbf{R}_{s-i} \mathbf{B}_i^{\mathrm{T}}, \qquad s = 1, \ldots, n.$$

Find an expression for $\boldsymbol{\rho}'$.

Chapter X | Modeling River Flows

10a. The Need and Scope of Modeling

We will consider several aspects of modeling a river flow series by a stochastic difference equation. There are a number of reasons for such a study. The river flow process is a very complex process that involves a number of primary as well as auxiliary variables such as rainfall, groundwater, and geomorphologic characteristics of the basin. Purely deterministic models of the rainfall-runoff process have not been very successful because of the inherent randomness of the variables affecting runoff. Many of the important variables, such as the geomorphological characteristics of watersheds, cannot be adequately incorporated into such models of runoff. Because of the complicated nature of the process, especially in large watersheds, it is hard to quantify the contribution of the rainfall (as measured by a single raingauge or even by a sparse raingauge network) to the flow. The time histories of related variables such as groundwater flows or rainfall are needed for computing the coefficients of the deterministic input–output model, and these are often not available. Finally, successful modeling of the river flow process by a stochastic difference equation is a good demonstration of the usefulness of the model building techniques discussed in this book.

Moreover, there have been suggestions in the literature that the river flow processes cannot be described by a difference equation model. This comment is based on the phenomenon of "persistency" in the flows which can be described as follows. In many annual flow or rainfall sequences it is not uncommon to see clusters of dry years and wet years; i.e., there appears to be a tendency for wet or dry years to persist. (A number of consecutive years during which the annual flows or rainfall are greater than the long-term average annual flow or rainfall defines a cluster or sequence of wet years. Similarly, if a sequence of annual flows or rainfall values is less than the corresponding long-term average values, it is defined as a dry sequence.) It has been suggested that a phenomenon with persistance cannot be described by any process obeying the stochastic difference equation models described in this book. It is further claimed that such processes can be described by the so-called fractional noise models described in Chapter II. Thus, we have an excellent opportunity for using the model validation methods described in Chapter VIII to check the validity of the assertion given above.

If a successful model can be developed for a river flow process, it will serve a variety of purposes:

(i) Single- and multiple-step-ahead forecasts. (The unit of time may be a day, month, or year.) Such forecasts are needed in the operation of reservoirs, agricultural planning, and the like.

(ii) Generation of synthetic flow data, i.e., generation of data by simulating the flow model on a digital computer such that the simulated or synthetic flow data possess all the dominant statistical characteristics such as spectra and correlograms, exhibited by the observed data. The synthetic data sequences are needed in the simulation studies for the storage design of dams and other devices and also for testing their reliability.

Statistical characteristics of flows in rivers in different climatic zones significantly differ from each other. A desirable property in the family of models chosen for modeling river flow sequences is that it must be general enough so that flows in rivers in different climatic zones can be modeled by a member in the family. Furthermore, it is interesting to inquire whether the models of flows in rivers which are geographically close to each other and which are in the same general climatic zone are substantially different from each other. In order to answer these questions, the model building techniques in this book were used to construct models of the monthly flows of the Krishna, Godavari, and Wabash rivers. The Krishna and Godavari rivers in India are geographically close to each other, and are in the tropical zone. The Wabash River in the United States is in the temperate zone.

For every one of the rivers mentioned above we will construct models based on the monthly and annual flow data. Further, for Wabash River data we will discuss a model based on daily flow data. The appropriate class of models needed in each time scale is different although the appropriate monthly model of all three rivers belong to the same class. The like holds true for yearly and daily data also. The best fitting models are validated by means of both residual testing and comparison of statistical characteristics of the output of the model and the observed data.

Next we will assess the predictive abilities of the models and compare the predictive ability of the monthly and yearly models to forecast the flow one year ahead.

Finally we will compare the best fitted models with other suggested models in the literature such as the seasonal ARIMA models. In particular, the best fitting difference equation model for the annual series will be compared with the corresponding fractional noise models in Section 10e.

10b. Discussion of Data

10b.1. Watershed and General Information

The first step in any model building problem is the inspection of the available time series data and the corresponding statistical characteristics such as correlograms and spectral densities. We will consider some of the relevant aspects of the flow data of the Krishna, Godavari, and Wabash rivers.

The Krishna and Godavari rivers are two of the major rivers in central and south India. They have their origins in the western "Ghats" (mountains) near Bombay. Although both of these rivers originate near Bombay, and run an approximately parallel course, they maintain their separate identities and discharge separately into the Bay of Bengal. The watersheds of both rivers are large and cover a variety of terrain. Some of the major tributaries of Krishna are the Bhima River in the north, and the Malaprabha and Tungabhadra in the south. The Godavari has two major tributaries: Penganga River in the north and the Manjra River in the south. Until about 1950, very few dams and other major retention structures existed on the Krishna, Godavari, and their tributaries. During the fifties, several major dams on the Krishna and its tributaries were designed and constructed. Thus, there is some flow modification in both of these rivers, although it is not very substantial.

The Wabash River, in Indiana in the United States, has a smaller watershed area than either the Krishna or Godavari rivers, although it drains the surface runoff from a large part of Indiana and from smaller parts of Illinois and Ohio. The most important tributary of the Wabash is the White River, and the Wabash itself drains into the Ohio River. Some details of these watersheds and the flow data from these rivers are shown in Table 10b.1.1.

Let $y_D(t)$ be the average flow on the tth day, $t = 1, 2, \ldots$. Then, from the daily data, we can construct the monthly data $\{y_M(\cdot)\}$ as follows:

$$y_M(t) = \text{mean monthly flow in the } t\text{th month} = \frac{1}{N_1} \sum_{j=1}^{N_1} y_D(t_1 + j),$$

where t_1 is the number of days in the previous $(t - 1)$ months, and $N_1 = 30$,

TABLE 10b.1.1. Some Details of Watershed and Flow Data

River and station	Duration of record	Type of data	Number of observations	Area of watershed (km^2)	Units	Data source
Krishna at	Jan. 1901–	Monthly	720		m^3/sec	[a]
Vijayawada	Dec. 1960	Yearly	60	251,355	m^3/sec	
Godavari at	Jan. 1902–	Monthly	708		m^3/sec	[a]
Dowleswaram	Dec. 1960	Yearly	59	299,320	m^3/sec	
Wabash at	Oct. 1928–	Monthly	504		m^3/sec	[b]
Mt. Carmel, Illinois	Nov. 1969	Daily	14,976	74,074	m^3/sec	
Mississippi at	Jan. 1861–	Yearly	99	444,200	m^3/sec	[b]
St. Louis, Missouri	Dec. 1960					

[a] "Discharge of Selected Rivers of the World," Vols. I and II, UNESCO, Paris, 1971.
[b] "Water Supply Papers" of the U.S.G.S., U.S. Dept. of the Interior, Washington, D.C.

TABLE 10b.1.2. Some Statistical Characteristics of River Flow Data

River	Daily[a]		Monthly		Annual		Histogram properties[b]	
	$\bar{y}$	$\bar{\sigma}$	$\bar{y}$	$\bar{\sigma}$	$\bar{y}$	$\bar{\sigma}$	$\gamma_1{}^c$	$\gamma_2{}^d$
Krishna	—	—	1795	2574	1795	455	1.56	4.79
Godavari	—	—	3208	5040	3208	1028	2.10	8.38
Wabash	740	937	740	790	740	303	2.03	9.80

[a] $\bar{\sigma}$ is standard deviation (m^3/sec), $\bar{y}$ is the mean (m^3/sec).
[b] Histogram properties are given for monthly data only.
[c] $\gamma_1 = 1/N \sum_{i=1}^{N} (y(i) - \bar{y})^3/\bar{\sigma}^3$.
[d] $\gamma_2 = 1/N \sum_{i=1}^{N} (y(i) - \bar{y})^4/\bar{\sigma}^4$.

31, 28, or 29, depending on the month and year. Similarly, we can construct the annual data sequence $\{y_A(t)\}$:

$$y_A(t) = \text{mean annual flow in the } t\text{th year} = \tfrac{1}{12} \sum_{j=1}^{12} y_M(12(t-1)+j).$$

Therefore, if the $\{y_D(\cdot)\}$ sequence is available, the corresponding $\{y_M(\cdot)\}$ and $\{y_A(\cdot)\}$ sequences may be easily constructed. Suppose we have data corresponding to N_A years which correspond to N_M months or N_D days. Let the mean daily flow $\bar{y}_D$ be defined as $\bar{y}_D = 1/N_D \sum_{j=1}^{N_D} y_D(j)$. Then $\bar{\sigma}_D$, the standard deviation associated with daily data, is defined as $\bar{\sigma}_D{}^2 = 1/N_D \sum_{j=1}^{N_D} (y_D(j) - \bar{y}_D)^2$. We can define $\bar{y}_A, \bar{\sigma}_A, \bar{y}_M$, and $\bar{\sigma}_M$ similarly. By definition $\bar{y}_A, \bar{y}_M$, and $\bar{y}_D$ are equal. Because of the averaging process involved in the definition of the monthly and annual sequences $\{y_M(\cdot)\}$ and $\{y_A(\cdot)\}$, we expect the empirical standard deviations to obey the relationship: $\bar{\sigma}_A < \bar{\sigma}_M < \bar{\sigma}_D$. The annual data have the least variance, and daily data the largest variance. The empirical mean and standard deviation of the three riverflow sequences are listed in Table 10b.1.2.

10b.2. Variation in Mean and Standard Deviation of Monthly Flows

Our next task is to see whether there are any drastic changes in the empirical mean and the standard deviation in the observed record. For simplicity, we consider the monthly data $\{y_M(\cdot)\}$:

$$\bar{y}_M(t) = \frac{1}{t} \sum_{j=1}^{t} y_M(j), \qquad \bar{\sigma}_M{}^2(t) = \frac{1}{t} \sum_{j=1}^{t} (y_M(j) - \bar{y}_M(t))^2.$$

Graphs of $\bar{y}_M(t)$ and $\bar{\sigma}_M(t)$ vs. t are indicated in Fig. 10b.2.1 for the three rivers. From Fig. 10b.2.1, the empirical means and standard deviations settle down to an almost constant value after about six years, or $t \geq 72$. Hence, there is no

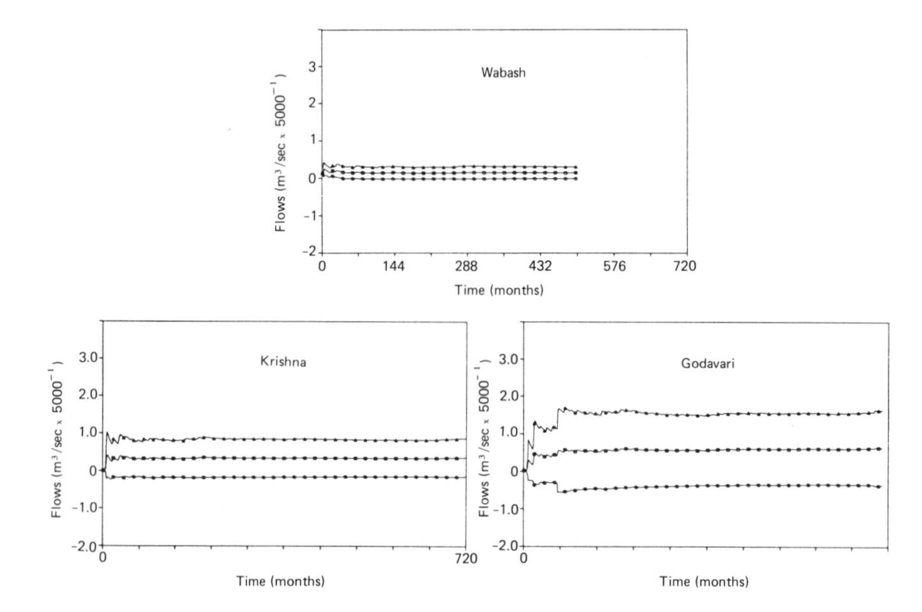

FIG. 10b.2.1. Variation of the means (middle curve) and the standard deviation (top and bottom curve) of monthly flows with sample size.

drastic change in the means of the processes. This fact is stressed because sometimes the construction of dams and other works in watersheds drastically alters the statistical flow characteristics of the river flow.

10b.3. Histograms of Monthly Flows

The histograms of the monthly flows of the three rivers in Fig. 10b.3.1. The strongly skewed nature of the histogram indicates that the distribution of monthly flows is not even approximately Gaussian. This feature may be important in the parameter estimation problems. The skewness and the non-centrality parameter corresponding to the three histograms of monthly data are listed in Table 10b.1.2.

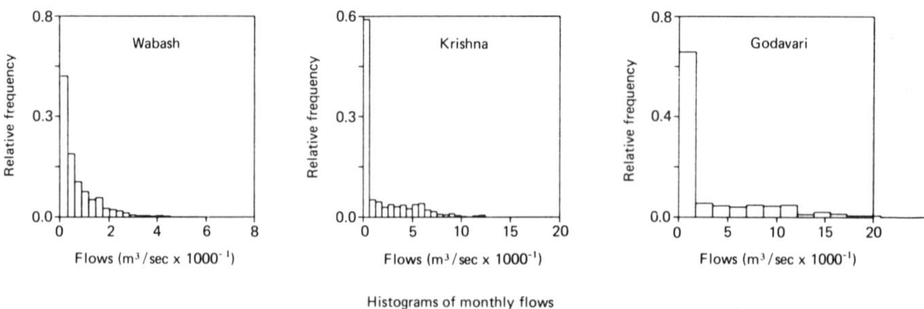

FIG. 10b.3.1. Histograms of monthly flows.

10b.4. Monthly Means and Standard Deviations of Monthly Flows

Let us consider the variation in the average flows in the different months of the year. The average flow in the ith month is $\bar{y}_M(i)$, where January, February, ..., December correspond, respectively, to the values of $i = 0, 1, \ldots, 11$. The corresponding standard deviation in the ith month is designated $\bar{\sigma}_M(i)$:

$$\bar{y}_M(i) = \frac{1}{N_A} \sum_{t=1}^{N_A} y_M(i + 12t), \qquad i = 0, 1, \ldots, 11$$

$$\bar{\sigma}_M{}^2(i) = \frac{1}{N_A} \sum_{t=1}^{N_A} (y_M(i + 12t) - \bar{y}_M(i))^2, \qquad i = 0, 1, \ldots, 11.$$

The graphs of $\bar{y}_M(i)$ and $\bar{\sigma}_M(i)$ vs. i are given in Fig. 10b.4.1 for the three rivers.

The behavior of the monthly means in Fig. 10b.4.1 indicates that in the Krishna and Godavari rivers, the greatest change occurs between the months of June and July because of the northeast monsoon. The flows in these two rivers in the nonmonsoon season are relatively low, showing the absence of major contribution by snowmelt or groundwater. In the Wabash River the

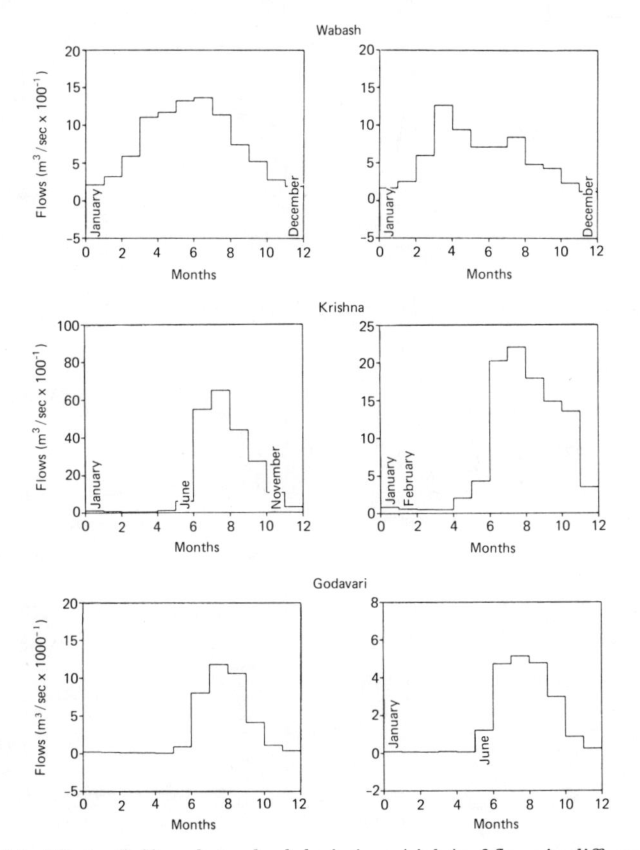

FIG. 10b.4.1. Means (left) and standard deviations (right) of flows in different months.

monthly means do not vary as steeply as in the Krishna and Godavari rivers since snowmelt and groundwater contribute substantially to the flows in the months of lower rainfall.

The variation in the monthly standard deviations is substantial in all three rivers. This variation in monthly standard deviations clearly indicates that any difference equation model excited by a noise with a constant variance will not be satisfactory since such models can lead to an output with constant variance. Hence the noise input may have to be multiplied by a time-varying term to account for the variation in the monthly standard deviation.

10b.5. Correlograms, Spectra, and R/σ Characteristics of Monthly Flows

The correlograms, the power spectral densities, and the R/σ characteristics of the monthly data are given in Figs. 10b.5.1 and 10b.5.2, for the three rivers.

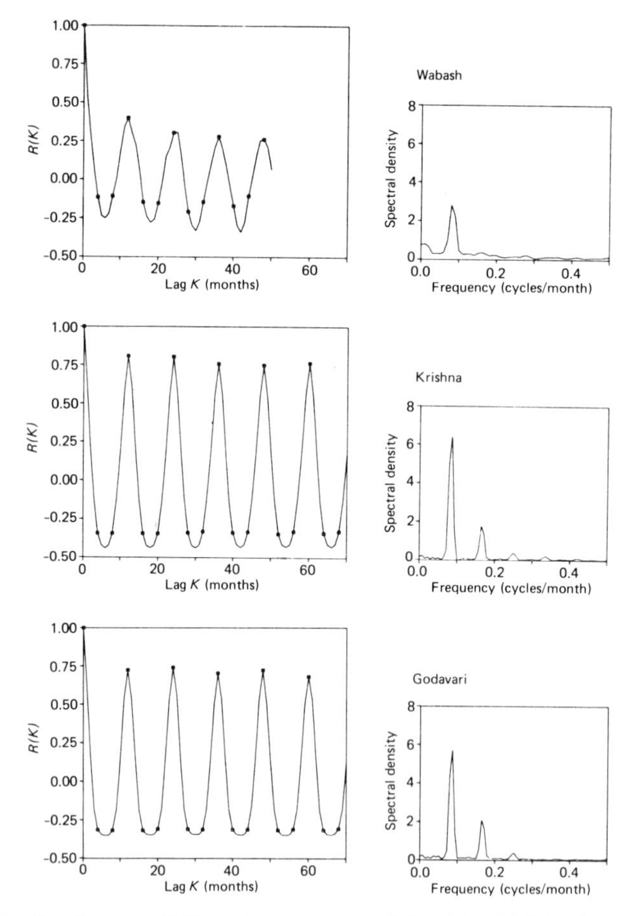

FIG. 10b.5.1. Correlograms (left) and power spectral densities (right) of monthly flows.

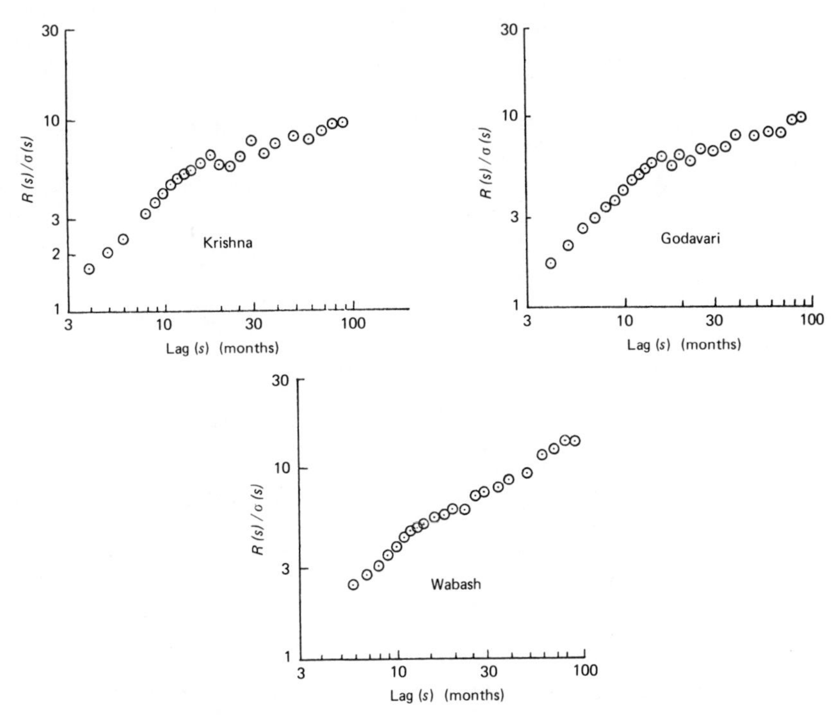

FIG. 10b.5.2. Rescaled range characteristics of observed monthly data.

We will briefly comment on the computation of power spectrum. The raw estimates of the power spectrum were obtained by using

$$PS(\omega_h) = \frac{2}{\pi} \sum_{k=0}^{M} E_k \gamma(k) \cos \frac{hk\pi}{M} \tag{10b.5.1}$$

where $\omega_h = h\pi/M$ is the frequency in radians per month, $h = 0, 1, 2, \ldots, M$; $E_k = 1, 0 < k < M$; $E_k = \frac{1}{2}, k = 0, M$; M is the integer nearest to $0.1N$, N being the number of observations; $\gamma(k)$ is the autocovariance at lag k in months; and $PS(\omega_h)$ is the raw estimate of the power spectrum at frequency ω_h.

These raw estimates were smoothed by using the Hamming window as shown below:

$$\begin{aligned}
k &= 0, & S(\omega_0) &= 0.54PS(\omega_0) + 0.46PS(\omega_1) \\
0 < k &< M, & S(\omega_k) &= 0.23PS(\omega_{k-1}) + 0.54PS(\omega_k) + 0.23PS(\omega_{k+1}) \\
k &= M, & S(\omega_M) &= 0.54PS(\omega_M) + 0.46PS(\omega_{M-1}).
\end{aligned} \tag{10b.5.2}$$

These estimates are divided by $\gamma(0)$ and plotted in Fig. 10b.5.1.

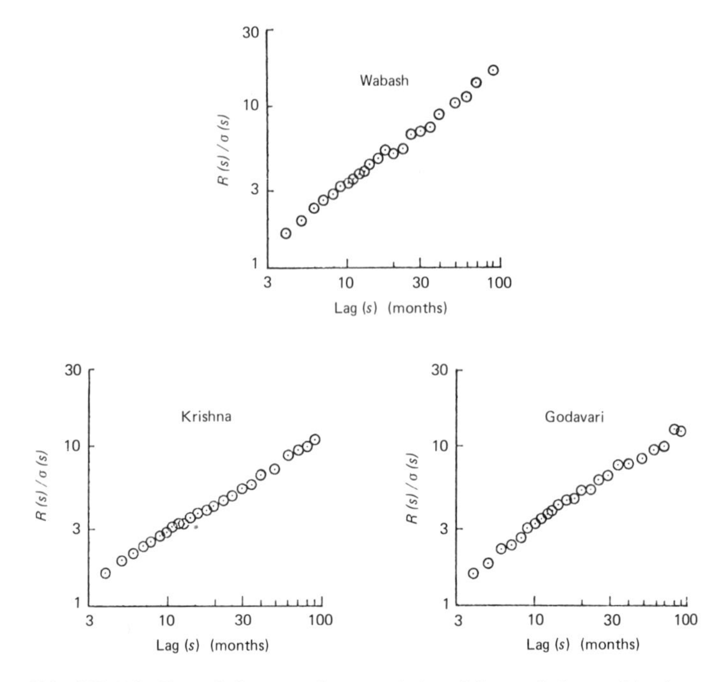

FIG. 10b.5.3. Rescaled range characteristics of detrended monthly data.

The correlograms especially those of the Krishna and Godavari rivers in Fig. 10b.5.1 are strongly oscillatory, with periods of 12 months, and show little damping. The power spectral density of the Wabash shows a sharp peak corresponding to the annual cycle. The Krishna and Godavari data show peaks corresponding to 12-, 6-, and 4-month periods. The Godavari data show an additional minor peak corresponding to the three-month period. The relative sharpness of the spectra indicates that covariance stationary models may be more appropriate than other models, e.g., seasonal ARIMA models, which lead to a smooth spectrum.

For the data from each river, the R/σ characteristic of the flows shown in Fig. 10b.5.2 is made up of two straight lines which show a sharp break at a lag slightly greater than 12 months. This is a common feature of the R/σ characteristic of pseudoperiodic series, and occurs at a lag slightly greater than the dominant period. If we remove the various components corresponding to the dominant frequencies in the spectrum and determine the R/σ characteristics for the detrended data, they are all approximately straight lines with slopes different from 0.5. These are shown in Fig. 10b.5.3. Asymptotically, white noise processes and other difference equation models driven by white noise should lead to a R/σ characteristic with a slope of about 0.5. It has been hypothesized that the significant deviation of the slope of the R/σ characteristic from 0.5 indicates that the river flows *may not obey* stochastic difference equation models excited by white noise. We will discuss this suggestion later.

10b.6. Daily Flow Data Characteristics

Let us call $\bar{y}_{\mathrm{D}}(t)$ and $\bar{\sigma}_{\mathrm{D}}(t)$ the empirical mean and standard deviation of all the daily flow observations until time t. The graph $\bar{y}_{\mathrm{D}}(t)$ and $\bar{\sigma}_{\mathrm{D}}(t)$ vs. t for the Wabash River is given in Fig. 10b.6.1. Considerable fluctuation is present in both the mean and standard deviation. This feature is absent in the monthly flows (Fig. 10b.2.1).

Second, consider the three empirical correlograms of the daily flows obtained from three different batches of data each having 1000 points in Fig. 10b.6.2. Considerable variability is present in the three correlograms. The variability in the correlograms again causes considerable problems in the construction of a suitable model for daily flows. This aspect is further discussed in Section 10d.

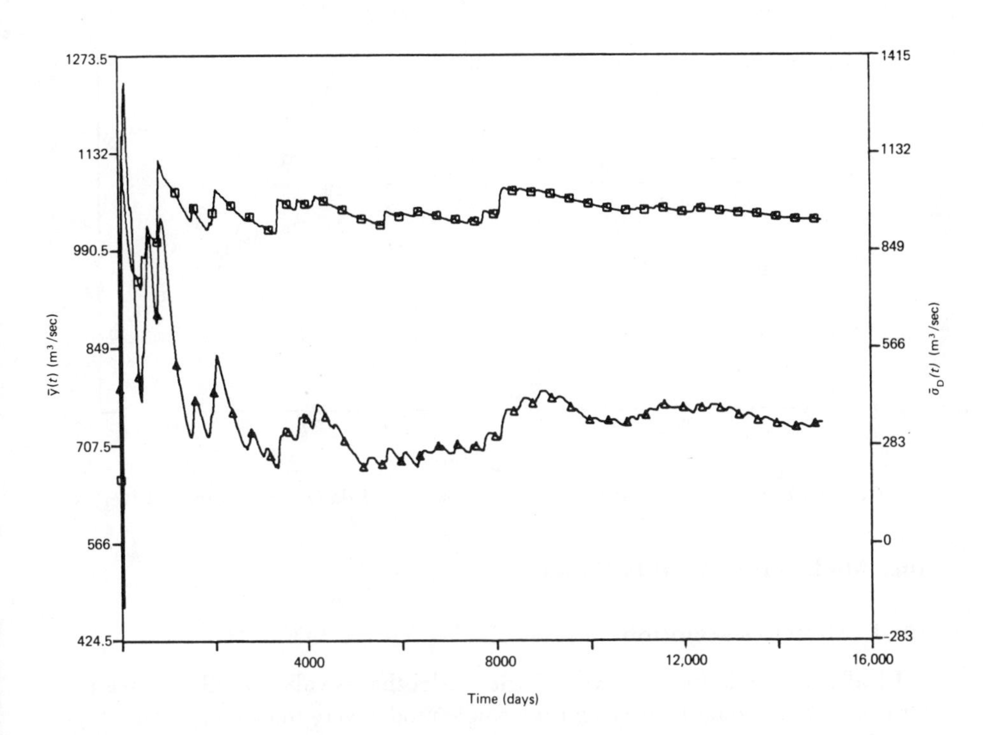

FIG. 10b.6.1. Variation of empirical mean (△) and standard deviation with time (days) (□) of daily Wabash River flows at Mt. Carmel, Illinois.

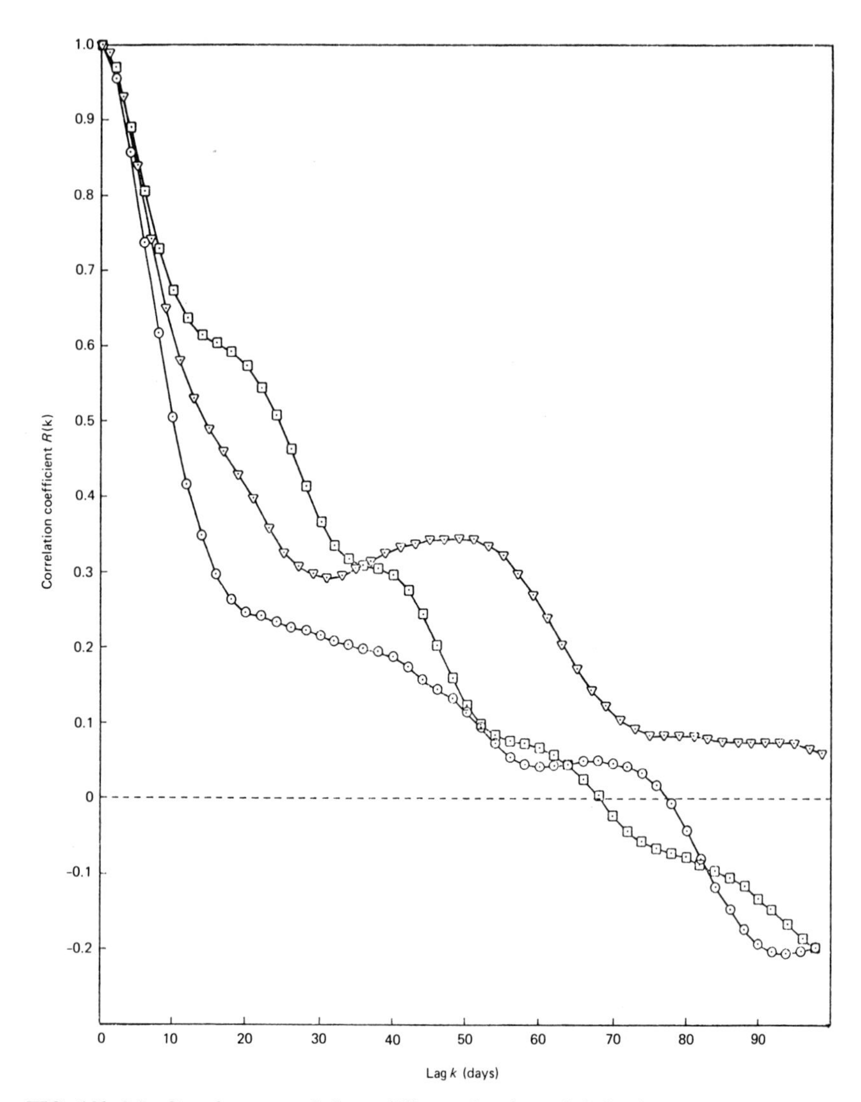

FIG. 10b.6.2. Correlograms of three different batches of daily data (Wabash River).

10c. Models for Monthly Flows

10c.1. Model Description

Ideally we would like to have a single model that is valid for all time scales; for example, it would be excellent if a single model were to give good one-day-ahead as well as one-year-ahead forecasts. Usually it is not possible to develop such general models. Accordingly, we will construct different models which are valid for different frequency ranges. In particular, we will construct a monthly flow model using the monthly data $\{y_M(\cdot)\}$. Similarly, we can construct models for the daily and annual flow data.

We cannot, a priori, state that the one-year-ahead forecasts given by the yearly model are always better than those given by the monthly model. We have to compare the corresponding properties of the monthly and yearly models before arriving at such a conclusion.

The strong variation in the monthly means and standard deviations is a clear indication of the fact that the monthly flow sequence cannot be modeled by weak stationary processes such as the AR or ARMA processes. Hence, we shall find the best model for each river in the class of nonstationary models. Even among nonstationary models, we shall restrict ourselves to the class of AR models with deterministic sinusoidal trend terms added to them since this class can be shown to be sufficient for our purposes.

Let $y(t)$ denote the mean flow in the month t. We will not use the subscript M in this section if there is no cause for confusion. We shall consider the following family of nonstationary models:

$$y(t) = \sum_{j=1}^{m_1} A_j y(t-j) + \psi(t-1) + v(t), \qquad v(t) = \phi(t)w(t), \quad (10c.1.1)$$

$$\begin{aligned} \psi(t) &= F_0 + \sum_{j=1}^{m_5} (F_{m_1+2j-1} \cos \omega_j t + F_{m_1+2j} \sin \omega_j t), \\ \phi(t) &= G_0 + \sum_{j=1}^{m_3} (G_{2j-1} \cos \omega_j t + G_{2j} \sin \omega_j t). \end{aligned} \quad (10c.1.2)$$

In (10c.1.1), $w(\cdot)$ is a zero mean white noise sequence with covariance ρ obeying assumption (A1). The time-varying multiplying term $\phi(\cdot)$ has been introduced to account for the systematic variation in the standard deviation of the monthly flows from month to month. Similarly, $\psi(t)$ is introduced to account for the systematic variation in the monthly means. $\omega_1, \omega_2, \ldots$, are the

TABLE 10c.1.1 Parameter Estimates of the Final Models for the Monthly Flows. The term $v(t)$ is defined in equations (10c.1.1) and (10c.1.2). $\omega_1 = 2\pi/12$, $\omega_2 = 2\pi/6$, $\omega_3 = 2\pi/4$, $\omega_4 = 2\pi/3$.

River and model no.	Final model
Krishna (K3)	$y(t) = 1327 + 0.161y(t-1) + 0.058y(t-2) + 0.059y(t-3) - 0.023y(t-4) - 1050 \sin \omega_1 t - 2050 \cos \omega_1 t + 1151 \sin \omega_2 t + 887 \cos \omega_2 t - 582 \sin \omega_3 t - 418 \cos \omega_3 t + 343 \sin \omega_4 t + 341 \cos \omega_4 t + v(t)$
Godavari (G2)	$y(t) = 2365 + 0.262y(t-1) - 2037 \sin \omega_1 t - 3454 \cos \omega_1 t + 2388 \sin \omega_2 t + 1387 \cos \omega_2 t - 1274 \sin \omega_3 t - 55 \cos \omega_3 t + v(t)$
Wabash (W2)	$y(t) = 435 + 0.408y(t-1) + 251 \sin \omega_1 t - 325 \cos \omega_1 t - 40 \sin \omega_2 t + 11.69 \cos \omega_2 t + v(t)$

TABLE 10c.1.2. The Estimates of G Coefficients for the Monthly Flow Models[b]

River and model	m_3	$\hat{G}_0$	$\hat{G}_1$	$\hat{G}_2$	$\hat{G}_3$	$\hat{G}_4$	$\hat{G}_5$	$\hat{G}_6$	$\hat{G}_7$	$\hat{G}_8$	$\hat{G}_9$	$\hat{G}_{10}$	$\hat{G}_{11}$	$\frac{1}{N}\sum_{i=1}^{N} w^2(i)$
Wabash W2	6	515.3	−369.6	57.0	−109.4	−79.5	140.9	−5.8	−34.0	66.700	−84.0	−23.5	66.5	1.000
Godavari G2	6	1726.4	−990.8	−2165.9	−681.6	836.5	143.5	−76.5	−104.0	171.300	255.3	−94.5	−97.1	1.000
Krishna K3	4	901.1	−260.4	−952.4	−152.9	166.9	140.0	−171.5	−191.5	0.548				1.055

[b] $\hat{G}_{2j-1}$ is the coefficient of $\cos \omega_j(t-1)$; $\hat{G}_{2j}$ is the coefficient of $\sin \omega_j(t-1)$. (See eq. 10c.1.2.)

TABLE 10c.1.3. Probability distributions of the Random Inputs $w(\cdot)$

Model K3:	$p_K(w) \sim N(-0.3, 1)$
Model G2:	$p_G(w) \sim N(-0.132, 1)$
Model W2:	$p_W(w) \sim 0.96N(-0.07, 0.8^2) + 0.04N(1.75, 10.65^2)$.
	$N(\mu, \rho) \equiv$ Normal distribution with mean μ, and variance ρ.

dominant frequencies in the corresponding spectra of monthly flows. The choice of the integer m_1 and the appropriate frequencies among $\omega_1, \omega_2, \ldots$, is discussed in Rao and Kashyap (1974). The coefficients $\{A_j\}$ obey assumption (A5) and the trend terms obey (A4). As both G_j in $\phi(t)$ and the covariance ρ of $w(t)$ cannot be arbitrary, we set ρ to unity.

As the distribution of flows is highly non-Gaussian, there is no rationale for computing the Gaussian CML estimates of the parameters $\{A_j\}$, $\{F_j\}$, and $\{G_j\}$ since they may not possess properties such as asymptotic minimum variance. Consequently, we restricted ourselves to finding the least squares estimates of the various coefficients in (10c.1.1), since these estimates appear to be quite adequate. Several models were obtained from (10c.1.1) and (10c.1.2) by using different values for m_1, m_3, and m_5, and these models were compared by using the decision rules discussed in Section 8b.2. The final models for the flows of the three rivers are given in Table 10c.1.1. The corresponding $\{G_i\}$ coefficients defined in (10c.1.2) are given in Table 10c.1.2. The appropriateness of the final models was confirmed by an examination of the characteristics of the one-month-ahead prediction errors, by the tests on residuals, and by a comparison of the statistical characteristics of the outputs of the models with the corresponding empirical characteristics of the observed flows. The probability distribution of the random inputs $\{w(\cdot)\}$ used for simulation of flows are listed in Table 10c.1.3. The details of the parameter estimation and validation of the fitted models may be found in Rao and Kashyap (1973, 1974).

10c.2. Comparison with Other Models

The basic model for monthly flows will be compared below with some of the other models suggested for monthly river flows. We will discuss only a few of them here; some others are mentioned in the bibliographical notes. The discussion is brief as numerical results pertaining to validation and simulation are not available for all the other models.

10c.2.1. Autoregressive and Autoregressive Moving Average Models

A. The Thomas–Fiering Model. The Thomas–Fiering (Thomas and Fiering; 1962) model may be written as shown below, where $y(k, i)$ are the monthly flows:

$$y(k, i) - \bar{y}_1(i) = \beta(i)[y(k, i-1) - \bar{y}_1(i-1)] + \bar{w}(k, i-1)\phi_1(i), \quad (10c.2.1)$$

where $\phi_1(i) = \sigma_i(1 - \gamma_{i-1}^2)^{1/2}$; $\beta(i)$ is the regression coefficient of the flows in months i and $i-1$; $\bar{w}(k, i)$ is a sequence of independent normal random variables with zero mean and unit variance; σ_i is the standard deviation of flows in the ith month; γ_{i-1} is the correlation coefficient between the flows of the ith and $(i-1)$th month; $\bar{y}(i)$ is the mean monthly flow during the ith month, $i = 1, 2, \ldots, 12$; and $y(k, i)$ is the flow during the ith month of the kth year, $k = 1, 2, \ldots$.

Since Eq. (10c.2.1) is a first-order Markov process, we will compare it with the first-order version of our model as given in Eq. (10c.1.1). The main difference between these two representations is that whereas there are a few autoregressive coefficients in our model in Eq. (10c.1.1), there are 12 corresponding parameters $\beta(1), \ldots, \beta(12)$, one for each month, in the model given in Eq. (10c.2.1). This increase in the number of parameters causes the estimates of the parameters to be less accurate than that of A_j in Eq. (10c.1.1). For instance, if N_1 years of data are available, A_j in Eq. (10c.1.1) can be estimated using $12N_1$ values of the monthly flows, whereas each of the $\beta(i)$, $i = 1, \ldots, 12$, can be estimated using only N_1 values. However, the factors multiplying the disturbance terms $w(t)$ and $w(k, i-1)$ are similar in both models. Both the multiplying factors $\phi(i)$ and $\phi_1(i)$ vary from month to month, $i = 1, \ldots, 12$. Further, if the Thomas–Fiering model is generalized to a higher order Markov model, the parameters $\beta(i)$ and $\phi_1(i)$ will no longer have the simple interpretation given above.

By expressing $\bar{y}(\cdot)$ as a combination of sinusoids in the model (10c.1.1) and by using the recursive estimation algorithm, any slow variation in the monthly means may be tracked. These variations in the monthly means are usually present in data covering a long time interval. In contrast, the Thomas–Fiering model is inflexible in this respect; it cannot be used to determine the changes in the monthly means $\bar{y}_M(t)$ over a period of time.

The basic objective and use of the Thomas–Fiering model is in simulation of flows, whereas the model in (10c.1.1) may be used for simulation as well as prediction.

The Thomas–Fiering model can maintain the monthly means and standard deviations, probability distribution of monthly flows, and the correlogram and spectral density functions. Its prediction performance appears to be inferior to that of model (10c.1.1). The model investigated in the present study can maintain all the above-mentioned characteristics along with the rescaled range characteristics of the flows.

10c.2.2. Other AR, ARMA, and Seasonal ARIMA Models

One can fit an autoregressive (AR) or an autoregressive moving average (ARMA) model such as Eq. (10c.2.2) to monthly flow data $y(k, i)$ without the trend terms $\psi(i)$. $\{w(\cdot)\}$ is a sequence of uncorrelated random variables:

$$y(k, i) = A_0 + \sum_{j=1}^{m_1} A_j y(k, i-j) + \sum_{j=1}^{m_2} A_{j+m_1} w(k, i-j) + w(k, i). \qquad (10c.2.2)$$

In view of the periodicities present in monthly flow data, the AR or ARMA model of the type given in Eq. (10c.2.2) gives large one-step-ahead prediction mean square errors. Model (10c.2.2) cannot reproduce the monthly mean and monthly variance characteristics of the observed flows. The AR models with suitable sinusoidal input terms have been shown in the present study to preserve adequately all the statistical characteristics of the observed flows.

Seasonal ARIMA models have been discussed by Clarke (1973) in modeling river flows and by Rao and Rao (1974) for rainfall processes. These models do not possess any advantage over the covariance stationary AR or ARMA models developed here. Additional drawbacks of the seasonal ARIMA models, such as the counterintuitive nature of the disturbance terms and the complexity of parameter estimation, were mentioned in Chapter III.

10c.2.3. The Models for Normalized Flows

In order to remove the periodicites present in the monthly flow data, two types of normalizations are performed on the data:

$$\tilde{y}_1(k, i) = \frac{y(k, i) - \bar{y}(i)}{\bar{\sigma}(i)} \tag{10c.2.3}$$

$$\tilde{y}_2(k, i) = y(k, i) - y_{\mathrm{F}}(i). \tag{10c.2.4}$$

In the first of these normalizations in (10c.2.3), the monthly means $\bar{y}(i)$ are subtracted from the monthly flows and the remainder is divided by the monthly standard deviations of flows (Singh and Lonnquist 1974). In the second type of normalization (10c.2.4) a Fourier analysis is conducted on the observed data $y(k, i)$ to obtain $y_{\mathrm{F}}(i)$ and the residuals from the Fourier analysis, $\tilde{y}_2(k, i)$, are analyzed further (Yevjevich, 1972). The observed monthly flows may also be transformed by using logarithmic, square root, or cube root transformations and then normalized. Autoregressive or autoregressive moving average models such as (10c.2.2) are fitted to the $\tilde{y}_j(k, i)$ series, $j = 1, 2$.

When monthly flows are analyzed, models such as (10c.2.2) with normalization such as (10c.2.3) will involve $24 + m_1 + m_2 + 1$ unknowns such as $\bar{y}(i)$, $\bar{\sigma}(i)$, and A_j. Similarly, $m_1 + m_2 + m_5 + 1$ unknowns are involved if normalization such as (10c.2.4) is used, where m_5 is the number of Fourier series terms in $y_{\mathrm{F}}(i)$. The variations in $\bar{y}(i)$ and $\bar{\sigma}(i)$ will greatly affect the prediction performance of the model and hence the transformed models may not be suitable for prediction.

The synthetic flows generated from the normalized models can reproduce the monthly means and standard deviations. Detailed information is not available about the R/σ properties of models based on normalized data, and about its capacity to reproduce histograms of flows.

The prediction errors with the normalized model may be larger than those of model (10c.1.1) since it is possible in the latter model to keep track of any variation in the estimation of the coefficient of the sinusoidal term via the real time parameter estimation algorithm (C1) of Chapter VI. In contrast one has to fix the values of these coefficients in the normalized flow models even though they may be slowly varying with time.

10c.2.4. The Fractional Noise (FN) Model

The rescaled range characteristic of the observed monthly flows is not a straight line when the R/σ values are plotted against lag s on logarithmic paper. As such, a direct FN model cannot be postulated for monthly flow data. However, the trend-free process, $y_1(t)$, which is obtained by subtracting the monthly mean $\bar{y}_M(t)$ from the flow $y(t)$ $[y_1(t) = y(t) - \bar{y}_M(t)]$, could be modeled by a FN model (Rao and Kashyap, 1974). Such a model of the trend-free process will reproduce the R/σ characteristics of the trend-free observed data.

One of the disadvantages of the FN model is that the flows simulated from the model cannot reproduce the observed monthly means and standard deviations in view of the stationarity of the FN model. Enough work has not yet been done with the FN models to ascertain their ability to preserve other characteristics such as correlation, spectral density, and the probability distribution of the detrended monthly flows. The prediction capability of the FN model is yet to be analyzed, even though preliminary results indicate that the mean square value of the one-step-ahead prediction error of the FN model is much greater than the corresponding value in the covariance stationary model investigated in the present study. These conclusions, however, are not surprising in view of the fact that the models used in the present study deal essentially with higher frequency aspects of the flows, whereas the FN model is designed to preserve the low-frequency aspects of the flows. A detailed discussion of the FN model for annual flow data is given in Section 10e.

10d. Modeling Daily Flow Data

There have been a number of modeling studies of daily data. In many of these studies the periodicites that are present in the daily flows are removed by fitting Fourier series to daily means and standard deviations. AR or ARMA models are fitted to the detrended data.

There have been relatively few studies in which the model is tested to determine whether it represents the given data in any sense. In many of these studies, even the whiteness tests on the residuals have not been reported. As mentioned earlier, the construction of a model that represents the data in some suitable sense is difficult with daily flow data because of the large number of observations available and the variability of the data.

In this section we will consider the problem of selection of an appropriate model for *forecasting* the daily flows. We would like to emphasize that we are not concerned with generation of synthetic data.

We will first compare the IAR and AR classes in order to select the best model. Obviously, other classes may also be similarly compared. Next, we will compare the different classes of models by forecasting. Although the intended purpose of the model is forecasting, we have also analyzed the characteristics of residuals and other properties of two different forecasting models. The daily flow data of the Wabash River are analyzed and discussed below.

10d.1. The Comparison of Classes by the Likelihood Criterion

We will first compare the various autoregressive (AR) models and the integrated autoregressive (IAR) models of various orders by means of the decision rule based on the likelihood discussed in Section 8b.1.3 The residual variances of the various models and the corresponding likelihood index are listed in Table 10d.1.1. In this table n_i is the number of terms estimated in the ith model and N is the total number of observations used, namely 14,976.

TABLE 10d.1.1. Variation in Residual Variance and Likelihood Index in AR and IAR Models of Different Orders

i	Model	n_i	$\hat{\rho}_i$	Index[a]
	AR models			
1	AR(2)	3	6398	63,874
2	AR(5)	6	6369	63,843
3	AR(10)	11	6341	63,816
4	AR(15)	16	6312	63,781
5	AR(20)	21	6297	63,775
6	AR(25)	26	6290	63,772
	IAR models			
1	IAR(2)	3	6599	64,099
2	IAR(5)	6	6513	64,006
3	IAR(10)	11	6450	63,945
4	IAR(15)	16	6397	63,885
5	IAR(20)	21	6372	63,862
6	IAR(25)	26	6355	63,847

[a] Index $= n_i + (N/2) \ln \hat{\rho}$.

The best model is AR(25) since it has the *smallest* value of the negative of the likelihood index listed in the last column of Table 10d.1.1. Also, the IAR models are consistently inferior to the AR models.

We have arbitrarily limited the order of AR models to 25. These conclusions are not altered when AR models of order higher than 25 are considered.

10d.2. Comparison of Classes by Forecasting Capability

We will use the first 10,000 observations to estimate the parameters in the chosen model. The model with the unknown parameters replaced by these estimates is used to obtain the one-step-ahead predictions of the remaining 4976 observations. Let $\hat{\rho}$ be the mean square error of these 4976 forecasts and $\bar{\rho}$ the

TABLE 10d.2.1. Variation in Residual and Prediction Error Variances in AR and IAR Models

Model	$\bar{\rho}$	$\hat{\rho}$
AR models		
AR(2)	5861	7471
AR(5)	5786	7608
AR(10)	5751	7622
AR(15)	5744	7682
AR(20)	5717	7708
AR(25)	5708	7714
IAR models		
IAR(2)	5989	7898
IAR(5)	5899	7817
IAR(10)	5847	7765
IAR(15)	5805	7838
IAR(20)	5780	7807
IAR(25)	5760	7814

residual variances from the model obtained by fitting 10,000 observations. We list the values of $\bar{\rho}$ and $\hat{\rho}$ for the various models in Table 10d.2.1.

The AR(2) model gives the least value of $\hat{\rho}$, and additional terms in the model increase the value of $\hat{\rho}$. Thus, from the prediction point of view, AR(2) model is the best fitting model among the models considered. This result conflicts with the decision given by the likelihood criterion. An explanation of this discrepancy is that the requirements for forecasting alone are not as stringent as the requirements for a reasonable representation of the data. The results of this section imply that we do not need a very complicated model if we are interested in forecasting only.

There is considerable discrepancy between the values of $\hat{\rho}$ and $\bar{\rho}$ in Table 10d.2.1. This difference may suggest that there may be other predictors which yield a value of $\hat{\rho}$ closer to that of $\bar{\rho}$. However, a detailed analysis does not warrant such a conclusion and the discrepancy between $\bar{\rho}$ and $\hat{\rho}$ can be explained by the fact that the residual variance of the model is not really a constant but varies considerably in time. In particular, the residual variance in the time interval $t = 10{,}001$ to 14,976 appears to be larger than that in the time interval $t = 1$ to 10,000. In order to see this we may recall that the residual variance for the AR(25) model constructed from the 14,976 observations is 6290 m^6/sec^2. From these 14,976 residuals, the last 4976 residuals were selected and their mean square value was 7400 m^6/sec^2. This value of the residual variance is consistent with the value of $\hat{\rho}$ equal to 7714 m^6/sec^2.

The observed variation in time of the residual variance is not entirely unexpected since the standard deviation of data also varies considerably with time. We emphasize that the AR(25) model discussed above cannot account for this observed variation in the variance.

10d.3. Validation of the Models

As mentioned earlier, although our emphasis in the development of a daily flow model is mainly on forecasting, we would also like to test the residuals and consider other properties of the model. With this objective in view, we consider the following three aspects.

- Possible variation of the coefficients of the model with time.
- Tests on the residuals to determine the extent of their "whiteness."
- Ability of the model to reproduce the characteristics of the data.

10d.3.1. Variation in the Coefficients with Time

We have already seen that the residual variance varies with time. We want to see whether the coefficients in the model also substantially vary in time. We will consider the AR(2) model. If $N = 10{,}000$, the fitted AR(2) model is

$$\begin{aligned} y(t) = \underset{(\pm 0.01)}{11.01} + \underset{(\pm 0.00006)}{1.78} y(t-1) - \underset{(\pm 0.00006)}{0.80} y(t-2) + w(t). \end{aligned}$$

If $N = 14{,}976$, the fitted AR(2) model is

$$\begin{aligned} y(t) = \underset{(\pm 0.007)}{11.73} + \underset{(\pm 0.00004)}{1.75} y(t-1) - \underset{(\pm 0.00004)}{0.77} y(t-2) + w(t). \end{aligned}$$

In these models the variation in the coefficients of $y(t-1)$ and $y(t-2)$ is about 1.5 to 4.0%, but the variation in the constant term is substantial—about 7%. This is presumably caused by the variation in time of the sample mean of the observed flow process.

In these two models, we have also included the corresponding standard deviations of the parameter estimates. These estimates of the standard deviations have been computed by assuming the validity of the given model. Since the validity of the model itself is in doubt, the standard deviations of the estimates are also doubtful. We stress this fact because, if we accept these standard deviations, then even the difference in the estimates of the coefficient of $y(t-1)$ in the two models is highly significant at the 95% level and implies that we should regard this coefficient also as varying in time.

We can conduct a similar analysis with the AR(25) model. The two AR(25) models obtained from the 14,976 and 10,000 observations are given in Table 10d.3.1. Again the two dominant terms, namely the coefficients of $y(t-1)$ and $y(t-2)$, show little variation, but the constant term shows considerable variation. There is considerable discrepancy between the corresponding AR coefficients of the terms $y(t-j)$, $j = 3, \ldots, 25$. However, the numerical values of these coefficients are small compared to the coefficients of $y(t-1)$ and $y(t-2)$.

TABLE 10d.3.1. Variations in the Coefficients of the AR(25) Model with Sample Size

	Model with 10,000 data points		Model with 14,976 data points	
Coefficient	Estimate	Standard deviation[b]	Estimate	Standard deviation[b]
A_0	6.997	0.0106	8.035	0.00757
A_1	1.845	0.00010	1.7717	0.000069
A_2	−0.893	0.00022	−0.7807	0.00014
A_3	−0.0086	0.00024	−0.01783	0.00015
A_4	0.0489	0.00024	−1.1604	0.00015
A_5	−0.0511	0.00024	0.00481	0.00015
A_6	0.0427	0.00024	0.0185	0.00015
A_7	0.0529	0.00024	0.0257	0.00015
A_8	−0.0664	0.00024	−0.0449	0.00015
A_9	−0.0315	0.00024	−0.0420	0.00015
A_{10}	0.0495	0.00024	0.144	0.00015
A_{11}	0.0348	0.00024	−0.144	0.00015
A_{12}	−0.0518	0.00024	0.0738	0.00015
A_{13}	−0.0029	0.00024	−0.00935	0.00015
A_{14}	0.063	0.00024	0.0119	0.00015
A_{15}	−0.0942	0.00024	−0.0415	0.00015
A_{16}	0.094	0.00024	0.0645	0.00015
A_{17}	−0.0817	0.00024	−0.0699	0.00015
A_{18}	0.044	0.00024	0.0302	0.00015
A_{19}	−0.0148	0.00024	0.0112	0.00015
A_{20}	−0.0072	0.00024	−0.03166	0.00015
A_{21}	0.0497	0.00024	0.05456	0.00015
A_{22}	−0.0157	0.00024	−0.02088	0.00015
A_{23}	−0.0464	0.00024	−0.0248	0.00015
A_{24}	0.0356	0.00022	0.0109	0.00014
A_{25}	−0.00391	0.00010	0.0065	0.0006

[b] The numbers have been rounded off.

10d.3.2. Residual Testing

We will consider the AR(25) model obtained with the aid of all 14,976 observations and ascertain the extent of whiteness of the corresponding residuals. If we consider batches of 1000 residuals, then each individual batch passes the cumulative periodogram test at the 95% significance level, showing that the residuals may be considered to be free of deterministic sinusoidal components; i.e., even though the periodogram shows bumps at the frequencies corresponding to periods such as seven days, these frequencies are still not significant. Presumably the AR coefficients are quite adequate to handle the observed frequency components of seven-day periods in the data. We should emphasize that we did not test all 15,000 residuals simultaneously by the cumulative periodogram test.

The extent of serial correlation among the residuals from different models is of interest. We will determine the largest integer l such that the residuals

$\bar{w}(t-j)$, $j = 0, 1, \ldots, l$, can be regarded as uncorrelated at the 95% level. The integer l can be determined by test 6 in Section 8c.2; its value is listed below for the various models:

Model	Integer l
AR(2)	2
AR(15)	29
AR(25)	41

Thus, the second-order AR model provides a very poor representation of the data. One could regard the AR(25) model as satisfactory if one is going to limit the use of the model to inferences involving several days, say 10–15 days.

We can cross-check the result above by considering the actual correlogram of the residuals. Since the standard deviation of any correlation coefficient of lag k of a white noise process computed from N observations is $1/\sqrt{N}$, we can accept a correlation coefficient to be true if its absolute value is less than two standard deviations, i.e., $2/\sqrt{N} = 0.016$, for $N = 14{,}976$. Among the first 41 correlation coefficients of the residuals of the AR(25) model there were five correlation coefficients that exceeded the bound 0.01. Based only on the correlogram, it is difficult to conclude that the true values of the first 40 correlation coefficients of the residuals are zero and that the deviations are caused only by the finiteness of the observations.

10d.3.3. Direct Comparison of Correlograms

We can compare the theoretical correlogram of the AR models with the corresponding empirical correlogram of the data. The AR(2) model is unsatisfactory. The AR(25) model correlogram is similar to that of the data up to about 40 lags.

If we compute the theoretical standard deviation of the correlation coefficients of the AR(25) model, the observed correlogram lies within four standard deviations of the theoretical correlogram of the model.

10d.4. Discussion

1. This example is an excellent illustration of the fact that a model that gives good forecasts does not necessarily represent the data well. The AR(2) model is better than the AR(25) model for forecasting, illustrating the parsimony principle.

2. The ratio of forecast mean square error to the signal mean square error is $7714/1.42 \times 10^6 = 0.0054$. The square root of this quantity is 0.074; i.e., the prediction error is about 7.4%.

3. The AR(25) model can claim to represent the data as given by the first 40 correlation coefficients.

10d.5. Deterministic Daily Flow Models

A number of deterministic, moisture accounting type of daily flow simulation models have been developed. The Stanford watershed model (Crawford and Linsley, 1966) is perhaps the most important of this type of model. The number of input parameters in these models is typically about 20. The parameters in these models are optimized to minimize the sums of squares of deviations between the observed and computed flow values or by another similar criterion. In order to simulate daily flows from these models, a rainfall model is necessary in addition to the input parameters.

A comparison of the stochastic models of the type discussed earlier and others available in the literature with the deterministic models reveals that the number of parameters is large in either case. The deterministic models are sometimes claimed to be better than the stochastic models as they are based on physical processes such as evapotranspiration which are involved in the transformation of rainfall to runoff. However, models of these processes are themselves rather crude and involve parameters that must be empirically estimated. A comparative study of the deterministic and valid stochastic models has not been reported to indicate clearly the advantages of the deterministic models. Data requirements of deterministic models are considerably greater than those of stochastic models. Finally, the development of daily rainfall generation models for large watersheds which is a prerequisite for deterministic riverflow models is a formidable problem.

In view of these considerations, stochastic models are to be preferred to deterministic models, especially for modeling daily flows from large watersheds.

10e. Models for Annual Flow Data

Construction and validation of AR or ARMA models for the annual flow data are relatively easy. However, it has been recently asserted that only fractional noise (FN) models can represent the observed annual river flow sequences in view of their ability to preserve the rescaled range characteristic h.

However, the ability of a model to represent one particular property of a process does not imply that the model can represent other characteristics of the data such as the correlogram and spectral density. Furthermore, the capability of the FN model to "preserve" the Hurst coefficient h does not imply its superiority to the best fitting AR or ARMA model according to direct tests such as the hypothesis tests.

In view of these considerations we will compare the best fitting AR and FN models of annual flows from the following points of view: (a) ability to reproduce correlation and rescaled range–lag characteristics, (b) classical hypothesis tests and Bayesian tests, (c) one-step-ahead prediction error properties, and (d) physical interpretation. The results are discussed in Section 10e.7. We will consider only the flow values normalized as follows

$$\frac{1}{N}\sum_{i=1}^{N} y(i) = 0; \qquad \frac{1}{N}\sum_{i=1}^{N} y^2(i) = 1.$$

10e.1. The Different Types of Models

Among the finite difference equation models, the class of autoregressive models is sufficient for modeling the annual flows. There is no need for considering IAR and other models since the empirical correlation functions decay fairly rapidly with increasing lags. Moving average terms do not also increase the quality of the fitted model. Consequently, AR models are developed for the annual flow data. We can determine the order of the AR models for each river as discussed in Chapter VIII. We will omit the details and give only the final models. They are denoted by the label M to distinguish them from the corresponding FN models, which are labeled M_1. $y(t)$ stands for normalized mean annual flows.

Godavari M_0:

$$y(t) = 0.2394y(t-2) + w(t)$$
$$w(t) \sim N(0, \rho_0 = 0.9053)$$

Krishna M_0:

$$y(t) = 0.2863y(t-1) + 0.1505y(t-3) - 0.2379y(t-4) + w(t)$$
$$w(t) \sim N(0, \rho_0 = 0.8278)$$

Mississippi M_0:

$$y(t) = 0.2967y(t-1) + w(t)$$
$$w(t) \sim N(0, \rho_0 = 0.970).$$

The term $y(t-1)$ in the Godavari M_0 model and the term $y(t-2)$ of the Krishna M_0 model are omitted since the estimates of the coefficients of these terms were insignificant according to test 5 of Section 8c.2.3. However, the assumption of normality cannot be validated because of the paucity of the number of observations.

Let us consider fitting the FN models for the same data. As discussed in Chapter II, a FN process y can be represented by the following moving average process:

$$y(t) = \sum_{t=1}^{M} w_1(t-k)/k^{(1.5-h)},$$

where M is a large integer, and $\{w_1(\cdot)\}$ is a zero mean independent sequence with a normal distribution $N(0, \rho_1)$.

Next we will estimate the parameters h and ρ_1 of the best fitting FN model for each river. A good estimate of the parameter h is the slope of the rescaled range (R/σ)–lag (s) characteristic of the given empirical sequence $\{y(\cdot)\}$ when it is plotted on log–log paper. The estimates of h were obtained from the log–log plots of R/σ versus lag characteristics which are shown in Fig. 10e.1.1.

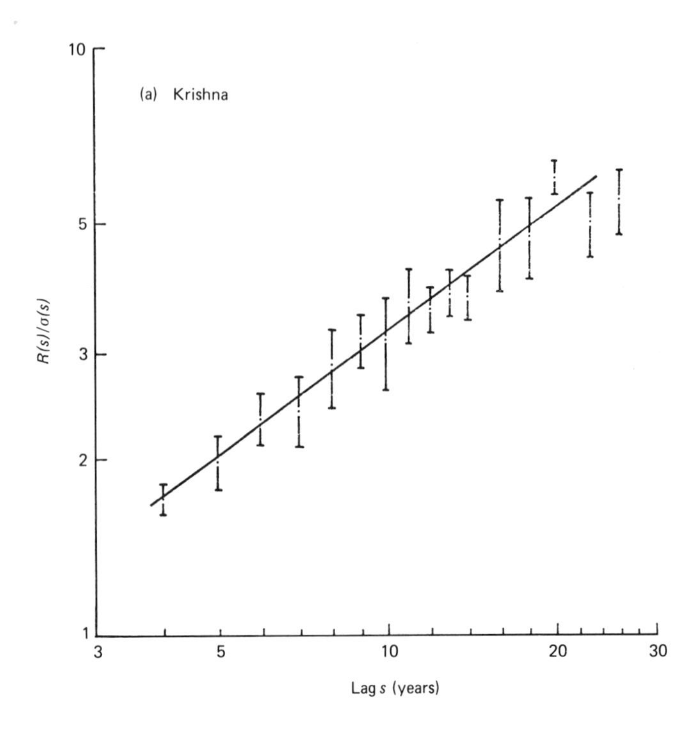

(a) Krishna
10
5
3
2
1
R(s)/σ(s)
3
5
10
20
30
Lag s (years)

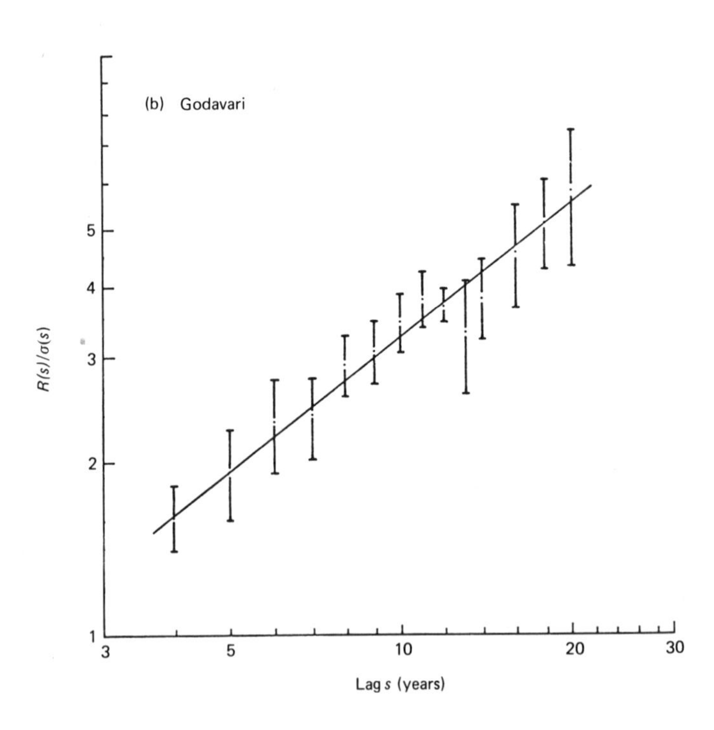

(b) Godavari
5
4
3
2
1
R(s)/σ(s)
3
5
10
20
30
Lag s (years)

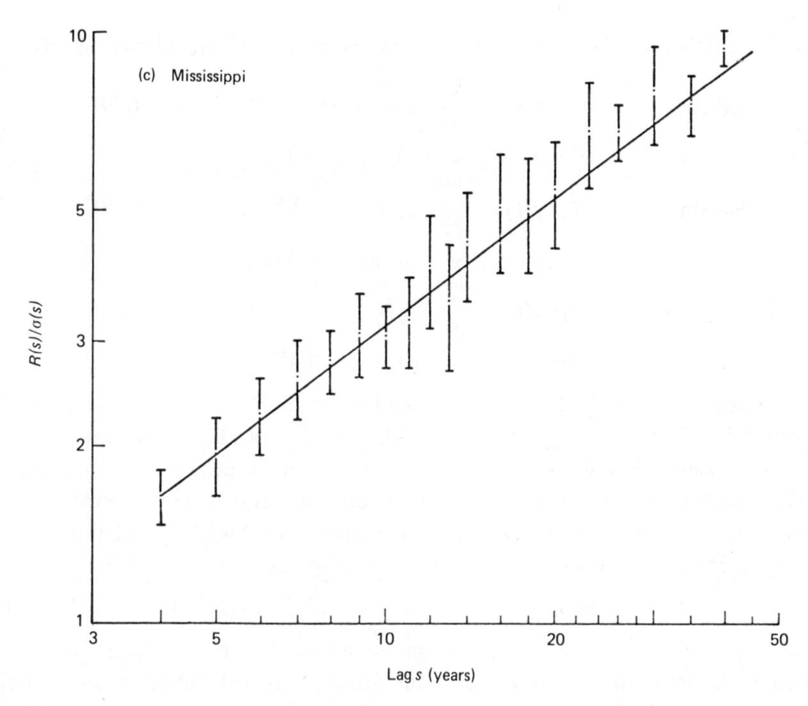

FIG. 10e.1.1. Annual flows of (c) Mississippi River at St. Louis, Missouri; (a) Krishna River at Vijayawada; and (b) Godavari River at Dowleswaram; mean $R(s)/\sigma(s)$ (points) and their standard deviations (bars).

In order to estimate ρ_1, we recall the following expression for the mean square value of y from Chapter II:

$$E[y^2] = \rho_1 \sum_{j=1}^{M} (1/j^{3-2h}).$$

Hence the estimate ρ_1 can be computed by

$$\rho_1 = \left(\frac{1}{N}\sum_{t=1}^{N} y^2(t)\right) \Big/ \sum_{j=1}^{M} (1/j^{3-2h}).$$

The estimates h and ρ_1 for the Krishna, Godavari, and Mississippi rivers are listed in Table 10e.1.1.

TABLE 10e.1.1. Parameters h and ρ_1

River	h	ρ_1
Krishna	0.704	0.441
Godavari	0.78	0.3755
Mississippi	0.72	0.422

The fitted FN models for the three rivers are given below, labeled as M_1:

$$\text{Godavari:} \quad M_1 : y(t) = \sum_{j=1}^{800} w_1(t-j)/j^{1.5-h}, \qquad h = 0.78$$
$$\text{where} \quad w_1 \sim N(0, \rho_1 = 0.3755)$$

$$\text{Krishna:} \quad M_1 : y(t) = \sum_{j=1}^{800} w_1(t-j)/j^{1.5-h}, \qquad h = 0.704$$
$$\text{where} \quad w_1 \sim N(0, \rho_1 = 0.441)$$

$$\text{Mississippi:} \quad M_1 : y(t) = \sum_{j=1}^{800} w_1(t-j)/j^{1.5-h}, \qquad h = 0.72$$
$$\text{where} \quad w_1 \sim N(0, \rho_1 = 0.422).$$

It is important to note that the fractional noise model M_1 is compared with only the "best" model M_0 in the ARMA family, and not *any* model. An *arbitrarily* chosen ARMA model cannot be used for comparison with the FN model M_1 as the results obtained from such comparison can be very misleading. For example, for the data of Godavari River flows, an ARMA(1, 1) model was fitted by using the methods discussed earlier. The resulting model is

$$y(t) = 125.5y(t-1) + w(t) - 125.502w(t-1). \qquad (10e.1.1)$$

The noteworthy feature of the model given above is the very high values of the two coefficients in the model and the consequent instability of the model. The reason for this feature is that the lag 1 empirical correlation coefficient $\hat{C}(1)$ for the Godavari data is very small and is almost insignificant, and the ratio $\hat{C}(2)/\hat{C}(1)$ is correspondingly very large (125.5). Consequently, in order to obtain valid results, an appropriate member of the ARMA family must be used for comparison. Although simulation studies have indicated that the ARMA(1, 1) model can preserve both the lag 1 correlation coefficient and the R/σ characteristics of some FN models, it is clear from the example presented above that the ARMA(1, 1) model itself may be a very bad model for the observed process.

10e.2. Comparison of the Models M_0 and M_1 Based on the Correlation Coefficients

The theoretical correlation coefficients $\{\hat{R}_0(j), j = 1, 2, \ldots, 10\}$ of model M_0 and $\{\hat{R}_1(j), j = 1, \ldots, 10\}$ of model M_1 may be compared with the corresponding empirical correlation coefficients $\hat{C}(j), j = 1, 2, \ldots$. In Table 10e.2.1, the first 10 correlation coefficients $\hat{C}(j)$, $\hat{R}_0(j)$, and $\hat{R}_1(j)$ are listed for the annual data from the Krishna, Godavari, and Mississippi rivers. These results reveal that in each case, the empirical correlation coefficients $\hat{C}(j), j = 1, 2, \ldots$, appear to be closer to the coefficients $\hat{R}_0(j)$ of model M_0 rather than the coefficients $\hat{R}_1(j)$ of M_1.

Even if the observed data did come from the process obeying M_0, the observed correlation coefficient $\hat{C}(j)$ will not be identically equal to the corresponding theoretical value $\hat{R}_0(j)$ for all j since $\hat{C}(j)$ has been computed by using

TABLE 10e.2.1. Correlation Coefficients of Observed Flows and Their Models

River	Item	LAG, j (years)									
		1	2	3	4	5	6	7	8	9	10
Krishna	$\hat{R}_1(j)$	0.7321	0.6109	0.5366	0.4847	0.4457	0.4148	0.3896	0.3685	0.3505	0.3368
	$\hat{C}(j)$	0.2696	0.0834	0.1063	−0.168	−0.0847	0.0169	0.0786	−0.0769	−0.0234	−0.0364
	$\hat{R}_0(j)$	0.2696	0.0836	0.1063	−0.168	−0.101	−0.03	−0.058	0.0087	0.0228	0.005
	σ_j	0.148	0.149	0.151	0.152	0.153	0.155	0.156	0.157	0.159	0.161
Godavari	$\hat{R}_1(j)$	0.7841	0.6802	0.6141	0.5666	0.5301	0.5007	0.4763	0.4555	0.4375	0.4217
	$\hat{C}(j)$	0.0019	0.2394	0.0261	0.0383	−0.1795	−0.1332	0.0435	−0.1716	0.1358	−0.0094
	$\hat{R}_0(j)$	0.0019	0.2394	0.0008	0.057	0.0003	0.0137	0.00008	0.003	0.00002	0.0008
	σ_j	0.139	0.141	0.142	0.143	0.144	0.146	0.147	0.148	0.150	0.151
Mississippi	$\hat{R}_1(j)$	0.7429	0.625	0.5523	0.5011	0.4625	0.4318	0.4067	0.3856	0.3675	
	$\hat{C}(j)$	0.2967	0.0185	−0.0426	−0.033	0.0088	−0.0093	−0.0104	−0.0766	−0.0922	
	$\hat{R}_0(j)$	0.2967	0.088	0.026	0.0077	0.0023	0.0007	0.0002	0.00006	0.00002	
	σ_j	0.111	0.111	0.112	0.113	0.113	0.114	0.115	0.115	0.116	

a finite amount of data. In order to determine whether the difference between $\hat{C}(j)$ and $\hat{R}_0(j)$ can be explained by sampling variations we can proceed as follows. Let us consider the estimate $\bar{C}(j) = (1/N) \sum_{i=j+1}^{j+N} y(i)y(i-j)$, which is computed by using the samples $y(1), \ldots, y(N)$ generated from the model M_0 simulated on a digital computer. Let σ_j be the standard deviation of this estimate, which has the following expression:

$$\sigma_j^2 = E[((\bar{C}j) - \hat{R}_0(j))^2] = \left(1 + 2 \sum_{j=1}^{\infty} (\hat{R}_0(j))^2\right) \Big/ N.$$

From Tables 10e.2.1 it is seen that $|\hat{R}_0(j) - \hat{C}(j)| < 2\sigma_j$. Hence we conclude that the difference between $\hat{R}_0(j)$ and $\hat{C}(j)$ is not significant for $j = 1, \ldots, 10$ and the observed difference between $\hat{R}_0(j)$ and $\hat{C}(j)$ is due to sampling variations only. Hence the given data can be considered to have been generated from model M_0 in all the rivers.

On the other hand, the theoretical correlation coefficients $\hat{R}_1(j)$ of model M_1 are very different from the corresponding empirical correlation coefficients $\hat{C}(j)$ in all three rivers, which casts doubt on the validity of model M_1 to represent the given data.

10e.3. Comparison of Models by the Classical Theory of Hypothesis Testing

The classical theory of hypothesis testing has been discussed in Chapter VIII. We will consider the model M_0 as the null hypothesis and M_1 as the alternate hypothesis. Any decision rule for the choice between the two models leads to two types of errors. The true model may be M_0 but it may be classified as M_1 (type I error), and vice versa (type II error). Our aim is to find a decision rule that minimizes the probability of type II error, for a specified value of the probability of type I error. Such a rule is the likelihood ratio test described below, based on the observation history $\mathbf{y}(N_1, N_2) \triangleq (y(N_1), \ldots, y(N_2))^{\mathrm{T}}$:

$$\frac{p(\mathbf{y}(N_1, N_2)|M_0)}{p(\mathbf{y}(N_1, N_2)|M_1)} \begin{cases} \geq c_1 \to \text{choose } M_0 \\ < c_1 \to \text{choose } M_1 \end{cases} \tag{10e.3.1}$$

The threshold c_1, which is discussed later, is to be chosen on the basis of the prespecified value of the probability of type I error.

Let $N_2 - N_1 + 1 = N_0$. By definition of the model M_0, we have

$$p\{\mathbf{y}(N_1, N_2)|M_0\} = \frac{1}{(2\pi)^{m/2}|\det \mathbf{H}_0|^{1/2}} \exp[-\tfrac{1}{2}\mathbf{y}(N_1, N_2)^{\mathrm{T}}\mathbf{H}_0^{-1}\mathbf{y}(N_1, N_2)]. \tag{10e.3.2}$$

Similarly,

$$p\{\mathbf{y}(N_1, N_2)|M_1\} = \frac{1}{(2\pi)^{m/2}|\det \mathbf{H}_1|^{1/2}} \exp[-\tfrac{1}{2}\mathbf{y}(N_1, N_2)^{\mathrm{T}}\mathbf{H}_1^{-1}\mathbf{y}(N_1, N_2)], \tag{10e.3.3}$$

where $\mathbf{H}_0$ is the $N_0 \times N_0$ matrix made up of the correlation coefficients $\hat{R}_0(j)$,

$j = 0, 1, 2, \ldots$, of model M_0 and $\mathbf{H}_1$ is the corresponding $N_0 \times N_0$ matrix for model M_1:

$$(H_0)_{ij} = \hat{R}_0(|i - j|), \qquad i, j = 1, \ldots, N_0$$
$$(H_1)_{ij} = \hat{R}_1(|i - j|), \qquad i, j = 1, \ldots, N_0.$$

The correlation coefficients $\hat{R}_0(\cdot)$ and $\hat{R}_1(\cdot)$ for models M_0 and M_1 can be obtained from the formulae in Chapter II. The numerical values of the $\hat{R}_0(j)$ and $\hat{R}_1(j)$ for $j = 1, \ldots, 10$ can be found in Tables 10e.2.1 for the Krishna, Godavari, and Mississippi rivers.

We can simplify the left-hand side of (10e.3.1) after substituting for $p(\mathbf{y}(N_1, N_2)|M_0)$ and $p(\mathbf{y}(N_1, N_2)|M_1)$ from (10e.3.2) and (10e.3.3). The simplified decision rule (10e.3.4) is

$$x \triangleq \mathbf{y}(N_1, N_2)^{\mathrm{T}}(\mathbf{H})\mathbf{y}(N_1, N_2)\begin{cases} \leq c \rightarrow \text{choose } M_0 \\ > c \rightarrow \text{choose } M_1, \end{cases} \tag{10e.3.4}$$

where $c = -2 \ln c_1 - \ln|\det \mathbf{H}_0| + \ln|\det \mathbf{H}_1|$ and $\mathbf{H} = \mathbf{H}_0^{-1}. - \mathbf{H}_1^{-1}$.

The threshold c should be chosen so that the probability of type I error is a prespecified value such as 0.05. To choose the value of c, the probability distribution of the statistic x in (10e.3.4) is needed under the condition that the observations $\mathbf{y}(N_1, N_2)$ are drawn from model M_0. A closed form expression for this probability distribution is not available, but it can be approximately determined via simulation as follows. Simulate the model M_0 on a digital computer to obtain values of $y(i)$, $i = 1, \ldots, 100$. The last N_0 of these $y(\cdot)$ values are then used to compute the statistic x which is labeled $x(1)$. Similarly, r statistically independent runs can be performed which give r values of x, say $x(1), \ldots, x(r)$. A histogram of these r values of x gives a good approximation to the probability density of the statistic x under the hypothesis H_0, which is designated $P(x|M_0)$.

The probability of type I error P_1 for a given c is then computed as follows:

$$P_1 = \int_{x=c}^{\infty} p(x|M_0)\, dx.$$

The threshold c may then be selected so that P_1 is equal to the probability of type I error which was selected to be 0.05.

We will summarize the results obtained from the procedure discussed above only for data from the Krishna and Godavari rivers, for which 59 observations were available. As the computation of the probability density $P(x|M_0)$ is extremely time-consuming and complex for N_0 equal to 59, N_0 was selected to be 19. The corresponding histograms and the cumulative frequency distributions for the variable x are given in Figs. 10e.3.1 and 10e.3.2 for the data from the Godavari and Krishna rivers, respectively. The values of the threshold c were obtained from these figures corresponding to cumulative probabilities of 0.95. These values are listed in Table 10e.3.1: for the Godavari River, $c = 15.14$, for the Krishna, $c = -9.94$. The data of each river can be divided into three

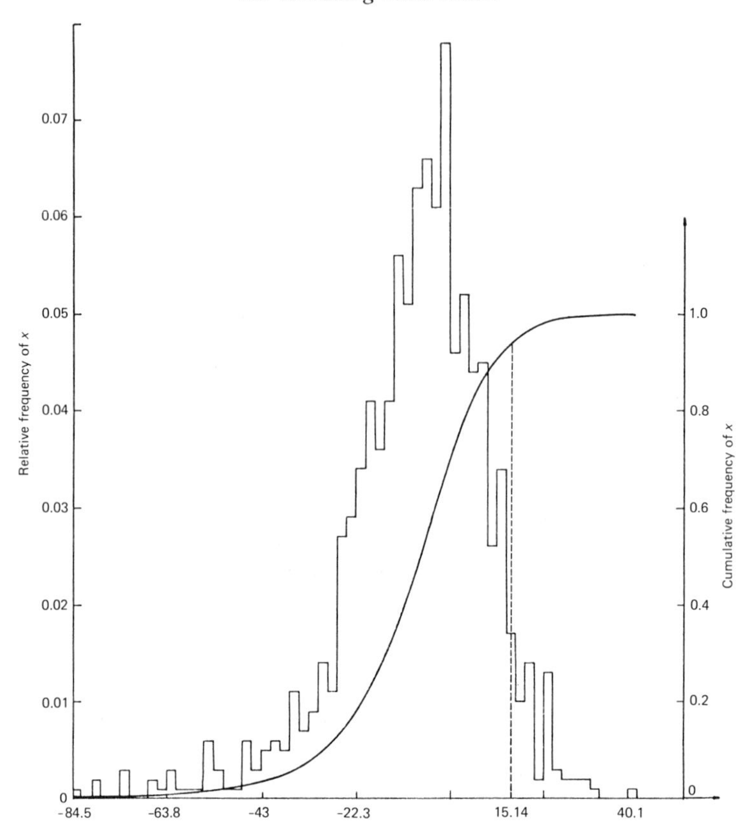

FIG. 10e.3.1. Histograms and cumulative frequency curves of the variable x for Godavari River data.

groups, each having 19 members, namely $\mathbf{y}(1, 19)$, $\mathbf{y}(20, 38)$, and $\mathbf{y}(39, 57)$. The value of the statistic x was separately computed for each group and these are also listed in Table 10e.3.1. The results for the Godavari River in Table 10e.3.1 indicate that the value of x is much smaller than c for all the groups, which indicates the overwhelming superiority of model M_0 over M_1. The results for the Krishna River show that the value of x in two of the groups is less than c,

TABLE 10e.3.1. Results of Comparison of Models by Classical Theory of Hypothesis Testing

River		x	c	$x < c$
Godavari	$\mathbf{y}(1, 19)^{\mathrm{T}}\mathbf{H}\mathbf{y}(1, 19)$	−70.3	15.14	Yes
	$\mathbf{y}(20, 38)^{\mathrm{T}}\mathbf{H}\mathbf{y}(20, 38)$	−20.17	15.14	Yes
	$\mathbf{y}(39, 57)^{\mathrm{T}}\mathbf{H}\mathbf{y}(39, 57)$	−32.68	15.14	Yes
Krishna	$\mathbf{y}(1, 19)^{\mathrm{T}}\mathbf{H}\mathbf{y}(1, 19)$	−47.45	−9.94	Yes
	$\mathbf{y}(20, 38)^{\mathrm{T}}\mathbf{H}\mathbf{y}(20, 38)$	−7.17	−9.94	No
	$\mathbf{y}(39, 57)^{\mathrm{T}}\mathbf{H}\mathbf{y}(39, 57)$	−24.32	−9.94	Yes

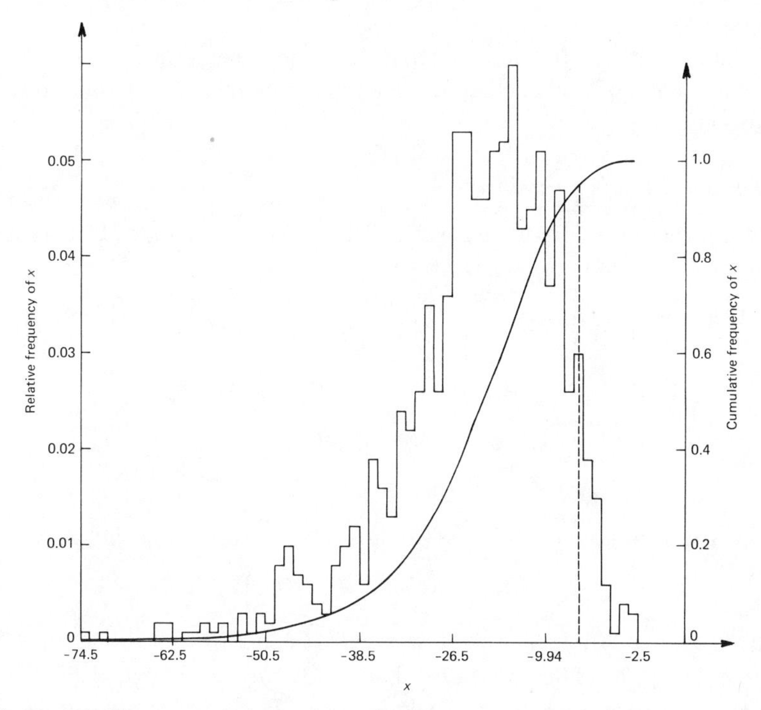

FIG. 10e.3.2. Histograms and cumulative frequency curves of the variable x for Krishna River data.

but the value of x for the other group is greater than c. The average of the three x values is -26.31. This is also substantially smaller than the value of c, which is equal to -9.94, indicating again the strong preference for model M_0 over M_1.

It is important to note that the above inference is not altered by small errors in the value c caused by small differences which may exist between the simulated histogram and the actual probability density of the variable x.

10e.4. The Relative Predictive Abilities of the Models from Classes M_0 and M_1

In this section we make use of the one-step-ahead prediction formulas given by the two models to obtain the actual one-step-ahead prediction of the annual flows and compute the corresponding observed mean square value of the prediction errors (abbreviated as prediction MSE). The model leading to the prediction formulas which yields the smaller of the two observed prediction MSE values should be more acceptable than the other model, especially if the difference between the two observed prediction MSE is substantial.

Second, we can inquire whether the observed MSE value of the optimal one-step-ahead forecasts of the flow variable computed from the prediction formulas, in each model, is close to the corresponding theoretical value of the mean square prediction error given by the model (ρ_0 in the case of M_0 and ρ_1

in the case of M_1). If the observed and theoretical mean square errors are very different from one another, then the corresponding model should be considered inconsistent.

10e.4.1. Prediction in a FN Process

We use the linear prediction method discussed in Chapter II. The optimal linear least squares predictor is $\bar{y}_r(t+1|t)$:

$$\bar{y}_r(t+1|t) = \sum_{j=0}^{r-1} d^*_{j+1,r} y(t-j), \tag{10e.4.1}$$

$$\mathbf{d}_r^* = (d^*_{1,r}, \ldots, d^*_{r,r})^{\mathrm{T}} = \mathbf{H}_1^{-1} \boldsymbol{\delta}_r.$$

$\mathbf{H}_{1,r}$ is an $r \times r$ matrix whose (i, j)th element is

$$(H_{1,r})_{i,j} = \hat{R}_1(|i-j|), \qquad i, j = 1, \ldots, r,$$

$\hat{R}_1(k)$ is the correlation coefficient of lag k for a process y obeying M_1:

$$\boldsymbol{\delta}_r = (\hat{R}_1(1), \ldots, \hat{R}_1(r))^{\mathrm{T}}.$$

To choose the integer r, we will compute the index $\bar{J}_r$ for different values of r, say 2, 3, ..., 20. The value of r that yields the smallest $\bar{J}_r$ defined in Eq. (10e.4.2) is chosen as the correct value. Let us say that r is equal to 20. Then the predictor in (10e.4.1) is used to obtain the one-step-ahead forecasts of $y(21), \ldots, y(N)$. The corresponding observed mean square prediction error is labeled $\bar{J}_r$:

$$\bar{J}_r = \frac{1}{N-20} \sum_{j=21}^{N} (y(j) - \bar{y}_r(j|j-1))^2. \tag{10e.4.2}$$

Since the estimate $\bar{J}_r$ was computed from $(N - 20)$ observations, the standard deviation of $\bar{J}_r$ is $[(2/(N-20))\bar{J}_r]^{1/2}$.

10e.4.2. Prediction in the AR Model M_0

The theoretical mean square value of the one-step-ahead prediction error is ρ_0. Let $y^*(i|i-1)$ be the least squares predictor, obtained from the AR model, of $y(i)$ based on $y(i-j), j > 1$:

$$y^*(i|i-1) = \sum_{j=1}^{m_1} A_j y(i-j). \tag{10e.4.3}$$

The empirical mean square prediction error J^* for the predictor $y^*(i|i-1)$ can be found by using the same set of observations $y(21), \ldots, y(N)$ previously used:

$$J^* = \frac{1}{N-20} \sum_{i=21}^{N} (y(i) - y^*(i|i-1))^2. \tag{10e.4.4}$$

10e.4.3. Results of Comparison of Prediction Errors

A. Godavari River

1. FN Model M_1. The empirical mean square errors $\bar{J}_r$ defined in (10e.4.2) were computed from optimal predictors of the type given in (10e.4.1). These are listed in Table 10e.4.1 for various values of r.

From Table 10e.4.1 we select r to be 2 since the corresponding $\bar{J}_2$ is the smallest value equal to 1.069. The theoretical value of the MSE is given by ρ_1, which is equal to 0.3755. Thus, the discrepancy between $\bar{J}_2$, the theoretically achievable value of mean square prediction error, and that given by the model, ρ_1, is very large, which indicates the poor prediction ability of the predictor $\bar{y}_2$ and the inadequacy of model M_1. Moreover, the empirical mean square prediction errors $\bar{J}_r$ in Table 10e.4.1 are all greater than unity, which is the empirical variance of the signal being predicted. This feature itself shows the poor performance of this predictor and model M_1.

TABLE 10e.4.1. Mean Square Prediction Error Variation with Lag r

River	r	2	3	4	5	10	15	20
Godavari	$\bar{J}_r$	1.069	1.111	1.077	1.104	1.083	1.076	1.078
Krishna	$\bar{J}_r$	0.6819	0.6409	0.6545	0.6435	0.64	0.646	0.6483
Mississippi	$\bar{J}_r$	1.10	1.08	1.07	1.065	1.055	1.065	1.095

2. AR Model. The theoretical value of the mean square prediction error is given by ρ_0, which is equal to 0.9053. The empirical mean square prediction error J^* as computed from (10e.4.3) and (10e.4.4) is equal to 0.724. The standard deviation of the estimate J^* is given by $(\frac{2}{39})^{1/2}\rho_0$, which is equal to 0.2045. Thus, the empirical mean square error J^* is within one standard deviation of the theoretical limit ρ_0. Moreover, the empirical mean square prediction error J^* computed by using the AR model is much smaller than the corresponding quantity $\bar{J}_2$ obtained from the FN model M_1. Thus, M_0 is clearly superior to M_1.

B. Krishna River

1. The FN Model M_1. The empirical mean square error $\bar{J}_r$ is listed in Table 10e.4.1 for various values of r. From the results given in Table 10e.4.1 the optimum value of r is found to be equal to 10, the corresponding MSE is equal to 0.64, whereas the theoretically achievable MSE is equal to 0.441. There is a wide discrepancy between the observed MSE and the theoretically achievable MSE, which is much smaller than the observed MSE in this case also.

2. The AR Model M_0. The theoretically achievable MSE, ρ_0, is equal to 0.8278. The observed MSE J^* from (10e.4.3) and model M_0 is equal to 0.611. The standard deviation of the estimate J^* is given by $(\frac{2}{39})^{1/2}\rho_0$ and is equal to 0.188. Thus, the observed MSE is within two standard deviations of the theoretically achievable minimum. Moreover, the observed minimum MSE with model M_0 is smaller than the observed minimum MSE with model M_1. Consequently, for the Krishna River flows also, the prediction ability of M_0 is superior to that of M_1.

C. Mississippi River

1. The FN Model M_1. The empirical mean square error $\bar{J}_r$ is listed in Table 10e.4.1 for various r. From the results given in Table 10e.4.1, the optimum value of r is about 10, the corresponding MSE being 1.055, which is greater than 1, the variance of signal. The predictor is not useful. As before, there is a wide discrepancy between the observed MSE and the theoretically achievable MSE of 0.422.

2. The AR Model M_0. The theoretically achievable MSE is 0.97. The observed MSE J^* equals 0.92. The standard deviation of the estimate J^* is $(\frac{2}{79})^{1/2}\rho_0$, and is equal to 0.1545. Thus, the observed MSE is within one standard deviation of the theoretically achievable minimum. Moreover, the observed minimum MSE with model M_0 is smaller than the observed minimum MSE with model M_1. Consequently, for Mississippi flows also, the predictive ability of model M_0 is superior to that of M_1.

It must be noted that the quantities $\bar{y}_r(t|t-1)$ in (10e.4.1) and $y^*(t|t-1)$ in (10e.4.3) are not strict predictors for $t < N$ because their coefficients are computed on the basis of the entire history of observations $y(1), \ldots, y(N)$. As such $\bar{J}_r$ and J^* can be described to be estimates of ρ_1 and ρ_0 instead of being labeled mean square prediction errors. But the thrust of the argument leading to the superiority of M_1 over M_0 is not affected by this consideration because it affects both predictors $\bar{y}$ and y^*.

We did not use split samples for parameter estimation and prediction tests because of the small number of available observations.

10e.5. Comparison Based on R/σ Characteristics

We will compare models M_0 and M_1 in their ability to preserve the R/σ characteristic of the observed flows. By definition and construction, the FN model reproduces the observed R/σ vs. s characteristic. Hence, we need to discuss model M_0 only. We will repeat the definition of rescaled range R.

By using the sequence of normalized observed flows $y(\cdot)$, the sequence $x(\cdot)$ is constructed in which t, s, and u are integers:

$$x(t, s, u) = \sum_{j=1}^{u} y(t+j) - \frac{u}{s}\sum_{j=1}^{s} y(t+j).$$

The rescaled range of the sequence $\{y(\cdot)\}$ at time t for lag s, is given by $R_3(t, s)$:

$$R_3(t, s) = \max_{0 \le u \le s} x(t, s, u) - \min_{0 \le u \le s} x(t, s, u).$$

The sample variance of the sequence $y(\cdot)$ in the interval $[t, t + s]$ is designated $\sigma_3{}^2(t, s)$:

$$\sigma_3{}^2(t, s) = \frac{1}{s} \sum_{u=1}^{s} y^2(t + u) - \left[\frac{1}{s} \sum_{u=1}^{s} (y(t + u))\right]^2.$$

Let $E[R_3(t, s)|M_0] = R_4(s)$, $E[\sigma_3{}^2(t, s)|M_0] = \sigma_4{}^2(s)$, $E[R_3(t, s)|M_1] = R_5(s)$, and $E[\sigma_3{}^2(t, s)|M_1] = \sigma_5{}^2(s)$. Our intention is to estimate the ratio $R_4(s)/\sigma_4(s)$ and $R_5(s)/\sigma_5(s)$.

For model M_0, an explicit expression cannot be obtained for the quantity $R_4(s)/\sigma_4(s)$ as a function of s. Hence, one has to simulate model M_0 on a computer and estimate the $R(s)/\sigma(s)$ vs. s characteristic from the generated data as discussed in Chapter II. The number of data points generated by model M_0 for subsequent use in plotting the R/σ characteristic is equal to 90.

In Figs. 10e.5.1a–c we have plotted the estimate $R(s)/\sigma(s)$ obtained from the data generated by a digital computer and model M_0 for the Krishna, Godavari, and Mississippi rivers, along with the bounds $R(s)/\sigma(s) + \mu(s)$, $R(s)/\sigma(s) - \mu(s)$, where $\mu(s)$ is the empirical standard deviation of the estimate $R(s)/\sigma(s)$ computed from the observed data by models as indicated in Chapter II.

For the data from the Krishna and Godavari rivers, the R/σ estimate of the river flow data is within one standard deviation of the R/σ estimate computed from samples from model M_0, for all lags less than 30, where 30 is roughly one-half of the sample size of 59. Similarly, for the Mississippi River the R/σ estimate of the river flow data is within one standard deviation of the R/σ estimate computed for the model for $s \leq 35$. Hence, model M_0 is consistent with the given data as far as the R/σ characteristic is concerned. Thus the R/σ characteristic by itself is unable to indicate the better model between M_0 and M_1.

There have been simulation studies of the AR models (Quimpo, 1967), and ARMA models (O'Connell, 1975). Results from these studies indicate that the slope h of the R/σ graph of AR models is always greater than 0.5. The results presented above demonstrate that the slope of the R/σ graph of the data simulated from the model reproduces the slope of the R/σ graph of the given flow data. The R/σ values computed by using the simulated data lie within the one standard deviation bound of the R/σ values computed by using the observed data. Further, the estimated standard deviations of the R/σ values computed from the observed flow data are comparable to the estimated standard deviations of the R/σ values computed from the simulated data.

It is often claimed in the literature (Mandelbrot and Wallace, 1969a) on "theoretical grounds" that AR models cannot reproduce the observed R/σ characteristic of the flow data. The argument runs as follows: Consider the range random variable $R_3(t, s)$ defined in Section 2i.1 constructed from the random sequence $\{y(t), \ldots, y(t + s)\}$. If the sequence $\{y(t), \ldots, y(t + s)\}$ obeys an AR model, then

$$E[R_3(t, s)|s] \triangleq R_4(t, s) = Ks^{0.5} \qquad \text{for large } s. \tag{10e.5.1}$$

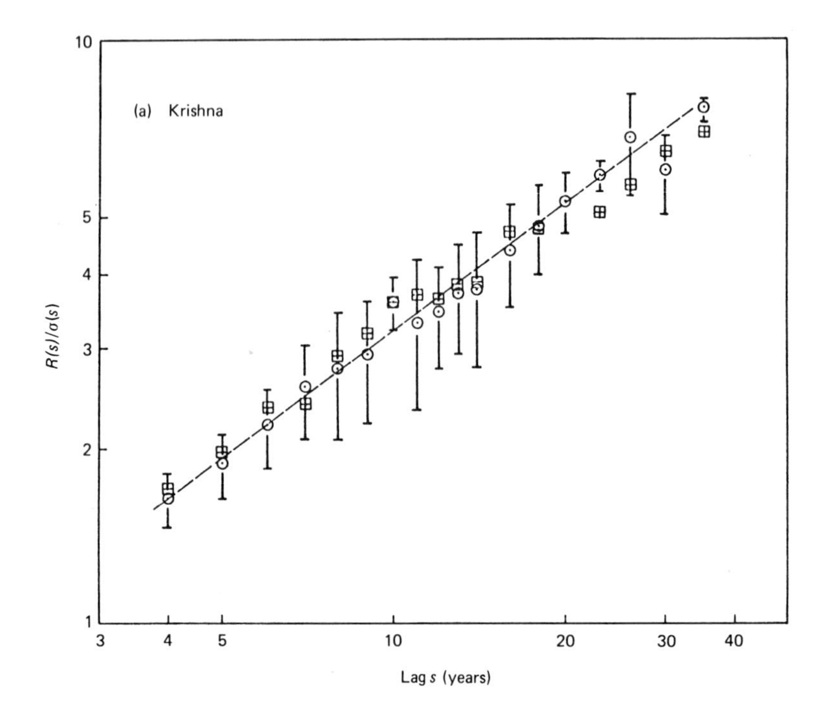
(a) Krishna
R(s)/σ(s)
10
5
4
3
2
1
3
4
5
10
20
30
40
Lag s (years)

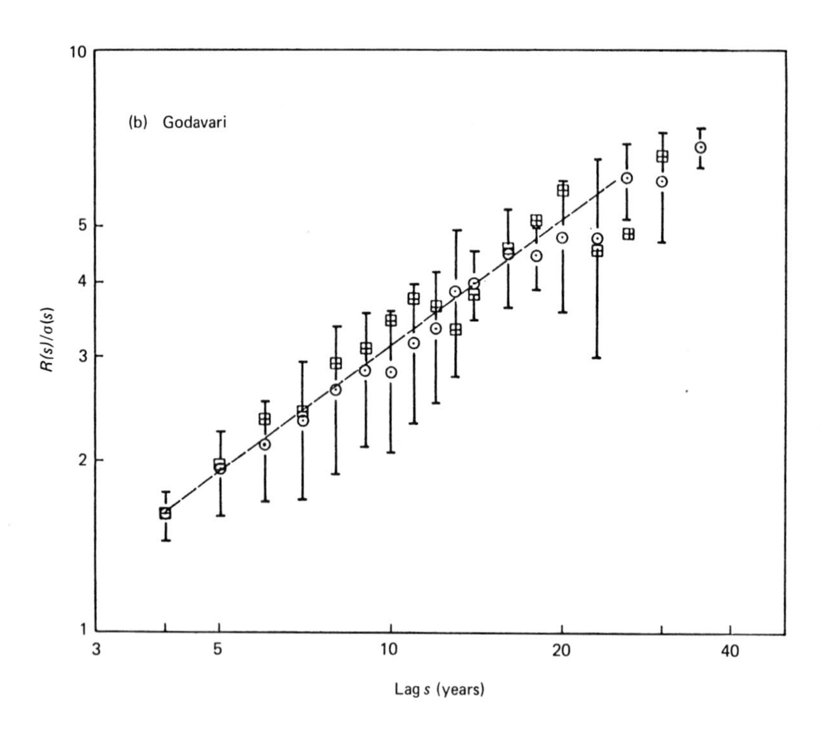
(b) Godavari
R(s)/σ(s)
10
5
4
3
2
1
3
5
10
20
40
Lag s (years)

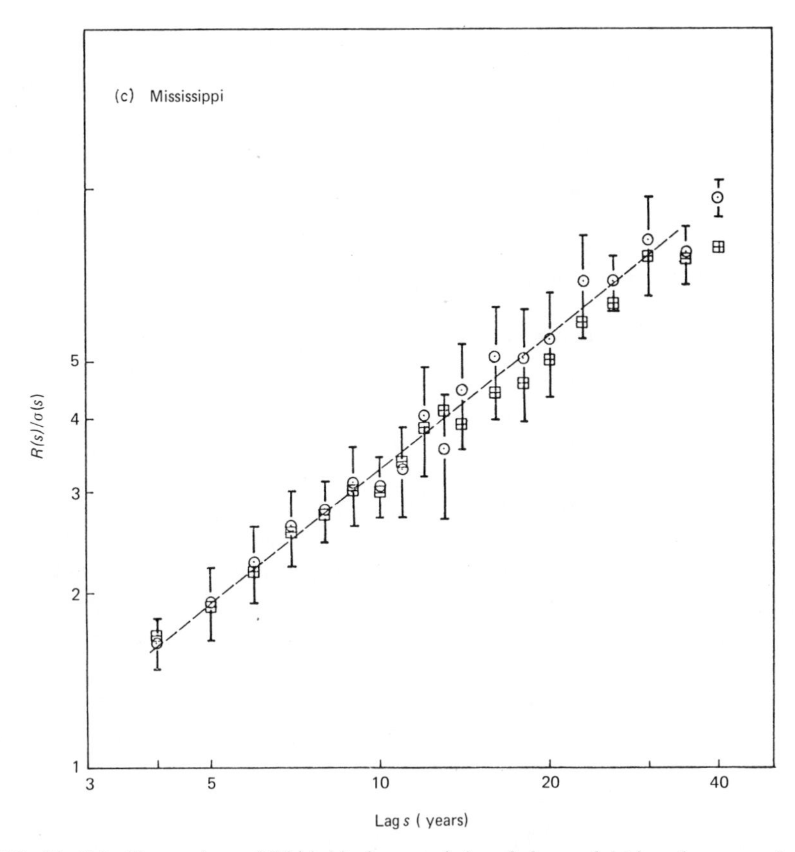

FIG. 10e.5.1. Comparison of $R(s)/\sigma(s)$ characteristics of observed (○) and generated (□) data [means (points) and standard deviations (bars)] on annual flows of (a) Krishna River at Vijayawada, (b) Godavari River at Dowleswaram, and (c) Mississippi River at St. Louis, Missouri.

Since the slope of the empirical R/σ graph obtained from the given data is distinctly greater than 0.5, it is claimed that the AR models are not appropriate. This argument deserves further analysis.

The relation (10e.5.1) is true only for large s. There is no theoretical or empirical evidence to believe that it is valid for small values of s, such as 100 or even several hundreds. At present, it is not known how large the value of s should be before (10e.5.1) is valid.

It is appropriate to mention that there are also some discrepancies in the explanation of the observed R/σ characteristics by the FN model as well. Consider the detrended monthly river flow data, i.e., the monthly river flow data from which the monthly sinusoidal trend components are removed. If the annual flows obey an FN model, the detrended monthly flows must also obey the same FN model (i.e., the h estimates should be the same or nearly the same in both of them). However, this is not so in practice. For example, for the

Godavari data, the h values are 0.88 and 0.78 for the detrended monthly and yearly data, respectively. The same is true for data from other rivers also.

One could argue that we are dealing with two different estimates of the same quantity h, and two arbitrary estimates of the same quantity cannot be expected to be identical. This argument would be reasonable if one could show that the observed difference between the two estimates is "typical," i.e., it is comparable with the expected value of the absolute difference between the two estimates. However, such a demonstration is not found in the literature.

10e.6. Bayesian Comparison of Models

The M_0 and M_1 models that are to be compared are

$$M_0: y(t) = \sum_{j=1}^{n} A_j y(t-j) + w(t)$$

$$M_1: y(t) = \sum_{j=1}^{M} w_1(t-j)/j^{1.5-h}.$$

$w(\cdot)$ and $w_1(\cdot)$ are sequences of normal, independently distributed (NID) random variables with zero mean and variances ρ_0 and ρ_1. The parameters n, $A_1, \ldots, A_n$, ρ_0, h, ρ_1 have been computed by using the techniques discussed earlier. M is a large integer, chosen here to be 800.

The empirical data for testing the validity of the models is the sequence $y(1), \ldots, y(N)$. As mentioned earlier, the observations have been normalized. Let

$$\mathbf{y}(N_1, N_2) = \begin{pmatrix} y(N_1) \\ \vdots \\ y(N_2) \end{pmatrix}.$$

The Bayesian decision theory has been discussed in Chapter VIII. The key feature in the Bayesian analysis is the posterior probabilities $P(M_0|\mathbf{y}(1, N))$ and $P(M_1|\mathbf{y}(1, N))$, where $P(M_0|\mathbf{y}(1, N))$ is the probability that model M_0 is the correct model on the basis of the observations $\mathbf{y}(1, N)$. $P(M_1|\mathbf{y}(1, N))$ is defined similarly.

Any decision involves errors, and the corresponding losses under the various contingencies can be represented by the matrix

		Action M_0	Action M_1
true model	M_0	d_{00}	d_{01}
	M_1	d_{10}	d_{11}

where d_{ij} is the loss caused by choosing model M_j to represent the given data when M_i is the correct model. Usually

$$d_{00} = 0, \quad d_{11} = 0$$
$$d_{01} = 1, \quad d_{10} = 1. \tag{10e.6.1}$$

Clearly we have to choose that decision rule which yields the least value of the average loss. The average losses for the two possible decisions are given by J_0, the average loss suffered by declaring model M_0 to be correct on the basis of data $\mathbf{y}(1, N)$:

$$J_0 = \{d_{00}P(M_0|\mathbf{y}(1, N)) + d_{10}P(M_1|\mathbf{y}(1, N))\}p(\mathbf{y}(1, N)); \quad (10e.6.2)$$

and J_1, the average loss suffered by declaring model M_1 to be correct on the basis of the data $\mathbf{y}(1, N)$:

$$J_1 = \{d_{01}P(M_0|\mathbf{y}(1, N)) + d_{11}P(M_1|\mathbf{y}(1, N))\}p(\mathbf{y}(1, N)). \quad (10e.6.3)$$

Hence the optimal decision rule, i.e., the decision rule that leads to the least average loss, is

$$J_0 \begin{cases} < J_1 \to \text{choose } M_0 \text{ as the model} \\ > J_1 \to \text{choose } M_1 \text{ as the model.} \end{cases} \quad (10e.6.4)$$

By Eqs. (10e.6.2) and (10e.6.3), we obtain

$$\frac{J_0}{J_1} < 1 \to \frac{P(M_0|\mathbf{y}(1, N))}{P(M_1|\mathbf{y}(1, N))} < \frac{d_{10} - d_{11}}{d_{01} - d_{00}}. \quad (10e.6.5)$$

Hence, combining (10e.6.4) and (10e.6.5) with the loss matrix in (10e.6.1), we get the following optimal decision rule:

$$\frac{P(M_0|\mathbf{y}(1, N))}{P(M_1|\mathbf{y}(1, N))} \begin{cases} < 1 \to \text{choose } M_1 \\ > 1 \to \text{choose } M_0, \end{cases} \quad (10e.6.6)$$

which is a reasonable rule. It simply asks us to choose the model that has the greatest posterior probability.

We now compute the posterior probabilities. Let $P(M_i)$, $i = 0, 1$, be the prior probability that M_i is the correct model:

$$P(M_0) + P(M_1) = 1.$$

The prior probabilities represent our degrees of belief in the validity of the two models *before* inspecting the data $\mathbf{y}(1, N)$. Let $P(M_0)$ and $P(M_1)$ be 0.5, implying that we have no prior preference for either model.

Let $P(M_i|\mathbf{y}(1, N))$ be the posterior probability that the model M_i is correct on the basis of the observations $\mathbf{y}(1, N)$. The posterior probabilities can be computed with the aid of Bayes law:

$$P(M_i|\mathbf{y}(1, N)) = \frac{p(\mathbf{y}(1, N)|M_i)P(M_i)}{p(\mathbf{y}(1, N)|M_0)P(M_0) + p(\mathbf{y}(1, N)|M_1)P(M_1)}.$$

By definition, the sum $P(M_1|\mathbf{y}(1, N)) + P(M_0|\mathbf{y}(1, N))$ is equal to one. By definition of model M_0, we can write the expression for $p(\mathbf{y}(1, N)|M_0)$:

$$p(\mathbf{y}(1, N)|M_0) = \frac{1}{(2\pi)^{N/2}|\det \mathbf{H}_0|^{1/2}} \exp[-\tfrac{1}{2}(\mathbf{y}(1, N))^{\mathrm{T}}\mathbf{H}_0^{-1}\mathbf{y}(1, N)]. \quad (10e.6.7)$$

$\mathbf{H}_0$ is the covariance matrix of $(y(1), \ldots, y(N))$ according to model M_0:

$$\mathbf{H}_0 = E\left[\begin{pmatrix} y(1) \\ y(2) \\ \vdots \\ y(N) \end{pmatrix} (y(1), y(2), \ldots, y(N))^{\mathrm{T}} \middle| M_0\right]$$

$$(H_0)_{ij} = \hat{R}_0(|i-j|), \qquad i, j = 1, \ldots, N.$$

$\hat{R}_0(k)$, $k = 0, 1, 2, \ldots$, are the correlation coefficients of model M_0 which may be computed as mentioned earlier.

Similarly, let $\mathbf{H}_1$ be the covariance matrix of the vector $(y(1), \ldots, y(N))$ according to model M_1:

$$(H_1)_{i,j} = \hat{R}_1(|i-j|), \qquad i, j = 1, 2, \ldots, N. \tag{10e.6.8}$$

$\hat{R}_1(k)$, $k = 0, 1, \ldots$, are the correlation coefficients of model M_1, which may be computed by the methods discussed earlier. The expression for $p(\mathbf{y}(1, N)|M_1)$ is

$$p(\mathbf{y}(1, N)|M_1) = \frac{1}{(2\pi)^{N/2}|\det \mathbf{H}_1|^{1/2}} \exp[-\tfrac{1}{2}\mathbf{y}(1, N)^{\mathrm{T}}\mathbf{H}_1^{-1}\mathbf{y}(1, N)]. \tag{10e.6.9}$$

Substituting the posterior probabilities (10e.6.7) and (10e.6.9) into (10e.6.6) we get the required decision rule.

Results. The numerical values of the posterior probabilities for the Krishna, Godavari, and Mississippi river data are given in Table 10e.6.1, and some of the numerical details are given in Table 10e.6.2. The results presented in Table 10e.6.1 indicate that the evidence is overwhelmingly in favor of model M_0 in all the rivers.

TABLE 10e.6.1. Results of Bayesian Comparison of Models

River	$P(M_1\|\mathbf{y}(1, N))$	$P(M_0\|\mathbf{y}(1, N))/P(M_1\|\mathbf{y}(1, N))$
Krishna	$3.6 \times 10^{-9.2}$	$2.75 \times 10^{8.2}$
Godavari	$10^{-22.4}$	$10^{22.4}$
Mississippi	$3.6 \times 10^{-18.8}$	$2.78 \times 10^{17.8}$

In the analysis given above both prior probabilities $P(M_0)$ and $P(M_1)$ were assumed to be 0.5 since we had no prior preference for either model. Even if a

TABLE 10e.6.2. Some Statistics Involved in Bayesian Comparison of Models

River	N	$\det \mathbf{H}_0$	$\det \mathbf{H}_1$	$\mathbf{y}^{\mathrm{T}}(1,N)\,\mathbf{H}_0^{-1}\mathbf{y}(1, N)$	$\mathbf{y}^{\mathrm{T}}(1, N)\mathbf{H}_1^{-1}\mathbf{y}(1, N)$
Krishna	59	3.29×10^{-4}	2.49×10^{-21}	58.89	138.0
Godavari	59	3.45×10^{-2}	3.48×10^{-26}	58.30	217
Mississippi	50	1.09×10^{-2}	8.3×10^{-22}	43.09	171.44

heavy preference is attached to model M_1, the evidence is strongly in favor of M_0. For instance, for Krishna, if $P(M_0) = 0.01$ and $P(M_1) = 0.99$,

$$P(M_0|\mathbf{y})/P(M_1|\mathbf{y}) = 2.75 \times 10^{6.2}:1.$$

Such a large ratio warrants some comment. From (10e.6.7) and (10e.6.8), we get

$$\frac{P(M_1|\mathbf{y}(1, N)]}{P(M_0|\mathbf{y}(1, N)]} = \frac{[\det \mathbf{H}_0]^{1/2}}{[\det \mathbf{H}_1]^{1/2}} \exp[-\tfrac{1}{2}\mathbf{y}^{\mathrm{T}}(1, N)\mathbf{H}_1^{-1}\mathbf{y}(1, N) + \tfrac{1}{2}\mathbf{y}^{\mathrm{T}}(1, N)\mathbf{H}_0^{-1}\mathbf{y}(1, N)], \tag{10e.6.10}$$

For the Krishna River, the quantity $\mathbf{y}^{\mathrm{T}}(1, N)\mathbf{H}_1^{-1}\mathbf{y}(1, N) = 138$, whereas the value $\mathbf{y}^{\mathrm{T}}(1, N)\mathbf{H}_0^{-1}\mathbf{y}(1, N)$ is only 58.89. Even though they may be considered to be of the same order, the exponent in (10e.6.10) makes the final value of the ratio in (10e.6.10) very small.

The only conclusion that can be drawn from the relative largeness of the quantity $\mathbf{y}^{\mathrm{T}}(1, N)\mathbf{H}_1^{-1}\mathbf{y}(1, N)$ as compared to $\mathbf{y}^{\mathrm{T}}(1, N)\mathbf{H}_0^{-1}\mathbf{y}(1, N)$ is that it is unlikely that the observations $\mathbf{y}(1, N)$ could have come from a model whose covariance matrix $E[\mathbf{y}(1, N)\mathbf{y}^{\mathrm{T}}(1, N)]$ is $\mathbf{H}_1$ defined in (10e.6.8).

Another interesting point is the relative values of the determinants $\mathbf{H}_0$ and $\mathbf{H}_1$. For the Krishna River, with $N = 59$,

$$\det \mathbf{H}_0 = 3.29 \times 10^{-4}, \qquad \det \mathbf{H}_1 = 2.49 \times 10^{-21}. \tag{10e.6.11}$$

By definition, the sum of all the eigenvalues of $\mathbf{H}_0$ is the sum of all the eigenvalues of $\mathbf{H}_1$, which equals 59. Equation (10e.6.11) implies that one or more eigenvalues of $\mathbf{H}_1$ must be vanishingly small and hence the matrix $\mathbf{H}_1$ is almost singular in comparison with $\mathbf{H}_0$. This feature again shows up the unsuitability of model M_1 since the covariance matrix in any reasonable model must be nonsingular and its determinant must be significantly different from zero.

10e.7. Discussion

In the preceding sections we discussed the comparison of models M_0 and M_1 from several points of view. The Bayesian test used for the comparison is computationally, as well as conceptually, the simplest test. The choice of prior probabilities for the correctness of the two models is arbitrary. However, this may not be very significant in the final inference as demonstrated by an example. The likelihood ratio test is free from this drawback, but it requires extensive computation for the determination of the threshold. Both of these tests involve the comparison of the probability distribution of all the observations with the two models. The importance of using these two tests for the choice of the model is self-explanatory when we consider that one of the principal purposes of a model is to generate synthetic flow data whose probabilistic characteristics should be close to those of the observed data. As one of the important uses of the model is in prediction, the comparison of models by means of their relative predictive abilities is also very natural. The comparison of the R/σ and correlation properties of the observed flow data with the corresponding characteristics obtained by models M_0 and M_1 is important as the data generated by the models must retain these characteristics.

Using these tests with the annual flow data from the Krishna, Godavari, and Mississippi rivers, we see that in each case the AR model M_0 is superior to the FN model M_1 from the point of view of the Bayesian test, classical hypothesis test, one-step-ahead predictive ability, and correlation characteristics. The R/σ characteristics of both models M_0 and M_1 are close to the corresponding characteristics of the actual flow data. Consequently, the AR model may be claimed to be superior to the FN model for all the rivers.

A criticism of the AR model is whether a constant coefficient model is appropriate for explaining the data ranging over 100 years. This question is discussed in some detail in Section 3a.4.3.

It is interesting to consider the physical significance of the random inputs $w(\cdot)$ in models M_0 and M_1. The random input sequence $w(i)$ in M_0 can be easily explained to be the rainfall input which occurs within the time interval $[i - 1, i]$ since it is well known that the annual rainfall series can be adequately modeled by a sequence of independent variables. But it is difficult to give a physical interpretation for the random input $w_1(\cdot)$ in model M_1. Interpretation of the random input $w_1(i)$ in model M_1 as a rainfall input is difficult to sustain since the river flow output in M_1 depends on the values of the input in the remote past. It is difficult to imagine that the present rainfall will influence the flow 10 or 20 years later in a substantial way. Both Scheidigger (1970) and Klemes (1974) have given vigorous critiques of the FN model on the ground that there does not exist any physical mechanism that can ensure such a long memory for the flow processes. Accordingly, the physical basis of the FN model is tenuous.

Further, it is claimed that the FN models have to be preferred since they can represent the long-term flow characteristics better than other models, and also because they explain the phenomenon of persistence. No precise definition of persistence is, however, available in the literature. As Monin (1970) and others have pointed out, the phrases long term, short term, etc., which have been used in discussions of persistence, lose their meaning unless the time constants of the process are specified. If the time constants are restricted from one to several years (less than about 20 years), as in this section, then a model of the ARMA family provides an adequate fit. Similarly, if persistence is assumed to be represented by the R/σ characteristic, even then the AR model investigated appears to represent the phenomenon of persistence adequately.

Finally, it has been claimed that the available data are not sufficient to discriminate between the FN and ARMA types of models. The available data base is insufficient if only the R/σ characteristic is used as the basis of judgment. However, if standard methods for comparing the different stochastic models such as the several tests discussed in the preceding sections are used, then the results can be definitive.

10f. Conclusions

We have discussed in detail the modeling of the river flows. The models for the daily, monthly, and yearly flows are quite different. The daily flow model is an AR equation, whereas the monthly flow model is a covariance stationary AR equation excited by a noise with time-varying variance. These two statements are not contradictory because the range of the validity of the models is different. The daily flow model can be used only for inferences involving several days (less than 30), whereas the monthly flow model is useful only for inferences up to a few months.

We have also analyzed in detail the hypothesis that the river flow process may not obey a finite stochastic difference equation model but a fractional noise model which can be represented by an infinite moving average equation. Similar hypotheses have been mentioned for many other meteorological processes. Our conclusion is that the available yearly and monthly observation sets do not lend support to such a hypothesis. Moreover, the mean square prediction error of the best fitting fractional noise model is much greater than that of the finite stochastic difference equation models treated here.

The models for the monthly and yearly flows are relatively simple and pass the validation tests. The daily flow model is relatively more complex and does not pass some of the validation tests. In particular, the daily flow models do not adequately explain the time-varying mean and standard deviation of the observed flows. The principal difficulty with the daily flow data is the availability of an extensive history of 15,000 observations. Such a lengthy history forces us to construct a very sensitive model for the given data so that the deviation between the characteristics of the output of the model and the data is of order $O(1/\sqrt{N})$, where N is the number of available observations. The appropriate model may not belong to the class of linear additive models considered here. We may have to investigate models with an entirely different structure, such as a multiplicative disturbance model. It should be emphasized that the problems encountered here have nothing to do with the daily flows per se. They have to do with the length of the data. If we take any 700 daily flow observations, we can construct an entirely satisfactory model as before. Similarly, we would have problems in modeling monthly flows if we were given 15,000 observations of monthly flows.

Finally, we note that the monthly models of rivers in the same climatic zone are similar to one another.

Notes

There is an extensive literature on the analysis of hydrologic time series and construction of models for them. The following references are only a small part cf the available literature.

The application of stochastic models of runoff series to water resources system design has been treated in detail in the books by Maass *et al.* (1962) and Hufschmidt and Fiering (1966). Dawdy and Matalas (1964) and Kisiel (1969) have reviewed the time series analysis of hydrologic data. Construction of stochastic models for hydrologic time series and simulation of these models are discussed by Chow (1964), Fiering (1967), Yevjevich (1972), and Fiering and Jackson (1971). Fiering and Jackson also briefly discuss several case studies in water resources engineering in which the stochastic models of hydrologic data play a prominent role. Clarke (1973) has dealt with some aspects of the application of ARIMA and similar other models to hydrologic processes. Additional references to hydrologic models may be found in the following conference proceedings:

"Systems Approach to Hydrology." Water Resources Publications, Fort Collins, Colorado (1971).

"Mathematical Models in Hydrology," *Proceedings of Warsaw Symposium*, International Association of Hydrological Sciences Pub. No. 101, (1974).

"International Symposium on Uncertainties in Hydrologic and Water Resource Systems." University of Arizona, Tucson, Arizona (1972).

"Symposium on the Design of Water Resources Projects with Inadequate Data," *Proceedings of the Madrid Symposium*, International Association of Hydrological Sciences (1973).

"Floods and Droughts." Water Resources Publications, Fort Collins, Colorado (1973).

Chapter XI | *Some Additional Case Studies in Model Building*

Introduction

In this chapter we consider some additional examples of modeling "real life" time series. There is considerable variety in the series considered here: some have growth, some possess almost periodic oscillations, some have both systematic oscillations and growth, and one involves smooth monotonic growth. Some are quite "ordinary," i.e., they have no noticeable features such as growth or systematic oscillations or smooth and monotonic growth or decay. All of them have observations numbering less than 300 so that there is no need for complex models similar to the daily river flow model discussed in Chapter X.

The principal purpose of the chapter is to illustrate the art of model building. In particular we demonstrate that different series which appear similar to one another may require entirely different models; hence, we emphasize the importance of considering several possible classes of models for the given series before selecting a model.

For instance, consider the three time series discussed in this book: the monthly river flow series in Chapter X, and the population time series of Canadian lynx and the annual sunspot series in this chapter. All three series have systematic periodic oscillations and no growth. Yet the appropriate models for them have completely different structures. The monthly flow series can be modeled by a first- or second-order AR model in y with sinusoidal terms added to it. A good model for the lynx series is an autoregressive model in the variable $\log y$ with sinusoidal terms added to it. Finally, the sunspot series is well represented by a high-order AR process with AR terms such as $y(t - [T - 1])$, where T is the period corresponding to dominant frequency and $[T - 1]$ is the nearest integer to $T - 1$, there being no need for sinusoidal terms. Similarly, the growth in the sales data has to be represented by means of the trend term t, whereas the IAR model with a constant term adequately explains the exponential-like growth in the U.S. population series and the whooping crane population.

Next we demonstrate how the validation tests indicate the limitations of the fitted model and thus help in choosing the correct model. For instance, consider both the lynx and the sunspot series. In each case, if we restrict ourselves to AR models, the class of second-order AR models is found to be the best. Now consider the best fitting AR(2) model in each case. The residuals from them pass all the whiteness tests, but the correlograms of the output of these models are bad fits to the corresponding empirical correlograms, showing the inadequacy of the AR(2) models. Hence, we have to try other classes until the correct class is identified.

We also consider testing an environmental hypothesis such as: "Urbanization increases rainfall." The validity of the hypothesis can be discussed only in a dynamical context because the rainfall time series is correlated in time. The fact that an invalid model for the rainfall can give erroneous inference about the validity of the hypothesis is demonstrated.

Next, we consider the problem of whether any purpose is served by the inclusion of moving average terms in a model. We will present a "real life" time series $E2$ (the first differences of the expenditure data) to show the superiority of a model with moving average terms over a model having only AR terms. The reason appears to be that the series exactly obeys an MA process. An AR model is only an approximation to this process and hence is inferior to the MA model regardless of the order of the AR model. We also note that the higher the order of the AR model, the poorer is the fit. Strangely enough, such an attempt at comparison has not been attempted earlier even though many general statements have been made on this problem. This example again shows the need for avoiding dogmatic approaches in model building.

Finally, we shall present an example of the construction of a multivariate model using the groundwater levels and rainfall at a number of stations. The causal relationships between the rainfall and groundwater levels are also discussed.

A large number of case studies of model building with varying degrees of detail can be found in the *IEEE Transactions On Automatic Control* (December 1974), Vansteenkiste (1975), Chisholm *et al.* (1971), and many other places. In many of these studies one finds the best fitting model for the given data in a prespecified class of models. There is usually no detailed comparison of the classes or comprehensive validation of the fitted model. Hence we will emphasize class comparison and validation aspects in these case studies.

One of the common errors in model building arises when assessing the need for models with time-varying coefficients. For instance, let $\hat{\theta}_i(t)$ be the estimate of a parameter θ_i in some model, say an AR model, based on t observations. Suppose we find that the difference between many pairs of members of the sequence $\{\hat{\theta}_i(\cdot)\}$ is very much greater than the standard deviation of these estimates. In such a case one often concludes that there is a need for a class of models with time-varying coefficients. Such an inference is often unwarranted. Often such behavior indicates that the class of models under consideration is inappropriate. If we consider the correct class of models, then the estimates of the coefficients in the best fitting model may not vary as drastically as those of incorrect models. This feature is well illustrated by the models of whooping crane population.

11a. Modeling of Some Biological Populations

11a.1. The Background

We consider three biological populations, namely the annual populations of the Canadian lynx (Fig. 3b.1.2), whooping crane (Fig. 11a.4.1), and the U.S. population (Fig. 3a.1.3). The U.S. population series is monotonically increasing at an exponential-like rate except for the war years 1943 and 1944. It has no systematic oscillations. The whooping crane series has an exponentially increasing component but does not increase monotonically. The population of Canadian lynx has no systematic growth or decay, but has systematic oscillations of approximately 9.7-year period.

There have been extensive attempts to model such series by deterministic differential equations such as those by Verhulst and Gompertz and there is an excellent discussion of such models in the review paper by Goel *et al.* (1971). The main criticism of these models is their poor prediction ability. This drawback manifests itself in the variation in the different estimates of the same coefficient in the differential equation based on different batches of data. One could argue that this is unavoidable while dealing with complex systems. But our experience is that such a feature, i.e., the drastic variation in the different estimates of the same coefficient based on different batches of data is due to the impropriety of the chosen structure of the model for the given data.

Goel *et al.* (1971) have suggested the stochastic versions of the Verhulst and Gompertz models by adding a multiplicative disturbance into the corresponding differential equations, i.e., the total disturbance is a product of the population variable and a white noise disturbance. Again there has not been any attempt to check the validity of the assumption involved, i.e., whether the disturbance is of the nature described above.

The models mentioned above are not the *only* models suggested by the physics of the problem. The only reliable method of testing the absolute necessity of an assumption for representing a process is to construct alternative models without involving the assumption and compare their relative performances by the techniques of Chapter VIII. For example, all the models described above have the property that the population (or the expected value of population in the stochastic case) tends to a constant as the time t tends to infinity. Such an assumption is difficult to prove or disprove with any amount of data. One could argue that the assumption is justified by the limited nature of resources. But any such assessment cannot be realistic because it ignores the adaptive possibilities of the species. Hence this type of argument is not fruitful. We should try to construct models that do not involve the assumption made above and compare their performances with those of the earlier models, by the methods of Chapter VIII.

Using the class selection methods outlined in Chapter VIII, we found that the most appropriate class of models for the U.S. population and the crane population are of the form

$$y(t) = x(t) \exp[c_1 t], \qquad c_1 > 0,$$

where $y(t)$ is the population variable and $\ln x(t)$ has zero mean and variance which increases linearly with t. The $\ln x(\cdot)$ obeys an IAR model with no constant term; i.e., $\ln y$ obeys an IAR equation with a constant term in it. Hence both the mean and the variance of the best fitting model for y increase exponentially with time. The best fitting model gives considerably better forecasts than both the deterministic and stochastic equations based on the Volterra equation. The forecast error in the one-year-ahead forecasts given by the best fitting model described here is only 0.1% of the predicted value, whereas it is about 5% with the deterministic model mentioned above.

The most appropriate model for the lynx data is a second-order autoregressive model in the variable $\log y$ which contains, in addition, the sinusoidal trend terms corresponding to the observed frequency of $(2\pi/9.7)$. This model gives better predictions and reproduces the statistical characteristics of the data better than many of the suggested models in the literature such as a second-order AR model in $\log y$.

11a.2. The Plausible Classes of Models

11a.2.1. The Deterministic Models†

The deterministic models are based on the following two assumptions: (i) The rate of change in population is proportional to the population. (ii) The rate of change is also proportional to a function $g(y)$ which is monotonically decreasing with time. This function is introduced to account for the possible limitations on the resources and the effect of predators that feed on the population. The differential equation is

$$\frac{dy}{dt} = yg(y). \tag{11a.2.1}$$

The function $g(y)$ is chosen so that the population tends to a fixed value as t tends to infinity. The stabilization of the population as t tends to infinity is an implication of the model which should be scrutinized.

We will consider two choices for the function $g(\cdot)$:

A. *Verhulst model:*

$$g(y) = c_3(1 - y/c_1), \qquad c_3, c_1 > 0. \tag{11a.2.2}$$

The corresponding solution for y from (11a.2.1) is

$$y(t) = \frac{c_1}{1 + [(c_1/y(0)) - 1] \exp(-c_3 t)}, \tag{11a.2.3}$$

† See Goel *et al.* (1971).

the classical logistic function. Starting from the value $y(0)$, where $y(0) < c_1$, the population steadily grows with time t and tends to c_1 as t tends to infinity.

B. *The Gompertz model:*

$$g(y) = c_4 - c_5 \ln y, \qquad c_4 \geq 0, \qquad c_5 > 0.$$

The corresponding solution for y is

$$\ln y(t) = (c_4/c_5) + (\ln y(0) - c_4/c_5) \exp[-c_5 t]. \tag{11a.2.4}$$

Note that the function $\ln y(t)$ monotonically *decreases* as t tends to infinity and $y(t)$ tends to the positive value $\exp[c_4/c_5]$.

Even though the deterministic models can give good fits to the data their predictive ability is rather poor. Moreover, it is difficult to imagine that a time series displaying considerable fluctuations can be successfully modeled by a purely deterministic model.

11a.2.2. The Stochastic Models

A natural stochastic extension of model (11a.2.1) can be obtained by adding a random input term signifying the combined effect of all the exogenous factors which affect the population, such as food, climate, and predators. As before, the intensity of the random input at any instant of time is proportional to the value of the population at that instant. Thus the stochastic version of (11a.2.1) is

$$\frac{dy}{dt} = yg(t) + y\eta(t). \tag{11a.2.5}$$

where $\eta(t)$ is another random process. The simplest assumption on $\eta(t)$ is that it is a zero mean "white noise." This is clearly unrealistic. It is difficult to imagine that the net effect of all the important factors that influence the population, such as climate and food, are uncorrelated with time. We will relax this assumption subsequently.

A. Stochastic Verhulst Model with $\eta(\cdot)$ White

$$\frac{dy}{dt} = c_3 y(1 - y/c_1) + y\eta. \tag{11a.2.6}$$

Equation (11a.2.6) becomes purely additive in η if we express it in terms of $\ln y(t)$:

$$\frac{d \ln y(t)}{dt} = c_3\left(1 - \frac{y(t)}{c_1}\right) + \eta(t). \tag{11a.2.7}$$

Since we have only observations at discrete instants of time, we will discretize the continuous equation (11a.2.7) by replacing the derivative by its estimate based on first difference:

$$\frac{d \ln y(t)}{dt} \approx \frac{\ln y(t + \delta) - \ln y(t)}{\delta}, \tag{11a.2.8}$$

where δ is a small interval of time. Substituting into (11a.2.7) the approximation for the derivative given in (11a.2.8) the following difference equation can be obtained:

$$\ln y(t + \delta) - \ln y(t) = \delta[c_3 - (c_3/c_1)y(t)] + w(t + \delta). \quad (11a.2.9)$$

Suppose our unit of time is one year. Therefore, we set $\delta = 1$, and Eq. (11a.2.9) can be rewritten as

$$\ln y(t) - \ln y(t - 1) = c_3 - c_4 y(t - 1) + w(t), \quad (11a.2.10)$$

where $c_4 = c_3/c_1$, and $\{w(t)\}$ is an IID sequence with zero mean and variance ρ, which is unknown. The parameters c_3, c_4, and ρ can be efficiently estimated as indicated in Chapter VI. These estimates have asymptotically minimum variance in the class of asymptotically unbiased estimates.

B. Stochastic Versions of the Gompertz Model with Nonwhite Disturbance. The stochastic version of the Gompertz model can be written as

$$\frac{dy}{dt} = y(c_4 - c_5 \ln y) + y\eta(t). \quad (11a.2.11)$$

Dividing (11a.2.11) throughout by y, we obtain

$$\frac{d \ln y}{dt} = c_4 - c_5 \ln y + \eta(t). \quad (11a.2.12)$$

In contrast to the earlier case, we assume that the random input term is not white. Instead, let us suppose that it obeys the first-order stochastic differential equation

$$\frac{d\eta}{dt} = a\eta + d(t) + v(t), \quad (11a.2.13)$$

where $a < 0$, $v(t)$ is a zero mean white noise sequence in continuous time, and $d(t)$ represents the sum of the constant term and trend terms such as t and $\cos \omega t$. The sinusoidal terms could represent the effect of climatic or metereological factors as in the river flow problem. We can eliminate $\eta(\cdot)$ from Eqs. (11a.2.12) and (11a.2.13). Differentiating (11a.2.12) with respect to t, we get

$$\frac{d^2 \ln y}{dt^2} = -c_5 \frac{d \ln y}{dt} + \frac{d\eta(t)}{dt}. \quad (11a.2.14)$$

In Eq. (11a.2.13), substitute for η in terms of y using (11a.2.12), and substitute for $d\eta/dt$ in terms of y using (11a.2.14) to obtain

$$\frac{d^2 \ln y}{dt^2} + c_5 \frac{d \ln y}{dt} = a \frac{d \ln y}{dt} - c_4 + c_5 \ln y + d(t) + v(t),$$

or

$$\frac{d^2 \ln y}{dt^2} + \alpha_1 \frac{d \ln y}{dt} + \alpha_2 \ln y = d_1(t) + v(t), \quad (11a.2.15)$$

where $\alpha_1 = c_5 - a$, $\alpha_2 = -ac_5$, and $d_1(t) = d(t) - ac_4$. Discretizing (11a.2.15) we obtain stochastic difference equation

$$\ln y(t) = \theta_1 \ln y(t-1) + \theta_2 \ln y(t-2) + d_1(t) + w(t), \quad (11a.2.16)$$

where $\{w(t)\}$ is a zero mean IID sequence with a common variance ρ.

We investigate three choices for $d_1(t)$: (i) $d_1(t) = \theta_0$, (ii) $d_1(t) = \theta_0 + \theta_3 t$, (iii) $d_1(t) = \theta_0 + \theta_3 \cos \omega_1 t + \theta_4 \sin \omega_1 t$. One should investigate case (ii) because it could account for the growth in the time series. Similarly, case (iii) may account for the systematic oscillations of approximate period $2\pi/\omega_1$ observed in the lynx time series among others. Thus we are led to three types of stochastic difference equations derived from the Gompertz models:

$$\ln y(t) = w(t) + \theta_0 + \theta_1 \ln y(t-1) + \theta_2 \ln y(t-2), \quad (11a.2.17)$$

$$\ln y(t) = w(t) + \theta_0 + \theta_1 \ln y(t-1) + \theta_2 \ln y(t-2) + \theta_3 t, \quad (11a.2.18)$$

$$\ln y(t) = w(t) + \theta_0 + \theta_1 \ln y(t-1) + \theta_2 \ln y(t-2) + \theta_4 \cos \omega_1 t + \theta_5 \sin \omega_1 t. \quad (11a.2.19)$$

C. Other stochastic models. All the models in Eqs. (11a.2.17)–(11a.2.19) are either stationary or covariance stationary in the variable $\ln y(t)$; i.e., the variance of the process $\ln y(t)$ tends to a constant as time t tends to infinity. In trying to model sequences that have strong exponential growth, such as the U.S. population, it may be useful to consider models such that the variable $\ln y(t)$ has both exponentially increasing mean and variance. This can be done by means of the following multiplicative model:

$$y(t) = v(t) \exp(b_1 t) \quad (11a.2.20)$$

where $\ln v$ obeys the IAR process

$$v_1(t) = \sum_{j=1}^{m_1} \theta_j v_1(t-j) + w(t) \quad (11a.2.21)$$

where $v_1(t) = \nabla \ln v(t) = \ln v(t) - \ln v(t-1)$. Using (11a.2.20) and (11a.2.21), we can obtain the following for $y(t)$:

$$y_1(t) = \theta_0 + \sum_{j=1}^{m_1} \theta_j y_1(t-j) + w(t), \quad (11a.2.22)$$

where $y_1(t) = \ln y(t) - \ln y(t-1)$. Thus, $\ln y$ obeys an IAR process of order m_1 with a constant term.

11a.3. Modeling the U.S. Annual Population Series

We consider the annual population series of the United States taken from *Statistical Abstracts of the United States* (1974).

11a.3.1. The Deterministic Model

Since we are dealing with a population that increases monotonically with time, the Verhulst model (11a.2.3) is the more appropriate of the two deterministic models. The three unknowns in model (11a.2.3), namely $y(0)$, c_1, and

c_3, can be evaluated using the population values in the years 1840, 1950, and 1960. The corresponding estimated model is

$$\hat{c}_1 = 246.5, \qquad \hat{c}_3 = 0.2984,$$

$$\text{Model } M_0\text{:} \quad y(t) = \frac{246.5}{1 + 2.243 \exp[-0.02984(t - 1900)]} \tag{11a.3.1}$$

where t is in calendar years and the population is in millions. The population graph of the fitted model is superposed on the actual population curve in Fig. 3a.1.3. The fit is very close in the years 1790–1930, but not very good (although satisfactory) in the years 1930–1960. We could have obtained a different fit by using the population of three other years to obtain the constants $y(0)$, c_1, and c_3. But there is no systematic and reasonable way to compare the different models in the deterministic framework. We could use a criterion such as least squares, but this is rather arbitrary. One of the advantages of dealing with the stochastic version of Eq. (11a.2.3) is that we have a systematic way of estimating its parameters.

Suppose we use the deterministic model (11a.3.1) to obtain one-step-ahead forecasts of the population in the years 1960–1971. The results are given in Table 11a.3.1. The predicted value is indicated as $\hat{y}$. The mean of the prediction errors is 7.796 and the mean square value of the prediction error is 73.162. The observed and predicted values obtained by model M_0 are given in Fig. 3a.1.4. Note that in Fig. 3a.1.4 the fractional forecast error, i.e., $(y - \hat{y})/y$, increases with t and is almost 6% in the year 1971. Thus we have an archetypical example of a deterministic model that fits the given data very well, but has poor prediction ability.

TABLE 11a.3.1. One-Step-Ahead Forecasts of U.S. Population Obtained by Using Deterministic Model (11a.3.1)

Year	y	$\hat{y}$	$y - \hat{y}$
1960	180.671	179.358	1.313
1961	183.691	180.806	2.884
1962	186.538	182.234	4.303
1963	189.242	183.641	5.600
1964	191.889	185.028	6.860
1965	194.303	186.395	7.910
1966	196.560	187.741	8.819
1967	198.712	189.066	9.646
1968	200.706	190.370	10.336
1969	202.677	191.653	11.024
1970	204.879	192.915	11.964
1971	207.049	194.156	12.893

11a.3.2. The Stochastic Models

We will consider the following four classes of stochastic models C_i, $i = 1, 2, 3, 4$, characterized by equations $\mathscr{s}_i$, $i = 1, \ldots, 4$. Let $y_1(t) = \ln y(t) - \ln y(t-1)$ and $y_2(t) = \ln y(t)$. Then

$$\begin{aligned}
\mathscr{s}_1 &: y_1(t) = w(t) + \theta_0 + \theta_1 y(t-1)\\
\mathscr{s}_2 &: y_2(t) = w(t) + \theta_0 + \theta_1 y_2(t-1) + \theta_2 y_2(t-2)\\
\mathscr{s}_3 &: y_1(t) = w(t) + \theta_0 + \theta_1 y_1(t-1)\\
\mathscr{s}_4 &: y_2(t) = w(t) + \theta_0 + \theta_1 y_2(t-1) + \theta_2 y_2(t-2) + (t-1)\theta_3.
\end{aligned}$$

Equation $\mathscr{s}_1$ is the stochastic version of the Verhulst model (11a.2.10). Equations $\mathscr{s}_2$ and $\mathscr{s}_4$ are derived from the stochastic versions of the Gompertz models. They are identical to (11a.2.17) and (11a.2.18), respectively. Equation $\mathscr{s}_3$ is identical to Eq. (11a.2.22) with $m_1 = 1$. We can create additional classes by defining other equations from (11a.2.22) with different values of m_1. But there is no need for additional classes since a model from class C_3 is found to be quite satisfactory.

We use the data in the years 1900–1971 to obtain the best fitting models in all classes listed below along with their residual variances. The best fitting models in class C_i are labeled M_i, $i = 1, \ldots, 4$:

$$\begin{aligned}
M_1 &: y_1(t) = w(t) + 0.0176 - 2.75 \times 10^{-4} y(t-1), \qquad \hat{\rho}_1 = 6.47 \times 10^{-5}\\
M_2 &: y_2(t) = w(t) + 0.0162 + 1.728 y_2(t-1) - 0.73 y_2(t-2),\\
& \qquad \hat{\rho}_2 = 2.65 \times 10^{-5}\\
M_3 &: y_1(t) = w(t) + 0.0034 + 0.7528 y_1(t-1), \qquad \hat{\rho}_3 = 8.77 \times 10^{-6}\\
M_4 &: y_2(t) = w(t) + 0.03836 + 0.000435(t-1)\\
& \qquad + 1.689 y_2(t-1) - 0.72 y_2(t-2), \qquad \hat{\rho}_4 = 5.54 \times 10^{-5}.
\end{aligned}$$

The classes are compared using the likelihood approach of Section 8b.1. Since all the models involve logarithmic transformations, we can compare them by using the following statistics, where n_i and $\hat{\rho}_i$ are the number of parameters and the residual variance, respectively:

$$\begin{aligned}
\hat{L}_i &= -\frac{N}{2} \ln \hat{\rho}_i - n_i\\
\hat{L}_1 &= -\tfrac{72}{2} \ln(6.47 \times 10^{-5}) - 2 = 345.25\\
\hat{L}_2 &= -\tfrac{72}{2} \ln(2.65 \times 10^{-5}) - 3 = 376.38\\
\hat{L}_3 &= -\tfrac{72}{2} \ln(8.77 \times 10^{-6}) - 2 = 417.19\\
\hat{L}_4 &= -\tfrac{72}{2} \ln(5.54 \times 10^{-5}) - 4 = 348.83.
\end{aligned}$$

$\hat{L}_3$ has the *maximum* value among $\hat{L}_i$, $i = 1, \ldots, 4$, and hence class C_3 has to be preferred. This is also expected since $\hat{\rho}_3$ is considerably smaller than $\hat{\rho}_1$, $\hat{\rho}_2$, and $\hat{\rho}_4$.

We will confirm this decision by means of the prediction approach of Section 8b.4.1. We will obtain the best fitting models using only the data in the years

1900–1947, and then use these fitted models to obtain one-year-ahead forecasts of the years 1948–1971. The models fitted to the data 1900–1947 are

$$M_1'\colon y_1(t) = w(t) + 0.0326 - 1.71 \times 10^{-4} y(t-1),$$
$$M_2'\colon y_2(t) = w(t) + 0.0969 + 1.519 y_2(t-1) - 0.527 y_2(t-2)$$
$$M_3'\colon y_1(t) = w(t) + 0.0038 + 0.717 y_1(t-1).$$

The forecasts and the forecast errors are listed in Table 11a.3.2, and the forecasts from M_3' are plotted in Fig. 11a.3.1. The mean and the mean square value of the 24 forecast errors are also given in Table 11a.3.2.

Model M_3' yields the least value of the mean absolute error and the mean square forecast error among all the models, indicating the superiority of class C_3. The typical forecast error with M_3' is about 0.1% of the population value,

TABLE 11a.3.2. Forecasts Obtained from Stochastic Models for U.S. Population

		Model M_1'		Model M_3'		Model M_2'	
Year	$y(t)$	$\hat{y}(t)$	$y(t) - \hat{y}(t)$	$\hat{y}(t)$	$y(t) - \hat{y}(t)$	$\hat{y}(t)$	$y(t) - \hat{y}(t)$
1948	147.2	145.8	1.4	147.3	−0.1	146.5	0.7
1949	149.8	148.8	1.0	149.6	0.2	148.9	0.9
1950	152.3	150.8	1.5	152.2	0.1	151.5	0.8
1951	154.9	153.3	1.6	154.7	0.2	153.9	1.0
1952	157.6	155.8	1.8	157.4	0.2	156.6	1.0
1952	160.2	158.4	1.8	160.1	0.1	159.3	0.9
1954	163.0	161.0	2.0	162.7	0.3	161.8	1.2
1955	165.9	163.8	2.1	165.7	0.2	164.8	1.1
1956	168.9	166.6	2.3	168.7	0.2	167.7	1.2
1957	172.0	169.5	2.5	171.7	0.3	170.7	1.3
1958	174.9	172.5	2.4	174.9	0.0	173.8	1.1
1959	177.8	175.4	2.4	177.7	0.1	176.6	1.2
1960	180.7	178.2	2.5	180.7	0.0	179.5	1.2
1961	183.7	181.0	2.7	183.4	0.3	182.3	1.4
1962	186.5	183.9	2.6	186.6	−0.1	185.4	1.1
1963	189.2	186.7	2.5	189.3	−0.1	188.1	1.1
1964	191.9	189.3	2.0	191.9	−0.6	190.7	0.6
1965	194.3	191.9	2.4	194.6	−0.3	193.3	1.0
1966	196.6	194.2	2.4	196.8	−0.2	195.6	1.0
1967	198.7	196.4	2.3	199.0	−0.3	197.7	1.0
1968	200.7	198.4	2.3	201.0	−0.3	199.8	0.9
1969	202.7	200.4	2.3	202.9	−0.2	201.7	1.0
1970	204.9	202.3	2.6	204.9	0.0	203.7	1.2
1971	207.0	204.4	2.7	207.3	−0.2	206.0	1.1
Mean forecast error		$\bar{e}_1 = 2.17$		$\bar{e}_1 = 0.19$		$\bar{e}_1 = 1.04$	
Mean square forecast error		$\bar{e}_2 = 4.91$		$\bar{e}_2 = 0.053$		$\bar{e}_2 = 1.12$	

$\bar{e}_k = \frac{1}{24} \sum_{t=1948}^{1971} |y(t) - \hat{y}(t|t-1)|^k.$

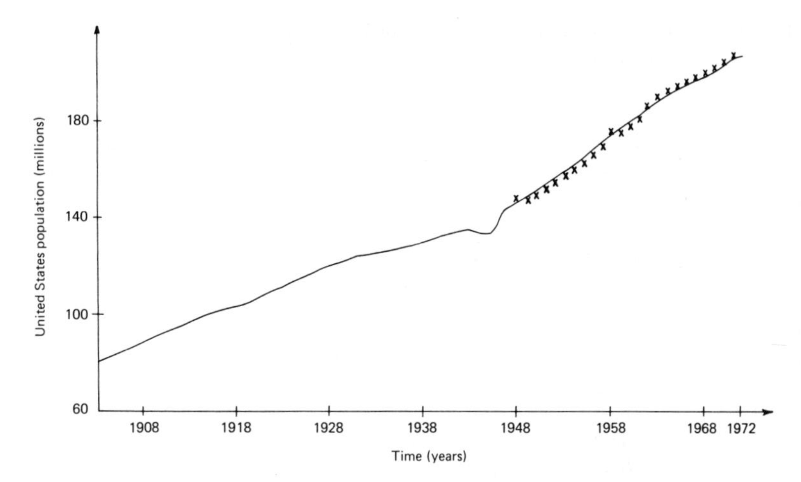

FIG. 11a.3.1. Prediction of United States population by model M_3'.

whereas it ranges from 5 to 1% in other models. Model M_3 passes all validation tests at the 95% significance level.

It is worth emphasizing that the best fitting model M_3 implies that both the mean and the variance of its output increase exponentially with time. There is little statistical support for the stochastic generalization of Verhulst and Gompertz models which indicate that the mean and variance of the population tend to fixed values as t tends to infinity. This conclusion has been verified by both the likelihood and the predictive approach. Let us consider several other aspects.

A. Long-Term Forecasting. We will show that model M_3 gives much better five- and 10-year-ahead forecasts than the deterministic model. Let $\hat{y}(t|t-k)$ denote the optimal forecasts of $y(t)$ based on $y(t-k)$, $t > k$, model M_3, and the loss function $E[(\ln y(t) - \ln \hat{y}(t|t-k))^2]$:

$$\hat{y}(t|t-k) = \exp \hat{y}_2(t|t-k),$$

where $\hat{y}_2(t|t-k) = E[\ln y(t)|t-k]$. To compute $\hat{y}_2(t|t-k)$ for various k, we proceed as follows: (i) Compute all the one-step-ahead predictors $\hat{y}_2(t|t-1)$ for the relevant t. (ii) Next compute $\hat{y}_2(t|t-2)$ for various t using the predictors found in step (i):

$$\hat{y}_2(t|t-2) = \hat{y}_2(t-1|t-2) + 0.0034 + 0.7528[\hat{y}_2(t-1|t-2) - y_2(t-2)].$$

(iii) We can compute $\hat{y}_2(t|t-3)$ for various t in a similar manner. In general, $\hat{y}_2(t|t-k) = \hat{y}_2(t-1|t-k) + 0.0034 + 0.7528[\hat{y}_2(t-1|t-k) - \hat{y}_2(t-2|t-k)]$.

In Table 11a.3.3, we list the population forecasts by the deterministic model M_0, the five-year-ahead forecasts $\hat{y}(t|t-5)$ and the ten-year-ahead forecasts $\hat{y}(t|t-10)$ for various t using M_3. These forecasts are graphed in Fig. 11a.3.2. Note that the deterministic model consistently and grossly underestimates the population, whereas the $\hat{y}(t|t-5)$ and $\hat{y}(t|t-10)$ forecasts are on either side of

TABLE 11a.3.3. Five- and Ten-Year Ahead Forecasts of U.S. Population

Year t	$y(t)$	$\hat{y}(t\|t-5)$	$e_1(t)$	$\hat{y}(t\|t-10)$	$e_2(t)$	$\bar{y}(t)$ forecast by M_0	$e_0(t)$
1954	163.03						
1955	165.93	164.06	1.87				
1956	168.90	166.39	2.51				
1957	171.98	168.93	3.05				
1958	174.88	171.65	3.23				
1959	177.83	174.28	3.55				
1960	180.67	177.24	3.43	175.97	4.70		
1961	183.69	180.29	3.40	178.26	5.43	180.81	2.88
1962	186.54	183.42	3.12	180.85	5.69	182.23	4.31
1963	189.24	186.79	2.45	183.75	5.49	183.64	5.60
1964	191.89	189.89	2.00	186.37	5.52	185.03	6.86
1965	194.30	193.11	1.19	189.46	4.84	186.40	7.90
1966	196.56	196.12	0.44	192.67	3.89	187.74	8.82
1967	198.71	199.45	−0.74	195.94	2.77	189.07	9.64
1968	200.71	202.59	−1.88	199.72	0.99	190.37	10.34
1969	202.68	205.55	−2.87	203.01	−0.33	191.65	11.03
1970	204.88	208.51	−3.63	206.57	−1.69	192.92	11.96
1971	207.05	211.19	−4.14	209.67	−2.62	194.16	12.89
1972	208.84	213.68	−4.84	213.42	−4.58	195.38	13.46
1973	210.40	216.07	−5.67	216.93	−6.53	196.57	13.83
1974	211.78	218.25	−6.47	220.17	−8.39	197.75	14.03
1975		220.32		223.57		198.91	
1976		222.74		226.60		200.04	
1977		225.22		229.38		201.16	
1978		227.26		232.11		202.25	
1979		229.34		234.49		203.33	
1980				236.53		204.38	
1981				239.17		205.41	
1982				242.17		206.42	
1983				244.59		207.41	
1984				247.12		208.38	
		$\bar{e}_1 = 0.074$		$\bar{e}_2 = 1.056$		$\bar{e}_0 = 9.54$	
		$\overline{e_1^2} = 11.34$		$\overline{e_2^2} = 22.374$		$\overline{e_0^2} = 103.25$	

$\overline{e_1^k} = \frac{1}{20}\sum_{t=1955}^{1974}(y(t) - \hat{y}(t|t-5))^k$,

$\overline{e_2^k} = \frac{1}{15}\sum_{t=1960}^{1974}(y(t) - \hat{y}(t|t-10))^k$, $\overline{e_0^k} = \frac{1}{14}\sum_{t=1961}^{1974}(y(t) - \bar{y}(t))^k$.

the actual observed population graph. We may mention that a criticism can be made of the $\hat{y}(t|t-5)$, $\hat{y}(t|t-10)$ forecasts for $t \leq 1970$, since these forecasts have been computed from model M_3, which was constructed using data up to the year 1971. However, this criticism is not valid for forecasts $\hat{y}(t|t-5)$ and $\hat{y}(t|t-10)$ for $t > 1971$. Even these estimates $\hat{y}(t|t-10)$ and $\hat{y}(t|t-5)$ for $t > 1971$ are much closer to $y(t)$ than the forecasts of the deterministic model. For instance, consider the forecast of the population for 1974 using the data

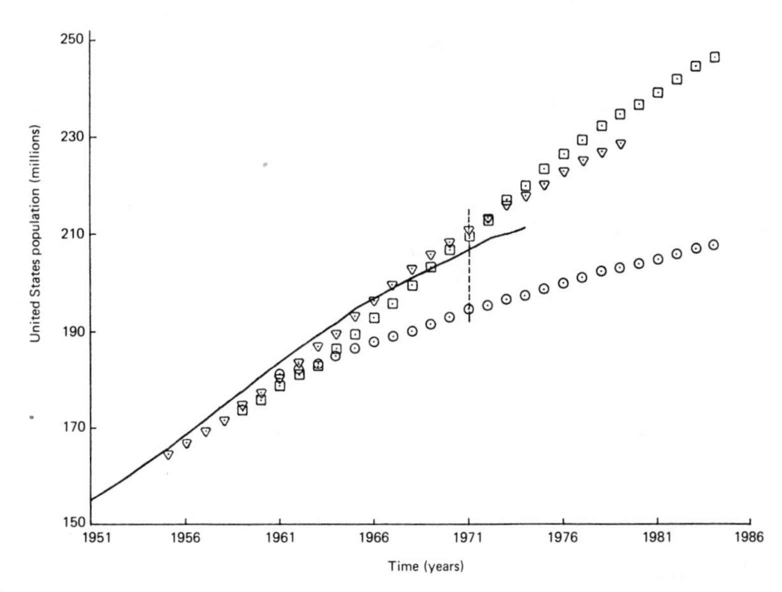

FIG. 11a.3.2. Five- (▽) and ten- (□) year-ahead forecasts of United States population using M_3; ○ = logistic curve forecasts.

up to 1969. The forecast by M_0 is 197.75, leading to an error of 14.03, whereas that by M_3 is 218.25 with an error of -6.47.

As mentioned earlier, we can construct a large number of different deterministic models by considering the population in any three different years; some of the models should give better forecasts than others. For instance, Croxton and Cowden (1939) suggest the following deterministic model based on the data of the years 1790–1930:

$$y(t) = \frac{190.830}{1 + \exp\{3.550671 - 0.3145945(t - 1800)/10\}}.$$

If we use this model to forecast population in the years 1960 and 1970, the forecast errors are about 14 and 20.1%.

B. Comparison of Models M_i and M_i', $i = 1, 2, 3, 4$. It is also interesting to compare the pair of models M_i and M_i' for $i = 1, \ldots, 4$, which are computed using the data of 72 and 48 years, respectively. Note that the corresponding coefficients in M_3 and M_3' are nearly the same; the difference between them can be ascribed to the sampling fluctuations. But the corresponding coefficients in the pairs M_1 and M_1', M_2, and M_2' are considerably different, showing the inappropriateness of classes C_1 and C_2. Comparing M_1 and M_1' or M_2 and M_2' we are not entitled to declare the need for a model with time-varying coefficients because both M_1 and M_2 are not valid models.

C. Comparison of Classes C_2 and C_3. Another feature of considerable interest in the general theory of modeling is the large difference in the predictive capabilities of the best fitting models from C_3 and C_2. Both appear to be second-order autoregressive equations in the variable $y_2(t) = \ln y(t)$ with the constant term added to them. But their residual variances and predictive capabilities are vastly different. Of course, the discrepancy in the performances of the two models is to be expected since the IAR model with the constant term leads to asymptotic growth, whereas the ordinary AR model leads to steady state behavior in the means and variances. This clearly shows the dangers in using an ordinary AR model for a process that obeys an IAR model having a constant term.

D. Comparison of Model M_1 with Deterministic Model M_0. Recall that model M_1 is the stochastic counterpart of the deterministic model M_0. We can therefore compare their coefficients. From Eq. (11a.3.1), we get the estimates $\hat{c}_1, \hat{c}_2$:

$$M_0: \hat{c}_1 = 246.5, \qquad \hat{c}_3 = 0.02984.$$

$$M_1: \hat{\theta}_0 = c_3 = 0.0326, \qquad \hat{\theta}_1 = -c_4 = -1.71 \times 10^{-4},$$

or

$$c_1 = c_3/c_4 = 0.0326/(1.71 \times 10^{-4}) = 190.64.$$

Note the discrepancy between the estimates of c_1 in M_0 and M_1. Also note the vast difference in the forecast errors between the deterministic model M_0 in (11a.3.1) and its stochastic version M_1, as given in Tables 11a.3.1 and 11a.3.2. The superiority of the stochastic model can be attributed to two causes: (i) The stochastic model M_1 uses the actual value of the observations in all the years prior to the forecast. This is not the case in forecasting by model M_0; i.e., forecasting with M_0 utilizes less information than M_1. (ii) The parameters are estimated in a systematic manner in M_1, the stochastic model. In the deterministic model M_0, parameter estimation is rather arbitrary because we estimated the three coefficients using three arbitrary population values.

11a.4. The Annual Population Series of the Whooping Crane

We consider the annual population of the whooping crane in the United States in the years 1938–1972 as given by Miller and Botkin (1974). The series is shown in Fig. 11a.4.1. There is a strong growth component in the series superposed on an irregular varying part.

In view of the strong irregular fluctuations there is no point in considering the deterministic models. We will consider only four classes of models: C_i, $i = 0, 1, 2, 3$. The corresponding equations are given below.

Let $y_1(t) = \nabla \ln y(t) = \ln y(t) - \ln y(t-1)$ and $y_2(t) = \ln y(t)$. Then

$$\mathcal{C}_0: y_1(t) = w(t) + \theta_0 + \theta_1 y(t-1)$$

$$\mathcal{C}_1: y_1(t) = w(t) + \theta_0 + \theta_1 y_1(t-1)$$

$$\mathcal{C}_2: y_2(t) = w(t) + \theta_0 + \theta_1 y_2(t-1) + \theta_2 y_2(t-2)$$

$$\mathcal{C}_3: y(t) = w(t) + \theta_0 + \theta_1 y(t-1).$$

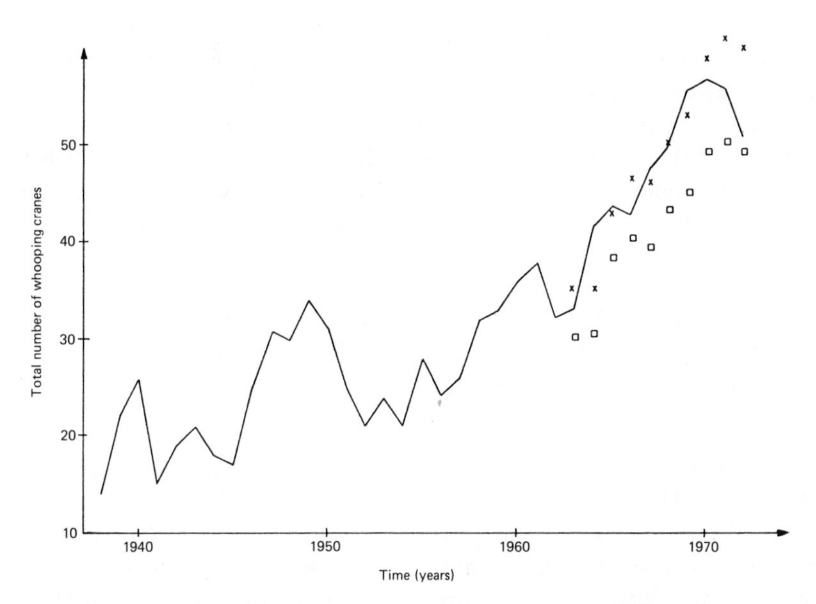

FIG. 11a.4.1. Observed and forecast whooping crane population; × = model M_1' forecast, □ = model M_3' forecast.

Equation $\mathscr{I}_0$ is the stochastic generalization of the Verhulst model of Eq. (11a.2.10). Equation $\mathscr{I}_2$ is the stochastic generalization of the Gompertz model in Eq. (11a.2.17). Equation $\mathscr{I}_1$ is the integrated autoregressive model that proved successful in the representation of the U.S. population. Equation $\mathscr{I}_3$, the ordinary AR process, was used in the earlier crane population studies (Miller and Botkin, 1974). We want to show the loss in the predictive ability caused by the use of such inappropriate models. We can say that $\mathscr{I}_3$ is inappropriate because it can represent only a weak stationary process, which is not the case with the crane population.

We use the 35 observations in the years 1938–1972 to estimate the parameters in the various equations and the corresponding residual variances. The fitted models M_i, $i = 0, 1, 2, 3$, are

$$\begin{aligned}
M_0\colon y_1(t) &= w(t) + 0.104 - 0.00247y(t-1), & \hat{\rho}_0 &= 0.0309\\
M_1\colon y_1(t) &= w(t) + 0.0312 - 0.136y_1(t-1), & \hat{\rho}_1 &= 0.0302\\
M_2\colon y_2(t) &= w(t) + 0.257 + 0.757y_2(t-1) + 0.174y_2(t-2), & \hat{\rho}_2 &= 0.0295\\
M_3\colon y(t) &= w(t) + 6.58 + 0.769y(t-1), & \hat{\rho}_3 &= 28.02.
\end{aligned}$$

We can compare the classes via the likelihood approach of Section 8b.1. The corresponding statistics $\bar{L}_i$, $i = 0, 1, 2, 3$, are given below, where n_i is the number of parameters in model M_i:

$$\bar{L}_i = -(N/2)\ln\hat{\rho}_i - n_i - N\hat{E}(\ln y(t)), \qquad i = 0, 1, 2$$

$$\bar{L}_0 = -60.08, \qquad \bar{L}_1 = -59.68, \qquad \bar{L}_2 = -60.27, \qquad \bar{L}_3 = -60.33.$$

Since $\bar{L}_1$ is the largest of the $\bar{L}_i$, $i = 0, \ldots, 3$, we prefer class C_1. Note how close the values of $\bar{L}_1$ and $\bar{L}_0$ are.

We can compare the classes via the prediction approach of Section 8b.4.1. We construct the best fitting models in each class based on the data 1938–1962 and use this model to obtain the one-year-ahead forecasts of the population in the years 1963–1971. The fitted models M_i', $i = 0, 1, 2, 3$, are

$$\begin{aligned} M_0'&: y_1(t) = w(t) + 0.317 - 0.0115y(t-1)\\ M_1'&: y_1(t) = w(t) + 0.031 - 0.15y_1(t-1)\\ M_2'&: y_2(t) = w(t) + 0.848 + 0.643y_2(t-1) + 0.0988y_2(t-2)\\ M_3'&: y(t) = w(t) + 9.1 + 0.81y(t-1). \end{aligned}$$

The forecasts, the forecast errors, and their mean and mean square values are given in Table 11a.4.1. The forecasts obtained from models M_1' and M_3' are shown in Fig. 11a.4.1.

We find that the mean square error of M_1' is lower than those of the other four values by a factor of 2.5 or more, showing the appropriateness of class C_1. There is a vast difference in the prediction capabilities of M_0' and M_1' even though the likelihood statistics $\bar{L}_0$ and $\bar{L}_1$ are close to each other. We see from Table 11a.4.1 that the mean value of the forecast errors of all the models except M_1' are also large, showing their inappropriateness. The discrepancy between the corresponding coefficients in the pair of models M_i and M_i', $i = 0, \ldots, 3$ is noteworthy. The discrepancy is least for M_1 and M_1'.

TABLE 11a.4.1. The One-Step-Ahead Forecasts Obtained from the Models for the Whooping Crane Data[a]

		Model M_0'		Model M_1'		Model M_2'		Model M_3'	
Year	$y(t)$	$\hat{y}(t)$	$y(t)-\hat{y}(t)$	$\hat{y}(t)$	$y(t)-\hat{y}(t)$	$\hat{y}(t)$	$y(t)-\hat{y}(t)$	$\hat{y}(t)$	$y(t)-\hat{y}(t)$
1963	33	32	1	35	−2	32	1	31	2
1964	42	32	10	35	7	32	10	32	10
1965	44	37	7	44	0	38	6	39	5
1966	43	38	5	47	−4	40	3	40	3
1967	48	38	10	46	2	40	8	40	8
1968	50	40	10	51	−1	42	8	44	6
1969	56	40	16	53	+3	44	12	45	11
1960	57	42	15	59	−2	47	10	50	7
1971	56	43	13	61	−5	49	7	50	6
1972	51	42	9	60	−9	48	3	50	1
Mean error $\bar{e}_1$			9.6		−1.10		6.8		5.9
Mean square error $\bar{e}_2$			110.6		19.3		57.6		44.5

$\bar{e}_k = \frac{1}{10}\sum_{t=1963}^{1972}(y(t) - \hat{y}(t|t-1))^k$.

[a] The numbers for $\hat{y}(t)$ have been rounded off.

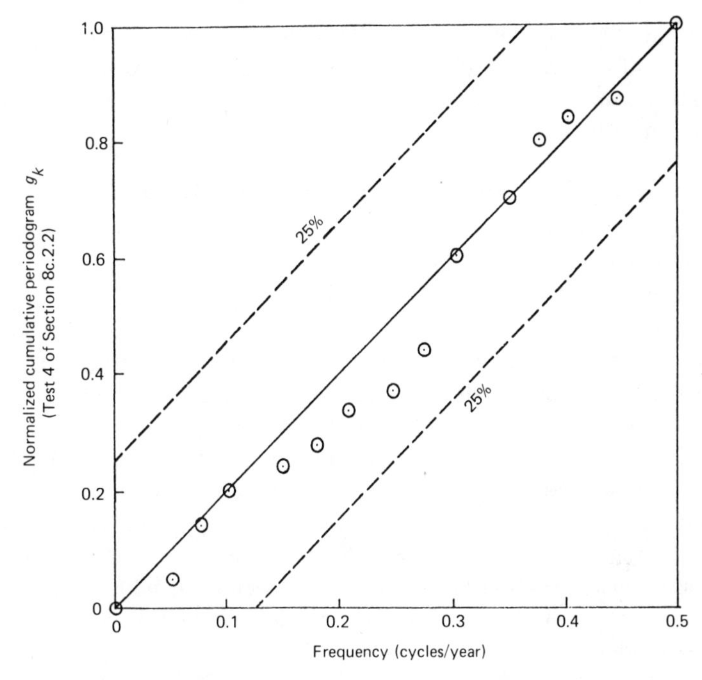

FIG. 11a.4.2. Normalized cumulative periodogram of residuals from model M_1 for the whooping crane data.

The residuals of the best fitting model M_1 pass validation tests at the 95% significance level (Figs. 11a.4.2 and 11a.4.3), showing that the assumptions in the corresponding model are approximately valid.

In conclusion model M_1 is more appropriate for the crane population compared to the Velhurst or Gompertz models. Both the mean and the variance

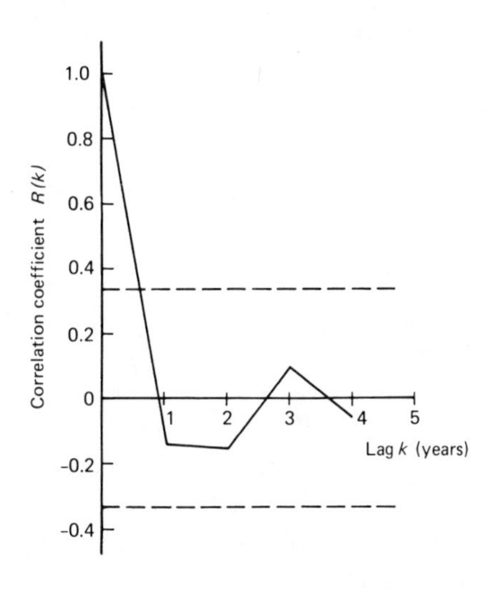

FIG. 11a.4.3. Correlogram of residuals from model M_1 for the whooping crane data; broken lines indicate the two standard error limits.

of the output of this model grow exponentially with time. The prediction mean square error of the AR model M_3 is about 2.5 times that of M_1.

We can obtain more sophisticated multivariate models by considering the population in different age groups.

11a.5. Analysis of the Population Cycle of the Canadian Lynx

We consider the empirical time series of the annual population of the Canadian lynx in the years 1821–1934 consisting of 114 observations. The time series is graphed in Fig. 3b.1.2 (Elton and Nicholson, 1942). It shows a strong cyclical component of period 9.7 years and displays no systematic growth or decay. One of the reasons for modeling this series is to determine the possible cause of the oscillations. This series has been analyzed a number of times by Moran (1953) and Hannan (1960), among others.

There is every reason to believe that the stochastic version of the Gompertz equation is quite adequate since the series has no growth components in it. The random input term η in Eq. (11a.2.11) can be regarded as the population of the hare, which is the principal source of food for the lynx. We could model the lynx population by treating hare population as an exogenous variable.

The discretized difference equation model derived from the stochastic Gompertz model in (11a.2.16) is

$$\ln y(t) = w(t) + \theta_1 \ln y(t-1) + \theta_2 \ln y(t-2) + d_1(t).$$

Additional lagged variables $\ln y(t-j)$ can be added if necessary. Our main interest is determining whether $d_1(t)$ is just a constant or a sum of the constant and sinusoidal terms of frequency $(2\pi/9.7)$. This comparison can be made with the aid of the class selection method of Chapter VIII. If we find that the appropriate model for $y(\cdot)$ needs the sinusoidal terms, then we can tentatively accept the hypothesis that the oscillations in the series are caused by an exogenous variable.

We will consider the five classes $C_0, \ldots, C_4$ of models whose difference equations $\mathscr{I}_i$, $i = 0, \ldots, 4$, are listed below.

Let

$$y_2(t) \triangleq \ln y(t), \qquad v(t) \triangleq w(t) + \theta_0 + \theta_1 y_2(t-1) + \theta_2 y_2(t-2)$$
$$y_3(t) \triangleq (\ln y(t-8) + \ln y(t-9))/2, \qquad \omega_1 = (2\pi)/9.7, \qquad \omega_2 = 2\pi/(112/33).$$

Then

$$\begin{aligned}
\mathscr{I}_0: y_2(t) &= v(t)\\
\mathscr{I}_1: y_2(t) &= v(t) + \theta_3 y_3(t)\\
\mathscr{I}_2: y_2(t) &= v(t) + \theta_4 \cos \omega_1 t + \theta_5 \sin \omega_1 t\\
\mathscr{I}_3: y_2(t) &= v(t) + \theta_3 y_3(t) + \theta_4 \cos \omega_1 t + \theta_5 \sin \omega_1 t\\
\mathscr{I}_4: y_2(t) &= v(t) + \theta_4 \cos \omega_1 t + \theta_5 \sin \omega_1 t + \theta_6 \cos \omega_2 t + \theta_7 \sin \omega_2 t.
\end{aligned}$$

We have included class C_1 because such models have been found to be successful in modeling time series with oscillations. (A model of the family C_1

is most appropriate for the sunspot series.) Classes C_0, C_2, and C_3 do not need explanation. We have included the sinusoidal trend terms of frequency ω_2 in $\mathscr{s}_4$ because this frequency appears to be dominant in the test on residuals of the best fitting model in C_0.

A. Class Selection. In Table 11a.5.1, we have listed the CML estimates of the parameters in each class along with the residual variance, obtained by using all 114 observations. We can compare the classes by means of the likelihood approach of Section 8b.1 using the residual variances $\hat{\rho}_i$. We find by inspection that the best class is C_2 according to the likelihood approach. We can confirm this decision by using the real time or recursive prediction approach of Section 8b.3. We will estimate the parameters in real time in each case, obtain the one-step-ahead forecasts, and thus compute the mean square value of the last 57 prediction errors. These values are listed as J_i in Table 11a.5.1.

The prediction index is smallest for the best fitting model M_2 from class C_2, confirming our preference for class C_2. We will validate the model M_2, the best fitting model in C_2. We will also see whether the model M_0 can be validated since this model has been mentioned often in the literature.

B. Residual Testing. We will test the residuals of the models M_0 and M_2 for whiteness by the tests of Chapter VIII. Tests 5 and 6 of Chapter VIII clearly indicate the lack of correlation in the residuals of both M and M_2 at the 95% level. Next we will test the residuals of M_0, say $\{x(i), i = 3, \ldots, 114\}$, for the absence of sinusoidal terms. The periodogram of the residuals shows relatively sharp peak at frequencies f_{33}, f_{34}, and f_{12}, defined below. Hence we want to check for the presence of such frequencies in the residuals. Let us denote by H_0 the hypothesis that the residuals are white. We will first apply the Fisher test (test 3). This is a multiple hypothesis test and is used to find the dominant frequencies among $f_k = k/112$ cycles per year, $k = 1, \ldots, 56$. Let us denote the periodogram corresponding to frequency f_k by $\gamma^2(k)$:

$$\gamma^2(k) = \frac{1}{112} \sum_{j=3}^{114} (x(j) \cos 2\pi f_k j)^2 + \frac{1}{112} \sum_{j=3}^{114} (x(j) \sin 2\pi f_k j)^2.$$

The test statistic corresponding to f_{33} is $g_{33} = \gamma^2(k = 33)/\sum_{j=1}^{56} \gamma^2(j) = 0.0712$. Recall that N is the number of residuals, which is 112. By the Fisher test

$$\text{prob}[g_{33} > 0.0712 | H_0] \approx \left(\frac{N}{2}(1 - 0.0712)\right)^{N/2} = 0.952.$$

Thus the existence of a sinusoidal trend term of frequency f_{33} in the $\{x(\cdot)\}$ sequence is highly improbable. The same is true with the other two frequencies f_{34} and f_{12}.

Next let us use test 2 of Chapter VIII and check the residuals of M_0 for the presence of trend terms of frequencies $f_{33}, f_{12}, \ldots$, *separately*. Consider frequency f_{33}. The test statistic $g = (\gamma^2(33) \times 112)/4\hat{\rho} = 4.86$. Under the null hypothesis H_0, g is distributed as $F(2, 112)$. Hence $P[g \geq 4.86 | H_0] \approx 0.01$. Hence, at the

TABLE 11a.5.1. The Mean Square Error and the Estimated Coefficients of Various Models for the Lynx Data

$$y_3(t) = (y_2(t-8) + y_2(t-9))/2$$

Model	Final values of estimates based on 114 observations									
	$\hat{\theta}_0$	$\hat{\theta}_1$	$\hat{\theta}_2$	$\hat{\theta}_3$	$\hat{\theta}_4$	$\hat{\theta}_5$	$\hat{\theta}_6$	$\hat{\theta}_7$		
M_i	1	$y_2(t-1)$	$y_2(t-2)$	$y_3(t)$	$\cos \omega_1 t$	$\sin \omega_1 t$	$\cos \omega_2 t$	$\sin \omega_2 t$	ρ_i	$J_i = \frac{1}{57} \sum_{k=58}^{114} e_i^2(k)$
M_0	1.059	1.384	−0.748						0.0519	0.0546
M_1	0.443	1.2112	−0.557	0.191					0.0469	0.06
M_2	0.888	1.067	−0.373		−0.236	−0.077			0.402	0.0473
M_3	0.758	1.049	−0.359	−0.044	−0.217	−0.014			0.0406	0.0493
M_4	0.751	0.951	−0.211		−0.300	−0.019	−0.056	−0.027	0.0394	0.0482

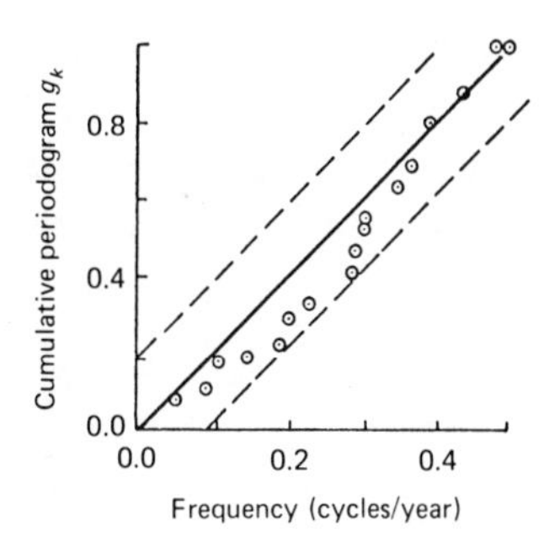

FIG. 11a.5.1. Cumulative periodogram of residuals of the model M_0 for lynx data.

95% significance level, the frequency term f_{33} is indeed significant! Similarly, at the same level, frequencies f_{34} and f_{12} are also significant. On the other hand, if we work at 99.5% significance level, then none of the frequencies are significant!

Next let us use test 4, the cumulative periodogram test. The cumulative periodogram of the residuals, i.e., the graph of $g_k = (\sum_{j=1}^{k} \gamma^2(j) / \sum_{j=1}^{56} \gamma^2(j))$ vs. f_k, $0 \leq f_k \leq 0.5$, is shown in Fig. 11a.5.1; it lies within the 95% band around the line joining (0, 0) to (0.5, 1) indicating that no sinusoidal trends are present in the residuals at the 95% significance level.

Thus we see that the results of the tests contradict each other at the 95% significance level. This discrepancy is heightened by the fact that test 3 is the most powerful symmetric test, and test 2 is the uniformly most powerful test among tests that depend on $(\gamma^2(k)/\hat{\rho})$. However, if we work at the 99.5% significance level, all the tests yield the same decision of acceptance of the null hypothesis H_0. Thus, one may be tempted to accept hypothesis H_0 for the residuals and hence model M_0 for the given data. The problem here is that the fitted AR model M_0 is a bad fit to the given process as shown by the comparison of the correlograms of the data and the AR process in Fig. 11a.5.2. Thus the

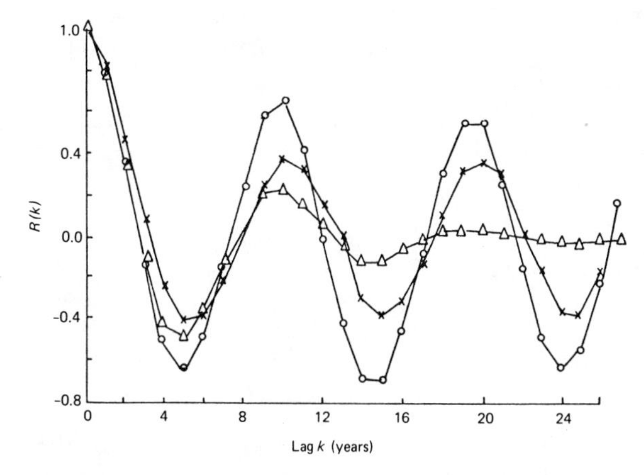

FIG. 11a.5.2. Comparison of empirical correlogram (○) with those of models $M_0(\triangle)$ and $M_2(\times)$ for the lynx data.

residual testing of model M_0 is inconclusive. The residuals of M_2 do not indicate the presence of sinusoidal terms at all.

C. Comparison of the Characteristics of M_0 and M_2. We can also compare the empirical correlogram of the data with the correlograms of models M_2 and M_0. The empirical correlogram is the graph of r_j vs. j where $r_j = c_j/c_0$,

$$c_j = (1/(114 - j)) \sum_{k=j+1}^{114} (y_2(k) - \bar{y}_2)(y_2(k - j) - \bar{y}_2), \bar{y}_2 = (1/114) \sum_{k=1}^{114} y_2(k).$$

The theoretical and empirical correlograms are given in Fig. 11a.5.2. Note that the correlogram of model M_2 is a better fit to the empirical correlogram than the correlogram of model M_0. Moreover, the correlogram of the data lies within two standard deviations of the correlogram of model M_2, whereas the same thing cannot be said of model M_0. The standard deviations of correlograms are computed with the aid of formulas in Chapter II. Comparison of the spectral characteristics also indicates that model M_2 adequately preserves the spectrum of the data.

In conclusion, we see that model M_2 involving the autoregressive terms and sinusoidal components is a good representation of the data. If M_2 is written as $A(D)y_1(t) = w(t) + c_1(t)$, then the zeros in $A(D)$ are not complex, showing that the oscillations in the series should be ascribed to an external factor. Further analysis is needed to ascertain external factor(s) which cause the oscillations.

11b. Analysis of the Annual Sunspot Series

The annual sunspot series has 222 observations in the years 1747–1969 (Waldmeier, 1961). The series has no growth in it, but only a systematic oscillation of approximate 11-year periods. This series has been analyzed a number of times and the autoregressive-type models have been fitted to it since the time of Yule (1927). However, the fitted second-order AR model does not pass all the validation tests proposed here. Hence, we would like to choose an appropriate model for the series using the class selection methods described in Chapter VIII. Even though the sunspot series resembles the lynx time series, the best fitting model is an AR process and does not involve sinusoidal trend terms.

We will consider the five classes C_i, $i = 0, \ldots, 4$, characterized by equations $\mathscr{s}_i$, $i = 0, 1, \ldots, 4$:

$$\mathscr{s}_0: y(t) = w(t) + \theta_0 + \theta_1 y(t-1) + \theta_2 y(t-2)$$
$$\mathscr{s}_1: y(t) = w(t) + \theta_0 + \theta_1 y(t-1) + \theta_2 y(t-2) + \theta_3 y(t-3)$$
$$\mathscr{s}_2: y(t) = w(t) + \theta_0 + \theta_1 y(t-1) + \theta_2 y(t-2) + \theta_4 y(t-10)$$
$$\mathscr{s}_3: y(t) = w(t) + \theta_0 + \theta_1 y(t-1) + \theta_2 y(t-2) + \theta_5 y(t-11)$$
$$\mathscr{s}_4: y(t) = w(t) + \theta_0 + \theta_1 y(t-1) + \theta_2 y(t-2) + \theta_6 \cos \omega_1 t + \theta_7 \sin \omega_1 t.$$

Class C_0 is the class currently used for fitting a model for the sunspot series. Class C_1 has been introduced just to show that nothing is gained by adding the third-order autoregressive term. We have introduced classes C_2 and C_3

involving the autoregressive terms $y(t-10)$ and $y(t-11)$. Such models have been suggested by Whittle (1954). Since the period of sunspot series is greater than 11 years, one would suspect that class C_3 is preferable to class C_2. But this is not the case. Class C_4 is the class of covariance stationary models involving sinusoidal terms of frequency ω_1.

We will compare the classes by using both the prediction method of Section 8b.4.1 and the likelihood approach of Section 8b.1. Using the first 160 observations of the series, we will obtain the maximum likelihood estimates of the parameters and the residual variance $\hat{\rho}$ corresponding to each class. Then using the fitted model in each class, we can obtain the one-step-ahead forecasts of the remaining 62 observations and compute p, the mean square value of the corresponding 62 prediction errors for each class. The fitted models M_i, $i = 0, 1, \ldots, 4$, are listed below. The corresponding values of $\hat{\rho}$ and p are given in Table 11b.1.1. We have included the standard deviations of the estimates for the parameter estimates of model M_2:

$$M_0: y(t) = w(t) + 14.17 + 1.352y(t-1) - 0.666y(t-2)$$

$$M_1: y(t) = w(t) + 14.59 + 1.334y(t-1) - 0.619y(t-2) - 0.035y(t-3)$$

$$M_2: y(t) = w(t) + \underset{(0.2)}{10.62} + \underset{(0.0054)}{1.284}y(t-1) - \underset{(0.0051)}{0.622}y(t-2) + \underset{(0.0032)}{0.1048}y(t-10)$$

$$M_3: y(t) = w(t) + 12.12 + 1.314y(t-1) - 0.666y(t-2) + 0.0864y(t-11).$$

Model M_2 yields the least value of the prediction index p, and hence C_2 is the preferred class by the prediction approach of Section 8b.4.1. This decision is confirmed by means of the likelihood approach of Section 8b.1 also, which involves the use of the residual variances $\hat{\rho}$ corresponding to the various classes.

Model M_2 chosen above has autoregressive terms $y(t-1)$, $y(t-2)$, and $y(t-10)$, which is better than one involving AR terms $y(t-1)$, $y(t-2)$, and $y(t-11)$. This is a little surprising since the oscillations in the series have periods of approximately 11 years. Unlike the lynx series, sinusoidal terms are not needed in the chosen model M_2.

We can validate the best fitting model M_2 and also compare it with model M_0 which is usually considered a good model for the series. We will first test the residuals of both models for whiteness using the standard tests (2–6) of Chapter VIII. The cumulative periodograms of the two sets of residuals, needed

TABLE 11b.1.1. The Residual Variance and Prediction MSE of Models for the Sunspot Series

Model	Residual variance $\hat{\rho}$	Prediction MSE p
M_0	230.6	393.5
M_1	226.7	386.5
M_2	218.5	333.1
M_3	223.8	372.5

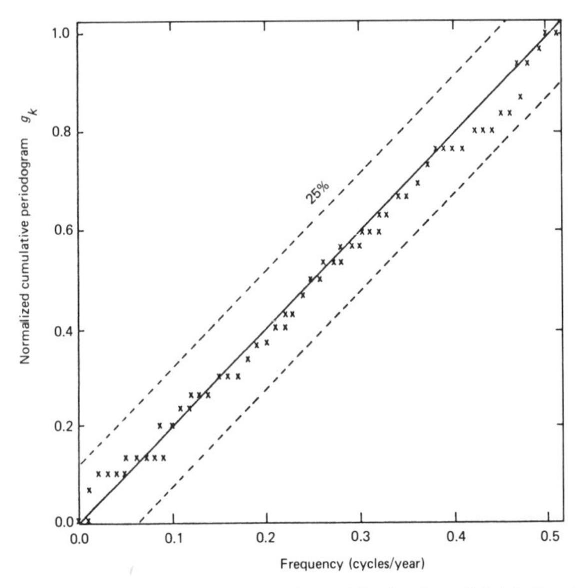

FIG. 11b.1.1. Cumulative periodogram of the residuals of model M_0 for the sunspot data.

in connection with test 4, are shown in Figs. 11b.1.1 and 11b.1.2. Tests 2–6 indicate the whiteness of the residuals of both the models at the usual 95% significance level. The correlograms of the residuals of M_0 and M_2 are given in Figs. 11b.1.3 and 11b.1.4. The residual testing does not clearly indicate the superiority of model M_2 over M_0.

We will directly compare the output characteristics of models M_0 and M_2

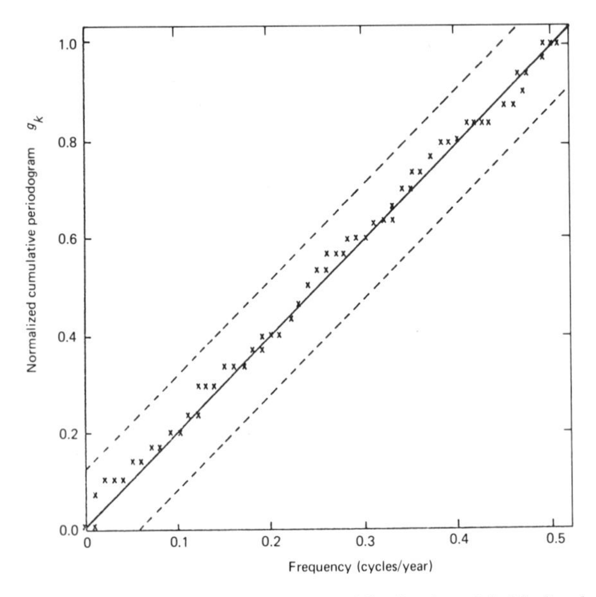

FIG. 11b.1.2. Cumulative periodogram of the residuals of model M_2 for the sunspot data.

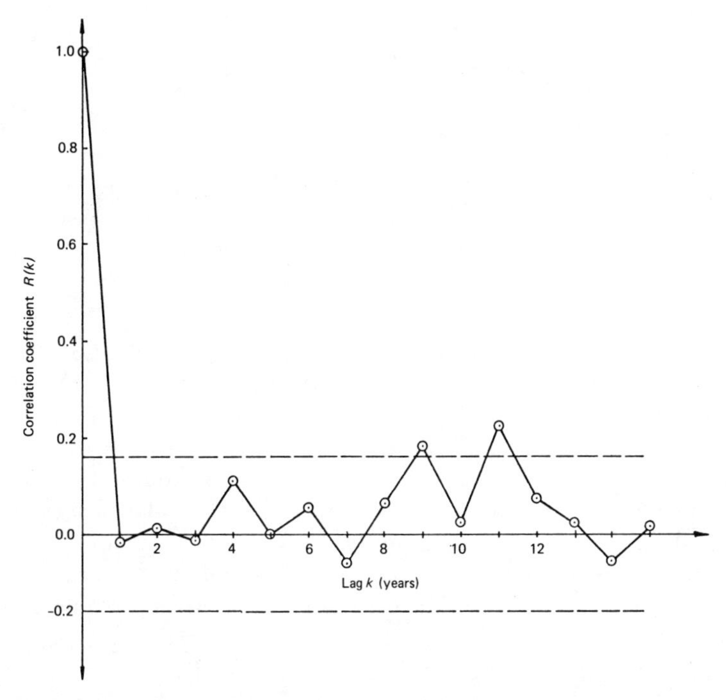

FIG. 11b.1.3. Correlogram of the residuals of model M_0 for the sunspot data; broken lines indicate two standard error limits.

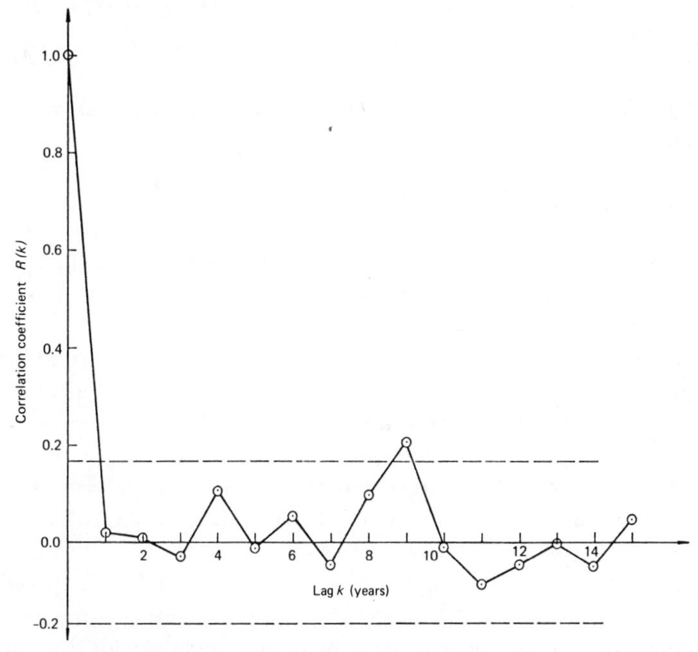

FIG. 11b.1.4. Correlogram of the residuals of model M_2 for the sunspot data; broken lines indicate two standard error limits.

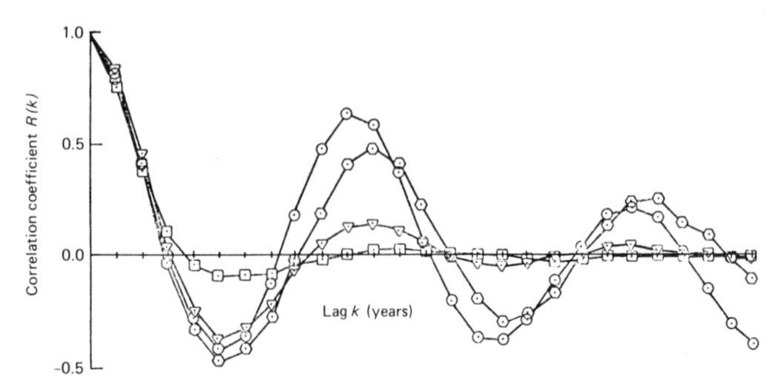

FIG. 11b.1.5. Correlogram of the data and the outputs of the models M_0 and M_2 (sunspot series); ○ = original data, □ = log AR(2) model; ▽ = AR(2) model M_0, ⬡ = model M_2. One unit on the abscissa equals one year.

with those of the empirical sunspot series. The correlograms of the outputs of models M_0 and M_2 and the empirical correlogram of the data are given in Fig. 11b.1.5. We see that the correlogram of model M_2 is a much better fit to the empirical sunspot correlogram than that of model M_0. In particular, the empirical correlogram lies within two standard deviations of the theoretical correlogram of M_2, where the standard deviation of the correlogram is 0.295 as calculated by the formulas in Chapter II. A similar statement cannot be made for M_0, showing the clear superiority of M_2 over M_0. The superior prediction capability of M_2 over M_0 has already been mentioned.

In conclusion, the AR model M_2 having three AR terms $y(t-1)$, $y(t-2)$, and $y(t-10)$ is an appropriate model.

11c. The Sales Data of Company X: An Empirical Series with Both Growth and Systematic Oscillations

We consider the empirical series of total monthly sales of a company in 77 months. The series is graphed in Fig. 11c.1.1, showing the presence of a periodic oscillation superposed on a growing component. The goal in modeling this series is forecasting. We are not interested in obtaining a model that reproduces the characteristics of the data.

Our main aim in choosing this example is to demonstrate that some time series which have growth and oscillatory components can be effectively represented by a covariance stationary model involving linear and sinusoidal trend terms; this model gives better forecasts than the best fitting seasonal ARIMA model. Moreover, the parameter estimation and forecasting formulas are simpler in covariance stationary models than in seasonal ARIMA models.

This time series was taken from a paper by Chatfield and Prothero (1973), who were also analyzing the capabilities of seasonal ARIMA models. These authors decided to construct a difference equation model for the log y instead of y, the observed sales variable. Following the methodology for the construction

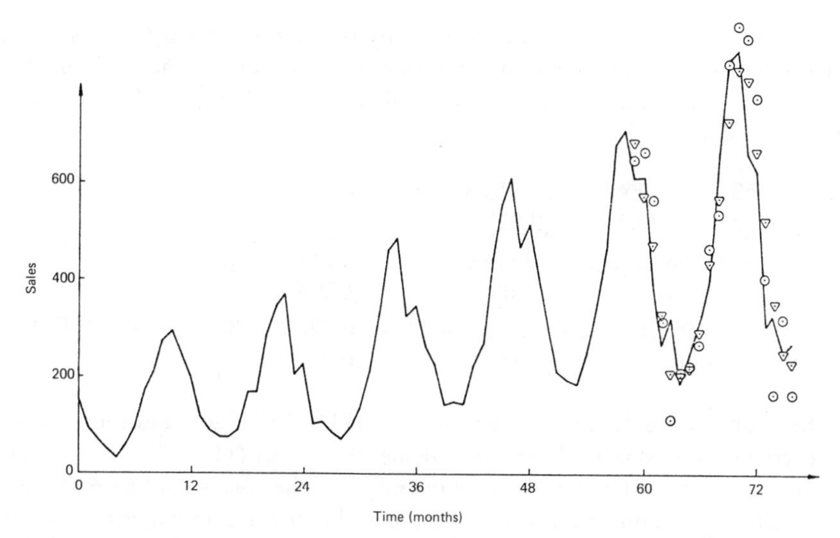

FIG. 11c.1.1. Observed sales and forecasts by seasonal ARIMA and covariance stationary models; ○ = forecasts from seasonal ARIMA model, ▽ = forecasts from covariance stationary model.

of ARIMA models discussed by Box and Jenkins (1970), they arrived at the model

$$(1 + 0.47D)x(t) = (1 - 0.81D^{12})w(t), \qquad x(t) = \nabla\nabla_{12} \log_{10} y(t). \qquad (11c.1.1)$$

The one-step-ahead forecasts from this model are relatively poor. However, one cannot conclude that for this series the seasonal ARIMA model is inappropriate until one can construct another model which should yield better forecasts than the model in (11c.1.1). It is also possible that the logarithmic transformation used here may be inappropriate. The fact that the seasonal ARIMA model in the variable $\log y$ yields bad forecasts does not imply that an appropriate seasonal ARIMA model for the variable y itself will also yield bad forecasts.

The last remark was emphasized by Wilson, one of the discussants of the Chatfield and Prothero (1973) paper, who suggested the following seasonal ARIMA model in terms of the variable y itself, which was obtained using the data $y(1), \ldots, y(59)$:

$$x(t) = 0.37x(t-1) + 0.44x(t-2) + w(t) - 0.56w(t-12) + 0.72w(t-24), \qquad (11c.1.2)$$

where $x(t) = \nabla_{12}^2 y(t) = (1 - D^{12})^2 y(t) = y(t) - 2y(t-12) + y(t-24)$. This model gives much better forecasts than model (11c.1.1).

To consider the possibility of improving the forecasts further, we consider the following class of covariance stationary models:

$$y(t) = \theta_0 + \theta_1 t + \theta_2 \cos \omega_1 t + \theta_3 \sin \omega_1 t + \theta_4 t \cos \omega_1 t + \theta_5 t \sin \omega_1 t + \sum_{j=1}^{m_1} \theta_{5+j} y(t-j) + w(t), \qquad (11c.1.3)$$

where $\omega_1 = 2\pi/12$.

In (11c.1.3), we have introduced not only the terms t, $\cos \omega_1 t$, and $\sin \omega_1 t$, but also the products of these terms. The integer m_1 was determined as indicated in Chapter VIII, and was fixed at the value $m_1 = 3$. The best fitting model based on the data $y(1), \ldots, y(59)$ is

$$\begin{aligned} y(t) = {} & \underset{(2.52)}{58.10} + \underset{(0.14)}{4.47t} - \underset{(2.82)}{24.81 \cos \omega_1 t} \\ & - \underset{(2.54)}{62.38 \sin \omega_1 t} + \underset{(0.09)}{2.05t \cos \omega_1 t} - \underset{(0.07)}{1.69t \sin \omega_1 t} \\ & - \underset{(0.02)}{0.27y(t-1)} + \underset{(0.02)}{0.1y(t-2)} - \underset{(0.02)}{0.12y(t-3)} + w(t). \end{aligned} \tag{11c.1.4}$$

The estimated residual variance from model (11c.1.4) is equal to 1568.9, with the corresponding standard deviation being 296.49. In (11c.1.4), the numbers in parentheses stand for the standard deviation of the corresponding estimates. In (11c.1.4) the numerical values of the estimates are much greater than that of the corresponding standard deviations. The coefficients of the terms $t \cos \omega_1 t$ and $t \sin \omega_1 t$ are quite significant.

We will compare the prediction capabilities of the seasonal ARIMA model in (11c.1.2) with that of (11c.1.4). Both models were constructed from the data $\{y(1), \ldots, y(59)\}$ and are used to compute the one-month-ahead forecasts of the values $y(60), \ldots, y(77)$, which were not used in the parameter estimation.

Let $\hat{y}(t|t-1)$ be the one-step-ahead forecast of $y(t)$ based on $y(t-j), j \geq 1$. Using (11c.1.2), the following equation for the forecast can be obtained:

$$\begin{aligned} \hat{y}(t|t-1) &= 2y(t-12) - y(t-24) + 0.37x(t-1) + 0.44x(t-2) + \hat{w}(t) \\ &\quad - 0.56\hat{w}(t-12) + 0.72\hat{w}(t-24), \qquad t = 60, \ldots, 77 \qquad (11c.1.5) \\ \hat{w}(t) &= y(t) - \hat{y}(t|t-1), \qquad x(t) = y(t) - 2y(t-12) + y(t-24). \end{aligned}$$

The forecasts $\hat{y}(60|59), \ldots, \hat{y}(77|76)$ can be computed recursively from (11c.1.5) provided we know the estimates $\hat{w}(j)$ of $w(j), j \leq 59$. They are automatically obtained in the course of the estimation of parameters in the seasonal ARIMA model. Otherwise they can be obtained from Eq. (11c.1.2), which can be rewritten as

$$w(t) = x_1(t) + 0.56w(t-12) - 0.72w(t-24), \qquad t = 27, \ldots, 59, \tag{11c.1.6}$$

where

$$x_1(t) = x(t) - 0.37x(t-1) - 0.44x(t-2). \tag{11c.1.7}$$

Since $x_1(t)$ depends on $y(t-26)$, we can only obtain the values of $x_1(27), \ldots, x_1(59)$ using $y(1), \ldots, y(59)$. Hence, we have to solve the 33 equations given by (11c.1.6) for $t = 27, \ldots, 59$ for the 57 unknowns $w(3), \ldots, w(59)$ in terms of the 33 knowns $x_1(27), \ldots, x_1(59)$. Since the number of unknowns is more than the number of equations, there is an infinite number of solutions for the set $\{w(3), \ldots, w(59)\}$. We will denote one such solution $\{\hat{w}(j), j = 3, \ldots, 59\}$, which can be easily evaluated as

$$\text{if } j = 3, \ldots, 26, \qquad \hat{w}(j) = 0, \tag{11c.1.8}$$

$$\text{if } j = 27, \ldots, 59, \qquad \hat{w}(j) = x_1(j) + 0.56\hat{w}(j-12) - 0.72\hat{w}(j-24). \tag{11c.1.9}$$

Thus we have all the required estimates $\hat{w}(j), j = 3, \ldots, 59$.

One may wonder whether the quality of the forecasts can be improved by a special choice of $\hat{w}(3), \ldots, \hat{w}(26)$ instead of choosing them arbitrarily as in (11c.1.8). This is not so, because the effect of the choice of these initial quantities on the forecasts $\hat{y}(60|59), \ldots, \hat{y}(77|76)$ is negligible.

We have given in Table 11c.1.1 the one-step-ahead forecasts $\hat{y}(60|59), \ldots, \hat{y}(77|76)$ obtained by the two models (11c.1.2) and (11c.1.4). In Fig. 11c.1.1, the forecasts have been superposed on the graph of the observed values. The MS prediction error by the covariance stationary model is 7763; the MS prediction error by the seasonal ARIMA model is 13,350. The superiority of

TABLE 11c.1.1 Forecasts of the Sales Data by the Seasonal ARIMA and Covariance Stationary Models

		Seasonal ARIMA		From Eq. (11c.1.4)	
t	$y(t)$	$\hat{y}(t\|t-1)$	$y(t) - \hat{y}(t)$	$\hat{y}(t\|t-1)$	$y(t) - \hat{y}(t)$
60	610	650.3	−40.3	686.8	−76.8
61	613	668.1	−55.1	574.3	38.7
62	392	566.8	−174.8	471.3	−79.3
63	273	313.9	−40.9	331.6	−58.6
64	322	113.0	209.0	206.5	115.5
65	189	203.9	−14.9	207.5	−18.5
66	257	223.5	33.5	217.2	39.8
67	324	268.5	55.5	290.3	33.7
68	404	468.0	−64.0	437.0	−33.0
69	677	539.8	137.2	567.3	109.7
70	858	845.3	12.7	727.2	130.8
71	895	929.1	−34.1	834.2	60.8
72	664	900.0	−236.0	812.6	−148.6
73	628	779.7	−151.7	663.4	−35.4
74	308	405.0	−97.0	522.7	−214.7
75	324	162.8	161.2	352.2	−28.2
76	248	320.8	−72.8	243.1	4.9
77	272	163.4	108.6	226.9	45.1
Mean forecast error $\bar{e}_1$			−14.61		−6.33
Mean square forecast error $\bar{e}_2$			13,350		7763

$\bar{e}_k = \frac{1}{18} \sum_{t=60}^{77} (y(t) - \hat{y}(t|t-1))^k$.

the covariance stationary model over the seasonal ARIMA model for forecasting is clear from the above results and those given in Table 11c.1.1 and Fig. 11c.1.1.

The covariance stationary model is appropriate for forecasting, but it has not been validated for use in other circumstances. The theoretical residual variance in the covariance stationary model is only 1568.9, whereas the MS prediction error is 7763.

We emphasize that this example does not imply that for every series a covariance stationary model always yields better forecasts than a seasonal ARIMA model. In time series modeling it is very difficult to make such general statements. However, the covariance stationary models often yield good performance and they are far easier to handle from the point of view of parameter estimation and forecasting than the seasonal ARIMA models.

11d. The Time Series $E2$: Role of Moving Average Terms

This time series is introduced to emphasize that the moving average terms may play a very important role in some models. The series $E2$ is the *first difference* of the series $E1$ which is the quarterly data of the total expenditures in producers' durables in the years 1947–01 (i.e., January 1974) until 1966–04 (Nelson, 1973). The series $E2$ is graphed in Fig. 3a.2.2. Since it has neither growth nor systematic oscillations, we need to consider only ARMA models. Let $y_i(t)$ be the observed variable in the $E1$ series and $y(t)$ be that in the $E2$ series. We consider the following six classes of models where the corresponding equations $\mathscr{I}_i$, $i = 0, \ldots, 5$, are

$$\mathscr{I}_0: y(t) = w(t) + \theta_1 w(t-1)$$

$\mathscr{I}_i$: AR model of order $(i + 1)$ with constant term added to it, $\quad i = 1, \ldots, 5$.

We will first compare them by the prediction approach of Section 8b.4. We use the first 56 observations to estimate the parameters in the model. The fitted models are labeled M_i, $i = 1, \ldots, 6$, respectively:

$$M_0: y(t) = w(t) + 0.257 + 0.306w(t-1)$$

$$M_1: y(t) = w(t) + 0.249 + 0.255y(t-1) - 0.1795y(t-2)$$

$$M_2: y(t) = w(t) + 0.235 + 0.257y(t-1) - 0.1732y(t-2) - 0.023y(t-3)$$

$$\begin{aligned} M_3: y(t) = {} & w(t) + 0.281 + 0.239y(t-1) - 0.2017y(t-2) + 0.035y(t-3) \\ & - 0.2394y(t-4) \end{aligned}$$

$$\begin{aligned} M_4: y(t) = {} & w(t) + 0.319 + 0.232y(t-1) - 0.188y(t-2) + 0.0136y(t-3) \\ & - 0.215y(t-4) - 0.082y(t-5) \end{aligned}$$

$$\begin{aligned} M_5: y(t) = {} & w(t) + 0.350 + 0.232y(t-1) - 0.225y(t-2) + 0.0094y(t-3) \\ & - 0.234y(t-4) - 0.0523y(t-5) - 0.126y(t-6). \end{aligned}$$

Using these fitted models, we can obtain the one-step-ahead forecasts of the remaining 24 observations. The forecasts, the forecast errors, and the mean square values of the 24 forecast errors are listed in Table 11d.1.1. The MA model M_0 has the least value of the mean square prediction errors, showing the appropriateness of class M_0. Note that among the AR models, AR(3) seems to

TABLE 11d.1.1. The Forecast Errors and Their Mean Square Value for the *E*1 Model

		$e(t) = y_1(t) - \hat{y}_1(t\|t-1)$					
t	$y_1(t)$	M_0	M_1	M_2	M_3	M_4	M_5
57	27.6	−2.25	−2.35	−2.30	−2.49	−2.35	−2.33
58	27.7	0.53	0.27	0.28	0.15	0.54	0.48
59	29.0	0.88	0.63	0.64	0.60	0.56	0.46
60	30.3	0.77	0.74	0.70	0.84	0.51	0.67
61	31.0	0.21	0.35	0.36	0.45	−0.22	−0.29
62	32.1	0.78	0.91	0.94	0.86	0.69	0.68
63	33.5	0.90	1.00	1.03	0.90	1.23	0.97
64	33.2	−0.83	−0.71	−0.69	−0.77	−0.36	−0.35
65	33.2	−0.00	0.08	0.11	0.01	0.26	0.42
66	33.8	+0.34	0.30	0.35	0.17	0.50	0.63
67	35.5	1.34	1.30	1.30	1.24	1.64	1.69
68	36.8	0.63	0.73	0.73	0.74	0.75	0.83
69	37.9	0.75	0.82	0.84	0.83	0.77	0.99
70	39.0	0.74	0.80	0.85	0.74	0.88	0.87
71	41.0	1.55	1.67	1.70	1.57	2.03	2.06
72	41.6	−0.13	0.04	0.07	−0.02	0.43	0.49
73	43.7	1.88	2.06	2.08	2.00	2.35	2.59
74	44.4	−0.13	0.02	0.08	−0.07	0.31	0.46
75	46.6	1.98	2.15	2.16	2.08	2.63	2.82
76	48.3	0.84	1.02	1.07	0.94	1.27	1.36
77	50.2	1.39	1.61	1.63	1.56	2.09	2.42
78	52.1	1.22	1.47	1.52	1.40	1.75	1.82
79	54.0	1.27	1.51	1.54	1.41	2.00	2.34
80	56.0	1.35	1.61	1.65	1.52	2.12	2.22
Mean square forecast error		1.239	1.46	1.49	1.39	1.99	2.27

be the best. The difference between the mean square errors of the MA model and the AR(3) model is more than 10% of the MSE of the MA model. The mean square prediction error increases considerably as the order of the AR model is increased beyond 3, showing the importance of keeping the number of parameters in a model as small as possible. The appropriateness of class M_0 can also be established by the likelihood approach. We will omit the details.

11e. Causal Connection between Increases in Rainfall Corresponding to Increased Urbanization

11e.1. Introduction

Considerable differences have been observed between the microclimates of urban and surrounding rural areas (Mathews *et al.*, 1971; WMO, 1970; Peterson, 1969; Lowry, 1971; Landsberg, 1970). Our interest here is in the conjecture regarding the increase in the average rainfall caused by urbanization (Rao and Rao, 1974). The physical reason behind the conjecture is that increased urbanization increases nucleation, the thermal and roughness effects which result in increased precipitation. To clarify the empirical basis of the conjecture, consider the rainfall in three urban stations, namely LaPorte, Indiana; Edwardsville, Illinois, and St. Louis, Missouri. The data are taken from the U.S. Weather Bureau publications.

The mean annual rainfall at LaPorte in the periods 1898–1928 and 1929–1968 was 36.4 and 44.59 in., respectively. The increase in the mean value in the second period (1929–1968), which is 22.5% of the mean value in the first period (1898–1928), is attributed to the effects of increased urbanization and industrial activity in the Chicago–Gary area, which is located to the west of LaPorte. Similarly, the mean annual rainfall at the Edwardsville station in the periods 1910–1940 and 1941–1970 was 38.21 and 40.48 in. The increase in the rainfall is about 4.25% and is attributed to the increased urbanization and industrialization of St. Louis, which is near Edwardsville. But there is some doubt about the validity of the conjecture because the mean annual rainfall at St. Louis itself in the periods 1837–1940 and 1941–1970 was 38.78 and 36.64 in.; i.e., there was an actual *decrease* in the average rainfall of the later (posturbanization) period which had greater urbanization than in the earlier (preurbanization) period.

The fundamental question is whether there is statistically significant changes in the magnitude of the rainfall in the preurbanization and posturbanization periods. Such a test for significance of the observed increases is important because if the observed increases are not statistically significant, then attributing the observed difference to urbanization or other causes would not be meaningful. Furthermore, as many annual rainfall series may be correlated (Rao and Rao, 1974), standard tests such as the t-tests used for testing the significance of differences in the mean values lose their power. Consequently, the tests developed in Chapters VIII and IX must be used to determine the significance of the observed increases.

In this section we discuss the use of AR models to investigate the significance of changes in the mean annual rainfall at LaPorte, Edwardsville, and St. Louis. Second, we discuss the importance of using validated models in these tests, and demonstrate that the use of an unvalidated model may yield a result that contradicts the evidence of observations. An invalid model may predict an increase in the mean values of the rainfall from the preurbanization to the posturbanization period, whereas, in fact, there may be a decrease in the mean annual rainfall.

11e.2. Testing the Significance of the Change in the Rainfall Means in the Two Periods by Generalized AR Models

11e.2.1. The Description of the Classes and Tests

Let $y(t)$ represent the total annual rainfall in year t. We have to compare two classes of models C_0 and C_1, $C_i = (\mathscr{I}_i, \mathscr{H}_i, \Omega)$, $i = 0, 1$, where

$$\mathscr{I}_0: y(t) = \theta_1 + \sum_{j=1}^{q} \theta_{j+1} y(t-j) + w(t),$$
$$\boldsymbol{\theta} = (\theta_1, \ldots, \theta_{q+1})^{\mathrm{T}} \in \mathscr{H}_0, \qquad \rho \in \Omega;$$
$$\mathscr{I}_1: y(t) = \theta_0 \eta(t) + \theta_1 + \sum_{j=1}^{q} \theta_{j+1} y(t-j) + w(t),$$
$$\boldsymbol{\theta} = (\theta_0, \theta_1, \ldots, \theta_{q+1})^{\mathrm{T}} \in \mathscr{H}_1, \qquad \rho \in \Omega.$$
$$\eta(t) = \begin{cases} 0 & \text{if} \quad t < m_1 \\ 1 & \text{if} \quad t \geq m_1. \end{cases}$$

Equation $\mathscr{I}_0$ is an ordinary AR model with a bias term. Thus class C_0 represents the class of models for the rainfall assuming that urbanization has no effect on rainfall. Equation $\mathscr{I}_1$ allows for a change in the mean annual rainfall caused by the urbanization for $t \geq m_1$. Thus, C_1 is a class of models that allow for the effect of urbanization on rainfall. In equation $\mathscr{I}_1$, we have allowed for a sharp change in the mean of rainfall in the year $t = m_1$. We may also provide for a gradual change. For simplicity, we assumed that the order of the AR model, q, is the same in both equations $\mathscr{I}$ and $\mathscr{I}_1$. We will choose q appropriately.

Our problem is to determine the appropriate class of models among C_0 and C_1 for representing the given rainfall. The appropriate class of models may conveniently be selected by the variance ratio test of Section 8b.3.

The variance test is as follows. By the least squares method, find the best model to fit the given data in class C_0 and let the corresponding value of ρ be ρ_0. Similarly, let the residual variance of the best fitting model in class C_1 be ρ_1. If the available observation history is $\{y(t), t = 1, \ldots, N\}$, the test statistic x_1 is

$$x_1 = \frac{N}{1} \frac{\rho_0 - \rho_1}{\rho_1}.$$

x_1 is $F(1, N-1)$ if class C_0 is the correct class. By using the table of F-distributions we can obtain the threshold x_0 for a significance level and accept C_0 if $x_1 < x_0$ and reject C_0 if $x_1 > x_0$.

q is also unknown but an appropriate value for q can be chosen by using the residual variance test for the class of model C_0. Let ρ_q and ρ_{q+1} be the residual variances of the best fitting qth- and $(q+1)$th-order AR models, respectively. If the qth-order AR model is satisfactory, then

$$x_2 = \frac{\rho_q - \rho_{q+1}}{\rho_{q+1}} \frac{N}{1} \qquad \text{is} \quad F(1, N-1).$$

We can devise a threshold test as before.

TABLE 11e.2.1. Parameter Estimates and Other Details of the AR Models. Numbers in parentheses are the standard deviations.

Station	$\hat{\theta}_1$	$\hat{\theta}_2$	$\hat{\theta}_3$	$\hat{\theta}_4$	$\hat{\theta}_5$	$\hat{\theta}_6$	Residual variance	$\hat{\theta}_0$	x_1	x_0 threshold	Decision
St. Louis	25.05	0.3346 (0.0883) (S)[a]	−0.0799 (0.0924)	0.0891 (0.0930)	0.1930 (0.093) (S)	−0.1915 (0.088) (S)	63.71	−1.309 (1.729)	0.463	3.92	C_0
Edwardsville	33.051	0.0767 (0.1343)					60.31	1.391 (2.041)	1.008	4.00	C_0
LaPorte	36.405	−0.0638 (0.1207)	0.1706 (0.1245) (S)				58.85	7.866 (2.285)	9.88	3.99	C_1

[a] S means significant at 95%.

11e.2.2. Results

We will consider the rainfall data of St. Louis, Missouri, LaPorte, Indiana, and Edwardsville, Illinois. We will choose the integer q by using the variance ratio test. The integer q is 2, 5, and 1 for the data from LaPorte, St. Louis, and Edwardsville (Table 11e.2.1). In addition, in the data from LaPorte, the first-order AR term is not significant and in the St. Louis data second- and third-order AR terms are not significant. In each case, we consider equations s_0 and s_1 so as to retain the significant terms only with the appropriate value of q. In Table 11e.2.1, we have listed the various estimated coefficients and their standard deviations.

In Table 11e.2.1, the statistic x_1 is also tabulated and compared with the corresponding threshold x_0 at the 95% significance level. Notice that only in LaPorte should we accept the class C_1. In both Edwardsville and St. Louis, we have to accept C_0. We have also listed in Table 11e.2.1 the estimated value of θ_0 for the best fitting model in class C_1. Note that, as expected, its numerical value is considerably less than its standard deviation for both the Edwardsville and St. Louis data. The sign of $\hat{\theta}_0$ for St. Louis data is negative, showing that there is a slight decrease in the rainfall in the posturbanization period. But the decrease is not significant.

We may also point out that the best fitted model from class C_0 for St. Louis and Edwardsville data and the best fitting model from class C_1 for the LaPorte data pass all the validation tests of Chapter VIII. This is the justification for considering only the two classes C_0 and C_1 in this problem. The validation of the fitted models and the comparison results mentioned earlier allow us to conclude that urbanization may sometimes increase rainfall, but not always.

11e.3. Inference with an Invalid Model

To investigate the significance of changes in mean values in the time series, Box and Tiao (1965) suggest the use of class $C_2 = \{\mathscr{s}_2, \mathscr{H}_2, \Omega\}$ of IMA(1, 1) models:

$$\mathscr{s}_2 \colon y(t) = \theta_0 + \theta_1 \eta(t) + \theta_2 \sum_{j=1}^{t-1} w(t-j) + w(t), \tag{11e.3.1}$$

where $w(\cdot)$ is the zero mean IID sequence $(0, \rho)$ and $\eta(\cdot)$ is the indicator variable defined earlier. In order to recognize that (11e.3.1) is an IMA equation, we can rewrite (11e.3.1) as

$$\begin{aligned} \nabla y(t) &= \theta_1 \eta'(t) + (\theta_2 - 1) w(t-1) + w(t), \\ \eta'(t) &= \begin{cases} 1 & \text{if } t = m_0 \\ 0 & \text{if } t \neq m_0. \end{cases} \end{aligned} \tag{11e.3.2}$$

where $\nabla y(t) = y(t) - y(t-1)$. Let us obtain the best fitting model for the rainfall data of St. Louis and Edwardsville. The corresponding estimates of

TABLE 11e.3.1. Parameter Estimates of the IMA(1, 1) Model[a]

City	$\hat{\theta}_0$	$\hat{\theta}_1$	$\hat{\theta}_2$
St. Louis	41.36	3.78	0.008
	(3.25)	(3.26)	
Edwardsville	38.82	1.69	0.01
	(1.57)	(1.96)	

[a] Numbers in parentheses are standard deviations.

θ_0, θ_1, and θ_2 are given in Table 11e.3.1. Let us consider the estimate of θ_1, the coefficient of the indicator variable $\eta(t)$. For the Edwardsville data $\hat{\theta}_1$ is numerically less than its standard deviation, and hence is not significant. For the St. Louis data $\hat{\theta}_1$ is positive, numerically greater than its standard deviation and should be termed significant. But since $\hat{\theta}_1$ is significant and positive, the fitted model implies that there is an increase in the mean value of the St. Lousis rainfall in the posturbanization period (1941–1970). But this inference directly contradicts the observations since the latter shows a *decrease* in the average rainfall after 1941.

This situation has come about because the IMA(1, 1) class is not appropriate for modeling the St. Louis rainfall data. This statement can be confirmed by the fact that the fitted IMA model for St. Louis data does not pass the validation tests (Rao and Rao, 1974). We can also see why the IMA model is not appropriate by considering the best fitting AR model for the St. Louis data given in Section 11e.2.2. The coefficients of the lag 4 and lag 5 terms in the AR model are significant and it is not possible to represent their effect in the IMA(1, 1) model. A similar erroneous inference would be obtained if we had used the AR(1) model for the St. Louis data.

However, we should mention that the IMA models have worked well in explaining changes in levels in other time series. This example clearly illustrates the need for validation of the fitted model. One cannot be sure of the validity of a model on purely a priori grounds.

11f. A Multivariate Model for Groundwater Levels and Precipitation

11f.1. Introduction

In this study we will illustrate the construction and validation of a multivariate model. We have considered the problem of modeling the response of the groundwater levels in an aquifer to rainfall. This study also illustrates another important feature, namely that strict causality (as used in experimental sciences) cannot be completely inferred from statistical considerations alone. Even though it is known from physical considerations that past values of groundwater levels in an aquifer will not influence rainfall in the future, it is still possible to have a validated multivariate model in which the equation for the rainfall may have lagged values of groundwater levels in them.

The groundwater in an aquifer is replenished by precipitation through the infiltration process. ["A geologic formation which contains water and transmits it from one point to another in quantities sufficient to permit economic development is called an aquifer," Linsley *et al.* (1958).] The extent of infiltration and replenishment is affected by several dynamic factors, such as the geology of the area, the time intervals between storms, and the evapotranspiration rates. It is difficult to model these processes deterministically and hence a stochastic model of rainfall and groundwater levels is appropriate. By knowing the aquifer constants, the volume of replenishment or loss from the aquifer can be easily computed if the water levels at several points in the aquifer are known.

In the present case study we have used the monthly water level data (1951–1970 corresponding to $N = 240$) from an unpumped observation well located in Wood County, Wisconsin (USGS No. Wd 29) and the total monthly rainfall data from Stevens Point, Marshfield, and Wisconsin Rapids, which are also located in Wisconsin. The data are taken from USGS publications. The water levels in the well were reported at irregular intervals which averaged out to about once a week, and the mean of these values over a month was assumed to be the monthly mean water level. Three rainfall stations were selected, as the data from only one station may not accurately reflect the effect of rainfall on the observed water level variations because of the inherent spatial variability of rainfall in a watershed.

Let $\mathbf{y}(t) = [y_1(t), y_2(t), y_3(t), y_4(t)]^{\mathrm{T}}$, where $y_1(t)$ represents groundwater levels, and $y_2(t), y_3(t)$, and $y_4(t)$, respectively, represent the rainfall at Wisconsin Rapids, Stevens Point, and Marshfield.

The various statistical characteristics such as the correlogram and spectral density, have been discussed elsewhere (Rao *et al.*, 1975). We will give the lag 1 correlation matrix of the vector $\mathbf{y}$ showing the strong negative correlation between the groundwater levels and rainfall:

$$\hat{E}[(\mathbf{y}(t) - \bar{\mathbf{y}})(\mathbf{y}(t-1) - \bar{\mathbf{y}})^{\mathrm{T}}] = \begin{bmatrix} 0.84 & -0.21 & -0.19 & -0.22 \\ -0.15 & 0.26 & 0.44 & 0.31 \\ -0.19 & 0.25 & 0.30 & 0.33 \\ -0.31 & 0.38 & 0.38 & 0.32 \end{bmatrix},$$

where

$$\hat{E}(\mathbf{y}(t)) = \frac{1}{N}\sum_{j=1}^{N} \mathbf{y}(j)\mathbf{y}^{\mathrm{T}}(j), \qquad \bar{\mathbf{y}} = \hat{E}[\mathbf{y}(t)].$$

The negative correlation between rainfall and groundwater levels is an indication of the time lag between rainfall and the arrival of the contribution of the rainfall to the groundwater storage. The cross correlation between different rainfall stations is relatively strong.

11f.2. Model Construction

The correlograms and power spectra of the water level and rainfall data indicated the strong presence of annual and semiannual cycles. These deterministic trends were removed from the data by Fourier analysis and the residuals from the Fourier analysis, called $y_i'(k)$, $i = 1, 2, 3, 4$, defined as

$$\begin{aligned} y_i'(t) &= y_i(t) - \theta_{i1}\sin(2\pi t/12) - \theta_{i2}\cos(2\pi t/12) \\ &\quad - \theta_{i3}\sin(2\pi t/6) - \theta_{i4}\cos(2\pi t/6) \qquad (11\text{f}.2.1) \\ \mathbf{y}'(t) &= [y_1'(t), y_2'(t), y_3'(t), y_4'(t)]^{\mathrm{T}} \end{aligned}$$

were fitted with the multivariate model. The θ_{ij} coefficients are given in Table 11f.2.1

An alternative procedure to detrend the data, which is widely used in hydrology, is to subtract the mean value of the corresponding month from the data and divide the resulting value by the standard deviation of the corresponding monthly value. This detrending procedure is given by

$$\begin{aligned} y_i''(t) &= \frac{y_i(t) - \bar{y}_i(j_t)}{\bar{s}_i(j_t)}, \qquad (11\text{f}.2.2) \\ \mathbf{y}''(t) &= [y_1''(t), y_2''(t), y_3''(t), y_4''(t)]^{\mathrm{T}} \end{aligned}$$

where $j_t = 1, \ldots, 12$ stands for January–December, respectively. This procedure results in a $y_i''(k)$ sequence that has zero mean and unit variance.

The mean, variance, and skewness coefficients of the detrended sequences are all smaller than the corresponding values of the observed sequence. More significantly, the lag 0 correlations between the different detrended rainfall sequences are considerably larger than the corresponding values obtained from the observed data, whereas the lag 1 correlations of detrended rainfall sequences are considerably smaller than those of the observed data. The cross correlation coefficients between water level and precipitation sequences of the detrended data are also smaller, in general, than the corresponding values of the observed data.

We will fit an autoregressive model for the vector sequences $\{\mathbf{y}'(\cdot)\}$ and $\{\mathbf{y}''(\cdot)\}$. Even though past values of groundwater level do not influence future rainfall, we did not incorporate this knowledge in the AR model in order to observe the corresponding consequences. We can use the two tests of Section 9c.3 for determining the appropriate order of the AR model.

TABLE 11f.2.1. Detrending Coefficients

Series	θ_{i1}	θ_{i2}	θ_{i3}	θ_{i4}
y_1'	0.2636	1.4210	0.4122	−0.3667
y_2'	−0.6187	−1.4222	−0.2979	−0.0038
y_3'	−0.6121	−1.2854	−0.1109	0.0478
y_4'	−0.7244	−1.5010	−0.1230	0.0419

The results of the F-test of Section 9c.3.1 indicate that a fourth-order model is adequate for the $\mathbf{y}'(k)$ series and a third-order model is adequate for the $\mathbf{y}''(k)$ series. The result from the partial correlation test of Section 9c.3.2 indicates that a second-order model is adequate for both the $\mathbf{y}'(k)$ and $\mathbf{y}''(k)$ series. As the latter test is the more powerful one, its results are accepted. The parameter matrices and the residual statistics are

$$\mathbf{y}'(t) = \mathbf{A}_1'\mathbf{y}'(t-1) + \mathbf{A}_2'\mathbf{y}'(t-2) + \mathbf{w}'(t), \tag{11f.2.3}$$

$$\mathbf{y}''(t) = \mathbf{A}_1''\mathbf{y}''(t-1) + \mathbf{A}_2''\mathbf{y}''(t-2) + \mathbf{w}''(t), \tag{11f.2.4}$$

where $\mathbf{w}'(\cdot) \sim N(0, \boldsymbol{\rho}'), \qquad \mathbf{w}''(\cdot) \sim N(0, \boldsymbol{\rho}'')$

$$\mathbf{A}_1' = \begin{bmatrix} 1.045 & -0.147 & -0.139 & -0.030 \\ -0.05 & 0.107 & 0.112 & 0.112 \\ -0.112 & 0.206 & -0.104 & -0.104 \\ -0.129 & 0.005 & 0.026 & 0.026 \end{bmatrix}, \qquad \mathbf{A}_2' = \begin{bmatrix} -0.255 & -0.108 & -0.055 & -0.022 \\ 0.113 & 0.044 & -0.016 & -0.038 \\ 0.198 & 0.074 & -0.042 & -0.102 \\ 0.308 & -0.094 & 0.123 & -0.152 \end{bmatrix},$$

$$\mathbf{A}_1'' = \begin{bmatrix} 1.003 & -0.077 & -0.125 & -0.001 \\ -0.072 & 0.004 & 0.175 & -0.103 \\ -0.069 & 0.221 & -0.101 & -0.006 \\ -0.106 & 0.056 & -0.024 & 0.032 \end{bmatrix}, \qquad \mathbf{A}_2'' = \begin{bmatrix} -0.248 & -0.019 & 0.001 & -0.051 \\ 0.228 & -0.006 & -0.026 & 0.012 \\ 0.253 & 0.072 & -0.111 & -0.033 \\ 0.269 & -0.051 & 0.023 & -0.074 \end{bmatrix},$$

$$\boldsymbol{\rho}' = \begin{bmatrix} 1.021 & -0.244 & -0.236 & -0.194 \\ -0.132 & 1.572 & 1.403 & 1.1 \\ -0.095 & 1.393 & 1.608 & 1.24 \\ -0.023 & 1.096 & 1.24 & 1.493 \end{bmatrix}, \qquad \boldsymbol{\rho}'' = \begin{bmatrix} 0.249 & -0.087 & -0.097 & -0.083 \\ -0.07 & 0.991 & 0.823 & 0.693 \\ -0.085 & 0.818 & 0.944 & 0.736 \\ -0.073 & 0.692 & 0.756 & 0.983 \end{bmatrix}.$$

The ratio of the mean square value of the residuals to the mean square value of the observations is about 0.23 for the residuals corresponding to y_1' and y_1'', and about 0.9 for other variates. This indicates that the multioutput model fitted to the Wisconsin data explains about 77% of the variation in the stochastic component of groundwater levels.

11f.3. Model Validation

We will first consider residual testing. The histograms of the individual residual sequences $w_i'(\cdot)$, $w_j''(\cdot)$ can be considered to be normally distributed according to the χ^2 goodness of fit tests. The noncentrality parameter was very close to 3 in all cases.

Next, we consider the cross correlation between the sequence $\{w_j'(\cdot)\}$, $j = 1, \ldots, 4$, using the tests of Section 9d.1. Typical results of the cross correlation test, coherence diagrams, integrated phase spectrum test, and the integrated sample cospectrum test are given in Figs. 11f.3.1–5 for some $\{w_j'\}$ and $\{w_j''\}$ sequences. The results from the tests indicate the absence of cross correlation between the individual sequences $\{w_j'(\cdot)\}, j = 1, 2, 3, 4$, at the 95% significance level.

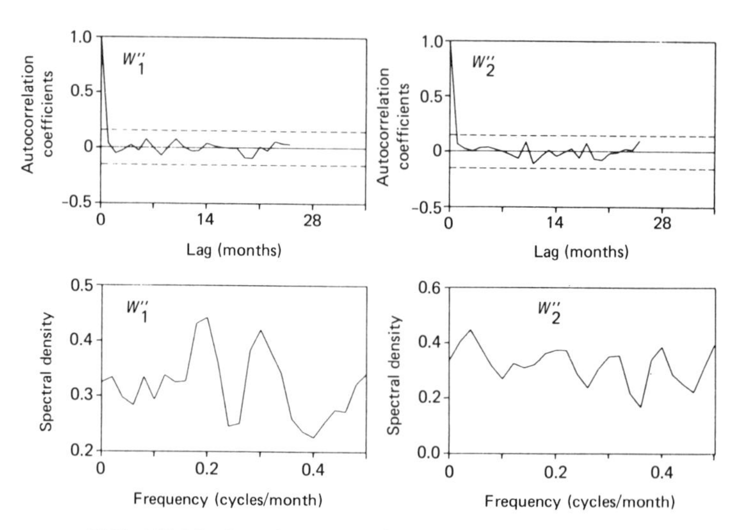

FIG. 11f.3.1. Correlograms and power spectra of residuals.

The model has also been validated by comparing the various characteristics of the empirical data such as the spectral density with the corresponding characteristics of the data generated by simulating the model on a digital computer. The characteristics of the observed data were found to be close to those of the generated data; the details can be found elsewhere (Rao *et al.*, 1975).

It is important to realize that in the validated model, the coefficient of the groundwater level variable $y_1(t-1)$ in the equation for rainfall, such as that for $y_2(t)$, $y_3(t)$ or $y_4(t)$ is significantly different from zero. Physically, it is absurd to believe that groundwater is a causal variable of rainfall. Thus, this study provides an excellent illustration of the idea that strict causality (in the sense of experimental sciences) can never be conclusively established by only looking at the data over whose generation we had no control. The fact that certain variables appear on the right-hand side of the equations of the validated model does not imply that these variables are causally related to the corresponding variables on the left-hand side.

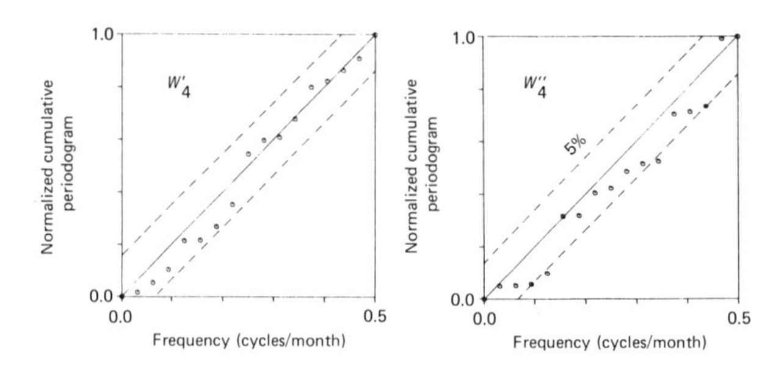

FIG. 11f.3.2. Results of cumulative periodogram test on residuals.

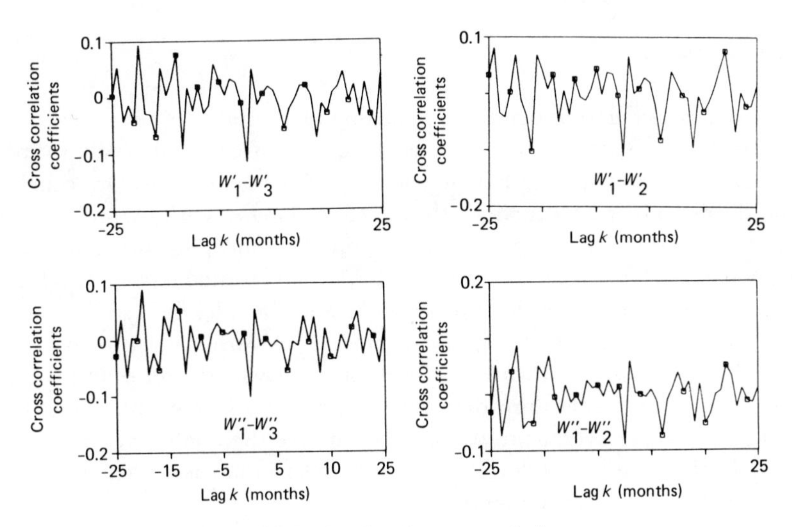

FIG. 11f.3.3. Results of cross correlation test.

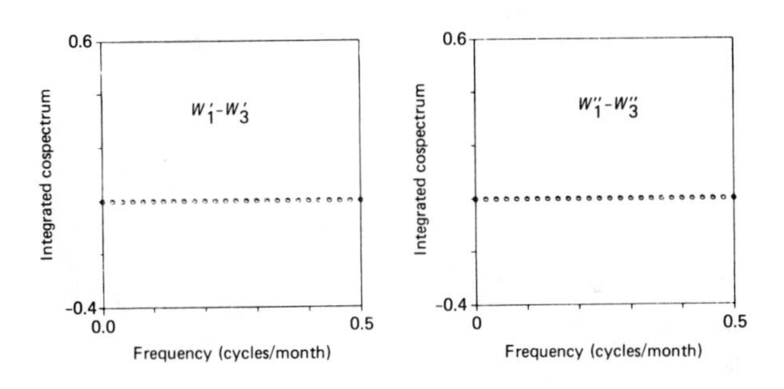

FIG. 11f.3.4. Examples of results of integrated cospectrum test.

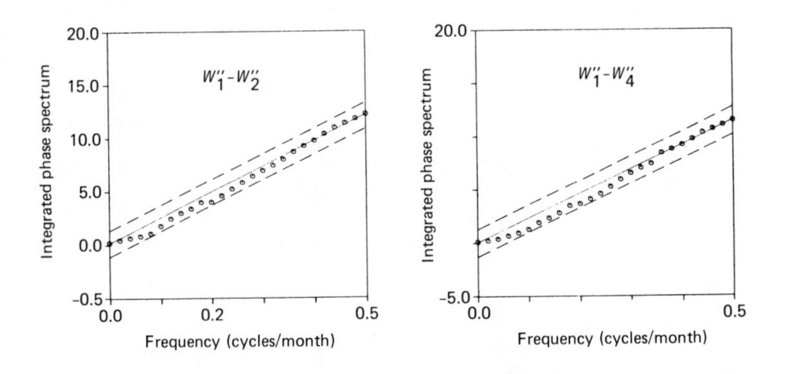

FIG. 11f.3.5. Results of integrated phase spectrum test.

11g. Conclusions

We have considered in detail the modeling of a number of time series. Three population time series were examined in detail to check the validity of the various assumptions in the usual models adopted for these series. Our conclusion is that the first-order IAR model is to be preferred to the Verhulst and Gompertz models for representing the U.S. or whooping crane population series. The modeling of the lynx series shows the limitations of residual testing in validating a model. The modeling of the sunspot series and the sales data series illustrates the need for looking into a variety of models. Covariance stationary models cannot be ruled out for modeling a series just because the series has a growth component. The expenditure data series provides an excellent example of a case where moving average terms greatly improve the quality of a model. The modeling of groundwater levels and rainfall variables is an example of the construction of multivariate models.

References

Some of the important references to the material used in the book are given here, although this list is not intended to be exhaustive. Wherever appropriate, we give a book or a review as a reference instead of the original paper so that additional references can be obtained from it.

Akaike, H. (1972). Information Theory and an Extension of the Maximum Likelihood Principle. *Int. Symp. Informat. Theory U.S.S.R.*

Akaike, H. (1973). Maximum likelihood identification of gaussian autoregressive moving average models. *Biometrika* **60**, 255–265.

Akaike, H. (1974). A new look at statistical model identification. *IEEE Trans. Automat. Contr.* **AC-19**, 716–722.

Andel, J., and Balek, J. (1971). Analysis of periodicity in hydrological sequences. *J. Hydrol.* **14**, 66–82.

Anderson, T. W. (1971). "The Statistical Analysis of Time Series." Wiley, New York.

Andrews, D. F. *et al.* (1972). "Robust Estimates of Location." Princeton Univ. Press, Princeton, New Jersey.

Astrom, K. J. (1967). On the achievable accuracy in identification problems. *IFAC Symp. Identification Automat. Contr. Syst., Preprints, Prague.*

Astrom, K. J., Bohlin, T., and Wensmark, S. (1965). Automatic Construction of Linear Stochastic Dynamic Models for Stationary Industrial Processes. Tech. Rep. TR-18-150, IBM Nordic Lab.

Bartlett, M. S. (1966). "An Introduction to Stochastic Processes." Cambridge Univ. Press, London and New York.

Bartlett, M. S., and Rajalakshman, D. V. (1953). Goodness-of-fit tests for simultaneous auto regressive series. *J. Roy. Stat. Soc.* (*B*) **15**, 107.

Birnbaum, A. (1969). Concepts of statistical evidence. *In* "Philosophy, Science and Method" (S. Morgenbesser *et al.*). St. Martins Press, New York.

Bohlin, T. (1971). On the problem of ambiguities in maximum likelihood identification. *Automatica* **7**, 199–210.

Box, G. E. P., and Jenkins, G. M. (1970). "Time Series Analysis—Forecasting and Control." Holden-Day, San Francisco, California.

Box, G. E. P., and MacGregor, J. F. (1972). The Analysis of Closed-Loop Dynamic Stochastic Systems. Dept. of Statistics, Univ. of Wisconsin, Tech. Rep. 309.

Box, G. E. P., and Tiao, G. C. (1965). A change in level of a non-stationary time series. *Biometrika* **52**, 181–192.

Box, G. E. P., and Tiao, G. C. (1973). "Bayesian Inference in Statistical Analysis." Addison Wesley, Reading, Massachusetts.

Caines, P. E., and Chan, C. W. (1974). Feedback Between Stationary Stochastic Processes. Contr. Syst. Rep. 7421, Univ. of Toronto, Electr. Eng. Dept.

Caines, P. E., and Rissanen, J. (1974). Maximum likelihood estimation of parameters in multivariate gaussian stochastic processes. *IEEE Trans. Informat. Theory* **IT-20**, 102–104.

Chatfield, C., and Prothero, D. L. (1973). Box–Jenkins seasonal forecasting: Problems in a case study. *J. Roy. Statistical Soc. Ser. A.* **136**, Part 3, 295–315 (with discussion).

Chisholm, M., Frey, A. E., and Haggett, P. (eds.) (1971). "Regional Forecasting." Archon Books, Hamden, Conn.

Chow, G. C. (1974). Identification and estimation in econometric systems: A survey. *IEEE Trans. Automat. Contr.* **AC-19**, 855–861.

Chow, V. T. (1964). "Sequential Generation of Hydrologic Information." Sec. 8, Part IV, *in* "Handbook of Applied Hydrology." Ed. V. T. Chow, McGraw-Hill, New York, N.Y.

Clarke, D. W. (1967). Generalized-least-square estimation of the parameters of a dynamic model. Presented at the *IFAC Symp. Identification Auto. Contr. Syst, Prague.*

Clarke, R. D. (1973). "Mathematical Models in Hydrology." Irrigation and Drainage paper #19, F.A.O., Rome, Italy.

Crawford, N. H., and Linsley, R. K. (1966). Digital Simulation in Hydrology: Stanford Watershed Model IV. Tech. Rep. No. 39, Dept. of Civil Eng., Stanford Univ., Stanford, California.

Croxton, F. E., and Cowden, D. J. (1939). "Applied General Statistics," pp. 452–458. Prentice-Hall, Englewood Cliffs, New Jersey.

Dhrymes, P. J. (1972). Simultaneous equations inference in econometrics. *IEEE Trans. Automat. Contr.* **AC-17**, 427–438.

Duda, R. O., and Hart, P. E. (1973). "Pattern Classification and Scene Analysis." Wiley, New York.

Durbin, J. (1959). Efficient estimation of parameters in moving average models. *Biometrika* **46**, 306–316.

Dawdy, D. R., and Matalas N. C. (1964). Statistical and Probability Analysis of Hydrologic Data, Sec. 8, Part III, *In* "Handbook of Applied Hydrology." Ed. V. T. Chow, McGraw-Hill, New York, N.Y. (1964).

Elton, C. S., and Nicholson, M. (1942). The ten year cycle in the numbers of lynx. *J. Anim. Ecol.* **1**.

Engle, R. F., and Ta-Chung Liu (1972). Effects of aggregation over time on dynamic characterization of an econometric model. *In* "Econometric Models of Cyclical Behavior" (B. G. Hickman, ed.). Columbia Univ. Press, New York.

Eykhoff, P. (1974). "System Identification: Parameter and State Estimation." Wiley, New York.

Feller, W. (1966). "An Introduction to Probability Theory and Its Applications," 2nd ed., Vol. 1. Wiley, New York.

Fiering, M. B. (1967). "Streamflow Synthesis." Harvard Univ. Press, Cambridge, Massachusetts.

Fiering, M. B., and Jackson B. B. (1971). "Synthetic Stream Flows," Water Resources Monograph 1, Am. Geophys. Union, Washington, D.C.

Fine, T. L. (1973). "Theories of Probability: An Examination of Foundations." Academic Press, New York.

Finigan, B. M., and Rowe, I. H. (1974). Strongly consistent parameter estimation by the introduction of strong instrumental variables. *IEEE Trans. Automat. Contr.* **AC-19**, 825–830.

Fisher, F. M. (1966). "The Identification Problem in Econometrics." McGraw–Hill, New York.

Gertler, J., and Banyasz, Cs. (1974). "A recursive (on-line) maximum likelihood identification method." *IEEE Trans. Automat. Contr.* **AC-19**, 816–819.

Gladysev, E. G. (1961). Periodically correlated random sequences. *Sov. Math.* **2**, 385–387.

Goel, N. S., Maitra, S. C., and Montroll, E. W. (1971). On the volterra and other nonlinear models of interacting populations. *Rev. Mod. Phys.* **43**, Number 2, Part I, 231–275.

Goldberger, A. S. (1968). "Topics in Regression Analysis." Macmillan, New York.

Granger, C. W. J. (1963). Economic processes involving feedback. *Informat. Contr.* **6**, 28–48.

Gustausson, I. (1972). Comparison of different methods for identification of industrial processes. *Automatica* **8**, 127–142.

Hannan, E. J. (1960). "Time Series Analysis." Methuen, London.

Hannan, E. J. (1969). The identification of vector mixed autoregressive–moving average systems. *Biometrika* **58**, 223–225.

Hannan, E. J. (1970). "Multiple Time Series." Wiley, New York.

Hannan, E. J. (1971). The identification problem for multiple equation systems with moving average errors. *Econometrica* **39**, No. 5, 751–765.

Hsia, T. C. (1975). On multistage least squares approach to system identification. Proc. VI World Congress of IFAC, Boston, Massachusetts.

Hufschmidt, M. M., and M. B. Fiering (1966). "Simulation Techniques for Design of Water Resource Systems." Harvard University Press, Cambridge, Mass. (1966).

Hurst, H. E., Black, R. P., and Simaika, Y. M. (1965). "Long Term Storage: An Experimental Study." Constable, London.

Jenkins, G. M., and Watts, D. G. (1968). "Spectral Analysis and Its Applications." Holden-Day, San Francisco, California.

Johnstone, J. (1963). "Econometric Methods." McGraw-Hill, New York.

Kailath, T. (1968). An innovations approach to least squares estimation—Part I. *IEEE Trans. Automat. Contr.* **AC-13**, 655–660.

Kalman, R. E. (1963). New methods in Wiener filtering theory. *Proc. Symp. Eng. Appl. Random Functions*, (J. Bogdanov and F. Kozin, eds.). Wiley, New York.

Kashyap, R. L. (1970a). A new method of estimation in discrete linear systems. *IEEE Trans. Automat. Contr.* **AC-15**, 18–24.

Kashyap, R. L. (1970b). Maximum likelihood identification of stochastic linear systems. *IEEE Trans. Automat. Contr.* **AC-15**, 25–34.

Kashyap, R. L. (1971). Probability and uncertainty. *IEEE Trans. Informat. Theory*, 641–650.

Kashyap, R. L. (1973). Validation of stochastic difference equation models for empirical time series. *Proc. IEEE Conf. Decision Contr., Miami.*

Kashyap, R. L., and Nasburg, R. E. (1974). Parameter estimation in multivariate stochastic difference equations. *IEEE Trans. Automat. Contr.* **AC-19**, 784–797.

Kashyap, R. L., and Rao, A. R. (1973). Real time recursive prediction of river flows. *Automatica* **9**, 175–183.

Kendall, M. G. (1971). Book review. *J. Roy. Statist. Soc. Ser. A* **134**, 450–453.

Kendall, M. G. (1973). "Time Series." Hafner, New York.

Kisiel, C. C. (1969). "Time Series Analysis of Hydrologic Data." Advances in Hydroscience, ed. V. T. Chow, Vol. 5, Academic Press, New York, N.Y.

Klein, L. R., and Evans, M. (1968). The Wharton Econometric Forecasting Model. Econ. Res. Unit., Univ. of Pennsylvania.

Klemes, V. (1974). The Hurst phenomenon, A puzzle? *Water Resources Res.* **10**, 675–688.

Landsberg, H. E. (1970). Man-made climate changes. *Science* **182**, 1265–1274.

Lee, T. C., Judge, G. G., and Zellner, A. (1970). "Estimating the Parameters of the Markov Probability Model from Aggregate Time Series Data." North-Holland Publ., Amsterdam.

Lehmann, E. L. (1959). "Testing Statistical Hypothesis." Wiley, New York.

Linsley, R. K., Kohler, M. A., and Paulhus, J. L. H. (1958). "Hydrology for Engineers." McGraw-Hill, New York.

Ljung, L., Gustavsson, I., and Soderstrom, T. (1974). Identification of linear multivariate system operating under linear feedback control. *IEEE Trans. Automat. Contr.* **AC-19**, 836–840.

Lowry, W. P. (1971). The climate of cities. *In* "Man and the Ecosphere." Freeman, San Francisco, California.

Maass, A., Hufschmidt, M. M., Dorfman, R., Thomas, H. A., Jr., Marglin, S. A., Fair, G. M., (1962). "Design of Water Resource Systems." Harvard University Press, Cambridge, Mass.

Malinvaud, E. (1970). The consistency of nonlinear regressions. *Annals Math. Statist.* **41**, 956–969.

Mandelbrot, B. B., and Wallis, J. R. (1968). Noah, Joseph, and operational hydrology, *Water Resource Res.* **4**(**5**), 900–918.

Mandelbrot, B. B., and Wallis, J. R. (1969a). Computer experiments with fractional Gaussian noises: 1. Averages and variances; 2. Rescaled ranges and spectra; 3. Mathematical appendix. *Water Resource. Res.* **5**(**1**), 228–267.

Mandelbrot, B. B., and Wallis, J. R. (1969b). Some long-run properties of geophysical records. *Water Resource Res.* **5**(**2**), 321–340.

Mandelbrot, B. B., and Van Ness, J. W. (1968). Fractional brownian motions, fractional noises and applications. *SIAM Rev.* **4**(**10**), 422–437.

Mann, A. B., and Wald, A. (1943). On the statistical treatment of linear stochastic difference equations. *Econometrica* **11**, 173–220.

Markel, J. D. (1972). Digital inverse filtering—a new tool for formant trajectory estimation. *IEEE Trans. Audio Electroacoustics* **AU-20**, 129–137.

Mathews, W. H., Kellog, W. W., and Robinson, G. D. (1971). "Man's Impact on the Climate." MIT Press, Cambridge, Massachusetts.

Mayne, D. Q. (1968). Computational procedure for the minimal realization of transfer function matrices. *Proc. IEEE* **115**, 1363–1368.

Medawar, P. B. (1969). "Induction and Intuition in Scientific Thought." Amer. Phil. Soc., Philadelphia, Pennsylvania.

Mehra, R. K. (1971). On-line identification of linear dynamic systems with applications to Kalman filtering. *IEEE Trans. Automat. Contr.* **AC-16**, 12–21.

Miller, R. B., and Botkin, D. B. (1974). Endangered species: models and predictions. *Amer. Sci.* **62**, 172–181.

Monin, A. S. (1972). "Weather Forecasting as a Problem in Physics." M.I.T. Press, Cambridge, Massachusetts.

Montroll, E. W. (1968). *In* "Lectures in Theoretical Physics" (A. O. Barut and W. E. Britten, eds.), vol. XA, Gordon and Breach, New York.

Moran, P. A. P. (1953). The statistical analysis of the Canadian lynx cycle, I. *Austr. J. Zool.* **1**, 163–173.

Nasburg, R. E., and Kashyap, R. L. (1975). Robust parameter estimation in dynamical systems. *1975 Conf. Informat. Sci. Syst.*, Johns Hopkins Univ., Baltimore, Maryland.

Nelson, C. R. (1973). "Applied Time Series Analysis for Managerial Forecasting." Holden-Day, San Francisco, California.

O'Connell, P. E. (1975). A simple stochastic modeling of Hurst's law. *Math. Models Hydrol.* vol. 1, IAHS Publ. 101 (Proc. of Warsaw Symp., 1971).

Pandya, R. N. (1974). A class of boot strap estimators and their relationship to the generalized 2 stage least squares. *IEEE Trans. Automat. Contr.* **AC-19**, 831–835.

Panuska, N. (1969). An adaptive recursive least-square identification algorithm. *Proc. 1969 IEEE Symp. Adaptive Proc.*, Pennsylvania State Univ.

Parzen, E. (1969). Multiple time series modeling. *In* "Multivariate Analysis II." Academic Press, New York.

Parzen, E. (1974). Some recent advances in time series modeling. *IEEE Trans. Automat. Contr.* **AC-19**, 723–729.

Peterson, J. T. (1969). "The Climate of the Cities, A Survey of Recent Literature." U.S. Dept. Health, Educ., Welfare, Nat. Air Pollut. Contr. Administration, Raleigh, North Carolina.

Popov, V. M. (1969). Some properties of the control systems with irreducible matrix-transfer functions. *In Lecture Notes in Math.*, 144, Springer Verlag, New York, pp. 169–180.

Popper, K. R. (1934). "The Logic of Scientific Discovery." Harper, New York.

Pratt, J. W., Raiffa, H., and Schlaifer, R. (1965). "Introduction to Statistical Decision Theory." McGraw-Hill, New York.

Quenouille, M. H. (1957). "The Analysis of Multiple Time Series." Hafner, New York.

Quimpo, R. G. (1967). Stochastic Model of Daily Riverflow Sequences, Hydrology Paper #18. Colorado State Univ., Fort Collins, Colorado.

Rao, C. R. (1965). "Linear Statistical Inference." Wiley, New York.

Rao, A. R., and Kashyap, R. L. (1973). Analysis, construction and validation of stochastic models for monthly river flows, Tech. Rept. CE-HYD-73-1, Purdue Univ.

Rao, A. R., and Kashyap, R. L. (1974). Stochastic modeling of river flows. *IEEE Trans. Automat. Contr.* **AC-19**, 874–881.

Rao, A. R., and Rao, R. G. S. (1974). Analyses of the Effect of Urbanization on Rainfall Characteristics—I. Tech. Rep. No. 50, Water Resources Res. Center, Purdue Univ., W. Lafayette, Indiana.

Rao, A. R., Rao, R. G. S., and Kashyap, R. L. (1975). "Stochastic Models for Ground Water Levels." Tech. Rept. 59, Water Resource Research Center, Purdue University, W. Lafayette. In.

Rosenbrock, H. H. (1970). "State Space and Multivariate Theory." Wiley, New York.

Rozanov, Yu. A. (1967). "Stationary Random Processes." Holden-Day, San-Francisco, California.

Savage, L. J. (1962). The foundations of statistical inference, Metheun, London.

Scheidegger, A. E. (1970). Stochastic models in hydrology. *Water Resource Res.* **6**, 750–755.

Singh, K. P., and Lonnquist, C. G. (1974). Two-distribution methods for modeling and sequential generation of monthly stream flows." *Water Resource Res.* **10**, 763–773.

Sims, G. A. (1972). Money, income, and causality. *Amer. Econ. Rev.* **62**, 540–552.

Slutsky, E. (1927). The summation of random causes as the source of cyclic processes. Reprinted in *Econometrica* **5**, 105.

Soderstrom, T. (1972). On the Convergence Properties of the Generalized Least Square Identification Method. Div. Automat. Contr. Lund Inst. Technol., Lund, Sweden, Rep. 7228.

Soderstrom, T. (1973). An On-Line Algorithm for Approximate Maximum Likelihood Identification of Linear Dynamic Systems. Div. Automat. Contr., Lund Inst. Technol., Lund, Sweden, Rep. 7308.

Statistical Abstract of the U.S. (1974). U.S. Dept. of Commerce, Social and Econom. Statist. Administration, Bur. of the Census, Washington, D.C.

Swamy, P. A. V. B. (1971). "Statistical Inference in Random Coefficient Regression Models." Springer-Verlag, Berlin.

Tao, P-C., Rao, A. R., and Rukvichai, C. (1975). Stochastic forecasting models of reservoir inflows for daily operation—Parts I and II. Tech. Rept. CEHYD-75-8, School of CE, purdue.

Tatarski, V. I. (1961). "Wave Propagation in a Turbulent Medium." McGraw-Hill, New York.

Teekens, R. (1972). "Prediction by Multiplicative Models." Netherlands Univ. Press, Amsterdam.

Thomas, H. A., and Fiering, M. B. (1962). Mathematical synthesis of streamflow sequences for the analysis of river basins by simulation. *In* "Design of Water Resources Systems" (A. Maass *et al.*, ed.), pp. 459–493. Harvard University Press, Cambridge, Massachusetts.

Toulmin, S. (1957). "Foresight and Understanding: An Enquiry into the Aims of Science." Harper, New York.

UNESCO (1971). "Discharge of the Selected Rivers of the World." Vol. II. UNESCO, Paris.

USGS (U.S. Geological Survey), "Water Supply papers" (misc. volumes), Dept. of the Interior, Washington, D.C.

U.S. Weather Bureau, "Climatalogical Data," (Miscellaneous volumes), U.S. Dept. of Commerce, National Oceanic and Atmospheric Administration, Environmental Data Service, Washington, D.C.

Vansteenkiste, G. C. (ed.) (1975). "Computer Simulation of Water Resource Systems." North Holland, Amsterdam.

Van Den Boom, A. J. W., and Van Den Enden, A. W. M. (1974). The determination of the orders of process and noise dynamics. *Automat.* **10**, 245–256.

Waldmeier, M. (1961). "Sunspot Activity in the Years 1610–1960." Zurich, Shultheiss and Co.

Walker, A. M. (1962). Large sample estimation of parameters for autoregressive processes with moving average residuals. *Biometrika* **49**, 117–131.

Whittle, P. (1951). "Hypothesis Testing in Time Series Analysis." Almqvist and Wiksell, Upsala.

Whittle, P. (1952). Tests of fit in time series. *Biometrika* **39**, 309–318.

Whittle, P. (1954). Statistical investigation of sunspot observations. *Astr. Phys. J.* **120**, 251–260.

Wittenmark, B. (1974). A self tuning predictor. *IEEE Trans. Automat. Contr.* **AC-19**, 848–851.

W. M. O. (1970). "Urban Climates." Proc. of the WMO Symposium on Urban Climates and Building Technology," held at Brussels, Oct. 1968, WMO-No. 254, T.P. 141, Geneva, Switzerland.

Wold, H. (1954). Causality and econometrics. *Econometrica* **22**, 162–177.

Wold, H. (1965). "Bibliography on Time Series and Stochastic Processes." Oliver and Boyd, Edinburgh.

Wolovich, W. A. (1974). "Linear Multivariable Systems." Springer-Verlag, New York.

Wong, K. Y., and Polak, E. (1967). Identification of linear discrete time systems using the instrumental variable method. *IEEE Trans. Automat. Contr.* **AC-12**, 707–718.

Wonnacott, R. J., and Wonnacott, T. H. (1970). "Econometrics." Wiley, New York.

Yaglom, A. M. (1958). Correlation theory of processes with stationary random increments of order *n*. *Amer. Math. Soc. Transl. Ser. 2* **8**.

Yaglom, A. M. (1967). Outline of some topics in linear extrapolation of stationary random processes. *Proc. Berkeley Symp. Math. Statist. Probl., 5th*, **2** Univ. of California, Berkeley, California.

Yevjevich, V. (1972). "Stochastic Processes in Hydrology." Water Resources Publ., Fort Collins, Colorado.

Young, P. C. (1970). An instrument variable method for real-time identification of a noisy process. *Automatica* **6**, 271–287.

Yule, G. U. (1927). On the method of investigating periodicities in disturbed series with special references to Wolfer's sunspot numbers. *Phil. Trans. A* **226**, 267.

Zellner, A. (1971). "An Introduction to Bayesian Inference in Econometrics." Wiley, New York.

Index

K

L

M

N

P

Q

R

S

T

U

V

W

Z

A 6
B 7
C 8
D 9
E 0
F 1
G 2
H 3
I 4
J 5